This book is a state-of-the-art look at combinatorial games, that is, games not involving chance or hidden information. It contains articles by some of the foremost researchers and pioneers of combinatorial game theory, such as Elwyn Berlekamp and John Conway, by other researchers in mathematics and computer science, and by top game players.

The articles run the gamut from new theoretical approaches (infinite games, generalizations of game values, two-player cellular automata, alpha-beta pruning under partial orders) to the very latest in some of the hottest games (Amazons, Chomp, Dot-and-Boxes, Go, Chess, Hex). Many of these advances reflect the interplay of the computer science and the mathematics. The book ends with an updated bibliography by A. Fraenkel and an updated version of the famous annotated list of combinatorial game theory problems by R. K. Guy, now in collaboration with R. J. Nowakowski.

Mathematical Sciences Research Institute
Publications

42

More Games of No Chance

Mathematical Sciences Research Institute Publications

Volumes 1–4 and 6–27 are published by Springer-Verlag

More Games of No Chance

Edited by

Richard Nowakowski
Dalhousie University

CAMBRIDGE
UNIVERSITY PRESS

CAMBRIDGE UNIVERSITY PRESS
Cambridge, New York, Melbourne, Madrid, Cape Town, Singapore,
São Paulo, Delhi, Dubai, Tokyo, Mexico City

Cambridge University Press
32 Avenue of the Americas, New York, NY 10013-2473, USA

www.cambridge.org
Information on this title: www.cambridge.org/9780521808323

First published 2002

A catalog record for this publication is available from the British Library

ISBN 978-0-521-80832-3 Hardback
ISBN 978-0-521-15563-2 Paperback

More Games of No Chance
MSRI Publications
Volume **42**, 2002

Contents

More Games of No Chance
MSRI Publications
Volume **42**, 2002

Preface

This volume arose from the second Combinatorial Games Theory Workshop and Conference, held at MSRI from July 24 to 28, 2000. The first such conference at MSRI, which took place in 1994, gave a boost to the relatively new field of Combinatorial Game Theory (CGT); its excitement is captured in *Games of No Chance* (Cambridge University Press, 1996), which includes an introduction to CGT and a brief history of the subject. In this volume we pick up where *Games of No Chance* left off.

Although Game Theory overlaps many disciplines, the majority of the researchers are in mathematics and computer science. This was the first time that the practioners from both camps were brought together deliberately, and the results are impressive. This bringing together seems to have formed a critical mass. There has already been a follow-up workshop at Dagstuhl (February 2002) and more are planned.

This conference greatly expanded upon the accomplishments of and questions posed at the first conference. What is missing from this volume are the reports of games that were played and analyzed at the conference; of Grossman's Dots-and-Boxes program beating everyone in sight, except for the top four humans who had it beat by the fifth move.

This volume is divided into five parts. The first deals with new theoretical developments. Calistrate, Paulhus, and Wolfe correct a mistake about the ordering of the set of game values, a mistake that has been around for three decades or more. Not only do they show that the ordering is much richer than previously thought, they open up a whole new avenue of investigations. Conway echoes this theme of fantastic and weird structures in CGT (2 being the cube root of ω), and he introduces the smallest infinite games.

The classical games are well represented. Elkies continues his investigations in Chess. There are many new results and tantalizing hints about the deep structure of Go. Moore and Eppstein turn one-dimensional solitaire into a two-player game, and conjecture that the $\mathcal{G}$−values are unbounded. (In attempting to solve this, Albert, Grossman and Nowakowski defined Clobber, a big hit at the Dagstuhl conference.)

Newer games (some of which were not represented in the first volume) appear here and form the basis of the third section. Grossman and Nowakowski consider the problem of one-dimensional phutball. Demaine et al. saw the game being played for the first time one night during the conference and by the following morning had a proof that just discovering whether the next player has a game-ending move was difficult. Amazons received much attention in the interval between the two conferences; Müller and Snatzke present some approaches.

The Puzzle section presents the results about puzzles which exhibit the same spirit as games. The game of Life has two papers devoted to it; Demaine et al. also present an entertaining look at solving coin-moving puzzles.

The proceedings is completed with updated versions of the "Unsolved Problems in Combinatorial Game Theory" by Guy and Nowakowski and Fraenkel's bibliography.

Many thanks must go to the MSRI staff, who helped make the Workshop a success. The facilities (and weather) were great. Thanks must also go to the Workshop Chairs, Elwyn Berlekamp and David Wolfe, who did much of the hard work. Thanks also go to the other organizers (David Blackwell, John Conway, Aviezri Fraenkel, Richard Guy, Jurg Nievergelt, Jonathan Schaeffer, Ken Thompson) who, together with Berlekamp and Wolfe, put together a wonderful program.

As with the last volume, credit must also go to Silvio Levy for his suggestions and expertise in making the volume look good.

Richard J. Nowakowski

The Big Picture

More Games of No Chance
MSRI Publications
Volume **42**, 2002

Idempotents Among Partisan Games

ELWYN BERLEKAMP

ABSTRACT. We investigate some interesting extensions of the group of traditional games, $\mathcal{G}$, to a bigger semi-group, $\mathcal{S}$, generated by some new elements which are idempotents in the sense that each of them satisfies the equation $G + G = G$. We present an addition table for these idempotents, which include the 25-year-old "remote star" and the recent "enriched environments". Adding an appropriate idempotent into a sum of traditional games can often annihilate the less essential features of a position, and thus simplify the analysis by allowing one to focus on more important attributes.

1. Introduction and Background

We assume the reader is familiar with the first volume of *Winning Ways* [Berlekamp et al. 1982], including Conway's axiomatization of $\mathcal{G}$, the group of partesan games under addition, which can also be found in [Conway 1976]. I now call the elements of this group *traditional games*. Each of the traditional games considered in this paper has only a finite number of positions. The identity of $\mathcal{G}$ is the game called 0, which is an immediate win for the second player. We investigate some interesting extensions of the group $\mathcal{G}$ to a bigger semi-group, $\mathcal{S}$, generated by some new elements which are idempotents in the sense that each of them satisfies the equation $G + G = G$. We also present an addition table for these idempotents.

Some of these idempotents have long been well-known in other contexts. The newer ones all fall into a class I have been calling *enriched environments*. A companion paper [Berlekamp 2002] shows how these idempotents prove useful in solving a particular hard problem involving a gallimaufry of checkers, chess, domineering and Go.

We begin with a review of definitions, modified slightly to fit our present purposes.

Moves. In Go, a *move* is the change on the board resulting from the act of a single player. In chess, this is commonly called a *ply*, and the term *move* is used to describe a consecutive pair of plys, one by White and one by Black. In

this paper, we use *move* as it is understood in Go. This is consistent with the
tradition of combinatorial game theory. This theory has been most successful
in analyzing positions which can be treated as sums of subpositions which are
relatively or completely independent of each other. Each of these subpositions,
as well as their sum, is called a *game*. Although the players alternate turns,
it is quite common that they may elect to play in different components, so
that within any particular game, the same player may make several consecutive
moves. Hence, the definition of a game, or any of its positions, *does not* include
any specification of whose turn it is to play next. Conway's traditional definition
of a game is written recursively as

$$G = \{G^L \mid G^R\},$$

where G^L and G^R are sets of previously defined games. G^L is the set of positions
to which Left can move immediately. These are also known as Left followers.
G^R is the set of Right followers.

Other axioms, with which the reader is assumed to be familiar, define *sum*,
negative, *greater-equal*, and *number*.

Outcomes. When played out, traditional games eventually yield outcomes.
There are two outcomes: Loutcome, which is the outcome if Left plays first,
and Routcome, which is the outcome if Right plays first. In the most general
case, either of these outcomes might assume any of the values LEFT, TIE, or
RIGHT, which Left prefers in the order LEFT > TIE > RIGHT. Ties and draws
are impossible within $\mathcal{G}$, but they can occur in some of the extensions we will
consider. In $\mathcal{G}$, play eventually terminates when the player to move is unable or
unwilling to do so. That happens to Right when the value of the position is a
nonnegative integer, or to Left when the value of the position is a nonpositive
integer. If the value of the position is 0, whichever player is next to move is the
loser.

Left plays to attain an outcome he prefers, while Right tries to thwart it.
Loutcome and Routcome are the results if both players play optimally. If both

$$\text{Loutcome}(G) \geq \text{Loutcome}(H) \quad \text{and} \quad \text{Routcome}(G) \geq \text{Routcome}(H),$$

we say that

$$\text{Outcomes}(G) \geq \text{Outcomes}(H).$$

Greater-Equal. Combinatorial games satisfy a partial order. One form of the
traditional definition of *greater-equal* states that

$$G \geq H \iff \text{For all } X, \ \text{Outcomes}(G + X) \geq \text{Outcomes}(H + X) \qquad (1\text{–}1)$$

If H and X have negatives, this is equivalent to the assertion that

$$\text{Outcomes}(G - H) \geq \text{Outcomes}(0)$$

But since Loutcome(0) = RIGHT, that half of the condition is trivially satisfied so a sufficient condition is that

$$\text{Routcome}(G - H) = \text{LEFT},$$

or, as more commonly stated, Left, playing second, can win on $G - H$.

Following Conway's original axioms, we say that

$$G = H \iff G \geq H \text{ and } H \geq G$$

and that

$$G > H \iff G \geq H \text{ but } H \not\geq G$$

Scores. For some purposes, it is convenient to define *scores* and work with them rather than with outcomes.

Play of any traditional combinatorial game must eventually yield a position whose value is a number. The value of the first such position is called the game's *score* or *stop*. If Left plays first and G is optimally played, the resulting number is called the Leftscore, denoted by Lscore(G). Similarly, if Right plays first and G is optimally played, the resulting number is Rscore(G). If G is any traditional game and x is any number, then

$$x > \text{Lscore}(G) \Rightarrow x > G,$$

$$\text{Lscore}(G) > x > \text{Rscore}(G) \Rightarrow x \text{ is confused with } G,$$

$$\text{Rscore}(G) > x \Rightarrow G > x.$$

It is known that if G is any game, then an optimal Left strategy for playing G ensures reaching a maximal score, and an optimal Right strategy for playing G ensures reaching a minimal score. However, the converse need not be true because several strategies might lead to the same score and some of them might yield a suboptimal outcome. This is due to the fact that when the score of a traditional game is 0, the outcome depends on who gets the last move.

Some real games have other scores, which are explicitly defined by the rules of the game. Go is such a game. It is an initially surprising and somewhat remarkable fact that these *official scores* imposed by any of several variations of the *official* Go rules are often identical to these *mathematical scores*. By appropriate choices of rules for "Mathematical Go" [Berlekamp and Wolfe 1994], we can attain agreement of scores in all but a few very rare positions, which are so exotic that different variations of the official Go rules then fail to agree with each other.

Dots-and-Boxes is another popular game in which scores are explicitly defined by the rules of the game. This pencil-and-paper game has very little in common with Go. But surprisingly, it again happens that the elegant mathematics of combinatorial game theory, when applied to an approximation of the popular game, yields decisive insights into how to play the popular game [Berlekamp 2000b].

In this paper, we treat *scores* in the mathematical sense: the value of the first position whose value is a number.

We next consider several idempotent games that have no negatives.

2. Definitions of Idempotents with Opening Ceremonies

Remote Star. The remote star ☆ is introduced in *Winning Ways*, Chapter 8 and plays a leading role there. Rather than rely on any of those results for a definition, I now propose a rule for playing

$$Y + ☆$$

Before moving on such a game, we require an *opening ceremony* during which each player submits a positive integer to the referee as a sealed bid. (From the mathematical perspective, there is no need for these bids to be sealed; public bids would work equally well. However, professionals and other serious competitors are loath to let the opponent know anything about their contingency plans prematurely. So it is easier to sell mathematical models to serious game-players when the rules ensure that losing bids remain unknown to the winning bidder.)

The referee selects the larger bid, n, and replaces ☆ by $*n$ before play begins. Ordering relations are determined in the usual way, using (1–1), with the understanding that the game X is specified before the opening ceremonies.

The play of $Y + ☆ + ☆$ begins with two successive auctions. Since either player can submit a bid to the second auction which is at least double the winning bid of the first auction, it follows that the sum of two remote stars is now again a remote star, whence

$$☆ + ☆ = ☆.$$

Ish. Traditionally, the term *ish* means *Infinitesimally SHifted*. It appears in such expressions as $\{1 \mid 1\} = 1* = 1\text{ish}$ and $\{1* \mid 1\} = 1\text{ish}$.

We can also manipulate ish as though it were another element of our semigroup S. To this end, we henceforth treat ish as a noun with the mnemonic Infinitesimal SHift. Its ordering relations might be defined as

$$G \leq H + \text{ish} \text{ and/or } G + \text{ish} \leq H \iff \text{Scores}(G + X) \leq \text{Scores}(H + X) \text{ for all } X,$$

However, to simplify the task of defining the sum of ish plus other idempotents, I prefer the following more intricate definition of ish:

At the opening ceremony of $G + X + \text{ish}$, each player submits a small positive number as a sealed bid to the referee, who announces the smaller such bid, which we will call ε. Then Left wins only if the score exceeds ε, and Right wins only if the score is less than $-\varepsilon$. A game which concludes with a net score of magnitude not exceeding ε declared to be a tie.

It is not hard to show that Left, going first, is able to *win* the game $G + X + \text{ish}$ if and only if $\text{Lscore}(G + X) > 0$. Left, going second, is able to *win* the game

$G + X + \text{ish}$ if and only if $\text{Rscore}(G + X) > 0$. Furthermore, it is easily verified that

$$\text{ish} + \text{ish} = \text{ish}.$$

Comment. *The sophisticated reader will recall several types of numbers that appear in [Conway 1976]: surreal, real, rational, dyadic rational. So when reading that each bid to determine ε must be a small positive number, she might ask which sort of number is required. It turns out that any type of number just listed is adequate for our present purposes. But only dyadic rationals are fully consistent with our focus on games with a finite number of positions.*

Positively Enriched Environment, $\mathcal{E}_t$. Enriched environments entail more elaborate opening ceremonies. Each positively enriched environment has a specified temperature, t, which is a positive number.

Every enriched environment contains an implicit ish, which is resolved first. This results in the specification of an ε. If other ishes are present, then this initial portion of the opening ceremonies continues until are all resolved into a single small positive ε. Then, to resolve the positively enriched environment $\mathcal{E}_t$, each player submits to the referee another small positive number as a sealed bid. The referee announces the winning (smaller) number, called δ. For simplicity, we restrict legal bids so as to ensure that δ is a divisor of t, so that t/δ is a positive integer. Then $\mathcal{E}_t$ is replaced by the sum of t/δ uniformly spaced switches, called *coupons*: $t|-t, (t-\delta)|(-t+\delta), \dots, \delta|-\delta$. This concludes the opening ceremonies. Then play begins. Play terminates when all coupons have been taken and the value of the position is a number. The net score is then taken to be this number plus all of Left's coupons minus all of Right's coupons. The outcome is declared to be a tie unless the magnitude of the score exceeds ε.

Fully Enriched Environment, $\mathscr{E}_t$. The temperature of a fully enriched environment can be any number not less than -1.

After resolving the implied and explicit ishes to an ε, each player submits a small positive bid for δ. Legal bids are constrainted to ensure that $(t + 1)/\delta$ is a nonnegative integer. The winning (smaller) bid is announced. Then $\mathscr{E}_t$ is replaced by a set of *coupons*, whose face values range from t down to $-1 + \delta$, with a constant decrement of δ. A very large number of coupons with value -1 is then placed at the bottom of the coupon stack.[1]

Play begins. At each turn, a player may either make a legal move from the current game, G, to one of his legal followers, *or* he may instead use his turn to

---

[1]Mathematically, one could do without the coupons with value -1, because it is possible for either player to pick an extremely small value of δ. However, the -1 point coupons make it easier to sell the concept of coupon stacks to serious competitive gamesmen. A good environment to accompany a 10×10 game of Amazons needs almost 80 coupons with value -1, even though δ of 0.1 or even 0.5 proves interesting. If there were few or no -1 point coupons, the appropriate δ might need to be reduced by nearly two orders of magnitude.

take the top coupon from the stack. Even if the value of the current position, G, becomes a number, play continues until after only -1 point coupons remain. Play terminates after each of the players has taken three -1 point coupons consecutively. Then there is a concluding ceremony, during which the referee gives a special $-\frac{1}{2}$ point *terminal komi* coupon to the player who did not take the last of the six consecutive -1 point coupons which caused the game to be terminated. Each player's score is then computed as the sum of all of the coupons he has taken. If the magnitude of the difference between these scores exceeds ε, the player with the higher score is declared the winner. Otherwise, the result is declared to be a tie.

Comment. *Why do we not end the game until after the sixth consecutive coupon of value -1 is taken? In part, this is modeled after a well-known rule in chess, which declares the game to be terminated with a drawn outcome after the same position occurs three times with the same player to move. Presumably this is intended to give each player multiple chances to consider other options. Theoretically, if both players are playing optimally, one might think that the first repetition of a position would be sufficient. However, in games like Go, which include a ko rule, there are situations in which a pair of consecutive coupons is taken as a kothreat and its response while the game is still quite active. So we might theoretically weaken the six consecutive coupons to four consecutive coupons but, at least in the case when the board positions include Go, two consecutive coupons is definitely not enough.*

3. Definitions of Idempotents Without Opening Ceremonies

On. Figure 1 shows two positions of a Black checker king. Although White has no pieces to move, Black can move to and fro between these two positions whenever he so desires. I call these positions *onto* and *onfro*. From *onto*, Black can move to *onfro*. From *onfro*, he can move back to *onto*. Formally,

$$\mathbf{onto} = \{\mathbf{onfro} \mid \}, \qquad \mathbf{onfro} = \{\mathbf{onto} \mid \}.$$

Taken together, **onto** and **onfro** can be viewed as the bifurcated components of an abstract game called **on**, which appears in the latter half of Chapter 11 of *Winning Ways*. In a sum of games including an **on**, Right will never be able to

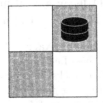 

Figure 1. Onto (left) and onfro (right).

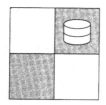

Figure 2. Off.

force the play of the game to terminate. Formally,

$$\mathbf{on} = \{\mathbf{on} \mid \}.$$

This game has infinite mean. If X is any traditional game, then our definition of equality implies that

$$\mathbf{on} + X = \mathbf{on}.$$

Off. Off differs from **on** only in that the lone checker king is now white instead of black. Now it is Right who can move to and fro at will. In a sum of games including an **off**, Left will never be able to prevent Right from playing. If Left has only a finite number of moves available elsewhere, he will eventually run out of options and lose the game.

This game has mean value of $-\infty$. Like **on**, it overpowers any traditional game to which it is added.

Dud. Although **on** and **off** superficially look like negatives of each other, every checker player knows that their sum is not zero (a win for the second player), but a *Deathless Universal Draw*, often realized on a single checkerboard when each player has only a single king, and the two kings are located in the double corners at opposite ends of the board.

Dud is not only a draw by itself, it also ensures that anything to which it is added also becomes a drawn game. **Dud** plays the role of a black hole. If a dud is present, no other features matter; the overall game behaves as a **dud**.

Comments on Outcomes with no Winner. Terminology has some history. In *Winning Ways*, we distinguished between *ties*, in which play terminated without either player winning, and *draws*, in which play could naturally be *drawn* out forever if not terminated by a special rule. In chess tournaments, ties due to stalemate and draws due to perpetual check are treated identically: each player receives $\frac{1}{2}$ win $+$ $\frac{1}{2}$ loss. However, Go has a quite different tradition. The natural rules allow certain positions to be repeated immediately in a two-move loop, one move by each player. Such global repetitions are universally banned by the so-called *Ko rule*. Positions involving ko are very common, occurring several times in most games. Lengthier repetitions, after 4 or 6 or more moves are also possible, but rather rare, occurring in only a small fraction of one percent of all professional games. The rules about how to handle such *superko* positions differ from time to time and from place to place. Today, only the North American rules and the New Zealand rules simply ban all superkos. Japanese rules explicitly allow them. Chinese and Taiwanese rules for superkos are more complicated. In many cases, one player or the other is permitted to repeat the position but the other player is not. Which player is banned depends on the details of the position.

If a game gets hung in a superko, the Japanese tournament rules do NOT treat it as a tied or drawn outcome. The official translation defines the outcome as *no result*. So I prefer the English word *hung*, as in a hung jury or a computer program that is hung in a loop. Like the hung jury, a Japanese Go game which hangs in a superko often leads to a new game in which the same two contestants begin again from scratch.

To further complicate the situation, some amateur Japanese Go games can end with a tied score. Many translators have called such an outcome a *draw*, in direct conflict with the terminology of *Winning Ways*. I call them ties, and try to avoid any use of the word *draw* in reference to Go. As we are primarily concerned with individual games, or sums of games, the question of how such outcomes are treated in tournaments need not concern us here. So chess games which draw in perpetual check and checker games which draw in a **dud** might also be said to have *hung*. Whatever the terminology, a win is surely better than either a tied or hung outcome, which in turn is better than a loss.

Loony. Some impartial games have what are called *complimenting moves*, and such games can have positions with a fascinating value called *loony*, and denoted by the lunar symbol, ☽. One of the best-known such games is Dots-and-Boxes, whose Impartial variation I shall now describe.

The game is played on a array of dots located on the integer points of a rectangular subset of the Cartesan plane. These dots appear at unit distances from each other in vertical and horizontal rows. A legal move for either player is to draw a new horizontal or vertical line of length one, joining two dots. Unless that line completes a unit square, it completes the mover's turn. However, if that line completes one or two unit squares (called boxes), the mover must continue to make another move. The game ends when no further legal moves remain, and the player who made the last move *loses*.

(Even though last player loses, this is regarded as a *normal* rather than a *misère* rule because the last move necessarily completes a box, so the turn is incomplete. The player loses because he is unable to fullfill the requirement that he make another move.)

Figure 3, left, shows the position of an impartial Dots-and-Boxes position. This position can be viewed as the sum of four positions: two squares in the northwest, one in the northeast, two in the southeast, and four in the southwest, whose respective values can be shown to be ☽, *, *, and *2. The figures in the middle and on the right show two quite different ways in which the player to

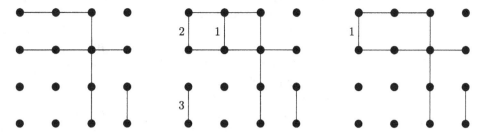

Figure 3. A Dots-and-Boxes position (left) and two possible continuations.

move can complete his turn. In the middle, he takes each of the boxes in the northwest and then completes his turn with a move elsewhere. In the right-hand figure, he completes his turn in another way, which forces his opponent to make the first move outside the northwest region. So although it might not be easy to determine whether or not one wishes to play first or second on the rest of the board, it is easily seen that the player to move from a sum which includes a loony component can win the game in either case. If he wants to make the first move elsewhere, he plays the loony component in a way which enables him to do that, as in the middle figure. On the other hand, if he wishes to force his opponent to make the first move elsewhere, he can achieve that objective by playing the loony component in the other way.

Loony values can also occur in games with entailing moves, as described in *Winning Ways*, Chapter 12. Entailing moves are special moves that require the opponent to move in a certain portion of the game. Complimenting moves are special moves (like completing a box in Dots-and-Boxes) which can be viewed as forcing the opponent to skip his next turn. Rather than attempt any general definition, for purposes of this paper it is sufficient to simply define $\mathfrak{D}$ very specifically as the northwest corner of the impartial Dots-and-Boxes position of the left panel in Figure 3.

4. The Addition Table

The addition table for the idempotents we have discussed is shown on the next page. The order in which they are listed may be viewed as an order of increasing *vim* (as in "vim and vigor"), with the understandings that **on** and **off** have equal vim, and that otherwise, whenever two of the idempotents are added, the one with the more vim predominates.

Another view is that adding in any of these idempotents destroys certain information, and the idempotents with more vim are more destructive.

In practice, adding in an appropriate idempotent can often be the key to the analysis of a particularly challenging position. The most helpful idempotent is the one which preserves just those features which are crucial to the winning line of play, while annihilating all of the less significant considerations.

Addition of Idempotents

	0	☽	☆	ish	$\mathcal{E}_\tau$	$\mathcal{E}_t$ ($t>\tau$)	$\mathscr{E}_t$	on	off	dud
zero	0	☽	☆	ish	$\mathcal{E}_\tau$	$\mathcal{E}_t$ ($t>\tau$)	$\mathscr{E}_t$	on	off	dud
loony	☽	☽	☆	ish	$\mathcal{E}_\tau$	$\mathcal{E}_t$	$\mathscr{E}_t$	on	off	dud
remote star	☆	☆	☆	ish	$\mathcal{E}_\tau$	$\mathcal{E}_t$	$\mathscr{E}_t$	on	off	dud
	ish	ish	ish	ish	$\mathcal{E}_\tau$	$\mathcal{E}_t$	$\mathscr{E}_t$	on	off	dud
enriched environ- ments $\Big\{$	$\mathcal{E}_\tau$	$\mathcal{E}_\tau$	$\mathcal{E}_\tau$	$\mathcal{E}_\tau$	$\mathcal{E}_\tau$	$\mathcal{E}_t$	$\mathscr{E}_t$	on	off	dud
	$\mathcal{E}_t$ ($t>\tau$)	$\mathcal{E}_t$	$\mathcal{E}_t$	$\mathcal{E}_t$	$\mathcal{E}_t$	$\mathcal{E}_t$	$\mathscr{E}_t$	on	off	dud
	$\mathscr{E}_t$	$\mathscr{E}_t$	$\mathscr{E}_t$	$\mathscr{E}_t$	$\mathscr{E}_t$	$\mathscr{E}_t$	$\mathscr{E}_t$	on	off	dud
on	on	on	on	on	on	on	on	on	dud	dud
off	off	off	off	off	off	off	off	dud	off	dud
dud	dud	dud	dud	dud	dud	dud	dud	dud	dud	dud

5. Other Properties of These Idempotents

Homomorphisms. Each idempotent I induces a mapping: $A \mapsto A + I$. If also $B \mapsto B + I$, it is evident that $A + B \mapsto A + B + I$. So each such mapping is a homomorphism on the semigroup $\mathcal{S}$.

We now give expanded descriptions of some of the idempotents:

Loony. Loony, or ☽, lies at the heart of a rich structure described in some detail in [Berlekamp 2000b]. This context, within which ☽ seems so powerful, is impartial; the only other games which appear there are nimbers. When compared with partesan games, we soon discover that ☽ has so little vim that it preserves *all* strict inequalities in the ordering relationships within the traditional group, $\mathcal{G}$. To see this, it is sufficient to verify that for every positive integer n,

$$\mathbf{miny}_n < ☽ < \mathbf{tiny}_n$$

where $\mathbf{tiny}_n$ is defined as $0 \,\|\, 0 \mid -n$ and $\mathbf{miny}_n = -\mathbf{tiny}_n$. It has long been known (as can be verified by an induction on zero-based canonical birthdays) that every positive game with a finite number of positions is greater than tinies which have sufficiently large n. So, if G and H are traditional games for which $G > H$, the inequality is preserved if ☽ is added to either or both sides.

Remote Star. Adding ☆, a remote star, into a game induces a homomorphism whose kernel contains all infinitesimals of higher order than ↑ = 0 | ∗.

Remote star plays the central role throughout Chapter 8 of *Winning Ways*, which presents the theory of atomic weights, including uppitiness and the Norton multiply operation. When the results of that chapter were first discovered, the main application was the game of Hackenbush hotchpotch, where the remote star, and its proximity to the red and blue kites, arises quite naturally. The generalization to other "all-small" games such as childish Hackenbush was fairly immediate, although we neglected to notice the relevance to the Toads-and-Frogs analysis in an earlier chapter of *Winning Ways*. Another decade passed before the discovery of the crucial role which atomic weight theory plays in getting the last one-point move in Go endgames [Berlekamp and Wolfe 1994; Moews 1996]. One must wonder why atomic weight theory, or even some approximate version of it, eluded Go players for so many centuries. I think the explanation is that, except for ∗, the nimbers (of which the remote star might be viewed as a limit) occur only very rarely in Go or in chilled Go.[2] So here we have a good example of how an issue in one game (How do I get the last one-point move in Go?) benefits from insights that originate in another quite different game (Hackenbush hotchpotch).

Enriched Environments. We shall see that adding a sufficiently hot enriched environment preserves the mean, but destroys all other information about the traditional game to which it is added. This is why it has become such an important tool in the study of Amazons [Berlekamp 2000a, Snatzke 2002] and Go: It eliminates all of the less essential features, allowing one to focus on the primary attribute of each position. At a relatively early stage of the endgame, it enables one to obtain a real-valued *count* which is precisely accurate in a well-defined mathematical sense. In many cases, the board contains numerous small battles which themselves serve as a plausible approximation to the enriched environment. For this reason, the line of play which is provably optimum in the enriched environment often provides an extraordinarily good guide as to how to play the actual endgame.

Coupons with positive values behave exactly like switches. The only real difference between $\mathcal{E}$ and $\mathscr{E}$ is that the former stack of coupons ends at 0, whereas the latter stack also includes negative coupons and a terminal komi. In other words, for positive t,

$$\mathscr{E}_t = \mathcal{E}_t + \mathscr{E}_0.$$

When $\mathcal{E}$ is played all by itself, the players alternately take coupons, effectively dealing the stack out between them. Each player ends with $n/2$ coupons, and each of the second player's coupons is δ less than the prior coupon just taken

---

[2]Even the existence of ∗2 was until very recently an open problem (listed for example in [Guy 1996, p. 487, problem 45]); its resolution came in [Nakamura and Berlekamp 2002] in the form of a chilled Go position of value ∗2.

by his opponent. So first player wins by a net score of $n\delta/2 = t/2$. One way to make the game fairer is to follow the common practice used in professional Go games, which assigns a specified number of points to the second player (White) as compensation for the advantage which Black gets from the first move. This compensation is called the *initial komi*. Evidently, the fair initial komi corresponding to $\mathcal{E}_t$ is $t/2$. Similarly, it can be verified that the terminal komi (of value $-\frac{1}{2}$) included in the coupon stack $\mathcal{E}_0$ is just enough to balance out the disadvantage of the last coupon (at temperature -1), so that when $\mathcal{E}_0$ is played out all by itself, the net score has magnitude less than ε, so the resulting outcome is a tie.

Thermographs. The thermograph of a given traditional game, G, consists of two functions of an independent variable, t, which is called the temperature. The functions are the thermograph's Leftwall and Rightwall, which may be defined as follows:

$$\text{Leftwall}(G, t) = \text{Lscore}(G + \mathcal{E}_t, t) - t/2,$$

$$\text{Rightwall}(G, t) = \text{Rscore}(G + \mathcal{E}_t, t) + t/2.$$

It can be shown that for traditional games, these definitions yield the same walls of the thermograph as the recursions based on the cooling homomorphism found in such references as Chapter 6 of *Winning Ways*. Alternatively, if one takes those recursions as the definition, then one can prove that the scores of $G + \mathcal{E}_t$ are the solutions of these same equations. In other words, the final score of a well-played game consisting of a traditional game, G, plus an enriched environment is equal to the fair initial komi, $t/2$, plus whichever wall of the thermograph corresponds to whoever plays first.

It is conventional to plot thermographs with the independent variable, t, running upwards along the vertical axis, starting from a minimal value which is often taken to be 0 or -1. Positive values are plotted as increasing towards the left along the horizontal axis, negative values to the right. This reversal of the signs normally used in analytic geometry facilitates a more direct comparison between thermographs and the expressions or graphs of the underlying games, in which Left seeks to move leftward, and Right seeks to move rightward.

If t is sufficiently large, then $\text{Leftwall}(G, t) = \text{Rightwall}(G, t) = \text{Mast}(G)$, a value independent of t. Traditionally, this mast is also known as the *mean value*, or $\text{mean}(G)$. The greatest lower bound on temperatures, τ, such that

$$\text{Leftwall}(G, t) = \text{Rightwall}(G, t) = \text{Mast}(G) \text{ for all } t \geq \tau,$$

is called the *temperature* of G.

In some applications, it is useful to color portions of the mast. If $t > \tau$ and if Left can attain the optimal Leftscore of $G + \mathcal{E}_t$ by playing on G, the mast of G at t can be colored *blue*. Similarly, if Right can attain the optimal Rightscore of $G + \mathcal{E}_t$ by playing on G, the mast of G at t can be colored *red*. A portion of the mast that is colored both red and blue is shown as *purple*. A portion that is neither red nor blue is shown as yellow (or gray). All traditional games have thermographs whose masts become gray at all sufficiently high temperatures.

One can also apply this reasoning to a sum of traditional games, $A + B + C$. For large t, one can study $A + \mathcal{E}_t$, $B + \mathcal{E}_t$, $C + \mathcal{E}_t$, and then add them up to obtain

$$
\begin{array}{c}
A + \mathcal{E}_t \\
B + \mathcal{E}_t \\
C + \mathcal{E}_t \\
\hline
A + B + C + \mathcal{E}_t,
\end{array}
$$

because $\mathcal{E}_t$ is an idempotent. Evidently, $\mathrm{Mast}(A+B+C) = \mathrm{Mast}(A)+\mathrm{Mast}(B)+\mathrm{Mast}(C)$, a result well-known in traditional thermography.

If G is any traditional game and n is any positive integer, then $\mathrm{Mast}(n\,G) = n\,\mathrm{Mast}(G)$ and $\mathrm{temp}(nG) \geq \mathrm{temp}(G)$. The scores of traditional games can be bounded by

$$-\tau + \mathrm{Mast}(G) \leq \mathrm{Rightscore}(G) \leq \mathrm{Mast}(G),$$

$$\mathrm{Mast}(G) \leq \mathrm{Leftscore}(G) \leq \tau + \mathrm{Mast}(G),$$

$$\mathrm{Mast}(G) - \frac{\tau}{n} \leq \frac{\mathrm{Scores}(nG)}{n} \leq \mathrm{Mast}(G) + \frac{\tau}{n},$$

from which is becomes reasonable to say that $\mathrm{Mast}(G)$ is the *mean* of G.

Thermographs of Games Including Kos. A common feature of Go positions is a situation called *ko*, which is a 2-move loop in the game graph of a local position. Globally, the immediate repetition of a position is banned in all dialects of the rules of Go. However, since the ban is global, it is quite common for the local position to be repeated locally in a so-called *kofight*. During the kofight, many moves elsewhere become worthwhile as *kothreats*. Locally, the game can stay in a two-cycle loop as long as each player is able to find worthwhile threats elsewhere.

The recursive definitions which formed the original basis of thermography were not designed to handle the loopiness of kos. This problem was addressed by Berlekamp [1996]. That paper introduced an extended thermography which was subsequently applied to the study of a collection of over eighty kos by Berlekamp, Müller and Spight [Berlekamp et al. 1996]. Many of those kos were taken from professional games studied by Müller.

Even though the play of actual kos can depend significantly on kothreats located in other regions of the board, the total environment consisting of all of these kothreats can often be usefully approximated by one of a small number of possibilities, depending on who is *komaster* or *komonster*, with a relatively small intermediate *hypersensitive* region. When the mast of a game depends on who is komaster of a ko occurring in one the game's positions, that game is said to be *hyperactive*. Other positions may contain kos which are called *placid*, because their means are independent of who is the master of the ko.

In any sum of games including at most one hyperactive ko position in which it is clear who is komaster, it remains true that the mast of the sum is still the sum of the masts. Because of this, extended thermographs prove very useful for analyzing a wide range of positions from actual Go games, which typically have at most one position depending on a hyperactive ko. However, sums of two or more games dependent on hyperactive kos are more complicated.

If a position's mast depends on a hyperactive ko, then its *mean* typically differs from either its mast with Left komaster or its mast with Right komaster. In many cases, the analysis of the game depends on one of these two masts rather than on the mean.

Spight [1998] has begun a theoretical investigation of sums of several hyperactive positions. The situation is complicated. Fortunately such sums are not very common in practice.

In some situations, an analysis of the whole board may be needed to distinguish whether a game including a ko position is best characterized as Left komonster, Left komaster, hypersensitive, Right komaster, or Right komonster. But often the answer is heavily biased by factors which depend only on a careful local analysis of the hyperactive ko.

Studies of this topic are continuing to progress at a rapid rate, as discussed in the section on further work at the conclusion of this paper.

Positive or Full Enrichment? Amazons [Berlekamp 2000a] is a hot game which, like Go, is primarily a territorial battle. However, unlike Go, positions with the traditional *number* values do occur in Amazons; Snatzke [2001] recently discovered a value of $\frac{1}{16}$. But even the best current players are often unable to determine whether the value of a position is a number or not. So enforcement of the traditional *stopping* rule of *Winning Ways*, pp. 145–162 is problematical. Negative coupons were introduced to address this issue.

The negative coupons in $\mathscr{E}$ have also proved helpful in teaching subzero thermographs [Berlekamp 1996]. Game positions with subzero thermographs actually occur in chilled Go [Berlekamp and Wolfe 1994], and some of their properties are more easily understood when chilled rather than when warmed [Takizawa 2001].

Environmental Go The concept of enriched environments grew out of a sequence of attempts to improve communication with professional Go players.

Historically, thermographs were first developed in terms of taxes rather than stacks of coupons. However, efforts to interest serious Go players in formal definitions of cooling and heating were very unsuccessful. The notion of determining tax rates by a competitive auction in [Berlekamp 1996] failed to catch on. Most people simply don't like to think about taxes. Changing the sign of the payoff (at positive temperatures) from negative to positive, and calling these payoffs *coupons* instead of *taxes* had a large impact. Jujo Jiang and NaiWei Rui, both famous 9-dan Go players, agreed to play the first demonstration game of environmental Go in 1998. An analysis of the endgame of this game appears in [Spight 2001]. Jiang and Rui played another game of environmental Go at MSRI on July 23, 2000. The analysis of that game is still underway. Many participants at the American Go Congress held in Denver in August 2000 expressed much interest in Environmental Go. As a means to provoke quantitative discussion, the concept of a stack of coupons must now be viewed as an enormous success.

The most popular current stack consists of 40 coupons, all positive, with values 20, 19.5, 19, ..., 0.5 This stack is placed next to the initially empty Go board. White receives a komi of 9.5 points, and Black moves first. There is universal agreement that Black should open by taking the 20-point coupon. White responds by taking the 19.5-point coupon. Excitement mounts as the play continues. Eventually, when the temperature of the coupon stack descends to somewhere in the low teens, someone plays the first stone on the board. Possibly the opponent replies by taking another coupon. Outstanding professionals such as Jiang and Rui often provide fascinating games. Unlike a conventional Go game, the environmental game forces the players to give us their expert opinions, at every move, as to whether or not the current move on the board is worth as much as the top coupon on the stack. In this way, we extract some interesting quantitative expert opinions about how big the various moves are, at least to within the $\delta = 0.5$ point difference between successive coupons.

Of course, the popular coupon stack is only a crude approximation to $\mathscr{E}_{20}$. One approximation is that $\delta = 0.5$ is relatively large. Another is that the popular initial komi is 9.5, which differs from the fair komi of $\mathscr{E}_{20}$, which is 10. These discrepencies are the results of an effort to maintain as much compatibility as feasible with the traditions and practices of conventional Go, where the score excluding the komi is necessarily an integer, and where the komi is a half-integer (to avoid ties) and generally one more than an *odd* integer (in order to minimize the risk that Chinese and Japanese scoring systems might disagree on the outcome of the game).

In Go, it is very difficult to realize a traditional game value of $\frac{1}{2}$. Bill Spight [2000] has constructed a position including a hyperactive ko and two sekis which, if there are no other kothreats on the board, behaves like the traditional mathematical value of $\frac{1}{2}$. Possibly the game value of $\frac{1}{2}$ cannot be realized on a Go board without a hyperactive ko position. I know no proof of this, but if such a position were to occur, it would almost certainly result in a dispute. After an appeal to the rules committees, this position would find its way into the collections of exotic cases covered by special scoring rules. Such collections now appear as appendices to the various dialects of official Go rules used in various Asian jurisdictions [Bozulich 1992]. The fact that none of the positions found in these appendices has traditional game value of $\frac{1}{2}$ provides strong evidence that such values are either impossible or can be constructed only with great effort.

Some of the more common complexities which can occur in Go scoring are described in Chapter 8 of [Mathews 1999], an excellent introductory book. Exotic subtleties substantially more intricate than have ever appeared in any known professional game have been composed by Harry Fearnsley [2000a; 2000b].

Most professional Go players are quite adept at calculating means and temperatures of latestage endgame positions. Their methodology differs somewhat from ours. It is faster but less accurate. Under appropriate assumptions which are usually satisfied in practice, it yields answers which agree with ours when the position is sufficiently simple and no kos are relevant. When the position is more complicated, professionals usually get approximate values quickly. Compared with mathematicians, the Go pros are vastly better at seeing the best local moves, but less patient and persistent in comparing the temperatures of different moves in different regions. Pros do approximations very quickly, but the mathematicians achieve more quantitative accuracy, and in many cases this improved accuracy translates back to improved lines of play which the players readily appreciate.

Orthodox Moves. Suppose that an enriched environment of sufficiently high temperature, t, is added to a game G, minus its mast value, and plus the appropriately signed fair initial komi, $t/2$. Then if both players play optimally, the net score at the end of the game will have magnitude less than ε, and the game will be declared a tie. The moves which a guru might play in the course of such a game are called *orthodox*. Other moves are called *unorthodox*.

As illustrated in [Berlekamp 2000a], there are many situations in which the traditional *canonical* methodology leads to a creeping growth of unmanageable complexity. So rather than consider all options of G which might be useful in playing $G + X$ for *some* X, orthodoxy focuses on playing $G + X$ for one particularly tractable value of X, namely $X = \mathcal{E}_t$. Thanks to the fact that $\mathcal{E}$ is idempotent, orthodox moves in G and H will, at the appropriate temperatures, remain orthodox moves in $G + H$.

An algorithm called *sentestrat* provides advice on how to play a sum of games. When a stack of coupons is present, the *ambient temperature* is defined as the value of the top coupon. If the opponent has just moved to a board position whose temperature is now higher than the ambient, then sentestrat tells you to respond in the same local game wherein your opponent has just moved. Otherwise, play in whichever region of the board or stack has the highest local temperature. Sentestrat is provably an orthodox strategy, assuming the only hyperactive ko (if any) has a clear komaster.

The *board temperature* is the temperature of the hottest region in which it is legal to move. (When a koban is applicable, the temperature of the ko is excluded.) The board temperature of a sum is the maximum of the temperatures of the summands.

One might choose to play an orthodox strategy, behaving as if coupons were present, even when there is no stack of coupons available. This requires a revised definition of the ambient temperature, which is defined historically. It is the minimum board temperature of any position which has yet occurred.

In the presence of an enriched environment, orthodox play by either player ensures that he will attain a result as least as desirable as the orthodox score, which is the mast value adjusted by the fair komi. When no enriched environment is present, then there are situations in which an unorthodox strategy can do better. Finding the optimal lines of play in such situations may require considerable search. The next section discusses an accounting technique for evaluating candidate lines of unorthodox play.

Orthodox Accounting. Orthodox accounting quantifies the benefits and costs of each move, whether the move is orthodox or not. The system is based on the forecasting methodology presented and illustrated in [Berlekamp and Wolfe 1994]. The present version includes refinements which handle kos.

Orthodox accounting attributes the final net score of a game to five different types of terms:

1. Mast value of the current position.
2. Fair current komi. The magnitude of this term is one half the ambient temperature, and the sign favors whoever's turn it is to move next.
3. $\frac{1}{2}$ Summation of signed temperature drops. Whenever the ambient temperature drops by Δt, then half that drop is awarded to whichever player got the last move at the old (higher) temperature.
4. Komaster adjustments. In many cases, these have already been correctly accounted for in term 1.
5. Komonster adjustments. This occurs only when one player finishes play on a hot ko; he then gains an adjustment equal to the difference between the temperature of his move and the ambient temperature.

In some sense, orthodox accounting is like a bank statement; it allows the observer, after the fact, to see exactly what credits came in and what debits were paid. It is a good accounting system in the sense that it always starts and ends with the correct balances, and the current *bottom line* does not undergo spurious big swings. In particular, suppose an orthodox move changes the mean by the current ambient temperature, which remains unchanged. In such a *transaction*, the changes in terms 1 and 2 cancel out, and the predicted net score remains unchanged.

When an enriched environment is present, Terms 3 and 5 are negligible. To see this, let m be an upper bound on the total number of moves that might be played on the board. For example, in a 19×19 Go game, most players would regard $m = 400$ as such a bound. Then since either player can place a bid which ensures that $\delta < \varepsilon/2m$, we can assume that at least one of them does so. In the course of the game, term 3 will provide many adjustments, but the number favoring one side cannot differ from the number favoring the other by more than m. A similar argument can bound the total effect of terms 5. When one player becomes komonster, he plays a kothreat and gets a response before retaking the ko. His opponent then takes a coupon as a kothreat, to which the komonster responds by taking another coupon after which the komonster's opponent retakes the ko. So the net affect of this typical sequence of six moves is two moves on the board and a decrease in ambient temperature of 2δ. So the sum of the magnitudes of all terms of type 5 cannot exceed $m\delta$.

The presence of an enriched environment ensures that the ambient temperature will decrease adiabatically. When no enriched environment is present, a decrease in the ambient temperature can be larger than δ. Such a decrease is called a *thermal shock*. Although an individual thermal shock can be significant, the sum of the magnitudes of all terms 3 is precisely the same as the magnitude of the fair initial komi, which is $t/2$. It is possible for the same player to get the benefits of all thermal shocks. But it is much more common for these benefits to be nearly evenly divided between the two players.

No matter how impoverished the environment, nor how poorly the game is played, orthodox accounting assigns a precise cost to each unorthodox move. In order for such a move to be a wise investment, it must lead to a future payoff via an item of type 3, 4, or 5. Even when an enriched environment is present, it is possible for the costs of the unorthodox move to be justified if it changes the balance of kothreats in such a way as to change the master of a forthcoming hyperactive ko. In an impoverished environment, there can also be other opportunities for returns on unorthodox investments (see the section on suspense and remoteness in Chapter 6 of *Winning Ways*). However, in the professional Go games we have analyzed, many of the unorthodox moves we have identified have simply turned out to be mistakes. Others subsequently breakeven in the sense that the cost is later recovered and the final score is the same as it would have been if an orthodox line had been played. Cases where unorthodox moves actually yield a profit seem to occur quite rarely in real play.

6. Suggestions for Further Work

Analysis of Real Go Endgames. The combination of local orthodox analysis, plus orthodox accounting, plus well-known search techniques such as alpha-beta pruning (which are well known in the artificial intelligence community) looks very promising but has yet to be thoroughly investigated and implemented.

Refinements of Ish. One can envision an expansion of the idempotent addition table to include more elements. Just below loony might be some sort of generic **tiny**. Tiny is a symbol which becomes completely defined only when followed by a subscript, G, which must be a game that is larger then some positive number.

Then $\mathbf{tiny}_G = 0 \,\|\, 0 \mid -G$. The smallest traditional tinies with finitely many positions are those for which $G = n$, a very large integer, and the largest are those for which $G = 2^{-k}$, a very small positive fraction. In some games we frequently encounter tinies whose subscripts are very complicated games which are definitely not numbers. Several of the positions in the complete analysis of Toads-and-Frogs in *Winning Ways*, Figure 12, p. 132 of are of this type. Chilling almost any professional Go game yields several such positions. So, it is often convenient to treat *tiny* generically, with the subscript unspecified. One may then need to specify whether the assertion one is making applies to all possible values of the subscript or to only some possible values of the subscript. In many cases, it doesn't matter, as in the assertion earlier in this paper that

$$\mathbf{miny} < \mathfrak{D} < \mathbf{tiny}$$

So I have yearned to insert **tiny** into the addition Table between $\mathfrak{D}$ and ☆, but I have not found any plausible opening ceremony with which to define a generic tiny that would be as nice as the other idempotents, and compatible with them.

In chilled Go, it is often desirable to distinguish between

$$0 \mid \mathbf{tiny} + \downarrow$$

and

$$0 \mid \mathbf{tiny} + \downarrow *$$

Both are common. The latter is a positive infinitesimal much bigger than tiny, but still small enough to be negligible for most purposes. The former is closely approximated by $*$. One may need to preserve the distinction between $*$ and 0 while neglecting $0|\mathbf{tiny} + \downarrow *$. The remote star is too crude for this purpose, because

$$☆ + *n = ☆$$

for all n, including $n = 1$.

An infinitesimal which preserves distinctions between 0 and $*$ needs not only to be of higher order than $\uparrow$, but also of higher order than $\uparrow^n$ [Conway 1976, pp. 199–200].

Near the big end of the range of ish, one might try to retrieve at least the biggest term in Norton's thermal dissociation (described in the Heating section of *Winning Ways*, Chapter 6). This term is nicely preserved by traditional *cooling*, but eradicated by enriched environments. In some composed problems, these terms can point the way to low-cost unorthodox plays which capture the benefit of a big thermal shock.

Kothreat Environments. In August 2000, Spight, Fraser, and I began studying several promising models of *kothreat environments*. These environments all behave like 0 when added to any traditional loopfree game, but they can have desirable simplifying effects on kos.

References

[Berlekamp 1996] E. Berlekamp, "The economist's view of combinatorial games", pp. 365–405 in *Games of no chance* (Berkeley, 1994), edited by R. Nowakowski, Math. Sci. Res. Publ. **29**, Cambridge Univ. Press, Cambridge, 1996.

[Berlekamp 2000a] E. Berlekamp, "Sums of $2 \times N$ Amazons", pp. 1–34 in *Game theory, optimal stopping, probability and statistics*, edited by T. Bruss and . LeCam, Lecture Notes – Monograph Series **35**, Inst. of Math. Stat., Hayward, CA, 2000.

[Berlekamp 2000b] E. Berlekamp, *The-dots-and boxes games: sophisticated child's play*, A K Peters, Natick (MA), 2000.

[Berlekamp 2002] E. Berlekamp, "The 4G4G4G4 problem and its solution", in *More games of no chance*, edited by R. Nowakowski, Math. Sci. Res. Publ. **42**, Cambridge Univ. Press, Cambridge, 2002.

[Berlekamp and Wolfe 1994] E. Berlekamp and D. Wolfe, *Mathematical Go: Chilling gets the last point*, A K Peters, Wellesley (MA), 1994.

[Berlekamp et al. 1982] E. R. Berlekamp, J. H. Conway, and R. K. Guy, *Winning ways for your mathematical plays*, Academic Press, London, 1982. Second edition, A K Peters, Natick (MA), 2001.

[Berlekamp et al. 1996] E. Berlekamp, M. Müller, and W. Spight, "Extended thermography for multiple kos in Go", Technical Report 96-030, Berkeley, 1996.

[Bozulich 1992] R. Bozulich, *The Go player's almanac*, Ishi Press, Mountain View, CA, 1992.

[Conway 1976] J. H. Conway, *On numbers and games*, London Math. Soc. Monographs **6**, Academic Press, London, 1976. Reprinted by A K Peters, Natick (MA), 2001.

[Fearnsley 2000a] H. Fearnsley, 2000. See http://www.goban.demon.co.uk/go/main.html.

[Fearnsley 2000b] H. Fearnsley, 2000. See http://www.goban.demon.co.uk/go/bestiary/zippersetc.html.

[Guy 1996] R. K. Guy, "Unsolved problems in combinatorial games", pp. 475–491 in *Games of no chance* (Berkeley, 1994), edited by R. Nowakowski, Math. Sci. Res. Publ. **29**, Cambridge Univ. Press, Cambridge, 1996.

[Mathews 1999] C. Mathews, *Teach yourself Go*, Hodder and Stoughton, London, 1999.

[Moews 1996] D. Moews, "Infinitesimals and coin-sliding", pp. 135–150 in *Games of no chance* (Berkeley, 1994), edited by R. Nowakowski, Math. Sci. Res. Publ. **29**, Cambridge Univ. Press, Cambridge, 1996.

[Nakamura and Berlekamp 2002] T. Nakamura and E. R. Berlekamp, "Analysis of composite corridors", ICSI technical report, 2002. See http://www.icsi.berkeley.edu/techreports/2002.abstracts/tr-02-005.html.

[Snatzke 2001] R. G. Snatzke, "Exhaustive search In Amazons", in *More games of no chance*, edited by R. Nowakowski, Math. Sci. Res. Publ. **42**, Cambridge Univ. Press, Cambridge, 2002.

[Spight 1998] W. Spight, "Extended thermography for multiple kos in Go", pp. 232–251 in *Computers and Games* (Tsukuba, Japan, 1998), edited by H. J. van den Herik and H. Iida, Lecture Notes in Comp. Sci. **1558**, Springer, Berlin, 1998.

[Spight 2000] W. Spight, 2000. Unpublished correspondence.

[Spight 2001] W. Spight, "Go Thermography: The 4/21/98 Jiang-Rui Endgame", in *More games of no chance*, edited by R. Nowakowski, Math. Sci. Res. Publ. **42**, Cambridge Univ. Press, Cambridge, 2002.

[Takizawa 2002] T. Takizawa, "An application of mathematical game theory to go endgames: some width-two-entrance rooms with and without kos", in *More games of no chance*, edited by R. Nowakowski, Math. Sci. Res. Publ. **42**, Cambridge Univ. Press, Cambridge, 2002.

ELWYN BERLEKAMP
MATHEMATICS DEPARTMENT
UNIVERSITY OF CALIFORNIA
BERKELEY, CA 94720-3840
    berlek@math.berkeley.edu

More Games of No Chance
MSRI Publications
Volume **42**, 2002

On the Lattice Structure of Finite Games

DAN CALISTRATE, MARC PAULHUS, AND DAVID WOLFE

ABSTRACT. We prove that games born by day n form a distributive lattice, but that the collection of all finite games does not form a lattice.

Introduction

A great deal is known about the partial order structure of large subsets of games. See, for instance, [BCG82] [Con76] for a complete characterization of games generated by numbers, and infinitesimals such as $\uparrow$ and $*n$. Linear operators applied to these games of temperature zero can often leverage this characterization to apply to hot games, such as positions occurring in Go [BW94] and Domineering [Ber88] [Wol93]. Some general results are known about the group structure of games, including a complete characterization of the group generated by games born by day 3 [Moe91], but surprisingly little has been written about the overall structure of the partial-ordering of games. Here we prove that the games born by day n form a distributive lattice, but that the collection of all finite games do not form a lattice.

We assume the reader is already familiar with combinatorial game theory definitions as in [BCG82] or [Con76]. In particular, we assume knowledge of the definitions of a game [BCG82, p. 21], sums and negatives of games [BCG82, p. 33], and the standard partial ordering on games [BCG82, p. 34].

The Lattices

Define the games born by day n, which we'll denote by $\mathcal{G}[n]$, recursively:

$$\mathcal{G}[0] \stackrel{\text{def}}{=} \{0\}$$
$$\mathcal{G}[n] \stackrel{\text{def}}{=} \{\{G^L \mid G^R\} : G^L, G^R \subseteq \mathcal{G}[n-1]\}$$

A *lattice*, $(S, \geq)$, is a partial order with the additional property that any pair of elements, $x, y \in S$ has a least upper bound or *join* denoted by $\vee$, and a greatest lower bound or *meet* denoted by $\wedge$. I.e., $x \geq x \vee y$ and $y \geq x \vee y$, and for any $z \in S$, if $z \geq x$ and $z \geq y$ then $z \geq x \vee y$. (Reverse all inequalities for $x \wedge y$.) In a *distributive lattice*, meet distributes over join (or, equivalently, join distributes over meet.) I.e, for all $x, y, z \in S$, $x \wedge (y \vee z) = (x \wedge y) \vee (y \wedge z)$.

We'll give a constructive proof that the games born by day n form a lattice by explicit construction of the join and meet operations. First, define

$$\lceil G \rceil \stackrel{\text{def}}{=} \{H \in \mathcal{G}[n-1] : H \not\leq G\}, \text{ and}$$

$$\lfloor G \rfloor \stackrel{\text{def}}{=} \{H \in \mathcal{G}[n-1] : H \not\geq G\}.$$

Then define the join and meet operations (over games born by day n) by

$$G_1 \vee G_2 \stackrel{\text{def}}{=} \left\{G_1^L, G_2^L \mid \lceil G_1 \rceil \cap \lceil G_2 \rceil\right\}, \text{ and}$$

$$G_1 \wedge G_2 \stackrel{\text{def}}{=} \left\{\lfloor G_1 \rfloor \cap \lfloor G_2 \rfloor \mid G_1^R, G_2^R\right\}.$$

Observe that $G_1 \vee G_2$ and $G_1 \wedge G_2$ are in $\mathcal{G}[n]$ since their left and right options are chosen from $\mathcal{G}[n-1]$.

Theorem 1. *The games born by day n form a lattice, with the join and meet operations given above.*

Proof. To verify these operations define a lattice, it suffices to show that

$$G_1 \vee G_2 \geq G_i \text{ (for } i = 1, 2\text{), and} \tag{0-1}$$

$$\text{if } G \geq G_1 \text{ and } G \geq G_2 \text{ then } G \geq G_1 \vee G_2. \tag{0-2}$$

($G_1 \wedge G_2$ can be verified symmetrically.)

To see (0–1), we'll show the difference game $(G_1 \vee G_2) - G_i$ (for $i = 1$ and $i = 2$) is greater or equal to 0, i.e., that Left wins moving second on this difference game. Left can respond to a Right move to $(G_1 \vee G_2) - G_i^L$ by moving to $G_i^L - G_i^L$. If, on the other hand, Right moves to $H - G_i$ where $H \in \lceil G_1 \rceil \cap \lceil G_2 \rceil$, then by definition of $\lceil G_i \rceil$, $H \not\leq G_i$, and hence Left wins moving first on $H - G_i$.

To see (0–2), suppose $G \geq G_1$ and $G \geq G_2$, and we'll show Left wins moving second on the difference game $G - (G_1 \vee G_2)$. Observe that any right option G^R of G is greater or incomparable to G, and hence is greater or incomparable to both G_1 and G_2. Therefore, $G^R \in \lceil G_1 \rceil \cap \lceil G_2 \rceil$. Thus, Left can respond to a Right move to $G^R - (G_1 \vee G_2)$ by moving to $G^R - G^R$. If, on the other hand, Right moves on the second component to some $G - G_i^L$ (for $i = 1$ or $i = 2$), Left has a winning response since $G \geq G_i$. $\square$

Theorem 2. *The lattice of games born by day n is distributive.*

Proof. First, observe the following identities:

$$\lfloor G_1 \vee G_2 \rfloor = \lfloor G_1 \rfloor \cup \lfloor G_2 \rfloor, \text{ and} \tag{0-3}$$

$$\lceil G_1 \wedge G_2 \rceil = \lceil G_1 \rceil \cup \lceil G_2 \rceil. \tag{0-4}$$

(To see the first, $\lfloor G_1 \vee G_2 \rfloor = \{X : X \not\geq G_1 \text{ or } X \not\geq G_2\} = \lfloor G_1 \rfloor \cup \lfloor G_2 \rfloor$.)

We wish to show $H \wedge (G_1 \vee G_2) = (H \wedge G_1) \vee (H \wedge G_2)$. Expanding both sides, call them S_1 and S_2, and rewriting S_2 using (0–3) and (0–4),

$$S_1 = \quad H \wedge (G_1 \vee G_2) \quad = \{ \quad \lfloor H \rfloor \cap \lfloor G_1 \vee G_2 \rfloor \quad | \quad H^R, \lceil G_1 \rceil \cap \lceil G_2 \rceil \quad \}$$
$$S_2 = (H \wedge G_1) \vee (H \wedge G_2) = \{ \lfloor H \rfloor \cap \lfloor G_1 \rfloor, \lfloor H \rfloor \cap \lfloor G_2 \rfloor \mid \lceil H \wedge G_1 \rceil \cap \lceil H \wedge G_2 \rceil \}$$
$$= \{ \quad \lfloor H \rfloor \cap \lfloor G_1 \vee G_2 \rfloor \quad | \quad \lceil H \rceil, \lceil G_1 \rceil \cap \lceil G_2 \rceil \quad \}$$

Clearly, $S_1 \geq S_2$, since S_2 has additional right options. To see that $S_2 \geq S_1$, we'll confirm Left wins second on the difference game $S_2 - S_1$. All right options match up except those moving S_2 to $X \in \lceil H \rceil$. By definition of $\lceil H \rceil$, $X \not\leq H$. Also, $H \geq S_1$, since S_1 is formed by the meet $H \wedge (G_1 \vee G_2)$. Hence $X \not\leq S_1$, and Left can win moving first on $X - S_1$.   □

Theorem 3. *The collection of finite games,* $\mathcal{G} = \bigcup_{n \geq 0} \mathcal{G}[n]$, *is not a lattice.*

Proof. We'll prove the stronger statement that no two incomparable games, G_1 and G_2, have a join in $\mathcal{G}$. We'll do this by arguing that if $G > G_1$ and $G > G_2$, then $G \mathbin{>=} H_n$ for some n, where

$$H_n \stackrel{\text{def}}{=} \{G_1, G_2 \parallel G_1, G_2 \mid -n\}$$

Since $H_0 > H_1 > H_2 > \cdots$, the theorem follows.

Suppose $G > G_1$ and $G > G_2$, and denote G's birthday by n. Note that all followers G' of G satisfy $-n < G' < n$. We'll confirm that Left wins moving second on the difference game $G - H_n$. Right cannot win by moving H_n to G_i (for $i = 1$ or $i = 2$), since $G > G_i$. When Right's initial move is on G, Left replies on the second component, $-H_n$, leaving $G^R - \{G_1, G_2 \mid -n\}$. Either Right plays on the first component, and Left wins by moving on the second component leaving $G^{RR} + n > 0$. Or Right moves the second component to some G_i and Left has a winning move since $G > G_i$.   □

Lattices up to Day 3

The specific structure of the distributive lattice of games born by day n remains elusive. We show the day 1 and day 2 lattices here; the day 2 lattice corrects errors found in [Guy96, p. 55] [Guy91, p. 15]. The lattice has 22 games divided among 9 levels. (Lattice edges need only be drawn between adjacent levels.)

By extending the software package, *The Gamesman's Toolkit* [Wol96] [Wol], we find the lattice born by day 3 has 1474 games and can be drawn on 45 levels, with the number of games on successive levels being 1, 2, 3, 5, 8, 9, 12, 14, 17, 20, 24, 26, 30, 34, 39, 45, 52, 58, 65, 72, 77, 81, 86, 81, ..., 3, 2, 1. As with the games born by day 2, the partial ordering appears to be composed of many sub-lattices which are hypercubes. In addition, the day-3 lattice has 44 *join-irreducible* elements whose partial order completely determines the lattice.

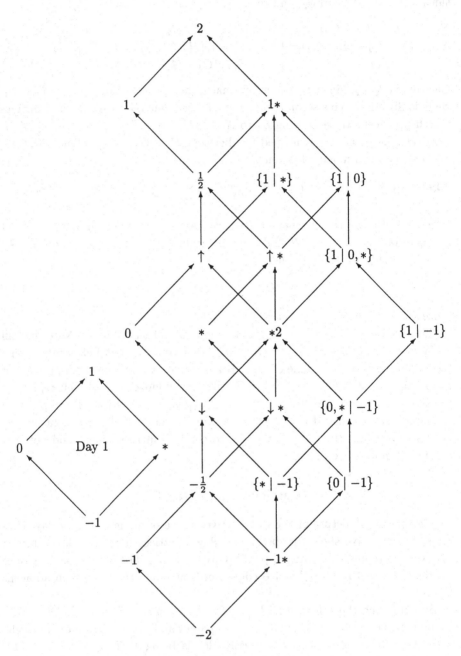

Figure 1. Games born by days 1 and 2.

These 44 elements are of the form g and $\{g \mid -2\}$, where g is one of the 22 games born by day 2. (Refer to a book on lattice theory such as [Bir67] or [DP90] for appropriate definitions and theorems.)

It would be interesting to describe the exact structure of the day 3 lattice, and (if possible) subsequent lattices.

Acknowledgements

We wish to thank Richard Guy for his thoughtful observations and his playful encouragement.

References

[BCG82] Elwyn R. Berlekamp, John H. Conway, and Richard K. Guy. *Winning Ways*. Academic Press, New York, 1982.

[Ber88] Elwyn R. Berlekamp. Blockbusting and domineering. *Journal of Combinatorial Theory*, 49(1):67–116, September 1988.

[Bir67] Garrett Birkhoff. *Lattice Theory*. American Mathematical Society, 3rd edition edition, 1967.

[BW94] Elwyn Berlekamp and David Wolfe. *Mathematical Go: Chilling Gets the Last Point*. A K Peters, Ltd., Wellesley, Massachusetts, 1994.

[Con76] John H. Conway. *On Numbers and Games*. Academic Press, London/New York, 1976.

[DP90] B. A. Davey and H. A. Priestly. *Introduction to Lattices and Order*. Cambridge University Press, 1990.

[Guy91] Richard K. Guy. What is a Game? *Combinatorial Games*, Proceedings of Symposia in Applied Mathematics, 43, 1991.

[Guy96] Richard K. Guy. What is a Game? In Richard Nowakowski, editor, *Games of No Chance: Combinatorial Games at MSRI, 1994*, pages 43–60. Cambridge University Press, 1996.

[Moe91] David Moews. Sum of games born on days 2 and 3. *Theoretical Computer Science*, 91:119–128, 1991.

[Wol] David Wolfe. Gamesman's Toolkit (C computer program with source) available at http://www.gustavus.edu/~wolfe; click on "For research on games".

[Wol93] David Wolfe. Snakes in domineering games. *Theoretical Computer Science*, 119(2):323–329, October 1993.

[Wol96] David Wolfe. The gamesman's toolkit. In Richard Nowakowski, editor, *Games of No Chance: Combinatorial Games at MSRI, 1994*, pages 93–98. Cambridge University Press, 1996.

DAN CALISTRATE
  calistrate@shaw.ca

MARC PAULHUS
  paulhusm@math.ucalgary.ca

DAVID WOLFE
DEPARTMENT OF MATHEMATICS AND COMPUTER SCIENCE
GUSTAVUS ADOLPHUS COLLEGE
800 WEST COLLEGE AVENUE
SAINT PETER, MN 56082
UNITED STATES
  wolfe@gustavus.edu

More Games of No Chance
MSRI Publications
Volume **42**, 2002

More Infinite Games

JOHN H. CONWAY

ABSTRACT. Infinity in games introduces some weird and wonderful concepts. I examine a few here.

My PhD thesis was on transfinite numbers and ordered types and when I got my first job in Cambridge it was as a resident mathematical logician. This was very fortunate for me since I am also interested in wasting time, professionally, and I have invented some very powerful ways of wasting time. One of these is combinatorial game theory. Most combinatorial game theorists automatically have a finite mind set when they look at games — a game is a finite set of positions. However, as a logician, developing surreal numbers, this was irrelevant. I just took whatever was needed to make the theory work. For additive games the notion of sum worked very well. One does not need finiteness just, essentially, the idea that you cannot make an infinite sequence of legal moves.

One of the main results, the existence of strategies for games with no infinite chain of moves, has two proofs one of which works well for finite games the other for all games. The first involves drawing the game tree and, starting from the leaves, marking a position with a P, for Previous, if the next player to play from this position cannot win and otherwise mark it with an N, for a Next player win. The other proof is a *reductio ad absurdum*. Suppose we have a game with no strategies. From this position look at the options: some will have strategies and some maybe not. If one option is marked with a P then we can move there and win so the original position is marked with an N. If all options are marked N then we can mark it as P. If there is no strategy then one of the options has no strategy and we iterate the argument and we get an infinite chain of moves which we supposing there wasn't. However, one of the interesting things about this argument is that a game can be perfectly well defined and computable but its winning strategy not be computable. The mathematical logician Michael Rabin studied this situation and found some interesting results.

There are games that are finite but the number of moves is not bounded. For example, choose an integer then reduce it by one on each turn afterward. Name

1000 and the game lasts for another 1000 moves. This is clearly unbounded but is boundedly unbounded — after one move you know how long the game will last. But there are games much worse than this. A boundedly unbounded game would have the property that at the beginning of the game, you could name a number n such that after n moves you know how long the game will last. An boundedly, unboundedly unbounded game would have a number n such that after n moves you could name a number m such that after m more moves you would know how long the game could last. It is easy to see that the order of unboundedness can be arbitrarily large, even infinite.

An example of this is Sylver Coinage, a game I invented when I wanted an example for a talk to a Cambridge Undergraduate Society. This game is played by two players who take turns in naming positive integers which cannot be made up from previously named numbers. The person who names 1 looses, otherwise the game is trivial. This game is finite, as can be shown by result of J. J. Sylvester. Naming 2 is a bad move because your opponent names 3 and now all even numbers cannot be named (since $4 = 2 + 2$, $6 = 2 + 2 + 2$, and so on) and neither can any odd number greater than 3 ($5 = 2 + 3$, $7 = 3 + 2 + 2$, and so on). Only naming 1 is left as a legal move, which loses.

Also, 4 and 6 are good replies to each. All the even numbers greater than 2 are eliminated and for $k \geq 1$, $4k + 1$, $4k + 3$ are good replies to one another. This seems like good evidence that the game is a loss for the first player.

However, a few weeks later, Hutchings came along and said "What was that game?" and we played. He started by naming 5 and he won. We played again, he named 5 and won. In fact, naming 5 can be shown to be a good move by a strategy stealing argument. So we know it is a good move but we don't know how. Indeed, this strategy stealing argument shows that all primes other than 2 and 3 are winning first moves for the first player. Since multiples of a prime are easy to answer, the only opening moves which are in doubt are those of the form $2^a 3^b$. We know that 2 and 3 are good answers to one another, so are 4 and 6 and so are 8 and 12. Is naming 16 a good move? Nobody knows.

It is easy to see that Sylver Coinage terminates; however it is unboundedly, unboundedly, ... , unboundedly unbounded! For example, naming 6^{1000} still leaves a game which is boundedly, unboundedly, unbounded since it could be followed by:

$$6^{999}, 6^{998}, \ldots, 6, \mathbf{3^n}, 3^{n-1}, \ldots, 3, \mathbf{2^m}, 2^{m-1}, \ldots, 2.$$

Is there an algorithm for Sylver Coinage that tells you, by looking over all (possibly infinite) options, what the status of the a position is and what if any are the winning moves? The answer is yes (see *Winning Ways*) but I do not know what the algorithm is.

What about other infinite games? Well, Sprague–Grundy theory (for impartial games) applies to heaps of infinite ordinals. Partizan theory has ordinary numbers and a mechanism for defining infinite numbers.

In Hackenbush, we can have strings of infinite length connected to the ground at one end. For example, a single blue edge would have value $\{0 \mid \} = 1$, two blue edges would have value $\{0, 1 \mid \} = 2$, etc., and an infinite beanstalk of blue edges would have value $\{0, 1, 2, \ldots \mid \} = \omega$, the first infinite ordinal. This is reminiscent of von Neumann's definition of the ordinals, that is, every ordinal is the set of all smaller ordinals. For us though, we have the vertical slash, Left membership and Right membership. For the ordinals we never have to use Right membership so the surreal numbers contain the reals. This is fantastic for a group. However, addition better not be Cantorian addition but I already knew about Hessenberg maximal addition which is the appropriate operation.

In ring theory, we get negatives, so we have $-\omega = \{ \mid 0, -1, -2, \ldots\}$. In game theory though, there are curious hybrids. We can take an infinite blue beanstalk and add a single red edge to get the game $\{0, 1, 2, \ldots \mid \omega\}$ and this game should be bigger than all the integers but smaller than ω. In fact, it is $\omega - 1$. So that ω is no longer the smallest infinite number but the smallest infinite ordinal. By adding an infinite red beanstalk on the top of the infinite blue beanstalk we get the game $\{0, 1, 2, \ldots \mid \omega, \omega - 1, \omega - 2, \ldots\} = \omega/2$.

Now every surreal number has a representation

$$\omega^{x_0} r_0 + \omega^{x_1} r_1 + \cdots + \omega^{x_\alpha} r_\alpha, \qquad \alpha < \beta$$

where x_i are surreal numbers, r_i are real numbers. This an analogue of Cantor's normal form. It doesn't really explain every number. Some are $\varepsilon = \omega^\varepsilon$ which is not explained by the normal form. No method of notation will explain every surreal number.

What kinds of infinities can occur? The first question is: how big can a game get? Well, a game can be as big as an ordinal but every game can be beaten by an ordinal For example, if $G = \{A, B, \ldots \mid D, E, \ldots\}$ then $G < \{A, B, \ldots \mid \}$, why give your opponent a move! But now, by induction, A, B, etc., are smaller than some ordinal $\alpha \in \mathbf{On}$. Now every surreal number is less than some $\alpha \in \mathbf{On}$ therefore we also have $1/x > 1/\alpha$. So a related question is how small can positive games be?

There is this game $\uparrow = \{0 \mid *\}$ where $* = \{0 \mid 0\}$. It is easy to show that $* \leq 1/4$ since $* = 0 \mid 0 \leq 0 \mid 1/2 \leq 1/4$. In fact, it is easy to show that $* \leq 1/\alpha$ for $\alpha \in \mathbf{On}$. Thus

$$\text{all negative numbers} < * < \text{all positive numbers}$$

Likewise

$$0 < \uparrow < \text{positive numbers}$$

so ↑ is very small. Note that ↑ is not a number: it is the value of a game, which is a more subtle concept. Also note that $1/{\uparrow}$ is not defined since it would be bigger than all surreal numbers and there are no such numbers. (In fact, it does exist but is one of the *Oneiric* numbers.)

Indeed, ${\uparrow} = 0\|0|0$ and $+_x = 0\|0| - x$ is a positive game, in fact it is in the smallest subgroup of positive games.

There is a small set of notes in *On Numbers and Games* about infinite games which no one has taken up and I wish somebody would.

Let's take a copy of the Real line and somewhere at the end is ω then ω^2, ω^ω, etc., as Cantor did 120 years ago, all the way to the end of the line, which doesn't exist (but it does which we shall soon see). But what is the smallest infinite number? There is a lot of space between the first infinite ordinal, ω, and the real numbers. Here are some surreal numbers that fit in the gap: $\omega - 1$, $\sqrt{\omega}$ and $\log \omega$. Numbers like ω^{1/ω^α} are the smallest infinite numbers but not the smallest infinite games.

The smallest appear to be an analogues of ↑, and are of the form $Z\|Z|Z = \infty$ and $Z\|Z|0 = \infty_0$. Moreover,

$$Z\|Z|0 - Z\|Z|Z = Z|0 - Z\|Z|0 = Z|0 - Z|Z.$$

So $Z \| Z|Z = Z|Z + Z|0$ is an analogue of the upstart inequality ${\uparrow} + {\uparrow} + * = \{0|{\uparrow}\}$.

So we can ask the question what is in between the reals and ω^{1/ω^α}. Certainly, $\infty_\alpha = Z\|Z|-\alpha$. Some of the questions were answered in *On Numbers and Games* but I would love to have a theory to cover the situation with Z. I think there should be some analogue with the small games — some infinite thermography in units of ∞_0.

Let's take a look at loopy games. Recall that $1 = \{0|\}$, $2 = \{0,1|\}$, $3 = \{0,1,2|\}$ etc. This is the way we generate all the ordinal numbers. Wouldn't it be nice to put yourself on the left. Well, we can: $G = \{G|\}$ or $G = \{\text{ pass }|\}$ and then by transfinite induction G is bigger than zero, all ordinals and, in fact, any game, The game is called On $= \{\text{On}|\}$. This is a loopy game, Left can move to On but Right cannot move. The existence of gaps in the line is solved by loopy games.

This game also gives $1/\text{On} = \{0|1/\text{On}\}$ and this game is the continuation of $1/2, 1/4, \ldots, 1/\omega, \ldots$. Therefore $1/\text{On}$ is absolutely the smallest positive game.

Each loopy game seems to have a left value and right value calling infinite number Left moves is a win for Left and an infinite number of moves for Right as a win for Right. We have some results but they are not true in full generality. The usual technique is to use sidling. Someone should follow this up.

Something else that intrigues me is multiplication of games. Addition of games is easy, $G + H = \{G^L + H, G + H^L | G^R + H, G + H^R\}$ Multiplication does not work with games in general but it does work for surreal numbers!

$$xy = \{x^L y + xy^L - x^L y^L, x^R y_x y^R - x^R y^R \mid x^L + xy^R - x^L x^R, x^R y + xy^L - x^R y^L\}.$$

That is why we could not define $1/\uparrow$. It took me two years to come up with the definition of multiplication. Surprisingly this definition works for Nimbers.

	*0	*1	*2	*3	*4	*5	*6	*7	*8
*0	*0	*0	*0	*0	*0	*0	*0	*0	*0
*1	*0	*1	*2	*3	*4	*5	*6	*7	*8
*2	*0	*2	*3	*1	*8	*10	*11	*9	*12
*3	*0	*3	*1	*2	*12	*15	*13	*14	*4
*4	*0	*4	*8	*12	*6	*2	*14	*10	*11

Let's drop the $*$. Note that $2 \times 2 = 3$ and $4 \times 4 = 6$. In fact, the nim-product of 2^{2^m} and 2^{2^m} is their normal product if $n \neq m$ and is $(3/2) \times 2^{2^n}$ otherwise. It all looks very finite. But to a logician it keeps on working into the infinite. In fact, I found that $\omega \times \omega \times \omega = 2$. So ω is the cube root of 2. This is a delightful part of the theory, that it still keeps surprising us with bizarre results, a wonderful mix of a coherent theory and hard calculation. I could only create artificial games with nim multiplication as part of their theory but Hendrik Lenstra has some quite natural games and knows more about these than anyone else. The ordinal that Cantor would call ω^3 let me write $[\omega^3]$ and our definition of cube inverts Cantor's multiplication, essentially $[\omega^3]^3 = \omega$. So our cube operation, moving from right to left, gives

$$2 \leftarrow \omega \leftarrow [\omega^3] \leftarrow [\omega^9] \leftarrow \cdots;$$

these tend to ω^ω and that turns out to be the fifth root of 4. As we now might expect, we have a chain of fifth powers

$$4 \leftarrow \omega^\omega \leftarrow [\omega^{\omega^5}] \leftarrow [\omega^{\omega^{2 \times 5}}] \leftarrow \cdots.$$

These ordinals tend to ω^{ω^2} which is the seventh root of $\omega + 1$. Now let $2 = \alpha_3$, $4 = \alpha_5$ and $\omega + 1 = \alpha_7$. I proved the theorem that α_p is the p-th root of the first number which did not already have a p-th root. For example, 0 and 1 have cube roots but 2 does not so $2 = \alpha_3$. Since 2 and 3 have fifth roots but 4 does not so $4 = \alpha_5$. All integers (and ω) have seventh roots so $\alpha_7 = \omega + 1$. Lenstra has shown that $\alpha_p^{1/p}$ is computable and computed several of them.

This is a wonderful theory, there are many ways in which games can go infinite and what is most surprising is the incredible structure at all levels. Usually, in mathematics, when things go infinite, things smooth out. Except here, when suddenly it is the cube root of 2.

JOHN H. CONWAY
DEPARTMENT OF MATHEMATICS
PRINCETON UNIVERSITY
FINE HALL
WASHINGTON ROAD
PRINCETON, NJ 08544

More Games of No Chance
MSRI Publications
Volume **42**, 2002

Alpha-Beta Pruning Under Partial Orders

MATTHEW L. GINSBERG AND ALAN JAFFRAY

ABSTRACT. Alpha-beta pruning is the algorithm of choice for searching game trees with position values taken from a totally ordered set, such as the set of real numbers. We generalize to game trees with position values taken from a partially ordered set, and prove necessary and sufficient conditions for alpha-beta pruning to be valid. Specifically, we show that shallow pruning is possible if and only if the value set is a lattice, and full alpha-beta pruning is possible if and only if the value set is a distributive lattice. We show that the resulting technique leads to substantial improvements in the speed of algorithms dealing with card play in contract bridge.

1. Introduction

Alpha-beta (α-β) pruning is widely used to reduce the amount of search needed to analyze game trees. However, almost all discussion of α-β in the literature is restricted to game trees with real or integer valued positions. It may be useful to consider game trees with other valuation schema, such as vectors, sets, or constraints. In this paper, we attempt to find the most general conditions on a value set under which α-β may be used.

The intuition underlying α-β is that it is possible to eliminate from consideration portions of the game tree that can be shown not to be on the "main line." Thus if one player P has a move leading to a position of value v, any alternative or future move that would let the opponent produce a value v' worse for P than v need not be considered, since P can (and should) always make choices in a way that avoid the value v'.

The assumption in the literature has been that terms such as "better" and "worse" refer to comparisons made using a total order; there has been almost no consideration of games where payoffs may be incomparable. As an example, imagine a game involving a card selected at random from a standard 52-card deck. If I make move m_1, I will win the game if the card is an ace. If I make

---

This work has been supported by DARPA/Rome Labs under contracts F30602-95-1-0023 and F30602-97-1-0294.

move m_2, I will win the game if the card is a spade. Since there are more spades in the deck than aces, m_2 is presumably the better move.

The conventional way to analyze this situation is to convert both conditions to probabilities, saying that m_1 has a payoff of $1/13$ and m_2 a payoff of $1/4$. The second payoff is better, so I make move m_2.

Now imagine, however, that I am playing a more complicated game. This game involves two subgames; in order to win the overall game, I need to win both subgames. The first subgame is as described in the previous paragraph. For the second subgame, m_1 wins if the card is an ace as before; m_2 wins if the card is a heart. It is clear that m_2 has no chance at all of winning the overall game (a card cannot be a heart and a spade both), so that m_1 is now to be preferred.

This example should make it clear that when we assign a number to a move such as m_1 or m_2, the assignment is only truly meaningful at the root of the game tree. At internal nodes, this game requires that we understand the context in which a particular move wins or loses, as opposed to simply the probability of that context. The contexts in which different moves win will not always be comparable, although they can be forced to be comparable by converting each context into a real number probabilistically. Doing so makes sense at the root of the game tree, but not at internal nodes or other situations where the contexts must be combined in some fashion.

Our goal in this paper is to analyze games while working with the most natural labels available. These may be real numbers, tuples of real numbers (natural if the payoff function involves multiple attributes), contexts (often natural in games of incomplete information), or other values. Given a game described in these terms, under what conditions can we continue to use α-β pruning to reduce the search space?

We will answer this question in the next three sections. Section 2 begins by introducing general operations on the value set that are needed if the notion of a game tree itself is to be meaningful. In Section 3, we go on to show that the validity of shallow α-β pruning is equivalent to the condition that the payoff values are taken from a mathematical structure known as a *lattice*. In Section 4, we show further that deep α-β pruning is valid if and only if the lattice in question is distributive. Section 5 describes an application of our techniques to card play problems arising in contract bridge. Related work is discussed briefly in Section 6 and concluding remarks are contained in Section 7.

2. Structure and Definitions

By a *game*, we will mean basically a set of rules that determine play from a given initial position I. We will assume that all games are two-player, involving a minimizer (often denoted 0) and a maximizer (often denoted 1). The "rules" of the game consist of a successor function s that takes a position p and returns the set $s(p)$ of possible subsequent positions. If $s(p) = \varnothing$, p is a terminal position and is assigned a value by an evaluation function e.

Definition 2.1. A *game* is an octuple

$$(P, V, I, w, s, e, f_+, f_-)$$

such that:

(i) P is the set of possible positions in the game.
(ii) V is the set of values for the game.
(iii) $I \in P$ is the initial position of the game.
(iv) $w : P \to \{0, 1\}$ is a function indicating which player is to move in each position in P.
(v) $s : P \to \mathcal{P}(P)$ gives the successors of each position in P.
(vi) $e : s^{-1}(\varnothing) \to V$ evaluates each terminal position in P.
(vii) $f_+ : \mathcal{P}(V) \to V$ and $f_- : \mathcal{P}(V) \to V$ are the combination functions for the maximizer and minimizer respectively.

Most games that have been discussed in the AI literature take V to be the set $\{-1, 0, +1\}$, with -1 being a win for the minimizer, $+1$ a win for the maximizer, and 0 a draw. The functions f_+ and f_- conventionally return the maximum and minimum of their set-valued arguments.

Note, incidentally, that the assumption that f_+ and f_- have sets as arguments is not without content. It says, for example, that there is no advantage to either player to have multiple winning options in a given position; one winning option suffices.

A game is finite if there is no infinite sequence of legal moves starting at the initial position. Formally:

Definition 2.2. A game $(P, V, I, w, s, e, f_+, f_-)$ will be called *finite* if there is some integer N such that there is no sequence $p_0, \ldots, p_N$ with $p_0 = I$ and $p_i \in s(p_{i-1})$ for $i = 1, \ldots, N$.

Given the above definitions, we can use the minimax algorithm to assign values to nonterminal nodes of the game tree:

Definition 2.3. Let $G = (P, V, I, w, s, e, f_+, f_-)$ be a finite game. By the *evaluation function for G* we will mean that function $\bar{e}$ defined recursively by

$$\bar{e}(p) = \begin{cases} e(p), & \text{if } s(p) = \varnothing; \\ f_+\{\bar{e}(p') | p' \in s(p)\} & \text{if } w(p) = 1; \\ f_-\{\bar{e}(p') | p' \in s(p)\} & \text{if } w(p) = 0. \end{cases} \tag{2-1}$$

The *value* of G will be defined to be $\bar{e}(I)$.

Proposition 2.4. *The evaluation function $\bar{e}$ is well defined for finite games.*

Proof. The proof proceeds by induction on the distance of a given position p from the fringe of the game tree. $\qquad \square$

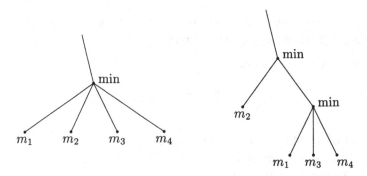

Figure 1. Equivalent games?

The expression (2–1) is, of course, just the usual minimax algorithm, reexpressed as a definition. In keeping with our overall goals, we have replaced the usual max and min operations with f_+ and f_-.

Consider now the two game trees in Figure 1, where none of the m_i are intended to be necessarily terminal. Are these two games always equivalent?

We would argue that they are. In the game on the left, the minimizer needs to select among the four options m_1, m_2, m_3, m_4. In the game on the right, he needs to first select whether or not to play m_2; if he decides not to, he must select among the remaining options. Since the minimizer has the same possibilities in both cases, we assume that the values assigned to the games are the same.

From a more formal point of view, the value of the game on the left is $f_-(m_1, m_2, m_3, m_4)$ and that of the game on the right is $f_-(m_2, f_-(m_1, m_3, m_4))$, where we have abused notation somewhat, writing m_i for the value of the node m_i as well.

Definition 2.5. A game will be called *simple* if for any $x \in v \subseteq V$,

$$f_+\{x\} = f_-\{x\} = x$$

and also

$$f_+(v) = f_+\{x, f_+(v - x)\} \quad \text{and} \quad f_-(v) = f_-\{x, f_-(v - x)\}.$$

We have augmented the condition developed in the discussion of Figure 1 with the assumption that if a player's move in a position p is forced (so that p has a unique successor), then the value before and after the forced move is the same.

Proposition 2.6. *For any simple game, there are binary functions $\wedge$ and $\vee$ from V to itself that are commutative, associative and idempotent*[1] *and such that*

$$f_+\{v_0, \ldots, v_m\} = v_0 \vee \cdots \vee v_m \quad \text{and} \quad f_-\{v_0, \ldots, v_m\} = v_0 \wedge \cdots \wedge v_m.$$

Proof. Induction on m. $\qquad\qquad\square$

---

[1] A binary function f is called *idempotent* if $f(a, a) = a$ for all a.

When referring to a simple game, we will typically replace the functions f_+ and f_- by the equivalent binary functions $\vee$ and $\wedge$. We assume throughout the rest of this paper that all games are simple and finite.[2]

The binary functions $\vee$ and $\wedge$ now induce a partial order $\leq$, where we will say that $x \leq y$ if and only if $x \vee y = y$. It is not hard to see that this partial order is reflexive ($x \leq x$), antisymmetric ($x \leq y$ and $y \leq x$ if and only if $x = y$) and transitive. The operators $\vee$ and $\wedge$ behave like greatest lower bound and least upper bound operators with regard to the partial order.

We also have the following:

Proposition 2.7. *Whenever $S \subseteq T$, $f_+(S) \leq f_+(T)$ and $f_-(S) \geq f_-(T)$.* $\square$

In other words, assuming that the minimizer is trying to reach a low value in the partial order and the maximizer is trying to reach a high one, having more options is always good.

3. Shallow Pruning

What about α-β pruning in all of this? Let us begin with shallow pruning, shown in Figure 2.

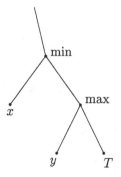

Figure 2. T can be pruned (shallowly) if $x \leq y$.

The idea here is that if the minimizer prefers x to y, he will never allow the maximizer even the possibility of selecting between y and the value of the subtree rooted at T. After all, the value of the maximizing node in the figure is $y \vee \overline{e}(T) \geq y \geq x$, and the minimizer will therefore always prefer x.

We will not give a precise definition of α-β pruning here, because the formal definition is fairly intricate if the game tree can be a graph (and we have nowhere excluded that). Instead, we observe simply that:

---

[2]We also assume that the games are sufficiently complex that we can find in the game tree a node with any desired functional value, e.g., $a \wedge (b \vee c)$ for specific a, b and c. Were this not the case, none of our results would follow. As an example, a game in which the initial position is also terminal surely admits pruning of all kinds (since the game tree is empty) but need not satisfy the conclusions of the results in the next sections.

Definition 3.1 (Shallow α-β pruning). A game G will be said to *allow shallow α-β pruning for the minimizer* if

$$x \wedge (y \vee T) = x \tag{3–1}$$

for all $x, y, T \in V$ with $x \leq y$. The game will be said to *allow shallow α-β pruning for the maximizer* if

$$x \vee (y \wedge T) = x \tag{3–2}$$

for all $x, y, T \in V$ with $x \geq y$. We will say that G *allows shallow pruning* if it allows shallow α-β pruning for both players.

As we will see shortly, the expressions (3–1) and (3–2) describing shallow pruning are identical to what are more typically known as *absorption identities*.

Definition 3.2. Suppose V is a set and $\wedge$ and $\vee$ are two binary operators on V. The triple $(V, \wedge, \vee)$ is called a *lattice* if $\wedge$ and $\vee$ are idempotent, commutative and associative, and satisfy the *absorption identities* in that for any $x, y \in V$,

$$x \vee (x \wedge y) = x, \tag{3–3}$$
$$x \wedge (x \vee y) = x. \tag{3–4}$$

Definition 3.3. A lattice $(V, \wedge, \vee)$ is called *distributive* if $\wedge$ and $\vee$ distribute with respect to one another, so that

$$x \vee (y \wedge z) = (x \vee y) \wedge (x \vee z), \tag{3–5}$$
$$x \wedge (y \vee z) = (x \wedge y) \vee (x \wedge z). \tag{3–6}$$

Lemma 3.4. *Each of (3–3) and (3–4) implies the other. Each of (3–5) and (3–6) implies the other.*

These are well known results from lattice theory [Grätzer 1978].

Proposition 3.5. *For a game G, the following conditions are equivalent:*

(i) *G allows shallow α-β pruning for the minimizer.*
(ii) *G allows shallow α-β pruning for the maximizer.*
(iii) *G allows shallow pruning.*
(iv) *$(V, \wedge, \vee)$ is a lattice.*

Proof. We show that the first and fourth conditions are equivalent; everything else follows easily.

If G allows shallow α-β pruning for the minimizer, we take $x = a$ and $y = T = a \vee b$ in (3–1). Clearly $x \leq y$ so we get

$$a \wedge (y \vee y) = a \wedge y = a \wedge (a \vee b) = a$$

as in (3–4).

For the converse, if $x \leq y$, then $x \wedge y = x$ and

$$x \wedge (y \vee T) = (x \wedge y) \wedge (y \vee T) = x \wedge (y \wedge (y \vee T)) = x \wedge y = x. \qquad \square$$

4. Deep Pruning

Deep pruning is a bit more subtle. An example appears in Figure 3.

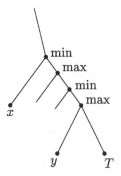

Figure 3. T can be pruned (deeply) if $x \leq y$.

As before, assume $x \leq y$. The argument is as described in the introduction: Given that the minimizer has a guaranteed value of x at the upper minimizing node, there is no way that a choice allowing the maximizer to reach y can be on the main line; if it were, then the maximizer could get a value of at least y.

Definition 4.1 (Deep α-β pruning). A game G will be said to *allow α-β pruning for the minimizer* if for any $x, y, T, z_1, \ldots, z_{2i} \in V$ with $x \leq y$,

$$x \wedge (z_1 \vee (z_2 \wedge \cdots \vee (z_{2i} \wedge (y \vee T))) \cdots) = x \wedge (z_1 \vee (z_2 \wedge \cdots \vee z_{2i}) \cdots).$$

The game will be said to *allow α-β pruning for the maximizer* if

$$x \vee (z_1 \wedge (z_2 \vee \cdots \wedge (z_{2i} \vee (y \wedge T))) \cdots) = x \vee (z_1 \wedge (z_2 \vee \cdots \wedge z_{2i}) \cdots).$$

We will say that G *allows pruning* if it allows α-β pruning for both players.

As before, the prune allows us to remove the dominated node (y in Figure 3) and all of its siblings.

Note that the fact that a game allows shallow α-β pruning does not mean that it allows pruning in general, as the following counterexample shows.

This game is rather like the game of the introduction, except only the suit of the card matters. In addition, the maximizer (but only the maximizer) is allowed to specify that the card must be one of two suits of his choosing.

The values we will assign the game are built from the primitives, "Win (for the maximizer) if the card is a club," "Win if the card is a diamond," and so on. The conjunction of two such values is a loss for the maximizer, since the card cannot be of two suits. The disjunction of two such values is a win for the maximizer, since he can require that the card be one of the two suits in question. The lattice of values is thus the one appearing in Figure 4. The maximizing function $\vee$ moves up the figure; the minimizing function $\wedge$ moves down.

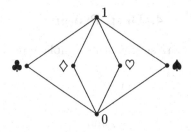

Figure 4. Values in a game where deep pruning fails.

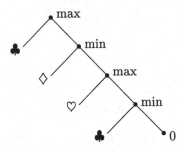

Figure 5. The deep pruning counterexample.

Consider now the game tree shown in Figure 5. If we evaluate it as shown, the backed up values are (from the lower right) 0, $\heartsuit$, 0, and $\clubsuit$, so that the maximizer wins if and only if the card is a club. But if we use deep α-β pruning to remove the 0 in the lower right (since its sibling has the same value $\clubsuit$ as the value in the upper left), we get values of $\clubsuit$, 1, $\diamondsuit$ and 1, so that the maximizer wins outright.

The problem is that we can't "push" the value $\clubsuit \wedge 0$ past the $\heartsuit$ to get to the $\clubsuit$ near the root. Somewhat more precisely, the problem is that

$$\heartsuit \vee (\clubsuit \wedge 0) \neq (\heartsuit \wedge \clubsuit) \vee (\heartsuit \wedge 0).$$

This suggests the following:

Proposition 4.2. *For a game G, the following conditions are equivalent:*

(i) *G allows α-β pruning for the minimizer.*
(ii) *G allows α-β pruning for the maximizer.*
(iii) *G allows pruning.*
(iv) *$(V, \wedge, \vee)$ is a distributive lattice.*

Proof. As before, we show only that the first and fourth conditions are equivalent. Since pruning implies shallow pruning (take $i = 0$ in the definition), it follows that the first condition implies that $(V, \wedge, \vee)$ is a lattice.

From deep pruning for the minimizer with $i = 1$, we have that if $x \leq y$, then for any z_1, z_2, T,

$$x \wedge (z_1 \vee (z_2 \wedge (y \vee T))) = x \wedge (z_1 \vee z_2)$$

Now take $y = T = x$ to get

$$x \wedge (z_1 \vee (z_2 \wedge x)) = x \wedge (z_1 \vee z_2). \tag{4-1}$$

It follows that each top level term in the left hand side of (4-1) is greater than or equal to the right hand side; specifically

$$z_1 \vee (z_2 \wedge x) \geq x \wedge (z_1 \vee z_2). \tag{4-2}$$

We claim that this implies that the lattice in question is distributive.

To see this, let $u, v, w \in V$. Now take $z_1 = u \wedge w$, $z_2 = v$ and $x = w$ in (4-2) to get

$$(u \wedge w) \vee (v \wedge w) \geq w \wedge ((u \wedge w) \vee v). \tag{4-3}$$

But $v \vee (u \wedge w) \geq w \wedge (v \vee u)$ is an instance of (4-2), and combining this with (4-3) gives us

$$
\begin{aligned}
(u \wedge w) \vee (v \wedge w) &\geq w \wedge ((u \wedge w) \vee v) \\
&\geq w \wedge w \wedge (v \vee u) \\
&= w \wedge (v \vee u).
\end{aligned}
$$

This is the hard direction; $w \wedge (v \vee u) \geq (u \wedge w) \vee (v \wedge w)$ for any lattice because $w \wedge (v \vee u) \geq u \wedge w$ and $w \wedge (v \vee u) \geq v \wedge w$ individually. Thus $w \wedge (v \vee u) = (u \wedge w) \vee (v \wedge w)$, and deep pruning implies that the lattice is distributive.

For the converse, if the lattice is distributive and $x \leq y$, then

$$
\begin{aligned}
x \wedge (z_1 \vee (z_2 \wedge (y \vee T))) &= (x \wedge z_1) \vee (x \wedge z_2 \wedge (y \vee T)) \\
&= (x \wedge z_1) \vee (x \wedge z_2) \\
&= x \wedge (z_1 \vee z_2),
\end{aligned}
$$

where the second equality is a consequence of the fact that $x \leq (y \vee T)$, so that $x = x \wedge (y \vee T)$. This validates pruning for $i = 1$; deeper cases are similar. $\square$

5. Bridge

To test our ideas in practice, we built a card-playing program for the game of contract bridge. Previous authors [Ginsberg 1996a; Ginsberg 1996b; Ginsberg 1999; Levy 1989] have too often considered only the perfect-information variant of this game; the problems with this approach have been pointed out by others [Frank and Basin 1998]. The proposed solution to this problem has been to search in the space of possible plans for playing a given deal or position, but the complexity of finding an optimal plan is NP-complete in the size (not depth) of the game tree, which is prohibitive. As a result, Frank et.al. report that it takes an average of 571 seconds to run an approximate search algorithm on problems of size 13 (running on a 300 MHz UltraSparc II) [Frank et al. 1998].

We built a search engine that is capable of solving problems such as these exactly. The value assigned to a fringe node is not simply the number of tricks that can be taken by one side or the other on a particular bridge hand, but a combination of the number of tricks and information that has been acquired about the hidden hands (e.g., West has the $\heartsuit 7$ but not the $\spadesuit Q$, and I can take seven tricks).

The defenders are assumed to be minimizing and to be playing with perfect information, so that their combination function is simple logical disjunction. As an example, if the nondefending side can take four tricks (and hence one, two or three as well) in one case, and only three tricks in another, the greatest lower bound of the two values is that the nondefending side can take three tricks – the disjunction.

The declaring side is assumed to be maximizing and is playing with incomplete information, so that they cannot simply take the best of two potentially competing outcomes. Instead, the maximizer works with values of the form $\mathtt{choose}(v_1, \ldots, v_n)$ where the v_i are the values assigned to the lines from which a choice must be made.

When these two operators are used to construct a (distributive) lattice in the obvious way, the ideas we have discussed can be applied to solve bridge problems under the standard assumption [Frank and Basin 1998] that the maximizer is playing with incomplete information while the minimizer is playing perfectly. The performance of the resulting algorithm is vastly improved over that reported by Frank. Endings with up to 32 cards are solved routinely in a matter of a few seconds (Frank et.al. could only deal with 13 cards), and endings with 13 cards are solved essentially instantly on a 500-MHz Pentium III. In addition, the results computed are exact as opposed to approximate.

6. Related Work

There appear to be two authors who have discussed issues related to those presented here.

Dasgupta et. al. examines games where values are taken from $\mathcal{R}^n$ instead of simply $\mathcal{R}$ [Dasgupta *et al.* 1996]. Since any distributive lattice can be embedded in $\mathcal{R}^n$ for some n, it follows from Dasgupta's results that if the values are taken from a distributive lattice, then α-β pruning can be applied. The converse, along with our results regarding nondistributive lattices, are not covered by this earlier work.

Müller considers games with values taken from arbitrary partially ordered sets, evaluating such games by taking a threshold value and then converting the original game to a 0-1 game by calling a position a win if its value exceeds the threshold and a loss otherwise [Müller 2000].

The problem with this approach is that you need some way to produce the threshold value; Müller suggests mapping the partial order in question into a discrete total order (the integers), and then using the values in that total order to produce thresholds for the original game.

While this appears to work in some instances (Müller examines capturing races in Go), it is unlikely to work in others. In bridge, for example, the natural mapping is from a context in which a hand can be made to the (discretized) probability that the context actually occurs. Unfortunately, it is possible to have three contexts c_1, c_2 and c_3 where the probability of c_1 or c_2 in isolation is less than the probability of c_3, but the probability of c_1 and c_2 together is greater than the probability of c_3. This suggests that there is no suitable linearization of the partial order discussed in the previous section, so that Müller's ideas will be difficult to apply.

7. Conclusion

If adversary search is to be effective in a setting where positional values have more structure than simple real numbers, we need to understand conditions under which proven AI algorithms can still be applied. For α-β pruning, this paper has characterized those conditions exactly: Shallow pruning remains valid if and only if the values are taken from a lattice, while deep pruning is valid if and only if the lattice in question is distributive. We also showed the somewhat surprising result that pruning is valid for one player if and only if it is valid for both.

We used the theoretical work described here to implement a new form of cardplay engine for the game of contract bridge. This implementation uses the standard bridge assumption that the defenders will play perfectly while the declarer may not, and avoids the Monte Carlo techniques used by previous authors. Compared to previous programs with this goal, we are able to compute exact results in a small fraction of the time previously needed to compute approximate ones.

References

[Dasgupta *et al.* 1996] Pallab Dasgupta, P. P. Chakrabarti, and S. C. DeSarkar. Searching game trees under a partial order. *Artificial Intelligence*, 82:237–257, 1996.

[Frank and Basin 1998] Ian Frank and David Basin. Search in games with incomplete information: A case study using bridge card play. *Artificial Intelligence*, 100:87–123, 1998.

[Frank *et al.* 1998] Ian Frank, David Basin, and Hitoshi Matsubara. Finding optimal strategies for imperfect information games. In *Proceedings of the Fifteenth National Conference on Artificial Intelligence*, pages 500–507, 1998.

[Ginsberg 1996a] Matthew L. Ginsberg. How computers will play bridge. *The Bridge World*, 1996.

[Ginsberg 1996b] Matthew L. Ginsberg. Partition search. In *Proceedings of the Thirteenth National Conference on Artificial Intelligence*, 1996.

[Ginsberg 1999]   Matthew L. Ginsberg. GIB: Steps toward an expert-level bridge-playing program. In *Proceedings of the Sixteenth International Joint Conference on Artificial Intelligence*, 1999.

[Grätzer 1978]   George Grätzer. *General Lattice Theory*. Birkhäuser Verlag, Basel, 1978.

[Levy 1989]   David N. L. Levy. The million pound bridge program. In D. N. L. Levy and D. F. Beal, editors, *Heuristic Programming in Artificial Intelligence*, Asilomar, CA, 1989. Ellis Horwood.

[Müller 2000]   M. Müller. Partial order bounding: A new approach to evaluation in game tree search. Technical Report TR-00-10, Electrotechnical Laboratory, Tsukuba, Japan, 2000. Available from http://web.cs.ualberta.ca/~mmueller/publications.html.

MATTHEW L. GINSBERG
CIRL
1269 UNIVERSITY OF OREGON
EUGENE, OR 97403-1269
UNITED STATES
ginsberg@cirl.uoregon.edu

ALAN JAFFRAY
jaffray@pobox.com

More Games of No Chance
MSRI Publications
Volume **42**, 2002

The Abstract Structure of the Group of Games

DAVID MOEWS

ABSTRACT. We compute the abstract group structure of the group **Ug** of partizan games and the group **ShUg** of short partizan games. We also determine which partially ordered cyclic groups are subgroups of **Ug** and **ShUg**.

As in [2], let **Ug** be the group of all partizan combinatorial games, let **No** be the field of surreal numbers, and for G in **Ug**, let $L(G)$ and $R(G)$ be the Left and Right sections of G, respectively. If $L(G)$ is the section just to the left or right of some number z, we say that z is the *Left stop* of G, and similarly for $R(G)$ and the *Right stop*. Let **ShUg** be the group of all short games in **Ug**; that is, **ShUg** is the set of all games born before day ω, or of all games which can be expressed in a form with only finitely many positions. For games U and integers n, we write

$$nU = n.U = \begin{cases} 0, & \text{if } n = 0; \\ U + \cdots + U \quad (n \text{ summands}), & \text{if } n \text{ is a positive integer;} \\ (-U) + \cdots + (-U) \ (-n \text{ summands}), & \text{if } n \text{ is a negative integer.} \end{cases}$$

Also, recall from [1, Chapter 8] the definition of Norton multiplication, for a game G and a game $U > 0$:

$$G.U = \begin{cases} \text{as above,} & \text{if } G \text{ equals an integer;} \\ \{G^L.U + U^L, G^L.U + (U + U - U^R)| & \\ \quad G^R.U - U^L, G^R.U - (U + U - U^R)\}, & \text{otherwise.} \end{cases} \quad (0\text{-}1)$$

Here, G^L, G^R, U^L, and U^R range independently over the left options of G, right options of G, left options of U, and right options of U, respectively. To define $G.U$, we must fix a form of G and sets of Left and Right options for U.

We will say that a subgroup $\mathfrak{X}$ of **Ug** has the *integer translation property* if it contains the integers and, whenever either Left or Right has a winning move in a sum $A_1 + \cdots + A_n$ of games from $\mathfrak{X}$, not all integers, he also has a winning move in an A_j which is not equal to an integer.

Lemma 1. *The real numbers have the integer translation property.*

Proof. Let x be an integer and G be a nonintegral real number, and set $G' = \{G^L + x | G^R + x\}$. It will do to show that $G' = G + x$. Let $L = \sup_{G^L} R(G^L)$ and $R = \inf_{G^R} L(G^R)$. Then since G is a number, $L < R$, and G is the simplest number satisfying $L < G < R$ [2, Theorem 56]. But $\sup_{G^L} R(G^L + x)$ equals $L + x$, $\inf_{G^R} L(G^R + x)$ equals $R + x$, and $L + x < R + x$, so G' will be the simplest number satisfying $L + x < G' < R + x$. Since $L + x < G + x < R + x$, to prove $G' = G + x$, we only need to show that no simpler number than $G + x$ satisfies $L + x < G + x < R + x$. Suppose S is born before $G + x$ and satisfies $L + x < S < R + x$. Since $G + x$ is real, and hence born on or before day ω, S must be a dyadic rational, and obviously $L < S - x < R$; but also, since G is the simplest number between L and R, G must be born before or at the same time as $S - x$, so G is a dyadic rational, $G = (2m+1)/2^n$, say, for integers $n > 0$ and m. Then for G to be the simplest number between L and R, we must have $m/2^{n-1} \leq L < (2m+1)/2^n$ and $(2m+1)/2^n < R \leq (m+1)/2^{n-1}$. Therefore,

$$(m+2^{n-1}x)/2^{n-1} \leq L+x < (2m+1+2^n x)/2^n$$

and

$$(2m+1+2^n x)/2^n < R+x \leq (m+1+2^{n-1}x)/2^{n-1},$$

so $(2m+1+2^n x)/2^n = G + x$ is in fact the simplest number between $L + x$ and $R + x$. $\qquad\square$

Theorem 2. *Suppose we have a subgroup $\mathfrak{X}$ of $\mathbf{Ug}$ with the integer translation property, and such that every $H \in \mathfrak{X}$ can be written in a form $\hat{H}$, where all positions of $\hat{H}$ are in $\mathfrak{X}$. Fix a game $U > 0$ and sets of Left and Right options for U, and define $G.U$ for each G in $\mathfrak{X}$ by using (0–1) with the form $\hat{G}$ for G and the given sets of options for U. Then, for all G and H in $\mathfrak{X}$, $(G+H).U = G.U+H.U$, and if $G \geq H$, then $G.U \geq H.U$.*

Proof. [1, Chapter 8]. $\qquad\square$

Let $\mathfrak{X}$ be the subgroup of real numbers. We fix forms for each real number by letting each dyadic rational have its canonical form; that is,

$$0 = \{ | \},$$

$$n = \{n-1 | \} \qquad \text{and} \qquad -n = \{ | -(n-1)\}$$

for integers $n > 0$, and

$$(2m+1)/2^n = \{m/2^{n-1} | (m+1)/2^{n-1}\}$$

for integers $n > 0$ and m. We let each real r that is not a dyadic rational have form

$$r = \{\lfloor r \rfloor, \lfloor 2r \rfloor/2, \lfloor 4r \rfloor/4, \ldots | \ldots, \lceil 4r \rceil/4, \lceil 2r \rceil/2, \lceil r \rceil\}.$$

By Lemma 1, the real numbers have the integer translation property, so we can now apply Theorem 2 to define $r.U$, where r is a real number and $U > 0$ is a game with specified sets of options.

Corollary 3. *For all real numbers r and s, and all games $U > 0$ with specified sets of options, $(r+s).U = r.U+s.U$, and if $r \geq s$, then $r.U \geq s.U$.*

Proof. Immediate.   □

Lemma 4. *If $n \geq 2$ is an integer and $x \in$ **No** is positive, then $G_{nx} = (2/n).\{2x|x\} - 3x/n$ has order n. The nonzero multiples of G_{nx} all have Left stops of x/n or larger.*

Proof. By Corollary 3, G_{nx} has order dividing n. Let $U = \{2x|x\}$; then $U^L = U+U-U^R = 2x$. Observe that $0.U$ has Right stop 0 and $1.U$ has Left stop $2x$. It follows by induction that for all dyadic rationals d in $(0,1)$, $d.U$ has Left stop $2x$ and Right stop 0, and then, for r real in $(0,1)$, $r.U$ also has stops $2x$ and 0. Similarly, since $1.U$ has Right stop x and $2.U = 3x$ has Left stop $3x$, $r.U$ has stops $3x$ and x for all r real in $(1,2)$, and $1.U = U$ clearly has stops $2x$ and x. This implies that $r.U$ is not a number for real r in $(0,2)$, so $m.G_{nx} \neq 0$ for $m = 1, \ldots, n-1$. Our claim on the Left stop of the multiples of G_{nx} follows from the computation of the stops of $r.U$.   □

No is the unique, up to isomorphism, universally embedding totally ordered field [2, Theorems 28 and 29]. We will prove a similar result about **Ug**.

An abelian group X is *universally embedding* if, given any abelian group G whose members form a set, and an embedding of a subgroup H of G in X, the embedding can be extended to an embedding of G in X. The members of such a group necessarily form a proper class.

Theorem 5. **Ug** *is a universally embedding abelian group.*

Proof. By Zorn's Lemma, it will do to show that if an abelian group G is generated by its subgroup H and its member $x \notin H$, and there is an embedding j of H in **Ug**, then there is an embedding of G in **Ug** extending j. Let M be the set of integers m with $mx \in H$. M is a subgroup of the integers. If $M = 0$, pick a large ordinal α, exceeding every element of $j(H)$, and embed G in **Ug** by sending x to α. Otherwise, M is cyclic, generated by $m > 1$, say. If $G_0 = j(mx)$, pick an ordinal $\beta > -G_0$ and sets of options for $G_0+\beta$, and set $G_1 = (1/m).(G_0+\beta)-\beta/m$. Obviously, $m.G_1 = G_0$. Let X be the subgroup of **Ug** generated by $j(H)$ and G_1, and let α be an ordinal such that $\alpha/2m$ exceeds every element of X. Now we can map G to **Ug** by sending x to $G_1+G_{m\alpha}$, and this will be an embedding if $q.(G_1+G_{m\alpha}) \neq j(h)$ for all $h \in H$ and $q \in \{1,\ldots,m-1\}$. But if $q.(G_1+G_{m\alpha}) = j(h)$, then $q.G_{m\alpha} \in X$, and since $q.G_{m\alpha}$ has Left stop at least α/m, $\alpha/2m \not\geq q.G_{m\alpha}$. This contradicts our choice of α. Hence we have embedded G into **Ug**.   □

Theorem 6. *Any universally embedding abelian group is isomorphic to **Ug**.*

Proof. Transfinite induction and a back-and-forth argument suffice to construct an isomorphism between any two universally embedding abelian groups.   □

Call a subgroup $\mathcal{G}$ of **ShUg** *odd-closed* if whenever G is a short game, n is an odd integer, and $n.G \in \mathcal{G}$, then $G \in \mathcal{G}$. Call it *position-closed* if whenever H is a position of the canonical form of $G \in \mathcal{G}$, then $H \in \mathcal{G}$.

Theorem 7. *Position-closed subgroups of* **ShUg** *are odd-closed.*

Proof. By a remark in [2], if G is short and n is odd, G is an integral linear combination of positions of (any form of) $n.G$. □

Theorem 8. [2, Theorem 92] *All short games have infinite order or order a power of 2.*

We now determine the abstract group structure of **ShUg**. Let $\mathcal{D}$ be the additive group of dyadic rationals.

Theorem 9. ShUg *is isomorphic to the direct sum of countably many* $\mathcal{D}s$ *and countably many* $\mathcal{D}/\mathbb{Z}s$.

Proof. We will find subgroups $\mathcal{S}_0, \mathcal{S}_1, \mathcal{S}_2, \ldots$ and $\mathcal{G}_0, \mathcal{G}_1, \mathcal{G}_2, \ldots$ of **ShUg** such that:

(i) Each $\mathcal{G}_l$ is a direct sum of $\mathcal{S}_0, \ldots, \mathcal{S}_l$.
(ii) $\bigcup_{l \geq 0} \mathcal{G}_l = $ **ShUg**.
(iii) Each $\mathcal{G}_l$ is position-closed (and hence odd-closed.)
(iv) Each $\mathcal{S}_l$ is isomorphic to either $\mathcal{D}$ or $\mathcal{D}/\mathbb{Z}$.

This will prove that **ShUg** is a countable direct sum of $\mathcal{D}s$ and $\mathcal{D}/\mathbb{Z}s$; this proves the theorem, unless possibly only finitely many $\mathcal{D}s$ or $\mathcal{D}/\mathbb{Z}s$ appear in the sum. If there were only finitely many $\mathcal{D}/\mathbb{Z}s$, k, say, then the subgroup of **ShUg** of games of order 2 would be $(\mathbb{Z}/2\mathbb{Z})^k$, which contradicts the existence of infinitely many games $(*, *2, *3, *4, \ldots)$ of order 2. Also, the tinies $+_1, +_2, +_3, \cdots$, generate a subgroup of **ShUg** isomorphic to the direct sum of countably many $\mathbb{Z}s$. Since this subgroup is torsion-free, it will map to an isomorphic subgroup of the quotient of **ShUg** by its torsion subgroup. If there are only finitely many $\mathcal{D}s$ in **ShUg**, k say, $\mathcal{D}^k$ will then have a subgroup isomorphic to $\mathbb{Z}^{k+1}$, which is impossible. Therefore, the direct sum must be as claimed.

We now proceed to the proof of 1–4. Well-order **ShUg** so that all options of H always precede H. (In this proof, by options and positions of a short game, we will always mean the options and positions of its canonical form.) We induce on l. Let $\mathcal{G}_0 = \mathcal{S}_0 = \mathcal{D}$. Clearly, 1, 3, and 4 are then true for $l = 0$. Otherwise, assume 1, 3, and 4 for $l = 0, \ldots, i$. Let q_i be the first short game not in $\mathcal{G}_i$, according to our order (so all options of q_i are in $\mathcal{G}_i$), and let r_i be an element of $q_i + \mathcal{G}_i$ with minimal order. Suppose that $2^b t r_i$ is in $\mathcal{G}_i$, where t is odd and $b \geq 0$. By odd-closure, $2^b r_i = z$, say, is in $\mathcal{G}_i$. Since $\mathcal{G}_i$ is 2-divisible, we see that there is y in $\mathcal{G}_i$ with $2^b y = z$. Then $2^b(r_i - y) = 0$, so $r_i - y$ has order dividing 2^b; by minimality of order, r_i also has order dividing 2^b, so $2^b r_i = 0$ and therefore $2^b t r_i = 0$. Hence $\mathcal{G}_i + \mathbb{Z}r_i$ is a direct sum. In fact, it is also position-closed; to see this, it will do to show that all positions of r_i are in $\mathcal{G}_i + \mathbb{Z}r_i$. Let $r_i = q_i + x$, $x \in \mathcal{G}_i$; all positions of r_i will equal $q' + x'$, where q' is a position of q_i and x' is a position of x. If q' isn't equal to q_i, then $q' + x'$ is already in $\mathcal{G}_i$; otherwise, $q_i + x' = r_i + (x' - x)$ is in $\mathcal{G}_i + \mathbb{Z}r_i$. This proves position-closure. Now for short games H, define

$$\phi(H) = \tfrac{1}{2}.(H + 2N_H) - N_H$$

where N_H is the minimal nonnegative integer such that $H + 2N_H > 0$. By our earlier remarks, $2\phi(H) = H$ for all H. Define

$$r_{ij} = \begin{cases} r_i, & j = 0, \\ \phi(r_{i(j-1)}), & j > 0. \end{cases}$$

Let $\mathcal{S}_{i+1} = \bigcup_{j \geq 0} \mathbb{Z}r_{ij}$. Evidently, $\mathcal{S}_{i+1}$ is isomorphic to $\mathcal{D}$ (if r_i has infinite order) or $\mathcal{D}/\mathbb{Z}$ (if r_i has order a power of 2.) Let $\mathcal{G}_{i+1} = \mathcal{G}_i + \mathcal{S}_{i+1}$. 4 is then certainly true. 1 will be true if the sum is direct. Let $2^k t r_{ij}$ be in $\mathcal{G}_i$, t odd, $k \geq 0$. By odd-closure, $2^k r_{ij}$ is in $\mathcal{G}_i$; if $k \leq j$, then $2^{j-k}2^k r_{ij} = 2^j r_{ij} = r_i$ is in $\mathcal{G}_i$, which is impossible. If $k > j$, then $2^k r_{ij} = 2^{k-j}r_i$ is in $\mathcal{G}_i$, and thus equals zero, since $\mathcal{G}_i + \mathbb{Z}r_i$ was direct. Hence $\mathcal{G}_i + \mathcal{S}_{i+1}$ is direct. For 3 to be true, we need $\mathcal{G}_{i+1}$ position-closed. It will do to show that for all j, all positions of r_{ij} are in $\mathcal{G}_{i+1}$. We induce on j. If $j = 0$, we have proved this above. Otherwise, we observe that any position of $\tfrac{1}{2}.K$, except $\tfrac{1}{2}.K$, is an integral linear combination of positions of K; therefore, any position of $r_{ij} = \phi(r_{i(j-1)})$ is either an integer translate of r_{ij} or an integer translate of an integral linear combination of positions of $r_{i(j-1)}$. The result then follows from the induction hypothesis.

This concludes the induction, proving that 1, 3, and 4 are true for all i. For 2, if some short game is not in $\bigcup_{l \geq 0} \mathcal{G}_l$, let K be the first such game, in our order. K will then eventually be chosen as some q_i; but $q_i \in \mathcal{G}_{i+1}$, which is a contradiction. This concludes the proof.   □

We would like to determine the abstract structure of **Ug** and **ShUg** as abstract partially ordered abelian groups. We have not done this, but we can approach the problem by first looking at cyclic subgroups of both groups. Any finite cyclic subgroup of **Ug** or **ShUg** must have all nonzero members incomparable with 0; so look at an infinite cyclic subgroup of either one, generated by G, say. We can't have $n.G > 0$ and $m.G < 0$ for positive m and n, since then $mn.G$ would have to be both positive and negative. Therefore either all positive multiples of G are positive or incomparable with 0, or all positive multiples of G are negative or incomparable with 0. By replacing G by $-G$ if necessary, we can assume that all positive multiples of G are positive or incomparable with 0. In this case, the set $\mathcal{S}$ of nonnegative integers n such that $n.G \geq 0$ must obviously be a submonoid of $\mathbb{Z}_{\geq 0}$. We will show that for $G \in$ **ShUg**, and hence also for $G \in$ **Ug**, all such submonoids can occur.

Lemma 10. $F = \{2|-1, \{0|-4\}\}$ has $n.F$ incomparable with 0, for all nonzero integers n.

Proof. First, we induce on n to show that $2+n.F \geq 0$ for all $n \geq 0$. If $n = 0$, this is clear. Otherwise, look at Right's first move. It can be to $1+(n-1).F$. Left has then won if $n = 1$; otherwise, he can respond on F to to $3+(n-2).F$, which is positive or zero by the induction hypothesis. Right's other first move is to $2+\{0|-4\}+(n-1).F$. In this case, Left should respond on $\{0|-4\}$, leaving $2+(n-1).F$, which is positive or zero by the induction hypothesis.

Now, it will do to show that both players have a winning first move from $n.F$ for all positive integers n. If $n > 0$, Left can move from $n.F$ to $2+(n-1).F$, and this is positive or zero by the above remarks. To show that Right has a winning first move, we induce on n. If $n = 1$, Right can move from F to -1 and win. If $n \geq 2$, Right's first move should be to $\{0|-4\}+(n-1).F$. If Left responds to $(n-1).F$, we have a good move by the induction hypothesis. Otherwise, Left must respond to $2+\{0|-4\}+(n-2).F$. If $n = 2$, Right can move to -2 and win. If $n = 3$, Right can move to $2+\{0|-4\}+\{0|-4\} = -2$ and win. Finally, if $n \geq 4$, Right can move to $-2+(n-2).F$. Left's only response is then to $(n-3).F$, and we can win this by the induction hypothesis. □

F has temperature 2 and mean value 0, so for all numbers $\varepsilon > 0$ and integers n, we have $-2-\varepsilon < n.F < 2+\varepsilon$.

Lemma 11. *All submonoids of $\mathbb{Z}_{\geq 0}$ are finitely generated.*

Proof. Let $\mathcal{S}$ be a submonoid of $\mathbb{Z}_{\geq 0}$. If it has no nonzero members, the result is obvious. Otherwise, let $n > 0$ be in $\mathcal{S}$, and for each $i > 0$, let $\mathcal{S}_i = \{j \in \{0, \ldots, n-1\} | j+ni \in \mathcal{S}\}$. Then $\mathcal{S}_1, \mathcal{S}_2, \ldots$ is a nondecreasing sequence of subsets of $\{0, \ldots, n-1\}$, so there must be some i_0 for which $\mathcal{S}_i = \mathcal{S}_{i_0}$ for all $i \geq i_0$. Then $\mathcal{S}$ is generated by $\mathcal{S} \cap \{1, 2, \ldots, n(i_0+1)-1\}$. □

Theorem 12. *If $\mathcal{S}$ is a submonoid of $\mathbb{Z}_{\geq 0}$, generated by positive integers $a_1, \ldots, a_n$, then for all integers $m > 0$ and $M > 6$,*

$$G = \{M, M+a_1.F, M+a_2.F, \ldots, M+a_n.F | -M-F\}$$

will have $2m.G > 0$ if m is in $\mathcal{S}$, and $2m.G \| 0$ otherwise.

Proof. Let $a_0 = 0$, and let $T = \{a_0, a_1, \ldots, a_n\}$. We make the following claims.

Claim 1. *For all integers b and nonnegative integers c, d, e, and q where $c+d+e \geq 2$, $V_{bcdeq} = (c+d+e).M+b.F-2c-d+e.\{0|-4\}+q.G$ is positive or zero.*

Proof of Claim 1. We induce on q. Let e' be the remainder when e is divided by 2. If $q = 0$, $V_{bcdeq} \geq b.F+(c+d+e).(M-2)+e'.\{2|-2\}$. But since $M > 6$, $(c+d+e).(M-2) > 8$, so this is positive. If $q > 0$, look at Right's first move in V_{bcdeq}. If it is in $\{0|-4\}$, we reply from G to M; we are then in a position equal to $V_{bcd(e+1)(q-1)}$, which is positive or zero by the induction hypothesis. If it is in F or $-F$, we reply from G to M; we are then in a position $V_{(b+\beta)(c+\gamma)(d+\delta)(e+\varepsilon)(q-1)}$, where β is 1 or -1, γ, δ, and ε are each 0 or 1, and $\gamma+\delta+\varepsilon = 1$. In any case, this is positive or zero by the induction hypothesis. The only other possibility for Right's first move is that it is in G. If $q \geq 2$, we reply from G to M. We are then at $V_{(b-1)cde(q-2)}$, which is positive or zero by the induction hypothesis. If $q = 1$, Right's move was to

$$(c+d+e-1).M+(b-1).F-2c-d+e.\{0|-4\}$$
$$\geq (c+d+e-1).(M-2)+(b-1).F-2+e'.\{2|-2\},$$

and since $M > 6$, $(c+d+e-1).(M-2) > 4$, so

$$(c+d+e-1).(M-2)-2+e'.\{2|-2\} > 0.$$

Since $(b-1).F$ is not negative, we have a position which is positive or incomparable with zero, which we can win. $\qquad \square$

Claim 2. *For all nonnegative integers m and n, not both zero, there is a winning strategy for Left playing first in $2m.G-n.F$.*

Proof of Claim 2. We induce on m. We know the claim already if $m = 0$. Otherwise, Left should open to $M+(2m-1).G-n.F$. Right may respond on G, to $(2m-2).G-(n+1).F$; we have a good move from this by the induction hypothesis. If $n > 0$, Right may also respond on $-F$, to $M+(2m-1).G-(n-1).F-2$. In this case, we should respond on G to $2M+(2m-2).G-(n-1).F-2$, which is positive or zero by Claim 1. $\qquad \square$

Claim 3. *For all integers b and nonnegative integers c, d, e, and q where $c+d+e \geq 1$, $W_{bcdeq} = -(1+c+d+e).M+b.F+2c+d+e.\{4|0\}+q.G$ is negative or zero.*

Proof of Claim 3. We induce on q. Let e' be the remainder when e is divided by 2. If $q = 0$, $W_{bcdeq} \leq b.F-M+(c+d+e).(2-M)+e'.\{2|-2\}$. Since $M > 6$, $-M+(c+d+e).(2-M) < -10$, so this is negative. If $q > 0$, look at Left's first move in W_{bcdeq}. If it is in $\{4|0\}$, we reply from G to $-M-F$; we are then in a position equal to $W_{(b-1)cd(e+1)(q-1)}$, which is negative or zero by the induction hypothesis. If it is in F or $-F$, we also reply in G; we are then in a position $W_{(b+\beta)(c+\gamma)(d+\delta)(e+\varepsilon)(q-1)}$, where β is 0 or -2, γ, δ, and ε are each 0 or 1, and $\gamma+\delta+\varepsilon = 1$. This is negative or zero by the induction hypothesis. The only other possibility is that it is in G. If $q \geq 2$, we reply from G to $-M-F$, leaving a position of $W_{b'cde(q-2)}$, for some integer b'. This is negative or zero by the induction hypothesis. If $q = 1$, Left's move was to

$$-(c+d+e).M+b'.F+2c+d+e.\{4|0\} \quad \text{(for some integer b')}$$
$$\leq (c+d+e).(2-M)+b'.F+e'.\{2|-2\},$$

and since $M > 6$, $(c+d+e).(2-M) < -4$, so this is negative. $\qquad \square$

Claim 4. *For all nonnegative integers m not in $\mathcal{S}$, there is a winning strategy for Right playing first in $2m.G$.*

Proof of Claim 4. We open by moving from G to $-M-F$, and we continue doing this as long as Left's reply to our play is also in G. If this goes on for $2m$ moves, we will end up moving from some position $(b_1+\cdots+b_m-m).F$, where $b_1, \ldots, b_m \in T$. This cannot be zero as $m \notin S$, so, by Lemma 10, we are moving from a game incomparable with 0 and will hence win. If this does not go on for $2m$ moves, Left responds in F or $-F$ at some point, leaving a position of the form $M+W_{bcdeq}$, where c, d, and e are nonnegative, $q \geq 1$, and $c+d+e = 1$. We should respond by moving from G to $-M-F$. This leaves the position $W_{(b-1)cde(q-1)}$, which is negative or zero by Claim 3. $\qquad\square$

Claim 5. *For $m \in S$, there is a winning strategy for Left playing second in $2m.G$.*

Proof of Claim 5. Since m is in S, we can express m as a sum of the positive a_i's; pad this with zeroes to make a sum of exactly m terms, so that

$$0 = (b_1-1)+(b_2-1)+\cdots+(b_m-1), \qquad b_1, \ldots, b_m \in T.$$

We may arrange these terms so that all initial partial sums are nonpositive. Then when Right opens, by moving from G to $-M-F$, our first response is on another copy of G, to $M+b_1.F$; if he moves on G again, our second response is from G to $M+b_2.F$, and so on. If this goes on for $2m$ moves, we will win, by moving to 0. Otherwise, Right responds on $-F$ at some point, leaving a position

$$(a+1).F+2q.G-2, \text{where } 1 \leq q < m \text{ and } a = b_1-1+\cdots+b_{m-q}-1 < 0. \quad (0\text{--}2)$$

We claim that we have a winning strategy from all positions (0–2). To prove this, induce on q. We should always respond to

$$M+(a'+2).F+(2q-1).G-2, \text{where } a' = b_1-1+\cdots+b_{m-q}-1+b_{m+1-q}-1.$$

Right must move from this position. If he moves on F or $-F$, respond from G to M; then the position is of the form $V_{b(c+1)de(2q-2)}$, where c, d, and e are nonnegative and $c+d+e = 1$. This is positive or zero by Claim 1. If he moves on G and $q > 1$, then his move is to a position (0–2) with q decreased by one, which we can win by the induction hypothesis. Finally, if he moves on G and $q = 1$, then $a' = 0$, so the position is now $F-2$, from which we move immediately to 0. $\qquad\square$

The theorem now follows immediately from Claims 2, 4, and 5. $\qquad\square$

References

[1]   E. R. Berlekamp, J. H. Conway, and R. K. Guy, *Winning Ways*, New York: Academic Press, 1982.

[2]   J. H. Conway, *On Numbers and Games*, London: Academic Press, 1976.

DAVID MOEWS
CENTER FOR COMMUNICATIONS RESEARCH
4320 WESTERRA COURT
SAN DIEGO, CA 92121
UNITED STATES
  dmoews@ccrwest.org

The Old Classics

More Games of No Chance
MSRI Publications
Volume **42**, 2002

Higher Nimbers in Pawn Endgames on Large Chessboards

NOAM D. ELKIES

Do ∗2, ∗4 and higher Nimbers occur on the 8 × 8 or larger boards?
— ONAE [Elkies 1996, p. 148]

It's full of stars!
— 2001: A Space Odyssey [Clarke 1968, p. 193]

ABSTRACT. We answer a question posed in [Elkies 1996] by constructing a
class of pawn endgames on $m \times n$ boards that show the Nimbers $*k$ for many
large k. We do this by modifying and generalizing T.R. Dawson's "pawns
game" [Berlekamp et al. 1982]. Our construction works for $m \geq 9$ and n
sufficiently large; on the basis of computational evidence we conjecture, but
cannot yet prove, that the construction yields $*k$ for all integers k.

1. Introduction

In [Elkies 1996] we showed that certain chess endgames can be analyzed by the
techniques of combinatorial game theory (CGT). We exhibited such endgames
whose components show a variety of CGT values, including integers, fractions,
and some infinite and infinitesimal values. Conspicuously absent were the values
$*k$ of Nim-heaps of size $k > 1$. Towards the end of [Elkies 1996] we asked whether
any chess endgames, whether on the standard 8 × 8 chessboard or on larger
rectangular boards, have components equivalent to $*2$, $*4$ and higher Nimbers.
In the present paper we answer this question affirmatively by constructing a new
class of pawn endgames on large boards that include $*k$ for many large k, and
conjecture — though we cannot yet prove — that all $*k$ arise in this class.

Our construction begins with a variation of a pawns game called "Dawson's
Chess" in [Berlekamp et al. 1982]. In § 3 we introduce this variation and show
that, perhaps surprisingly, all quiescent (non-entailing) components of the modified game are equivalent to Nim-heaps (Theorem 1). We then determine the

value of each such component, and characterize non-loony moves (Theorem 2). In §4 we construct pawn endgames[1] on large chessboards that incorporate those components. These endgames do not yet attain our aim, because the values determined in Theorem 2 are all 0 or ∗1. In §5 we modify one of our components to obtain ∗2. In §6 we further study components modified in this way, showing that they, too, are equivalent to Nim-heaps (Theorem 3). We conclude with the numerical evidence suggesting that all Nim-heaps can be simulated by components of pawn endgames on large rectangular chessboards.

Before embarking on this course, we show in §2 a pair of endgames on the standard 8 × 8 board in the style of [Elkies 1996] that illustrate the main ideas, specifically the role of "loony moves". Readers who are much more conversant with CGT than with chess endgames will likely prefer to skim or skip §2 on first reading, returning to it only after absorbing the theory in §3. Conversely, chessplayers not fluent in CGT will find in §2 motivation for the CGT ideas central to §3 and later sections.

2. An Illustrative Pair of Endgames

We introduce the main ideas of our construction by analyzing the following pair of composed endgames on the standard 8 × 8 chessboard:[2]

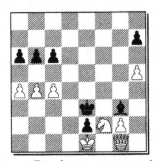

Diag. 1A: whoever moves loses   Diag. 1B: whoever moves wins

Diagram 1A consists of two components. On the Kingside, seven men are locked in a mutual Zugzwang that we already used in [Elkies 1996]. Both sides can legally move in the Kingside, but only at the cost of checkmate (Qh1(2) Bxf2) or ruinous material inferiority. Thus the outcome of Diagram 1A hinges on the Queenside component, with three adjacent pawns on each side. We have

---

[1]More precisely, King-and-pawn endgames; the first two words are usually suppressed because every legal chess position must have a White and a Black King.

[2]These are not pawn endgames, but all units other than pawns are involved in the Kingside Zugzwang, and are thus passive throughout the analysis. We can construct plausible positions where that Zugzwang, too, is replaced by one using only Kings and pawns, but only at the cost of introducing an inordinate number of side-variations tangential to the CGT content of the positions. For instance, replace files e–g in Diag. 1A by White Kg1, Ph2 and Black Kh3, Pg2; in Diag. 1B, do the same and move the h5/h7 pawns to e3/e5.

not seen such a component in [Elkies 1996], but it turns out to be a mutual Zugzwang: whichever side moves first, the opponent can maneuver to make the last pawn move on the Queenside, forcing a losing King move in the Kingside component. Thus[3] a5 can be answered by bxa5 bxa5 c5, likewise c5 by bxc5 bxc5 a5, and b5 by axb5 axb5 c5. In this last line, it is no better to answer axb5 with cxb5, since then c5 wins: if played by Black, White will respond a5 and promote first, but Black c1Q will be checkmate; while if White plays c5 and Black answers a5, White replies c6 and promotes with Black's pawn still two moves away from the first rank, winning easily.[4] Note the key point that a5 or c5 must not be answered by b5?, since then c5 (resp. a5) transfers the turn to the second player and wins — but not cxb5? cxb5 (or axb5? axb5), regaining the Zugzwang.

Diagram 1B is Diag. 1A with the h5 and h7 pawns added; these form an extra component which we recognize from [Elkies 1996] as having the value $*$. Thus we expect that Diag. 1B is a first-player win, and indeed either side can win by playing h6, effectively reducing to Diag. 1A. The first player can also win starting on the Queenside: a5 bxa5 bxa5 ($* + * = 0$), and likewise if a5 is answered by b5 (c:b5 etc. wins, but not c5? h6). The first player must not, however, start b5?, when the opponent trades twice on b5, in effect transforming $* + *$ into $0 + *$, and then wins by playing h6.

Note the role of the move b5 in the analysis of both Diagrams 1A and 1B. In the terminology of [Berlekamp et al. 1982], this is an "entailing move": it makes a threat (to win by capturing a pawn) that must be answered in the same component. But, whether the rest of the position has value 0 (Diag. 1A) or $*$ (Diag. 1B), the move b5 loses, because the opponent can answer the threat in two ways, one of which passes the turn back (advancing the threatened pawn to a5 or c5), one of which in effect retains the turn (capturing on b5, then making another move after the forced re-capture). Since the first option wins if the rest of the position has value 0, and the second wins if the rest of the position has value $*$ (or any other nonzero Nimber), b5 is a bad move in either case. In [Berlekamp et al. 1982] such an unconditionally bad move is called "loony" (see p. 378). Since b5 is bad, it follows in turn that, in either Diag. 1A or 1B, the

---

[3]Note that we do not specify whether White or Black begins the sequence. Fortunately the algebraic chess notation for these moves, and for most of the pawn moves that occur in this paper, is the same whether they are played by White or Black. When we exploit this circumstance, we refrain from the usual practice giving move numbers, which would specify who made which move: White begins 1 a5 bxa5 2 bxa5 c5, while Black begins 1...a5 2 bxa5 bxa5 3 c5.

[4]This part of the analysis explains why we chose this Kingside position from [Elkies 1996]: the position of White's King, but not Black's, on its first rank makes it more vulnerable to promoted pawns, exactly compensating for the White pawns being a step closer to promotion than Black's. In [Elkies 1996] this Kingside Zugzwang had the Kings on f1 and f3, not e1 and e3; here we shifted the position one square to the left so as not to worry about a possible White check by a newly promoted Queen on a8. Thus 1 b5 can be answered by 1...cxb5 as well as 1...axb5.

entailing move a5 may be regarded as the non-entailing "move" consisting of the
sequence a5 bxa5 bxa5: the only other reply to a5 is b5, which is loony and so
can be ignored.

We next show that this analysis can be extended to similar pawn configurations on more than three adjacent files. We begin our investigation with a
simplified game involving only the relevant pawns.

3. A Game of Pawns

3.1. Game Definition. Our game is played on a board of arbitrary length n
and height 3. At the start, White pawns occupy some of the squares on the
bottom row, and Black pawns occupy the corresponding squares on the top row.
For instance, Diag. 2 shows a possible starting position on a board with $n = 12$:

Diag. 2: A starting position

The pawns move and capture like chess pawns, except that there is no double-
move option (and thus no *en passant* rule). Thus if a file contains a White pawn
in the bottom row and a Black pawn in the top row, these pawns were there in
the initial position and have not moved; we call such a file an "initial file".

White wins if a White pawn reaches the top row, and Black wins if a Black
pawn reaches the bottom row. Thus there is no need for a promotion rule because
whoever could promote a pawn immediately wins the game. But it is easy enough
to prevent this, and we shall assume that neither side allows an opposing pawn
to reach its winning row. The game will then end in finitely many moves with
all pawns blocked, at which point the winner is the player who made the last
move. We shall sometimes call this outcome a "win by Zugzwang", as opposed
to an "immediate win" by reaching the opposite row.

For instance, from Diag. 2 White may begin 1 c2. Since this threatens to
win next move by capturing on b3 or d3, Black must capture the c2-pawn; if
Black captures with the b-pawn, we reach Diag. 3A. Now Black threatens to win
by advancing this pawn further, so White must capture it; but unlike Black's
capture, White's can only be made in one way: if 2 dxc2?, threatening to win
with 3 cxd3, Black does not re-capture but instead plays d2, producing Diag. 3B.

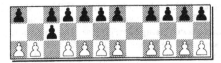

Diag. 3A: After 1 c2 bxc2

Diag. 3B: If then 2 dxc2? d2, winning

Black then wins, since touchdown at d1 can only be delayed by one move with 3 exd2 exd2 (but not 3...cxd2??, when 4 c3 wins instead for White!), and then 4...d1.

Diag. 3C: After 2 bxc2 dxc2 3 dxc2

Diag. 3D: Instead 2...d2

Diag. 3E: Then 3 exd2 exd2

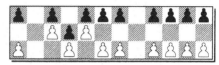

Diag. 3F: or 3 e2

Thus White must play 2 bxc2 from Diag. 3A. This again threatens to win with 3 cxd3, so Black has only two options. One is to re-capture with 2...dxc2, forcing White in turn to re-capture: 3 dxc2, reaching Diag. 3C. Alternatively, Black may save the d3-pawn by advancing it to d2 (Diag. 3D). This forces White to move the attacked pawn at e1. Again there are two options. White may capture with 3 exd2, forcing Black to reply 3...exd2 (Diag. 3E), not cxd2? when White wins immediately with 4 c3. Alternatively White may advance with 3 e2 (Diag. 3F), when Black again has two options against 4 exf3, etc. Eventually the skirmish ends either with mutual pawn captures (as in Diag. 3C or 3E) or when the wave of pawn advances reaches the end of the component (3...f2 4 g2), leaving one side or the other to choose the next component to play in.

As noted in the Introduction, this game is reminiscent of the game called "Dawson's Chess" in [Berlekamp et al. 1982, pp. 88 ff.]; the only difference is that in Dawson's Chess a player who can capture a pawn must do so, while in our game, as in ordinary chess, captures are optional.[5] Because of the obligation to capture, Dawson's Chess may appear to be an entailing game, but it is quickly seen to be equivalent to a (non-entailing) impartial game, called "Dawson's Kayles" in [Berlekamp et al. 1982]. Our game also has entailing moves, and features a greater variety of possible components; but we shall see that it, too, reduces to a non-entailing impartial game once we eliminate moves that lose immediately and loony moves.

---

[5]Since Dawson was a chess problemist, we first guessed that the game analyzed here was Dawson's original proposal, before the modification in [Berlekamp et al. 1982]. But in fact R.K. Guy reports in a 16.viii.1996 e-mail that Dawson did want obligatory captures but proposed a misère rule (last player loses). Guy also writes: "I am aware of some very desultory attempts to analyze the game in which captures are allowed, but little was achieved, to my knowledge." I thank Guy for this information, and John Beasley who more recently sent me copies of pages from the 12/1934 and 2/1935 issues of *The Problemist* Fairy Supplement in which Dawson proposed and analyzed his original game.

3.2. Decomposition into Components. We begin by listing the possible
components. We may ignore any files in which no further move may be made.
These are the files that are either empty (such as the h-file throughout Diag. 3,
the b-file in Diag. 3C–3F, and the d- and e-files in 3C and 3E respectively) or
closed. We say a file is "closed" if it contains one pawn of each color, neither
of which can move or capture (the c-file in 3C–3F and the d-file in 3E–3F).
One might object that such a currently immobile pawn may be activated in the
future; for instance in Diagram 3D if White plays 3 exd2 then the dormant c-
file may awake: 3...cxd2. But we observed already that this Black move loses
immediately to 4 c3. Since we may and do assume that immediately losing moves
are never played, we may ignore the possibility of 3...cxd2?, and regard the c-file
in Diag. 3D and the d-file in Diag. 3F as permanently closed.

By discarding empty or closed files we partition the board into components
that do not interact except when an entailing move requires an answer in the
same component. Thus at each point there can be at most one entailing component (again assuming no immediately losing moves). We next describe all
possible components and introduce a notation for each.

A non-entailing component consists of m consecutive initial files, for some
positive integer m. We denote such a component by $[m]$. For instance, Diag. 3A
is $[7] + [4]$; Diag. 3C is $[1] + [3] + [4]$; and Diag. 3E is $[1] + [2] + [4]$. An entailing
component contains a pawn that has just moved (either vertically or diagonally)
to the second rank, threatening an immediate win, and can be captured. We
denote this pawn's file by ":". There are four kinds of entailing component,
depending on whether this pawn can be captured in one or two ways and on how
many friendly pawns defend it by being in position to re-capture.

• If the pawn is attacked once and defended once, then the attacking and
defending pawns are on the same file (else the side to move can win immediately
as in Diag. 3B). Thus the ":" file is at the end of a component each of whose
remaining files is initial. We denote the component by $[:m]$, where m is the
number of initial files in the component. For instance, Diag. 4A shows $[:5]$. For
the mirror-image of "$[:m]$", we use either the same notation or "$[m:]$". Either
side can move from $[m + 1]$ to $[:m]$ by moving the left- or rightmost pawn. Faced
with $[:m]$, one must move the attacked pawn, either by capturing the ":" pawn or
by advancing it. Advancing yields $[:(m - 1)]$ (see Diag. 4B), unless $m = 1$ when
the advance yields 0 since all files in the component become blocked. Capturing
yields $[:.]+[m - 1]$ (see Diag. 4C), where $[:.]$ is the the component defined next.

• If the pawn is attacked once and not defended, then it has just captured,
and is subject to capture from an unopposed pawn. We denote the file with
one unopposed pawn by ".". The capture is obligatory, and results in a closed
file. Therefore the adjacent ":" and "." files do not interact with any other
components, even if they are not yet separated from them by empty or closed
files. We may thus regard these files as a separate component $[:.]$, which entails a

move to 0. For instance, files b, c in Diag. 4C constitute [∴], which is unaffected by the [4] on files d through g.

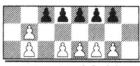

| Diag. 4A: [:5] | Diag. 4B: [:4] | Diag. 4C: [∴]+[4] |

• If the pawn is attacked twice and defended twice, then the component of the "∶" file consists of that file, m initial files to its left, and m' initial files to its right, for some positive integers m and m'. We denote such a component by $[m{:}m']$. For instance, Diag. 5A shows [2:4]. We already encountered this component after the move c2 from Diag. 2. The component $[m{:}m']$ entails a capture of the "∶" pawn. This yields either $[m-1]+[.{:}m']$ or $[.{:}m] + [m' - 1]$, where $[.{:}m]$, defined next, is our fourth and last kind of entailing component, and $[m-1]$ (or $[m'-1]$) is read as 0 if $m = 1$ (resp. $m' = 1$).

• If the pawn is attacked twice and defended once, then it is part of a component obtained from [∴] by placing m initial files next to the "∶" file, for some positive integer m. Naturally we call such a component $[.{:}m]$ (or $[m{:}.]$). As explained in the paragraph introducing [∴], an initial file next to the "." file cannot interact with it, and thus belongs to a different component. For instance, the two possible captures from [2:4] yield [1]+[.:4] (Diag. 5B, also seen in Diag. 3A) and [2:.]+[3] (Diag. 5C). Faced with $[.{:}m]$, one has a single move that does not lose immediately: capture with the "." pawn, producing $[:m]$, as seen earlier in connection with Diag. 3C.

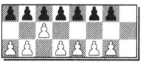

| Diag. 5A: [2:4] | Diag. 5B: [1]+[.:4] | Diag. 5C: [2:.]+[3] |

We summarize the available moves as follows. It will be convenient to extend the notations $[m]$, $[:m]$, $[.{:}m]$, $[m{:}m']$ by allowing $m = 0$ or $m' = 0$, with the understanding that

$$[0] = [{:}0] = [0{:}0] = 0, \quad [.{:}0] = [∴], \quad [m{:}0] = [0{:}m] = [{:}m].$$

We then have:

• From $[m]$, either side may move to $[m_1 : m_2]$ for each $m_1, m_2 \geq 0$ such that $m_1 + m_2 = m - 1$.

• If $m > 0$ then $[:m]$ entails a move to either $[∴] + [m - 1]$ or $[:(m - 1)]$.

• $[.{:}m]$ entails a move to $[:m]$. In particular (taking $m = 0$), [∴] entails a move to $[:0] = 0$.

- If $m, m' > 0$ then $[m : m']$ entails a move to either $[m - 1] + [. : m']$ or $[m' - 1] + [.:m]$.

For instance, Diag. 3D shows $[1] + [:3] + [4]$, moving either to $[1] + [:.] + [2] + [4]$ and thence to $[1] + [2] + [4]$ (Diag. 3E), or to $[1] + [:2] + [4]$ (Diag. 3F).

Our list of possible moves confirms that $[m]$, $[:m]$, $[:.]$, $[.:m]$, $[m:m']$ are the only possible components: the initial position is a sum of components $[m_i]$, and each move from a known component that does not concede an immediate win yields a sum of 0, 1, or 2 known components.

3.3. Analysis of Components.
Since in each component both sides have the same options, we are dealing with an *impartial* entailing game. We could thus apply the theory of such games, explained in [Berlekamp et al. 1982], to analyze each component. But it turns out that once we eliminate loony moves the game is equivalent to an ordinary impartial game, and thus that each component $[m]$ is equivalent to a Nim-heap of size depending on m.

Consider the first few m. Clearly $[1]$ is equivalent to $*1$, a Nim-heap of size 1. At the end of § 2 we have already seen in effect that a move from $[2]$ to $[:1]$ is loony: if the rest of the game has value 0, then the reply $[:1]\rightarrow 0$ wins; otherwise, the reply $[:1]\rightarrow[:.]$ forces $[:.]\rightarrow 0$, again leaving a forced win in the sum of the remaining components. (In particular, $[2]$ is mutual Zugzwang, equivalent to 0.) Thus also $[3]\rightarrow[1:1]$ is loony, because the forced continuation $[1:1]\rightarrow[.:1]\rightarrow[:1]$ again leaves the opponent in control. On the other hand, $[3]\rightarrow[:2]$ is now seen to force $[:2]\rightarrow[:.]+[1]$, since the alternative $[:2]\rightarrow[:1]$ is known to lose. We thus have the forced combination $[3]\rightarrow[:2]\rightarrow[:.]+[1]\rightarrow[1]$, which amounts to a "move" $[3]\longrightarrow[1]$.[6] Moreover, we have shown that this is the only reasonable continuation from $[3]$. Since $[1] \cong *1$, we conclude that $[3]$ is equivalent to a Nim-heap of size $\mathrm{mex}(\{1\}) = 0$, i.e. a mutual Zugzwang, as we already discovered in the analysis of Diags. 1A,1B.

What of $[4]$ and $[5]$? From $[4]$, there are again two options, one of which can be eliminated because it leads to the loony $[:1]$, namely $[4]\rightarrow[2:1]$ (after $[2:1]\rightarrow[1]+[.:1]\rightarrow[1]+[:1]$). This leaves $[4]\rightarrow[:3]$, which in turn allows two responses. One produces the sequence $[4]\rightarrow[:3]\rightarrow[2]+[:.]\rightarrow[2]$, resulting in a value of 0. The other response is $[4]\rightarrow[:3]\rightarrow[:2]$, which we already know forces the further $[:2]\rightarrow[:.]+[1]\rightarrow[1] \cong *1$. In effect, the response to $[4]\rightarrow[:3]$ can interpret the move either as $[4] \longrightarrow [2] \cong 0$, or as $[4] \longrightarrow [1] \cong *1$ *with the side who played* $[4] \rightarrow [:3]$ *on move again*. We claim that the latter option can be ignored. The reason is that the first interpretation wins unless the remaining components of the game add to 0; but then the second interpretation leaves the opponent to move in a nonzero position, and thus also loses. We conclude that $[4]$ is equivalent

---

[6]Here and later, we use an arrow "$\rightarrow$" for a single move, and a long arrow "$\longrightarrow$" for a sequence of 3, 5, 7, ... single entailing moves considered as one "move".

to an impartial game in which either side may move to 0, and is thus equivalent to a Nim-heap of size mex($\{0\}$) = 1. As to [5], there are now three options, only one excluded by [:1], namely [5]→[3:1]. The option [5]→[2:2] (move the center pawn) forces the continuation [2:2]→[1]+[.:2]→[1]+[:2]→[1]+[:.]+[1]→[1]+[1], and is thus tantamount to [5] $\longrightarrow$ [1] + [1] $\cong$ 0. This leaves [5]→[:4], which we show is loony for a new reason. One continuation is [:4]→[:.]+[3]→[3], interpreting the move as [5] $\longrightarrow$ [3] $\cong$ 0. The other is [:4]→[:3], which as we have just seen is equivalent to [:4]$\longrightarrow$[2]. Since [2] $\cong$ 0, this continuation interprets [5]→[:4] as a move to 0 *followed by an extra move*. Thus a move to [:4] always allows a winning reply, namely [:4]→[:3] if the remaining components add to 0, and [:4]→[:.]+[3] if not. Hence the move to [:4] is loony as claimed, and [5] $\cong$ *(mex$\{0\}$) = *1.

The analysis of [1] through [5] shows almost all the possible behaviors in our game; continuing by induction we prove:

Theorem 1. (i) *For each integer $m \geq 0$, the component* [m] *is equivalent to a Nim-heap of some size* $\varepsilon(m)$.

(ii) *A move to* [: 1] *is loony. For each integer $m > 1$, a move to* [: m] *is either loony or equivalent to a move to* $[m - 1] \cong *(\varepsilon(m - 1))$. *The move is loony if and only if a move to* [:(m − 1)] *is* <u>not</u> *loony and* $\varepsilon(m - 1) = \varepsilon(m - 2)$.

(iii) *For any positive integers m_1, m_2, a move to* $[m_1 : m_2]$ *is either loony or equivalent to a move to* $[m_1 - 1] + [m_2 - 1] = *(\varepsilon(m_1 - 1) \overset{*}{+} \varepsilon(m_2 - 1))$. *The move to* $[m_1 : m_2]$ *is non-loony if and only both m_1 and m_2 satisfy the criterion of* (ii) *for a move to* [:m] *to be non-loony.*

(iv) *We have* $\varepsilon(0) = 0$, $\varepsilon(1) = 1$, *and for $m > 1$ the values $\varepsilon(m)$ are given recursively by*

$$\varepsilon(m) = \max_{m_1, m_2} \left(\varepsilon(m_1 - 1) \overset{*}{+} \varepsilon(m_2 - 1)\right).$$

Here the mex runs over pairs (m_1, m_2) of nonnegative integers such that $m_1 + m_2 = m - 1$ and a move to $[m_1 : m_2]$ is not loony, as per the criteria in parts (ii) *[for $m_1 m_2 = 0$] and* (iii) *[for $m_1 m_2 > 0$]. For this equation we declare that $\varepsilon(-1) = 0$.*

In parts (iii) and (iv), "$\overset{*}{+}$" denotes the Nim sum: $(*k) \overset{*}{+} (*k') = *(k \overset{*}{+} k')$. Thus, once parts (i)–(iii) are known for all components with fewer than m initial files, (iv) is just the Sprague-Grundy recursion for impartial games. Once (iv) is known for all $m \leq m_0$, so is (i). The arguments for (ii) and (iii) are the same ones we used for components with up to 5 initial files. For instance, for (iii) the move $[m_1 + 1 + m_2] \to [m_1 : m_2]$ forces a choice among the continuations

$$[m_1 : m_2] \to [m_1 - 1] + [.:m_2] \to [m_1 - 1] + [:m_2],$$
$$[m_1 : m_2] \to [m_1 :.] + [m_2 - 1] \to [m_1 :] + [m_2 - 1].$$

If a move to $[:m_1]$ or $[:m_2]$ is loony then one or both of these continuations wins. Otherwise by (ii) both continuations are tantamount to $[m_1 + 1 + m_2] \longrightarrow [m_1 - 1] + [m_2 - 1]$. ▲

Carrying out the recursion for $\varepsilon(m)$, we quickly detect and prove a periodicity:

Theorem 2. *For all $m \geq 0$ we have $\varepsilon(m) = 0$ if m is congruent to 0, 2, 3, 6, or $9 \bmod 10$, and $\varepsilon(m) = 1$ otherwise. A move to $[:m]$ is loony if and only if $m = 5k \pm 1$ for some integer k.*

Proof. Direct computation verifies the claim through $m = 23$, which suffices to prove it for all m as in [Berlekamp et al. 1982, pp. 89–90], since 23 is twice the period plus the maximum number of initial files lost by a "move" $[m] \longrightarrow [m-2]$ or $[m_1 + 1 + m_2] \longrightarrow [m_1 - 1] + [m_2 - 1]$. ▲

So, for instance, Diagram 2 is equivalent to $*(\varepsilon(7)) + *(\varepsilon(4)) = *1 + *1 = 0$ and is thus a mutual Zugzwang, a.k.a. $\mathcal{P}$-position or previous-player win. The next player thus might as well play a loony move such as 1 c2, in the hope of giving the opponent Enough Rope [Berlekamp et al. 1982, p. 17]; the only correct response is 1... bxc2 (Diag. 3A) 2 bxc2 d2! (Diag. 3D), maintaining the win after either 3 exd2 exd2 (Diag. 3E) or 3 e2 (Diag 3F) fxe2! 4 fxe2.

4. Embedding into Generalized Chess

Consider Diag. 6A, a pawn endgame on a chessboard of 9 rows by 12 files:

Diag. 6A: whoever moves wins (c5!)

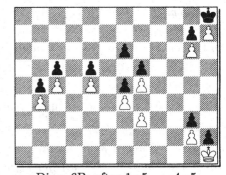

Diag. 6B: after 1 c5 ... 4 e5

There are four components. In each of the top right and bottom right corners, a King and three pawns are immobilized. Near the middle of the board (on the g- and h-files), we have a mutual Zugzwang with three pawns on a side; a player forced to move there will allow an opposing pawn to capture and soon advance to Queen promotion, giving checkmate. In the leftmost five files we have a pawn game with initial position [5], arranged symmetrically about the middle rank. An "immediate win" in this game is a pawn that can promote to Queen in three moves, ending the game by checkmate. We may thus assume that, as in our pawn game of the previous section, both sides play to prevent an "immediate win", and the leftmost five files will eventually be empty or blocked. This is why we have chosen a chessboard with an odd number of rows: with an even number, as on the orthodox 8×8 board, one side's pawns would be at least one move closer to promotion, and we would have to work harder to find positions in which, as in Diags. 1A and 1B, an "immediate win" in the pawns game by either player yields a chess win for the same player. Once play ends in the [5] component, we see why the component in the g- and h-file is needed: the Zugzwangs arising from the [m] components all end with blocked pawns, and if those were the only components on the board then the chess game would end in stalemate, regardless of which side "won" the pawns game. But, with the g- and h-files on the board, the side who lost the pawns game must move in the central Zugzwang and lose the chess game.

To see how this happens, suppose that White is to move in Diag. 6A. White must start 1 c5, the only winning move by the analysis in the previous section. Play may continue dxc5 2 dxc5 b5 (Black is lost, so tries to confuse matters with a loony move) 3 axb5 (declining the rope 3 a5? e5) axb5 4 e5 (Diag. 6B). With all other pawns blocked, Black must now play 4...g6 5 hxg6 g5. If now 6 h7? g4 ends in stalemate, so White first plays 7 gxh5 (or even 7 h4), and then promotes the pawn on g6 and gives checkmate in three more moves.

This construction clearly generalizes to show that any instance of our pawn game supported on a board of length n can be realized by a King-and-pawn endgame on any chessboard of at least $n + 6$ files whose height is an odd number greater than 5.

5. Stopped Files

By embedding our pawn game into generalized chess, we have constructed a new class of endgames that can be analyzed by combinatorial game theory. But we have still not attained our aim of finding higher Nimbers, because by Theorem 2 all the components of our endgames have value 0 or *1. To reach *2 and beyond, we modify our components by *stopping* some files. We illustrate with Diag. 7A:

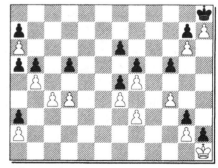

Diag. 7A: whoever moves wins   Diag. 7B: after 1 b5 cxb5 2 axb5

We have replaced the component [5] of Diag. 6A by two components. One is familiar: on the j-file (third from the right) we see [1] $\cong$ *1. On the leftmost

four files we have a new configuration. This component looks like [4], but with
four extra pawns on the a-file. These pawns are immobile, but have the effect of
stopping the file on both sides, so that a White pawn reaching a6 or a Black pawn
reaching a4 can no longer promote. Without these extra pawns, Diag. 7A would
evaluate to $[4] + [1] \cong *1 + *1 = 0$ and would thus be a mutual Zugzwang. But
Diag. 7A is a first-player win, with the unique winning move b5!. Indeed, suppose
White plays b5 from Diag. 7A. If Black responds 1... axb5 then White's reply
2 axb5 produces $[:2]+[1]$ and wins. So Black instead plays 1... cxb5, expecting
the loony reply 2 cxb5. But thanks to the stopped a-file White can improve with
2 axb5!. See Diag. 7B. If now 2... a5, this pawn will get no further than a4, while
White forces a winning breakthrough with 3 c5, for instance 3... dxc5 4 dxc5
bxc5 5 b6 c4 6 b7 (Diag. 7C) and mates in two. Notice that the extra pawns
on the a-file do not stop the b-file: if Black now captures the pawn on b7 then
the pawn on a7 will march in its stead. We conclude that in Diag. 7B Black
has nothing better than 2... axb5 3 cxb5, which yields the same lost position
$([1] + [1] \cong 0)$ that would result from 1... axb5. After 1 b5 Black could also try
the tricky 1... c5, attempting to exploit the a-file stoppage by sacrificing the
a6-pawn. After the forced 2 dxc5 (d5? axb5 3 axb5 j5 wins) dxc5 (Diag. 7D),
White would indeed lose after 3 bxa6? j5, but the pretty 3 a5! wins. However
Black replies, a White pawn will next advance or capture to b6, and three moves
later White will promote first and checkmate Black.

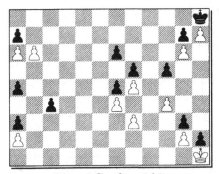

Diag. 7C: after 6 b7

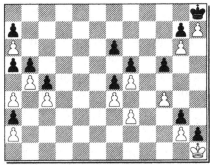

Diag. 7D: after 1... c5 2 dxc5 dxc5

Diag. 7A remains a first-player win even without the [1] component on the
j-file (Diag. 7E). The first move d5 wins as in our analysis of [4]: either cxd5 cxd5
or c5 bxc5 bxc5 a5 produces a decisive Zugzwang. In fact, d5 is the only winning
move in Diag. 7E. The move c5 is loony as before (bxc5 bxc5 d5/dxc5). With
the a-file stopped, a5 is loony as well. The opponent will answer b5 (Diag. 7F),
and if then cxb5 axb5!, followed by c5 and wins while the pawn left on a5 is

useless as in Diag. 7B. This leaves (from Diag. 7F) c5, again producing the loony [:1]. Thus a5 is itself loony as claimed.

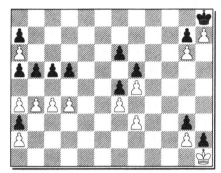
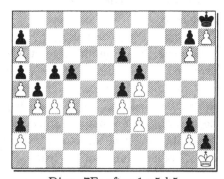

Diag. 7E: whoever moves wins   Diag. 7F: after 1 a5 b5

Therefore the component in files a–d of Diag. 7A and Diag. 7E is equivalent to an impartial game in which either side may move to either 0 (with d5) or *1 (with b5). Hence this component has the value *2! On longer boards of odd height ≥ 9, we can stop some of the files in $[m]$ for other m. We next show that each of the resulting components is equivalent to a Nim-heap, some with values *4, *8 and beyond.

6. The Pawns game with Stopped Files

6.1. Game Definition and Components. We modify our pawn game by choosing a subset of the n files and declaring that the files in that subset are "stopped". A pawn reaching its opposite row now scores an immediate win only if it is on an unstopped file.[7] *We require that no two stopped files be adjacent.* This requirement arises naturally from our implementation of stopped files in King-and-pawn endgames on large chessboards. As it happens, the requirement is also necessary for our analysis of the modified pawns game. For instance, if adjacent stopped files were allowed then a threat to capture a pawn might not be an entailing move.

In the last section we already saw the effects of stopped files on the play of the game. We next codify our observations. We begin by extending our notation for quiescent components. In § 3, such a component was entirely described by the number m of consecutive initial files that the component comprises. In the modified game, we must also indicate which if any of these m files is stopped. We denote a stopped initial file by 1, and an unstopped one by 0. A string of m binary digits, with no 1's adjacent, then denotes a quiescent component of m initial files. To avoid confusion with our earlier notation, we use double instead of single brackets. For instance, the component we called $[m]$, with no stopped

---

[7]If the file is stopped, the pawn does not "promote": it remains a pawn, and can make no further moves. Recall that this was the fate of Black's a5-pawn in Diag. 7C.

files, now becomes $[[000\cdots0]] = [[0^m]]$; the component with value $*2$ on files a–d of Diag. 7A is denoted $[[1000]]$. An initial file that may be either open or closed will be denoted i (or i_1, i_2, etc.); an arbitrary "word" of 0's and 1's will be denoted w (or w', w_1, w_2, etc.).

A component comprising just one initial file, stopped or not, still has value $\{0|0\} = *1$. In a component of at least two initial files, every move threatens to capture and is entailing. This is true even if the file(s) of the threatened pawn(s) is or are stopped, because an immediate win is then still threatened by advancing in that stopped file, as we saw in Diag. 7D where White wins by 3 a5!. (Here we need the condition that two adjacent files cannot both be stopped: if in Diag. 7D the b-file were somehow stopped as well then 3 a5 would lose to either 3...bxa5 or 3...axb5.) This remains true even in the presence of other components. For instance, if more components were added to Diagram 7B and Black played in another component, White could win immediately by capturing once or twice in that component until it became quiescent and then proceeding with c5.

Consider first a move by the pawn on the first or last file of the component (without loss of generality: the first), attacking just one pawn. As in §3, the opponent must move the attacked pawn on the second file, either advancing it or capturing the attacking pawn. In the latter case, the pawn must be re-captured, and the sequence has the effect of removing the component's first two files. In the latter case, the component becomes quiescent if it had only two files (in which case the first move in the component was loony, as before); otherwise the advanced pawn in turn attacks a third-file pawn, which must capture or advance. But now a new consideration enters: if the *first* file is stopped, then the capture loses immediately since the opponent will re-capture from the first file and touch down on the second, necessarily unstopped, file. (See Diag. 7F after 2 cxb5 axb5.) Note that the first file, though closed, can still affect play for one turn after its closure if it is stopped. We thus need a notation for such files, as well as stopped ":" files, which may become closed. We use an underline: a stopped ":" file will be denoted "$\underline{:}$", and a stopped blocked file will simply be denoted "$\_$". Thus the moves discussed in this paragraph are as follows, with each w denoting an arbitrary word *of positive length*:

- From $[[0]]$ or $[[1]]$, either side may move to 0.

- From $[[0w]]$ or $[[1w]]$, either side may move to $[[:w]]$ or $[[\underline{:}w]]$ respectively.

- a move to $[[:0]]$, $[[:1]]$, or $[[\underline{:}0]]$ is loony.

- $[[:0w]]$ entails a move to $[[:.]]+[[w]]$ or $[[:w]]$; $[[\underline{:}0w]]$ entails a move to $[[\underline{:}.]]+[[w]]$ or $[[\_:w]]$; and $[[:1w]]$ entails a move to $[[:.]]+[[w]]$ or $[[\underline{:}w]]$. Each of $[[:.]]$ and $[[\underline{:}.]]$ entails a move to 0.

- $[[\_:0]]$ or $[[\_:1]]$ entails a move to 0; and $[[\_:0w]]$ or $[[\_:1w]]$ entails a move to $[[:w]]$ or $[[\underline{:}w]]$ respectively.

It remains to consider a pawn move in the interior of a quiescent component. Such a move attacks two of the opponent's pawns, and entails a capture. If neither of the attacked pawns is on a stopped file, then either of them may capture, forcing a re-capture from the same file, just as in the pawn game without stopped files. If both pawns are on stopped files (see Diag. 8A), then a capture from either of these files can be met by a capture from the other file (Diag. 8B), forcing a further capture and re-capture to avoid immediate loss. The first player may also choose to make the first re-capture from the same file (Diag. 8C), but we can ignore this possibility because the opponent can still re-capture again to produce the same position as before, but has the extra option of advancing the attacked pawn.

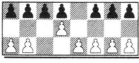

Diag. 8A (c, e files blocked)   Diag. 8B   Diag. 8C

Finally, if just one of the two attacked pawns lies on an unstopped file, it may as well capture, forcing a re-capture in the same file: capturing with the other pawn lets the first player capture from the stopped file, forcing a further capture and re-capture to avoid immediate loss, and thus denying the opponent the option to capture with one attacked pawn and then advance the other. We next summarize the moves discussed in this paragraph that we did not list before. Here w, w_1, w_2 again denote arbitrary words, which may be empty (length zero) except for the first item:

- From $[[w_1 0 w_2]]$ or $[[w_1 1 w_2]]$ with w_1, w_2 of positive length, either side may move to $[[w_1 {:} w_2]]$ or $[[w_1 {:} w_2]]$ respectively.

- $[[w_1 i_1 {:} i_2 w_2]]$ entails a move to $[[w_1]] + [[{.:} i_2 w_2]]$ or $[[w_1 i_1 {:.}]] + [[w_2]]$; likewise, $[[w_1 010 w_2]]$ entails a move to $[[w_1 0]] + [[{.:} w_2]]$ or $[[w_1 {:.}]] + [[0 w_2]]$.

- $[[{.:} w]]$ entails a move to $[[{:} w]]$; likewise, $[[{.:} w]]$ entails a move to $[[{:} w]]$.

- A move to $[[w_1 1 {:} 1 w_2]]$ is equivalent to a move to $[[w_1]] + [[w_2]]$.

- A move to $[[w_1 1 {:} 0 w_2]]$ is equivalent to a move to $[[w_1]] + [[{:} 0 w_2]]$.

Only the last two cases are directly affected by stopped files.

6.2. Reduction to Nim. Even with stopped files it turns out that our pawn game still reduces to an impartial game, and thus to Nim, once immediately losing and loony moves are eliminated:

Theorem 3. (i) *Each component $[[w]]$ is equivalent to a Nim-heap of some size $\varepsilon(w)$.*

(ii) *A move to $[[{:} i]]$, or $[[{:} 0 i]]$ is loony. For each w of positive length, a move to $[[i w]]$ is either loony or equivalent to a move to $[[w]] \cong *\varepsilon(w)$. The move to*

$[[:0w]]$ or $[[:1w]]$ is loony if and only if a move to $[[:w]]$ or $[[:w]]$ respectively is not loony and is equivalent to a move to $*\varepsilon(w)$. For each w of positive length, a move to $[[:0iw]]$ is either loony or equivalent to a move to $[[iw]] \cong *\varepsilon(iw)$. The move to $[[:00w]]$ or $[[:01w]]$ is not loony if and only if a move to $[[:w]]$ or $[[:w]]$ respectively is not loony and is equivalent to a move to $*(0w)$ or $*(1w)$.

(iii) For any words w_1, w_2, a move to $[[w_1 0 : 0 w_2]]$ or $[[w_1 0 : 0 w_2]]$ is either loony or equivalent to a move to $[[w_1]] + [[w_2]] = *(\varepsilon(w_1) \overset{*}{+} \varepsilon(w_2))$. The move to $[[w_1 0 : 0 w_2]]$ is non-loony if and only both w_1 and w_2 satisfy the criterion of (ii) for a move to $[[:0w]]$ to be non-loony. Likewise, the move to $[[w_1 0 : 0 w_2]]$ is non-loony if and only both w_1 and w_2 satisfy the criterion of (ii) for a move to $[[:0w]]$ to be non-loony.

(iv) The function ε from strings of 0's and 1's with no consecutive 1's to non-negative integers is recursively determined by (ii) and (iii): $\varepsilon(i) = 1$, and for w of length > 1 the value $\varepsilon(w)$ is the mex of the values of the Nim equivalents of all non-loony moves as described in (ii), (iii).

Proof. This is proved in exactly the same way as Theorem 1. Note that we do not evaluate moves to $[[:w]]$. Such a move is available only if the opponent just moved to $[[:0w]]$. If that move was loony then we win, capturing unless the other components sum to $\varepsilon(w)$ in which case we advance, forcing the opponent to advance in return and winning whether that forced advance was loony or not. If the opponent's move to $[[:0w]]$ was not loony then capturing or advancing our attacked pawn yields equivalent positions. ▲

6.3. Numerical Results. Theorem 3 yields a practical algorithm for evaluating $\varepsilon(w)$. If w has length m, the recursion in (iv) requires $O(m^2 \log m)$ space, to store $\varepsilon(w')$ as it is computed for each substring w' of consecutive bits of w, and $O(m^3)$ table lookups and nim-sums, to recall each $\varepsilon(w')$ as it is needed and combine pairs. Unlike the situation for the simple game with no files stopped, where we obtained a simple closed form for $\varepsilon(m)$ (Theorem 2), here we do not know such a closed form. We can, however implement the $O(m^3)$ algorithm to compute $\varepsilon(w)$ for many w. We conclude this paper with a report on the results of several such computations and our reasons for believing that $\varepsilon(w)$ can be arbitrarily large.

We saw already that $\varepsilon(1000) = 2$; this is the unique w of minimal length such that $[w]$ has value $*2$, except that the reversal $[[0001]]$ of $[[1000]]$ has the same value. Clearly $[[0]]$ and $[[1]]$ are the smallest instances of $*1$. We first find $*4$, $*8$ and $*16$ at lengths 9, 20, and 43, for $w = 101001000$, 10100100010100001000, $1010010001000000101000100000001010001001001$, among others. The following table lists for each $k \leq 16$ the least m such that $\varepsilon(w) = k$ for some word w of length m:

k	1	2	3	4	5	6	7	8	9	10	11	12	13	14	15	16
m	1	4	6	9	11	14	16	20	22	25	27	30	32	37	39	43

It seems that, for each k, instances of $*k$ are quite plentiful as m grows. The following table gives for each $35 \leq m \leq 42$ the proportion of length-m words w with $\varepsilon(w) = 0, 1, 2, \ldots, 9$, rounded to two significant figures:[8]

m	0	1	2	3	4	5	6	7	8	9
35	24%	26%	19%	15%	5.4%	5.7%	2.7%	2.5%	.51%	.25%
36	22%	27%	18%	15%	5.5%	5.7%	2.6%	2.8%	.54%	.27%
37	26%	22%	14%	19%	5.8%	5.5%	2.8%	2.8%	.55%	.31%
38	25%	23%	16%	17%	5.7%	5.7%	3.1%	2.7%	.56%	.35%
39	22%	26%	19%	14%	5.6%	5.9%	3.0%	3.0%	.59%	.37%
40	24%	24%	16%	18%	5.9%	5.7%	3.0%	3.2%	.61%	.40%
41	26%	22%	15%	19%	5.9%	5.8%	3.3%	3.1%	.61%	.44%
42	22%	24%	18%	15%	5.8%	6.0%	3.3%	3.2%	.63%	.47%

Especially for $*0$ through $*3$, the proportions seem to be bounded away from zero but varying quite erratically with m. The small proportions of $*6$ through $*9$ appear to rise slowly but not smoothly. We are led to guess that for each k there are length-m components of value $*k$ once m is large enough — perhaps $m \gg k$ suffices — and ask for a description and explanation of the proportion of components of value $*k$ among all components of length m. In particular, is it true for each k that this proportion is bounded away from zero as $m \to \infty$?

It is well known that the number of binary words of length m without two consecutive 1's is the $(m + 2)$-nd Fibonacci number. This number grows exponentially with m, soon putting an end to exhaustive computation. We do not expect to be able to extend such computations to find the first $*32$, which probably occurs around $m = 90$. Nevertheless we have reached $*32$ and much more by computing $\varepsilon(w)$ for periodic w of small period p. This has the computational advantage that for each $m' < m$ we have at most p substrings of length m' to evaluate, rather than the usual $m + 1 - m'$.[9]

We have done this for various small p. Often the resulting Nim-values settle into a repeating pattern, of period p or some multiple of p. This is what happened in Theorem 2 for $p = 1$, with period $10p$. Usually the multiplier is smaller than 10, though blocking files $14r$ and $14r + 5$ produces a period of $504 = 36 \cdot 14$.

All repeating patterns with $p \leq 4$ soon become periodic, but for larger p some choices of pattern yield large and apparently chaotic Nim-values. For instance, we have reached $*4096$ by blocking every sixth file in components of length up to $2 \cdot 10^5$. For each $\alpha = 3, 4, \ldots, 12$ the shortest such component of value $*(2^\alpha)$

---

[8]To compute such a table takes only $O(m)$ basic operations for each choice of w, rather than $O(m^3)$, because $\varepsilon(w')$ has already been computed for each substring w' — this as long as one has enough space to store $\varepsilon(w')$ for all w' of length at most $m - 2$.

[9]Once this is done for some period-p pattern, one can also efficiently evaluate components with the same repeated pattern attached to any initial configuration of blocked and unblocked files. We have not yet systematically implemented this generalization.

has files $6r + 4$ blocked, with length n given by the following table:

α	3	4	5	6	7	8	9	10	11	12
n	51	111	202	497	1414	3545	8255	21208	61985	187193

This again suggests that all $*k$ arise: even if the Nim-values for $p = 6$ ultimately become periodic, we can probably re-introduce chaos by blocking or unblocking a few files. Of course we have no idea how to prove that arbitrarily large k appear this way.

Finally, for a few repeating patterns we observe behavior apparently intermediate between periodicity and total chaos. Blocking every fifth or tenth file yields Nim-values that show some regularity without (yet?) settling into a period. Indeed in both cases the largest values grow as far as we have extended the search (through length 10^5), though more slowly and perhaps more smoothly than for $p = 6$. Such families of components seem the most likely place to find and prove an arithmetic periodicity or some more complicated pattern that finally proves that all $*k$ arise and thus fully embeds Nim into generalized King-and-pawn endgames.

Acknowledgements

This paper was typeset in LaTeX, using Piet Tutelaers' chess fonts for the Diagrams. The research was made possible in part by funding from the Packard Foundation. I thank the Mathematical Sciences Research Institute for its hospitality during the completion of an earlier draft of this paper, and Andrew Buchanan for a careful reading of and constructive comments on that draft.

References

[Berlekamp et al. 1982] E.R. Berlekamp, J.H. Conway, R.K. Guy: *Winning Ways For Your Mathematical Plays, I: Games In General.* London: Academic Press, 1982.

[Clarke 1968] A. Clarke: *2001, A Space Odyssey.* London: Legend, 1968 (reprinted 1991).

[Elkies 1996] N.D. Elkies: *On numbers and endgames: Combinatorial game theory in chess endgames.* Pages 135–150 in *Games of No Chance* (R.J. Nowakowski, ed.; MSRI Publ. **29**, 1996 via Cambridge Univ. Press; proceedings of the 7/94 MSRI conference on combinatorial games).

NOAM D. ELKIES
DEPARTMENT OF MATHEMATICS
HARVARD UNIVERSITY
CAMBRIDGE, MA 02138
UNITED STATES
elkies@math.harvard.edu

More Games of No Chance
MSRI Publications
Volume **42**, 2002

Restoring Fairness to Dukego

GREG MARTIN

ABSTRACT. In this paper we correct an analysis of the two-player perfect-information game Dukego given in Chapter 19 of *Winning Ways*. In particular, we characterize the board dimensions that are fair, i.e., those for which the first player to move has a winning strategy.

1. Introduction

The game of Quadraphage, invented by R. Epstein (see [3] and [4]), pits two players against each other on a (generalized) $m \times n$ chess board. The Chess player possesses a single chess piece such as a King or a Knight, which starts the game on the center square of the board (or as near as possible if mn is even); his object is to move his piece to any square on the edge of the board. The Go player possesses a large number of black stones, which she can play one per turn on any empty square to prevent the chess piece from moving there; her object is to block the chess piece so that it cannot move at all. (Thus the phrase "a large number" of black stones can be interpreted concretely as $mn - 1$ stones, enough to cover every square on the board other than the one occupied by the chess piece.) These games can also be called Chessgo, or indeed Kinggo, Knightgo, etc. when referring to the game played with a specific chess piece.

Quadraphage can be played with non-conventional chess pieces as well; in fact, if we choose the chess piece to be an "angel" with the ability to fly to any square within a radius of 1000, we encounter J. Conway's infamous angel-vs.-devil game [2]. In this paper we consider the case where the chess piece is S. Golomb's Duke, a Fairy Chess piece that is more limited than a king, in that it moves one square per turn but only in a vertical or horizontal direction. In this game of Dukego, we will call the Chess player $\mathcal{D}$ and the Go player $\mathcal{G}$.

Berlekamp, Conway, and Guy analyzed the game of Dukego in [1], drawing upon strategies developed by Golomb. As they observe, moving first is never a disadvantage in Dukego; therefore for a given board size $m \times n$, either the first

1991 *Mathematics Subject Classification.* 90D46.

player to move has a winning strategy (in which case the $m \times n$ board is said to be *fair*), or else $\mathcal{D}$ has a winning strategy regardless of who moves first, or $\mathcal{G}$ has a winning strategy regardless of who moves first. In [1] it is asserted that all boards of dimensions $8 \times n$ ($n \geq 8$) are fair, while $\mathcal{D}$ has a winning strategy on boards of dimensions $7 \times n$ even if $\mathcal{G}$ has the first move. This is not quite correct, and the purpose of this paper is to completely characterize the fair boards for Dukego. The result of the analysis is as follows:

The only fair boards for Dukego are the 8×8 board, the 7×8 board, and the $6 \times n$ boards with $n \geq 9$. On a board smaller than these, $\mathcal{D}$ can win even if $\mathcal{G}$ has the first move, and on a board larger than these, $\mathcal{G}$ can win even if $\mathcal{D}$ has the first move.

We make the convention throughout this paper that when stating the dimensions $m \times n$ of a board, the smaller dimension is always listed first. With this convention, the winner of a well-played game of Dukego is listed in the table below: the entry $*$ denotes a fair board, on which the first player to move can win, while the entries "D" and "G" denote boards on which the corresponding player always has a winning strategy independent of the player to move first.

	$n \leq 5$	$n = 6$	$n = 7$	$n = 8$	$n \geq 9$
$m \leq 5$	D	D	D	D	D
$m = 6$		D	D	D	$*$
$m = 7$			D	$*$	G
$m = 8$				$*$	G
$m \geq 9$					G

As a variant of these Quadraphage games, we can allow $\mathcal{G}$ to have both white (wandering) and black (blocking) stones, where the white stones can be moved from one square of the board to another once played. In this variant, $\mathcal{G}$ has the following options on each of her turns: place a stone of either color on an empty square of the board, move a white stone from one square of the board to any empty square, or pass. With a limited number of stones, it might be the case that $\mathcal{G}$ cannot completely immobilize the Duke, yet can play in such a way that the Duke can never reach any of the edge squares. For instance, it is shown in [1] that $\mathcal{G}$ can win (in this sense of forcing an infinite draw) against $\mathcal{D}$ on an 8×8 board with only three white stones, or with two white stones and two black stones, or with one white stone and four black stones. In this paper we show the following:

If $\mathcal{G}$ has at most two white stones and at most one black stone, then $\mathcal{D}$ can win this variant of Dukego on a board of any size, regardless of who has the first move. On the other hand, if $\mathcal{G}$ has at least three white stones, or two white stones and at least two black stones, then the winner of this variant of Dukego is determined by the size of the board and the player with the first move in exactly the same way as in the standard version of Dukego (as listed in the table above).

In the analysis below, we consider the longer edges of the board to be oriented horizontally, thus defining the north and south edges of the board, so that an $m \times n$ board (where by convention $m \leq n$) has m rows and n columns. Also, if one or both of the dimensions of the board are even, we make the convention that the Duke's starting position is the southernmost and easternmost of the central squares of the board.

2. How $\mathcal{D}$ Can Win

In this section we describe all the various situations (depending on the board size, the player to move first, and the selection of stones available to $\mathcal{G}$) in which the chess player $\mathcal{D}$ has a winning strategy.

We begin with the simple observation that if the Duke is almost at the edge of the board—say, one row north of the southernmost row—and $\mathcal{G}$ has (at most) one stone in the southernmost two rows, then $\mathcal{D}$ can win even if it is $\mathcal{G}$'s turn. For after $\mathcal{G}$ plays a second stone, one of the stones must be directly south of the Duke to prevent an immediate win by $\mathcal{D}$. By symmetry, we can assume that the second stone is to the west of the Duke, whereupon $\mathcal{D}$ simply moves east every turn; even if $\mathcal{G}$ continues to block the southern edge by placing stones directly south of the Duke on each turn, $\mathcal{D}$ will eventually win by reaching the east edge. If $\mathcal{D}$ is in this situation, we say that $\mathcal{D}$ has an Imminent Win on the south edge of the board (and similarly for the other edges).

It is now easy to see that $\mathcal{D}$ can always win on a $5 \times n$ board even if $\mathcal{G}$ has the first move. The Duke will be able to move either north or south (towards one the long edges) on his first turn, since $\mathcal{G}$ cannot block both of these squares on her first turn; and then $\mathcal{D}$ will have an Imminent Win on the north or south edge of the board, correspondingly. Of course, this implies that $\mathcal{D}$ can win on any $3 \times n$ or $4 \times n$ board as well, regardless of who moves first (Dukego on $1 \times n$ and $2 \times n$ boards being less interesting still). Similarly, $\mathcal{D}$ can win on a $6 \times n$ board if he has the first move, since he can move immediately into an Imminent Win situation along the south edge of the board (recalling our convention that the Duke starts in the southernmost and/or easternmost central square).

We can also see now that $\mathcal{D}$ can win on any size board if $\mathcal{G}$ is armed with only two white (wandering) stones. $\mathcal{D}$ selects his favorite of the four compass directions, for instance south, and pretends that the row directly adjacent to

the Duke in that direction is the edge of the board. By adopting an Imminent
Win strategy for this fantasy edge row, $\mathcal{D}$ will be able either to reach the east or
west edge of the board for a true win, or else to move one row to the south for
a fantasy win. But of course, repeating this Fantastic Imminent Win strategy
will get the Duke closer and closer to the south edge of the board, until his last
fantasy win is indeed a win in reality. Similarly, $\mathcal{D}$ can always win if $\mathcal{G}$ adds a
single black (blocking) stone to her two white stones. $\mathcal{D}$ plays as above until $\mathcal{G}$ is
forced to play her black stone (if she never does, then we have just seen that $\mathcal{D}$
will win); once the black stone is played, $\mathcal{D}$ rotates the board so that the black
stone is farther north than the Duke, and then uses this Fantastic Imminent Win
strategy towards the south edge.

We now describe a slightly more complicated situation in which $\mathcal{D}$ has a
winning strategy. Suppose that the Duke is two rows north of the southernmost
row and three rows west of the easternmost row, and that there are no stones in
the two southernmost rows or in the two easternmost rows of the board, nor is
there a stone directly east of the Duke (see the left diagram of Figure 1). Then
we claim that $\mathcal{D}$ can win even if it is $\mathcal{G}$'s turn to move. For (referring to the
second diagram of Figure 1) $\mathcal{G}$ must put a stone in square **A** to block $\mathcal{D}$ from
moving into an Imminent Win along the south edge. $\mathcal{D}$ moves east to square
1, whereupon $\mathcal{G}$ must put a stone in square **B** to block an Imminent Win along
the east edge. But this is to no avail, as $\mathcal{D}$ then moves to square **2** to earn an
Imminent Win along the south edge anyway. If $\mathcal{D}$ is in this situation described
in Figure 1, we say that $\mathcal{D}$ has a Corner Win in the southeast corner of the board
(and similarly for the other corners).

Notice that it is necessary for the shaded area to be empty for the Corner Win
to be in force. If $\mathcal{G}$ has a stone in the second-southernmost row somewhere to
the west of the Duke, then she can place a stone on square **2** to safely guard the
south edge of the board. Similarly, if $\mathcal{G}$ has a stone in the second-easternmost
row somewhere to the north of the Duke, then she can defend both edges of the
board by playing a stone at square **A** on her first turn and one at square **3** on
her second turn.

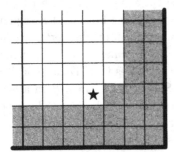 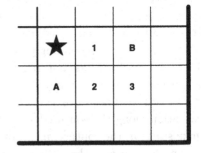

Figure 1. The Corner Win: If $\mathcal{G}$ has no stones in the shaded region (left), then
$\mathcal{D}$ can win from position ⋆ (right).

We can now see that $\mathcal{D}$ can win on a 6×8 board regardless of who has the first move, since the Duke starts the game in a Corner Win position. This implies that $\mathcal{D}$ can win on 6×6 and 6×7 boards regardless of who has the first move. Also, we can argue that $\mathcal{D}$ can win on a 7×7 board even if $\mathcal{G}$ has the first move. By symmetry, we can assume that $\mathcal{G}$'s first stone is placed to the northwest of the Duke or else directly north of the Duke, whereupon $\mathcal{D}$ can move to the south and execute a Corner Win in the southeast corner of the board. Similarly, $\mathcal{D}$ can win on an 8×8 board (and thus on a 7×8 board as well) if he has the first move, since he can again gain a Corner Win situation by moving south on his first turn.

3. How $\mathcal{G}$ Can Win

We have now shown all we claimed about $\mathcal{D}$'s ability to win; it's time to give $\mathcal{G}$ her turn. We start by exhibiting strategies for $\mathcal{G}$ to win on a 7×8 board and on a 6×9 board with the first move. To begin, we assume that $\mathcal{G}$ possesses only three white (wandering) stones; since having extra stones on the board is never disadvantageous to $\mathcal{G}$, the strategy will also show that $\mathcal{G}$ can win with a large number of black stones and no white stones.

The squares labeled with capital letters in Figure 2 are the strategic squares for $\mathcal{G}$'s strategies on these boards. The key to reading the strategies from Figure 2 is as follows: whenever the Duke moves on a square labeled with some lowercase letters, $\mathcal{G}$ must choose her move to ensure that the squares with the corresponding capital letters are all covered. Other squares may be covered as well, as this is never a liability for $\mathcal{G}$. If the Duke is on a square with a + sign as well as some lowercase letters, $\mathcal{G}$ must position a tactical stone on the edge square adjacent to the Duke (blocking an immediate win) as well as having strategic stones on the corresponding uppercase letters. All that is required to check that these are indeed winning strategies is to verify that every square marked with lowercase

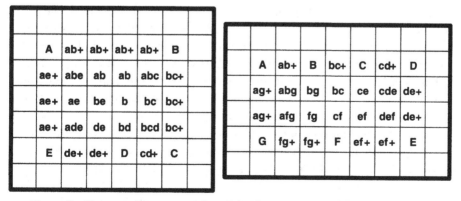

Figure 2. $\mathcal{G}$'s strategies on a 7×8 board (left) and on a 6×9 board (right) with three white stones.

letters (counting + as a lowercase letter for this purpose) contains all of the letters, save at most one, of each of its neighbors, so that $\mathcal{G}$ can correctly change configurations by moving at most one white stone.

In the case of the 7×8 board, the Duke begins on the square marked "b" in the left-hand diagram of Figure 2 (recalling our convention about the precise beginning square for $\mathcal{D}$ on boards with one or both dimensions even), so $\mathcal{G}$'s first move will be to place a stone on the square marked **B**. Notice how $\mathcal{G}$ counters the instant threat of a Corner Win by $\mathcal{D}$ in the southeast corner, by playing her first stone on the second-easternmost column of the board at **B**, a move which also begins the defense of the north and east edges of the board against direct charges by the Duke.

In the case of the 6×9 board, we need to make another convention about the beginning of the game. The Duke begins on the more southern of the two central squares (the square marked "cf" in the right-hand diagram of Figure 2), and we stipulate that $\mathcal{G}$'s first stone be played at square **F**. We may also assume that $\mathcal{D}$'s first move is not to the north, for in this case $\mathcal{G}$ may rotate the board 180 degrees and pretend that $\mathcal{D}$ is back in the starting position, moving her white stone from the old square **F** to the new square **F**. Eventually, $\mathcal{D}$ will move east or west, to a square marked either "ef" or "fg", and at this point $\mathcal{G}$ begins to consult the right-hand diagram for her strategy, playing her second stone at square **E** or **G**, respectively.

$\mathcal{G}$'s strategy on the 7×8 board can be converted into a strategy using two white stones and two black stones without too much difficulty. As before, $\mathcal{G}$ begins by placing a white stone on the square **B**. The goal of $\mathcal{G}$ is to establish her two black stones on squares **A** and **C** (or on **B** and **E**), and then use her white stones both on the strategic squares **B** and **E** (or **A** and **C**, respectively) and as tactical stones blocking immediate wins. The strategic square **D** is only used to keep $\mathcal{D}$ in check until the two black stones can be established. The conversion is straightforward and we omit the details.

Less straightforward, however, is showing that $\mathcal{G}$ also has a strategy for winning on the 6×9 board with the first move, if she has two white stones and two black stones. We include in Figure 3 a full strategy for $\mathcal{G}$ in this situation. $\mathcal{G}$'s goal is to coerce $\mathcal{D}$ into committing to one of the two sides of the board, corresponding to diagram 2 or diagram 3 in Figure 3. Then she will be able to establish her black stones on squares **A** and **F** (or **D** and **F**), and use her white stones both as strategic stones on squares **G** and **H** (or **E** and **H**, respectively) and as tactical stones blocking immediate wins for $\mathcal{D}$.

$\mathcal{G}$ begins by reading the topmost diagram (labeled 1). Since the Duke starts on the square marked "f", $\mathcal{G}$ places a stone on the square **F**; since square **F** is shaded black in the diagram, this stone must be a black stone. When the Duke moves to a square marked with a number, $\mathcal{G}$ immediately switches to the corresponding diagram in Figure 3 and moves according to the Duke's current position. For example, suppose that the Duke's first move is to the east, onto the

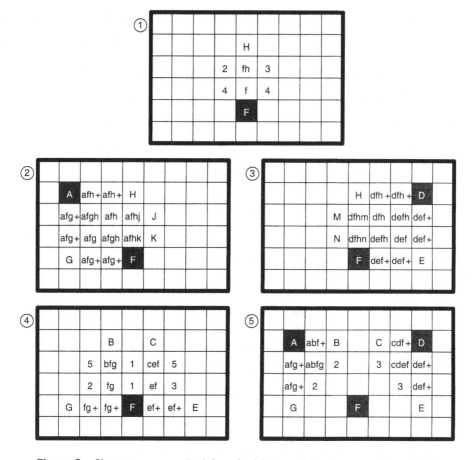

Figure 3. 𝒢's strategy on a 6×9 board with two white stones and two black stones.

square marked **4**; in this case 𝒢 switches to diagram 4, where the Duke's square is marked "ef", indicating that 𝒢 must add a stone to square **E** (a white stone, since square **E** is not shaded black) in addition to her existing black stone on square **F**. We remark, to ameliorate one potential source of confusion, that diagram 5 is really a combination of two smaller diagrams, one for each side of the board; in particular, there will never be a need to have black stones simultaneously at squares **A**, **D**, and **F**.

Of course, any opening move for 𝒢 other than placing a stone on square **F** would lead to a quick Imminent Win for 𝒟 along the south edge. It turns out, though, that even if 𝒢's first move is to play a *white* stone at **F**, a winning strategy exists for 𝒟 (assuming still that 𝒢 has exactly two stones of each color at her disposal). A demonstration of this would be somewhat laborious, and so we leave the details for the reader's playtime.

To conclude this section, we are now in a position to argue that 𝒢, armed with a large number of black stones or with three white stones or with two stones of each color, can win on a 7×9 board (and thus on any larger board) even if $\mathcal{D}$ has the first move. By symmetry we may assume that $\mathcal{D}$'s first move is either to the south or to the east. If $\mathcal{D}$ moves east on his first move, then $\mathcal{G}$ ignores the westernmost column of the board and adopts the above 7×8 strategy on the remainder of the board; alternatively, if $\mathcal{D}$ moves south on his first turn, then $\mathcal{G}$ ignores the northernmost row of the board and adopts the above 6×9 strategy on the rest of the board.

4. Afterthoughts

It is not quite true that we have left no stone unturned (pun unintended) in our analysis of Dukego. For instance, it is unclear exactly how many black stones $\mathcal{G}$ needs to win the original version of Dukego (no white stones) on the various board sizes; determining these numbers would most likely involve a fair amount of computation to cover $\mathcal{D}$'s initial strategies.

Somewhat more tractable, however, would be to determine how many black stones $\mathcal{G}$ needs to win when she also possesses a single white stone. The strategy given in [1] for $\mathcal{G}$ on an 8×8 board works perfectly well when $\mathcal{G}$ has one white stone (to be used tactically) and four black stones (to be placed strategically), and this is the best that $\mathcal{G}$ could hope for. On the other hand, $\mathcal{G}$'s strategies on the 7×8 and 6×9 boards as described above require five and seven black stones, respectively, to go along with the single white stone (and it requires some care to show that seven black stones suffice for the 6×9 board, since we need to account for $\mathcal{D}$ moving north on his first move).

We believe that $\mathcal{G}$ *cannot* win on either a 7×8 board or a 6×9 board with a single white stone and only four black stones. If this is the case, then the least number of black stones that $\mathcal{G}$ can win with, when supplemented by a white stone, would be five on a 7×8 board; but we don't know whether the analogous number on a 6×9 board is five, six, or seven.

References

[1] E. R. Berlekamp, J. H. Conway, and R. K. Guy, *Winning ways for your mathematical plays. Vol. 2*, Academic Press Inc. [Harcourt Brace Jovanovich Publishers], London, 1982, Games in particular.

[2] J. H. Conway, *The angel problem*, Games of no chance (Berkeley, CA, 1994), Cambridge Univ. Press, Cambridge, 1996, pp. 3–12.

[3] R. A. Epstein, *The theory of gambling and statistical logic*, revised ed., Academic Press [Harcourt Brace Jovanovich Publishers], New York, 1977.

[4] M. Gardner, *Mathematical games: Cram, crosscram, and quadraphage: new games having elusive winning strategies*, Sci. Amer. **230** (1974), no. 2, 106–108.

GREG MARTIN
DEPARTMENT OF MATHEMATICS
UNIVERSITY OF TORONTO
CANADA M5S 3G3
gerg@math.toronto.edu

More Games of No Chance
MSRI Publications
Volume **42**, 2002

Go Thermography:
The 4/21/98 Jiang–Rui Endgame

WILLIAM L. SPIGHT

Go thermography is more complex than thermography for classical combinatorial games because of loopy games called kos. In most situations, go rules forbid the loopiness of a ko by banning the play that repeats a position. Because of the ko ban one player may be able to force her opponent to play elsewhere while she makes more than one play in the ko, and that fact gives new slopes to the lines of ko thermographs. Each line of a thermograph is associated with at least one line of orthodox play [Berlekamp 2000, 2001]. Multiple kos require a new definition of thermograph, one based on orthodox play in an enriched environment, rather than on taxes or on composing thermographs from the thermographs of the followers [Spight 1998]. Orthodox play is optimal in such an environment.

Reading a Thermograph

Many go terms have associations with thermographs. They are not defined in terms of thermographs, but I will be using them in talking about the game, and it will be helpful to be able to visualize the thermographs when I do. The inverse of the slope of a thermographic line indicates the net number of local plays. If one player makes 2 local plays while her opponent makes only plays in the environment, the slope will be plus or minus $\frac{1}{2}$. The color of a line indicates which player can afford to play locally at a certain temperature. The player does not necessarily wish to play at that temperature, but she can do so without loss.

In the simple thermograph in Figure 1, the black mast at the top extends upward to infinity. It indicates a region of temperature in which neither player can afford to play locally without taking a loss. The mast starts at temperature, $t = 2$, which we call the temperature of this game, at the top of the hill. It also indicates the local count (or mast value), which is -5. Just below the top of the hill, the blue line of the Left wall indicates that Black (or Left) will not be unhappy to make a local play in this region of temperature. And it shows what

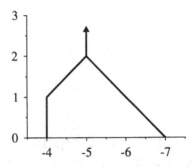

Figure 1. Simple Thermograph.

the local score at each temperature would be when Black plays first. Similarly, the red line of the Right wall indicates an initial play by White (or Right). A vertical slope, whether of a mast or a wall, indicates that each player has made the same number of local plays. The vertical section of the Left wall indicates that Black initiated play and that the whole sequence of play contained an equal number of local plays.

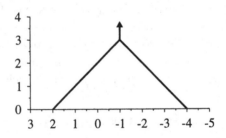

Figure 2. Gote.

Figure 2 shows the prototypical gote thermograph. Two inclined slopes meet at the top of the hill, which means that each player is inclined to play below the temperature of the game, which is 3. (The mast value is −1.) She initiates a sequence of play in which she makes one more play locally than her opponent. Above the temperature of the game the mast is black, which means that neither player is inclined to play locally.

Figure 3 shows the prototypical sente thermograph. Its mast value is 2 and its temperature is 1. It has one vertical wall at the top of the hill, which indicates that, just below the temperature of the game, one player will be able to play locally and force her opponent to make an equal number of local plays. This is White's sente because the Right wall is vertical. Also, the mast is colored red, which means that White will not be unhappy to initiate local play in the region between the temperature of the game and the temperature above which the mast

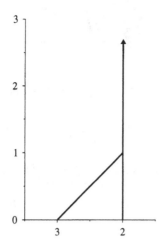

Figure 3. Sente.

is black. So when $1 < t < 2$, White will be able to play in the game with sente, as Black does best to reply locally.

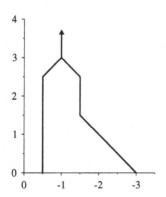

Figure 4. "Double sente".

A common go term is double sente. At first it does not seem to make much sense in terms of thermography, because it suggests two distinct vertical lines, which can never meet to form a mast. However, it does make sense as a temperature relative term. In the thermograph in Figure 4, both walls are vertical when $1\frac{1}{2} < t < 2\frac{1}{2}$. At those temperatures, each player will be able to play in the game with sente, and will be eager to do so, as the gain from playing costs nothing (versus allowing the opponent to play first locally). Hence the go proverb, "Play double sente early." Of course, earlier, when $t > 2\frac{1}{2}$, a local play is gote, but double sente tend to arise when the ambient temperature is lower than the local temperature.

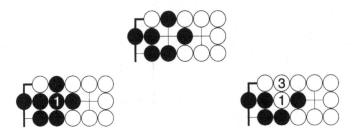

Figure 5. Ko.

The top central position in Figure 5 is a ko. The single Black stone on the upper side has only one adjacent free point (liberty) and may be captured by a play on that point (W 1). Then the stone at 1 has only one liberty, and could be captured except for the ko ban.

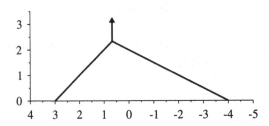

Figure 6. Ko Thermograph (White komaster).

Figure 6 shows the thermograph for the ko in Figure 5 when White is komaster, which means that he can take and win the ko if he plays first [Berlekamp 1996]. If Black wins the ko by filling with B 1, she gets 3 points locally, 2 points of territory plus 1 point for the dead White stone on the 2-1 point. The Left wall indicates the single Black play to a local score of 3. If White captures the ko with W 1 and then, after Black plays elsewhere, wins it with W 3, he gets 2 points of territory plus 1 point for the dead Black stone on the 2-4 point plus 1 point for the stone captured by W 1, for a total of 4 points. The Right wall indicates the 2 White moves to a local score of -4. The mast value of the ko is $\frac{2}{3}$, and its temperature is $2\frac{1}{3}$.

Figure 7 shows the thermograph of the same ko when Black is komaster. Even if White plays first, Black can win the ko. We do not see a separate Right wall, because it coincides with the Left wall. Since the t wo walls coincide, they form a mast. The inclined section of the mast when $t < 2\frac{1}{3}$ is purple, as it combines blue and red lines. When White takes the ko, Black makes a play that carries a threat which is too large for White to ignore. White replies to this ko threat, and then Black takes the ko back and wins it on the following move. The local result is the same as when Black simply wins the ko. The purple mast indicates that either player can play without a local loss at each temperature in

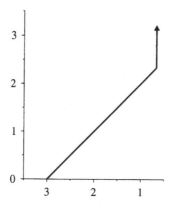

Figure 7. (Black komaster).

that range, but that fact does not take into account the dynamic aspects of the ko. If Black can afford to wait, as the temperature drops the value of the game increases, and that favors Black. So Black will be inclined not to win the ko yet. Therefore the koloser, White, should take the ko and force Black to win it before the temperature drops. It costs Black one move to win the ko, and that allows White to make a play in the environment. The hotter the ambient temperature, the more White gains from that play. This idea is counterintuitive, and most amateur go players have no conception of starting a ko when they cannot win it. They shy away from it.

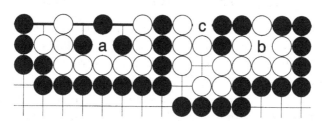

Figure 8. Ko and Gote.

Figure 8 shows a ko on the right and a gote on the left. In the gote, if Black plays at a she kills the White group for 23 points; if White plays at a he saves the group for a local score of −8. The gote has a mast value of $7\frac{1}{2}$ and a temperature of $15\frac{1}{2}$. If Black takes the ko at b and then connects 1 point above b, she kills the White group for 28 points; if White plays at c to take the 3 Black stones, he gets 8 points. The ko has a mast value of 4 and a temperature of 12.

Figure 9 illustrates the unusual thermographs that can occur with sums including kos. This is the thermograph of the sum of the gote and ko in Figure 8, when neither player has a ko threat The gote is hotter than the ko, so that when $t > 12$, each player will prefer the gote to the ko, and the thermograph looks like the standard gote thermograph. But when $t < 12$, both plays are

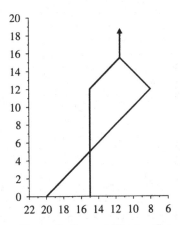

Figure 9. Sum of Ko plus Gote.

viable, and their interaction produces a curious loopback and osculation. When $5 < t < 12$, Black to move will play in the gote, allowing White to win the ko. That exchange produces the vertical segment of the Left wall. White to move will also play in the gote, allowing Black to take and win the ko while White makes a play elsewhere. That exchange produces the backward loop in the Right wall. Below the osculation point, when $t < 5$, each player will prefer to play in the cooler component, the ko.

The Jiang–Rui Environmental Go Game

On April 21, 1998, Jiang Zhujiu and Rui Nai Wei, 9-dan professional players, played the first environmental go game. Jiang played Black and Rui played White. Jiang won by $\frac{1}{2}$ point. The environment for environmental go is a stack of coupons. Instead of making a play on the board, a player may take the top coupon. This coupon is worth t points, the ambient temperature. For this game the values of the coupons ranged from 20 to $\frac{1}{2}$ in $\frac{1}{2}$ point decrements. To offset the advantage of playing first, White received compensation (komi) of $9\frac{1}{2}$ points. (In theory the komi should be 10 points, half the ambient temperature, but a non-integral komi is customary to avoid ties.) For go regions of even moderate size, exhaustive search is not possible, even with computer assistance, and we cannot absolutely guarantee all of our results. However, with the collaboration of t wo of the best go players in the world, we feel confident that we know the main lines of orthodox play from this point on.

Figure 10 shows a position from the game indicating marked territories and the temperatures of some regions. (The temperatures assume orthodox play with White to play in this position.) The hottest region is the Southwest corner, extending into the central South region. The top coupon is now worth 4 points, and this is a double sente at that temperature. We do not need to know the local temperature precisely, only that it is greater than 4. It is approximately 5.

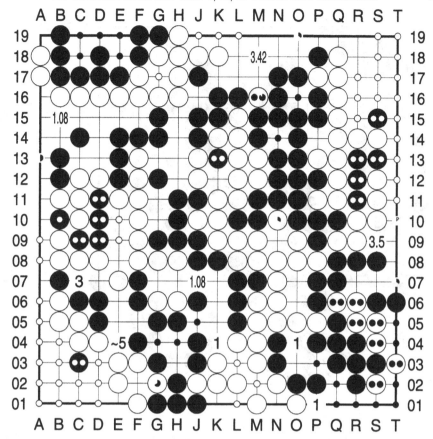

Figure 10. Territories and Temperatures.

If White plays at E04, she threatens to cut off the area just above that west of the E coordinate, which is fairly large. That threat is worth 6 points or so. If Black plays there, he threatens to cut at D03. This cut does not win the cut off stones, but it allows a sizable incursion into the corner. The cut is also hotter than 4 points. The next hottest play, worth $3\frac{1}{2}$ points, is in the East. The next hottest play, a White sente in the North, is worth $3\frac{5}{12}$. Then Black has a 3 point sente at C07 in the Southwest. Next there are 2 plays worth $1\frac{1}{12}$ points in the Northwest and Center, which are related. Finally there are some 1 point plays in the South.

Orthodox Play

Now let's take a look at what we believe to be orthodox play. There are several orthodox variations which all produce the same result: White wins by $2\frac{1}{2}$ points. White actually lost the game by $\frac{1}{2}$ point. Considering that the ambient temperature was only 4, White took a sizable loss of 3 points versus orthodox play.

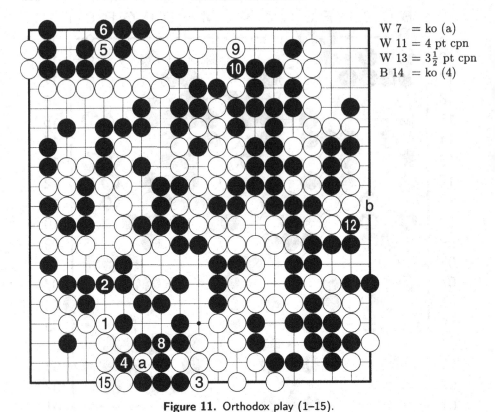

W 7 = ko (a)
W 11 = 4 pt cpn
W 13 = $3\frac{1}{2}$ pt cpn
B 14 = ko (4)

Figure 11. Orthodox play (1–15).

Figure 11: First White plays the double sente at 1. After B 2 White should play W 3 in the South. This is like a double sente, because the mast is purple at the ambient temperature. W 3 threatens to take 4 Black stones by a second play at 8. (Stones which may be taken at the opponent's next play are said to be in atari.) Black does not submit meekly, but takes the ko with B 4, threatening to invade White's corner. This threat is also hotter than the ambient temperature. However, W 5 threatens to kill Black's group in the Northwest corner. Since White would gain a sizable territory, Black replies. Then W 7 takes the ko and forces Black to play B 8, saving his stones by connecting them to living neighbors. Black has a ko threat but nothing that would be bigger than White's taking these 4 stones, because that capture would put Black's whole group in the Southwest and Center in jeopardy. If White waits to play at 3, Black can take the ko first. Then, even though White can still force Black to connect at 8, it costs an extra ko threat to do so.

Next, White takes her sente in the North with W 9. This play has a nominal temperature of less than 4, but it raises the local temperature to around $4\frac{1}{3}$, so Black replies. If White omits this play Black has the possibility of making a gain through unorthodox play. White plays sente while she still can. In general, with combinatorial games, it is a good idea to play sente early. In go, it is usually advisable to wait, because a sente may be used as a ko threat.

After B 10, the mast value of the game is $\frac{1}{4}$. Since the temperature is 4, with a perfectly dense environment the final result with orthodox play would be 2 points better for White, $-1\frac{3}{4}$. The sparseness of the environment happens to favor White in this case, by $\frac{3}{4}$ point. W 11 takes the 4 point coupon.

Now the mast value is $-3\frac{3}{4}$. Black can make a $3\frac{1}{2}$ point play in the East or take the $3\frac{1}{2}$ point coupon. These plays have the same temperature.[1] In go terms, they are miai. Whichever one Black chooses, White gets the other. Usually it is a matter of indifference which one is chosen, and that is the case here. In this variation, Black plays the board play at 12 and White takes the coupon. (Black has an alternative local play at b, also worth $3\frac{1}{2}$ points. The final result is the same if Black plays there, as well.)

Next, B 14 takes the ko in the Southwest at 4. This ko has a low temperature, only $1\frac{2}{3}$. Despite the fact that White is komaster, she meekly replies, because Black's threat to invade the corner is so large, over 4 points, that the mast is blue up to that temperature. Even if White fights the ko, its temperature is so low that she cannot afford to win it yet. Black does not need to play a ko threat, but can simply continue with his normal play, and take the ko back on his next turn. Black can continue to force White to waste ko threats in this fashion. Rather than do so, she connects at 15.

This kind of ko, in which one player takes the ko, forcing the opponent to connect, and then later the other player takes the ko back, forcing the first player to connect, is not uncommon. At first the mast is purple, but after Black connects at 8, the resulting ko carries no further threat for White, only one for Black, and its mast is blue.

Figure 12: At this point the mast value is still $-3\frac{3}{4}$, and the ambient temperature has dropped to 3. In a perfect environment with orthodox play the result would be $-2\frac{1}{4}$. The sparseness of the environment between temperatures 3 and 4 has benefited White by $\frac{1}{2}$ point.

B 16 is a 3 point Black sente. If Black cuts off White's Southwest corner, White will need to protect it from an inside attack. After White's response, Black takes the 3 point coupon. If Black had taken the coupon without playing the sente, White could have made a reverse sente play at 16 and gotten the last 3 point move. In the sparse environment of a real game, it is normally best to get the last play before the temperature drops.

The mast value is now $-\frac{3}{4}$. The local temperature in the North is $2\frac{2}{3}$. After W 19 Black responds with B 20. W 19 is not a prototypical sente, because the mast is black. The temperature of B 20 is the same as that of W 19, so White cannot play sente early and raise the local temperature to force Black's reply.

---

[1] By the more common traditional form of go evaluation, these plays are 7 point gote. The less common form, called miai valuation, corresponds to temperature. Whenever I refer to the value of a play, I am using the miai value.

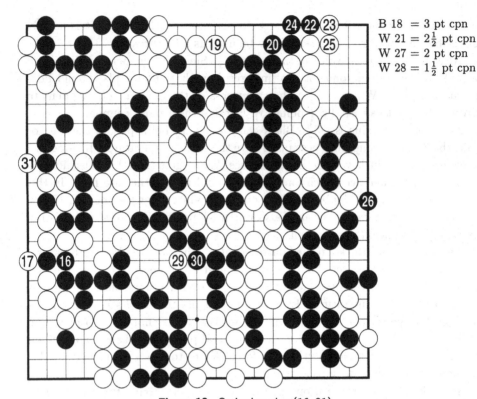

$$B\ 18\ = 3\ \text{pt cpn}$$
$$W\ 21\ = 2\tfrac{1}{2}\ \text{pt cpn}$$
$$W\ 27\ = 2\ \text{pt cpn}$$
$$W\ 28\ = 1\tfrac{1}{2}\ \text{pt cpn}$$

Figure 12. Orthodox play (16–31).

Next White takes the $2\frac{1}{2}$ point coupon. Now B 22 initiates a sente sequence through W 25, with temperature $2\frac{1}{3}$. Black threatens a large invasion of White's corner.

The mast value is $-3\frac{1}{4}$, and the ambient temperature is 2. Again we have a miai between the 2 point coupon and a play at 26 in the East. And again it does not seem to matter which alternative Black chooses. Here Black takes the board play. After White takes the 2 point coupon Black takes the $1\frac{1}{2}$ point coupon. The mast value is now $-1\frac{3}{4}$.

White has a clever play at W 29 in the center, with a temperature of $1\frac{1}{12}$. Black must respond at 30 or lose his group in the South. W 31 now completes the play. If White plays first at 31, the temperature in the center rises to $1\frac{1}{6}$ (!), and Black gets to make the hotter play at 29. These plays are related by White's threat to cut at 41 in Figure 13.

Figure 13: The mast value is now $-2\frac{5}{6}$, and the ambient temperature is 1. B 32 threatens an iterated ko, with 3 steps between winning and losing, raising the local temperature to about $1\frac{2}{5}$. Before replying at 35, White interjects a 1 point sente at W 33, which threatens Black's group in the corner. B 36 is another 1 point sente, and then B 38 takes the 1 point coupon. Now W 39 takes the $\frac{5}{6}$ point play in the North. The mast value is now $-2\frac{2}{3}$, and the ambient

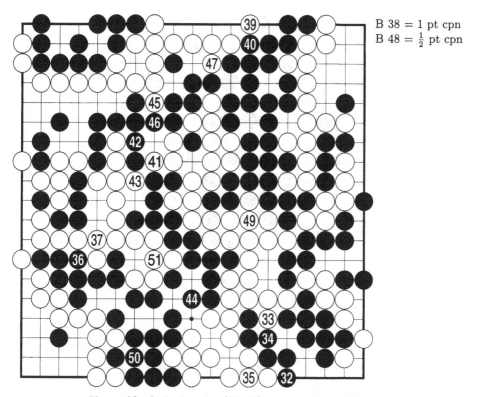

B 38 = 1 pt cpn
B 48 = $\frac{1}{2}$ pt cpn

Figure 13. Orthodox play (32–51) — one point and below.

temperature is $\frac{1}{2}$. In addition to the $\frac{1}{2}$ point coupon there are 2 board plays worth $\frac{1}{2}$, one in the North and one in the North Central section. Again, it does not matter who gets these plays. B 40 takes the play in the North, and then White takes the play in the North Central. This play is a gote, but the mast is red up to a very high temperature, because White's cut with W 41–43 carries a huge threat (Figure 14).

If Black omits B 4, White can play W 5. W 7 puts Black's stones in the center in atari and threatens to kill Black's large group. B 8 takes White's stone at a and puts the 2 neighboring White stones in atari. W 9 saves them and threatens to cut off Black's stones on the left with a play at 10. B 10 saves them, but now W 11 takes the ko at a, renewing the threat against Black's whole group. If Black connects at b, White takes at c, killing the group. Black cannot afford to fight this ko, and avoids it with B 44.

Such a ko is called a hanami ko, or flower-viewing ko, in reference to the Cherry Blossom festivals held each spring in Japan, when people gather to watch cherry blossoms bloom and later fall in clouds.

Black's group is originally alive, and if Black loses the ko he loses a huge amount by comparison. In contrast, if White loses the ko she loses only a couple of points. Black needs huge ko threats to fight the ko, while White needs only

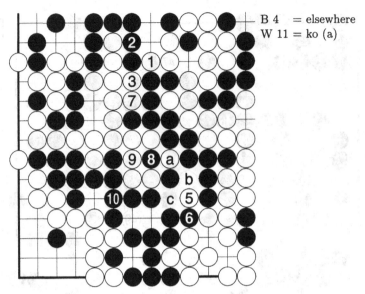

B 4   = elsewhere
W 11 = ko (a)

Figure 14. Hanami ko.

small ones. It is difficult for Black to be the komaster of White's hanami ko. We are just beginning the theoretical study of hanami kos, and of other kos whose structure affects the question of which player is komaster for that ko.

After B 48 in Figure 13 the mast value is $-2\frac{1}{6}$. There are 3 ko positions with temperature $\frac{1}{3}$ on the board. White gets two of them with W 49 and W 51 and Black gets one with B 50. (W 51 does not look like a ko, but if Black plays first he makes one with the same sequence as B 8, W 9, B 10 in Figure 14.) The result is a $2\frac{1}{2}$ point win for White.

Actual Play

Figure 15: White takes the 4 point coupon, but that is a costly mistake. After B 2 – W 3, White has lost 1 point on the South side by comparison with Figure 13. Then after B 4 – W 5, White has lost an additional $\frac{5}{8}$ point in the Southwest versus playing the double sente herself. Since Black plays B 2 and B 4 with sente, while White could have played with sente, these are differences between vertical thermographic lines, which means that Black's gains have come at no cost. "Play double sente early."

After W 5 the mast value is $-2\frac{1}{8}$, and the ambient temperature has dropped to $3\frac{1}{2}$. But the top coupon and the East side are miai, and Black should play his reverse sente sequence in the North (B 14 – W 11, B a), which is worth $3\frac{5}{12}$. Instead, B 6 is worth only $2\frac{5}{8}$ points. B 6 does aim at B 10, which gives Black several ko threats against White's Southwest corner, but with otherwise orthodox play, these threats do not seem to gain anything, nor did Black gain from them in the game.

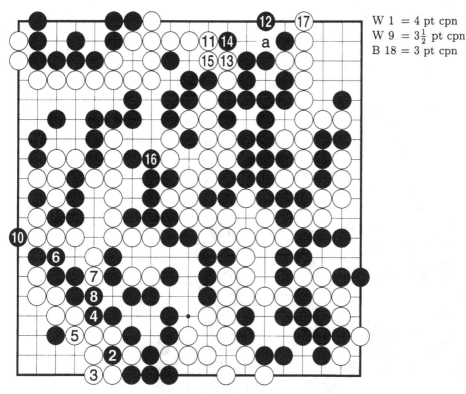

Figure 15. Actual play (1–18).

W 1 = 4 pt cpn
W 9 = 3½ pt cpn
B 18 = 3 pt cpn

After W 9 takes the $3\frac{1}{2}$ point coupon, Black should take the $3\frac{1}{2}$ point play in the East. B 10 gives Black several ko threats in the Southwest corner to make Black komaster in general, but B 10 is worth only $2\frac{1}{4}$ points, plus $\frac{1}{2}$ the value of the threats. That value appears to be $\frac{1}{6}$ point, increasing the value of B 10 to $2\frac{1}{3}$ points.

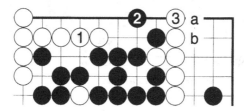

Figure 16.

After B 10 White should play her sente in the North, reaching the position in Figure 16. When White is komaster Black replies to W 1 at 1 point below 2, but since Black is komaster, B 2 is orthodox. If White omits W 3, Black can play there, and then if W a, B b makes a large ko. The mast value when Black is komaster is $\frac{1}{6}$ point greater than when White is komaster. There are some small kos and potential small kos on the board, but it appears that Black can realize no gain in their mast values from being komaster.

Returning to Figure 15, the fact that Black is now komaster after B 10 makes a play at 14 in the North double sente. White should play W 14 – B 13 in the North and then make the $3\frac{1}{2}$ point play in the East. Both players overlook the play in the East.

B 12 is sente, but may not be best. Black could try the aggressive B 17 – W a, B 12, making a large ko. Jiang 9-dan judged not to make that play in the game. The ko fight is difficult and has not yet been analyzed. Assuming that Black should not play at 17, his orthodox play is B a, a gote worth $3\frac{1}{6}$ points. In that case B 12 gives up $\frac{1}{3}$ point versus the mast value.

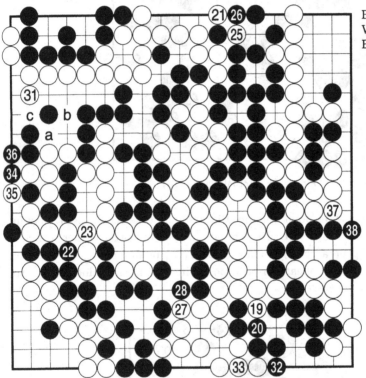

B 24 = $2\frac{1}{2}$ pt cpn
W 29 = 2 pt cpn
B 30 = $1\frac{1}{2}$ pt cpn

Figure 17. Actual play (19–38).

Figure 17: The mast value is now $-\frac{1}{3}$. Both players continue to ignore the East side, the hottest area on the board. W 21 is worth $2\frac{1}{2}$ points. W 25 is a $2\frac{1}{3}$ point sente. The exchange, W 27 – B 28, is unorthodox. It eliminates White options in some variations (See Figure 14 for an example.), and loses $\frac{1}{6}$ point versus the mast value, which is now $-\frac{1}{6}$.

After B 30 the mast value is $-\frac{2}{3}$. W 31 loses $\frac{1}{2}$ point versus the mast value. Before W 31, B 34 is worth $1\frac{1}{2}$ points, but afterwards it is worth 2 points.

If Black plays elsewhere after W 31, White can play at 35, threatening W a – B b, W c, which wins the 2 Black stones on the side. B 32 is a 1 point sente, but W 33 is worth less than a play in either the East or West. After B 36 the mast value is $-\frac{1}{6}$. White plays W 37 in the East, but loses 2 points by comparison with correct play.

Figure 18: In post-game discussion Rui Nai Wei spotted the correct play. After the skillful play of W 1, the local temperature drops to 3. But since the ambient temperature is less than that, Black replies. W 3 is another skillful play, threatening the Black stones on the side. Through B 8 the sequence is sente for White. White has the same territory as in the game, but Black has 2 points less. W 37 was White's last chance to win the game. If White had played as in Figure 18 the mast value would have been $-\frac{2}{3}$, but instead it is $1\frac{1}{3}$.

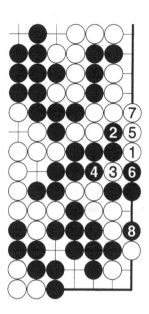

Figure 18.

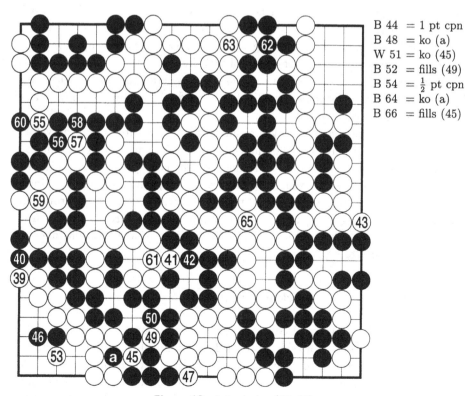

B 44 $= 1$ pt cpn
B 48 $=$ ko (a)
W 51 $=$ ko (45)
B 52 $=$ fills (49)
B 54 $= \frac{1}{2}$ pt cpn
B 64 $=$ ko (a)
B 66 $=$ fills (45)

Figure 19. Actual play (39–66).

Figure 19: There are some slight deviations from orthodox play at the end. W 39 should take the ko at 45 in the South. The ko has a temperature of $1\frac{1}{3}$. W 39 is a 1 point sente, and a possible threat for this ko. B 44 misses a chance when he takes the 1 point coupon instead of filling at 45. If Black fills at 45, White takes the 1 point coupon, and Black can play at 47 with sente. Then Black takes the $\frac{1}{2}$ point card. At that point there would be only 2 kos on the board. White would have to win the ko fight to get the same result as in the game. If Black won the ko fight, he would score 1 point more.

B 46 is an oversight. White does not have to answer this apparent ko threat right away. White plays atari at 47, and then when Black takes the ko back, White plays at 49. If Black fills the ko White can capture Black's 7 stones, so Black must take at 50. Then White takes the ko back and Black has to connect at 49. Since White is komaster, Black should simply play B 46 at 47. That gets the last 1 point play, but then White gets the $\frac{1}{2}$ point card instead of Black, so there is no difference in the score.

Summary

When White took the 4 point coupon, she overlooked the double sente in the South and Southwest, allowing Black to make the game close. Then Black overestimated the value of his plays in the West, allowing White to play first in the North. White chose the wrong spot in the North, ending up with gote. Then for a long time both players overlooked White's skillful option in the East and underestimated its temperature. White lost several chances to win, while Black failed to secure the game. Finally, White made the wrong play in the East, losing 2 points and the game.

Traditional evaluation of go plays produces the temperatures and mean values of classical thermography. Thermographs, however, yield additional information about orthodox play in different environments. The concept of komaster allows us to find the mast values and temperatures of all positions involving single kos. Defining thermographs in terms of play in a universal enriched environment extends thermography to positions with multiple kos. Research continues into positions in which neither player is komaster and into what conditions allow a player to be komaster.

Acknowledgments

This analysis has been, and continues to be, a collaborative effort. It is the result of many discussions involving Elwyn Berlekamp, Bill Fraser, Jiang Zhujiu, Rui Nai Wei, and myself. This year Wei Lu Chen also contributed to the analysis. It would have been impossible without the aid of Bill Fraser's Gosolver program. For instance, the North is particularly complex, and we have entered more than

20,000 North positions into the program's database. I wish to thank Elwyn Berlekamp and Martin Müller for their helpful comments on this work.

References

[Berlekamp 1996] E. R. Berlekamp, "The economist's view of combinatorial games", pp. 365–405 in *Games of No Chance*, R. Nowakowski, ed., Cambridge University Press, Cambridge (1996).

[Berlekamp 2000] E. R. Berlekamp, "Sums of N x 2 amazons", pp. 1–34 in *Game Theory, Optimal Stopping, Probability and Statistics*, T. Bruss and L. Le Cam, eds., Lecture Notes – Monograph Series **35**, Institute of Mathematical Statistics, Hayward, CA, (2000).

[Berlekamp 2001] E. R. Berlekamp, "Idempotents among partisan games", in this volume.

[Spight 1998] W. L. Spight, "Extended thermography for multiple kos in go", pp. 232–251 in *Computers and Games*, J. van den Herik and H. Iida, eds., Lecture Notes in Computer Science **1558**, Springer, Berlin (1998).

WILLIAM L. SPIGHT
729 55TH ST.
OAKLAND, CA 94609
UNITED STATES
bspight@pacbell.net

More Games of No Chance
MSRI Publications
Volume **42**, 2002

An Application of Mathematical Game Theory to Go Endgames: Some Width-Two-Entrance Rooms With and Without Kos

TAKENOBU TAKIZAWA

ABSTRACT. The author is part of a research group under the supervision of Professor Berlekamp. The group is studying the late-stage endgame of Go and has extended the theory of quasi-loopy games using the concept of a ko-master. This paper discusses some width-two-entrance rooms with and without kos that have arisen in this study.

1. Easy Examples without Kos

Figure 1.1 shows a simple width-two-entrance 5-room position. Though it is easily analyzed by mathematicians, it is difficult even for strong Go players to determine the best move. Black's t wo options from Figure 1.1 are shown in Figures 1.2 and 1.3, and White's t wo options are shown in Figures 1.4 and 1.5.

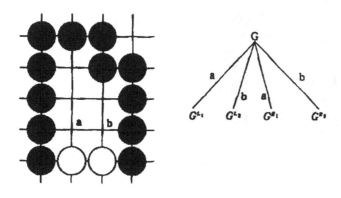

Figure 1.1

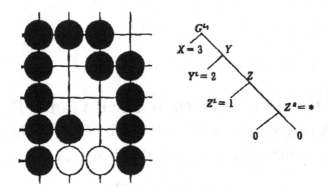

Figure 1.2

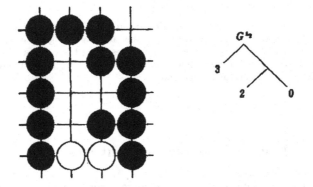

Figure 1.3

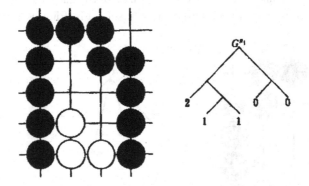

Figure 1.4

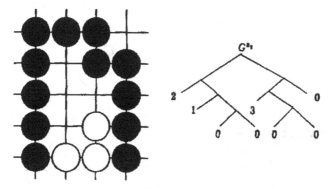

Figure 1.5

We get the following results: $G_1^{L_1} = 2 + \frac{1}{8}$, $G_1^{L_2} = 2\uparrow$, $G_1^{R_1} = \frac{3}{4}$ and $G_1^{R_2} = \frac{1}{2}$, thus $G^{L_1} > G^{L_2}$ and $G^{R_2} < G^{R_1}$. Therefore, Black should play a but White should play b from the position in Figure 1.1, and $G_1 = \{G_1^{L_1} - 1 | G_1^{R_2} + 1\} = 1 + \frac{1}{4}$ is the chilled game of G (Figure 1.6).

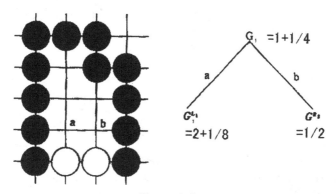

Figure 1.6

Figure 1.7 shows an even easier game than the game in Figure 1.1. Black and White should play in the same place, and these moves are analyzed in Figures 1.8 and 1.9. The canonical form of G and its analysis using the chilling method are shown in Figure 1.10. We get $G_1 = 2 + \frac{1}{8}$.

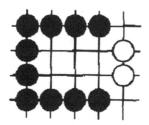

Figure 1.7

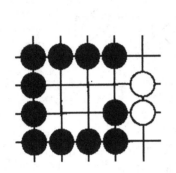

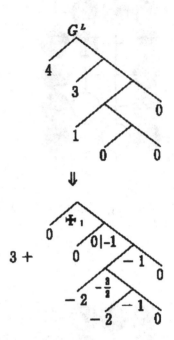

Figure 1.8

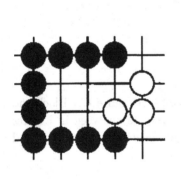

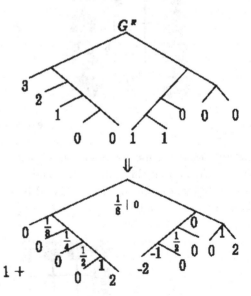

Figure 1.9

2. A Simple 1-point Ko

Figure 2.1 X shows a simple one-point ko. If Black plays A in X then the position moves to Y. If White can play B in Y then the position moves back to X. The ko rule bans each player from taking back a ko directly after another player takes it, so that a game position which includes a ko is not a loopy game but a quasi-loopy game.

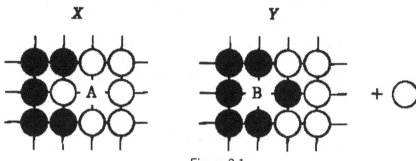

Figure 2.1

White has to play in some other place if Black takes the ko. If Black responds in the other place, then White may take back the ko. White's play in the other place is called a ko-threat.

Figure 2.2 (left) shows the game tree of X. It can also be expressed by a graph as in Figure 2.2 (right). Let game G be the game X in Figure 2.1. We can make three copies of X and name them X_1, X_2 and X_3. $X_1 + X_2 + X_3 = 3 \cdot G$ (Figure 2.3).

Figure 2.2

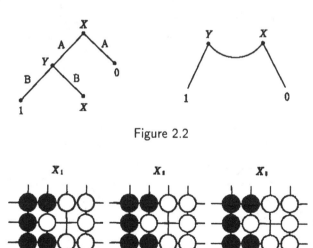

Figure 2.3

If Black plays first, he plays X_1 to $X_1^L + X_2 + X_3$. Because $(X_1^L - 1) + X_2 = 0$, White then plays X_3 to $X_1^L + X_2 + X_3^R$ and the value of such a game is 1. If White plays first, she plays X_3 and then Black plays X_1 to reach the same position. Therefore, $3 \cdot G = 1*$. And the mean value of G is $\frac{1}{3}$.

When there is a ko, Black or White eventually fills the ko. The filling side is the winner of the ko and called the ko-master. Note that the ko-master must fill the ko. When Black is the ko-master, then the game G is denoted by $\hat{G}$; when White is the ko-master, then G is denoted by $\check{G}$.

Figure 2.4 shows the game graph where Black is the ko-master and Figure 2.5 where White is the ko-master. If White is the ko-master, Black may take the ko by playing at A in Figure 2.1, but when White takes back the ko, Black cannot take back the ko again. So the position is the same but the situation is different. The ko-master White should fill the ko, and we denote this position as $\dot{X}$ rather than $\check{X}$.

3. An Example Containing a Hidden Ko

Figure 3.1 shows an example of a Go position which contains a hidden ko. Black's options are shown in Figure 3.2, while White's options are shown in Figure 3.3.

Figures 3.4 through 3.6 show game tree analyses of chilled games:

$$\hat{G}_1^{L_1} = 3 + \{0|\uparrow\} = 3 + \Uparrow *, \check{G}_1^{L_1} = 3 + +_1, G_1^{L_2} = 2 + \{1|*\}, G_1^{L_3} = 2.$$

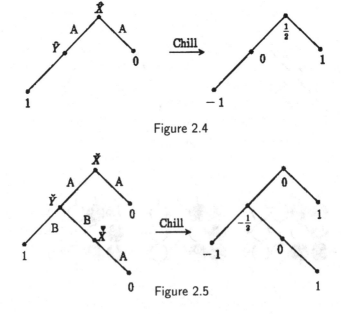

Figure 2.4

Figure 2.5

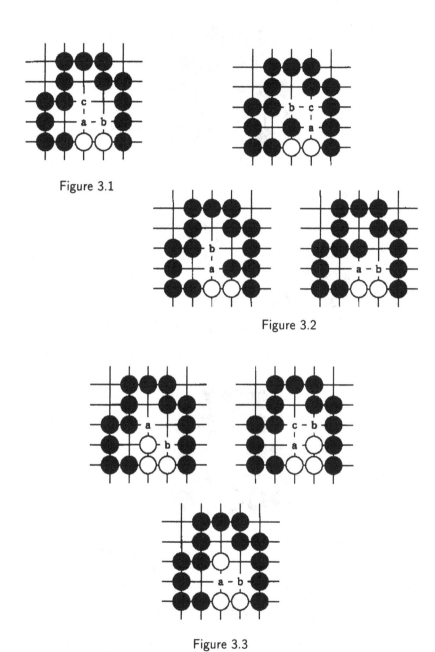

Figure 3.1

Figure 3.2

Figure 3.3

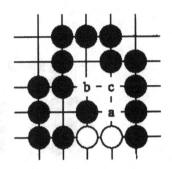

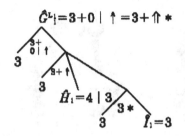

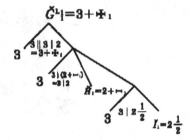

Figure 3.4

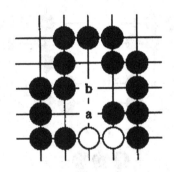

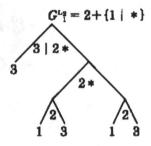

Figure 3.5

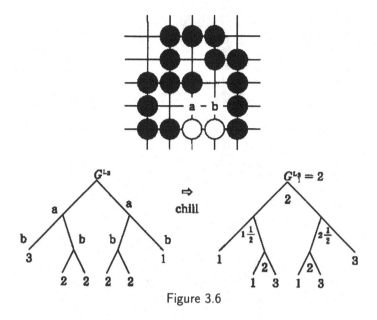

Figure 3.6

And Figures 3.7 through 3.9 show game tree analyses of chilled games:

$$G_1^{R_1} = \tfrac{3}{4}, \quad \hat{G}_1^{R_2} = (1+\uparrow)|1 = 1*,$$
$$\check{G}_1^{R_2} = (1|0, \tfrac{1}{2})\|1 = \tfrac{3}{4},$$
$$\hat{G}_1^{R_3} = 3\|2|1\||1 = 2|1, \quad \check{G}_1^{R_3} = 3|0\|1 = 0.$$

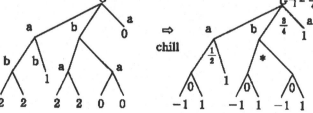

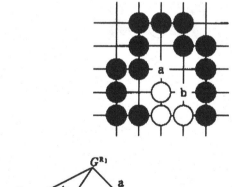

Figure 3.7

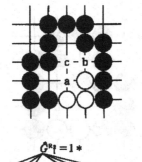

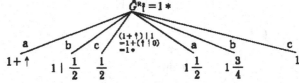

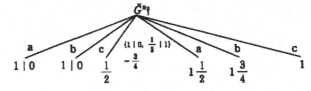

Figure 3.8

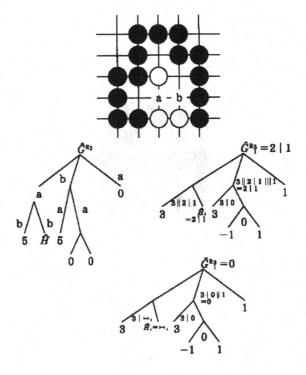

Figure 3.9

Finally, we get the analysis of the chilled game G itself as shown in Figure 3.10:

$$\hat{G}_1 = \{(2+\Uparrow *)|1\tfrac{3}{4}\}, \quad \check{G}_1 = \{(2++_1)|1\} = 2|1.$$

Thus, Black should play a in Figure 3.1 regardless of who the ko-master is, but White should play a when Black is the ko-master and play c when White is the ko-master.

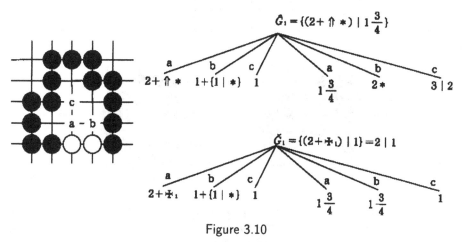

Figure 3.10

4. The Rogue Positions

Elwyn Berlekamp and David Wolfe have found an interesting Go position and named it the rogue position. This is now called Wolfe's Rogue Position and is shown in Figure 4.1. It is a width-one-entrance 7-room position. We then found a quite interesting width-two-entrance 7-room position and named it Takizawa's Rogue Position. This position is shown in Figure 4.2. There are two hidden kos in Takizawa's Rogue Position.

Black and White's options are shown in Figures 4.3 and 4.4. All but one, Black's option G^{L3}, contain at least one ko in the game tree. The chilled game

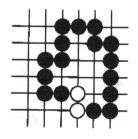

Figure 4.1

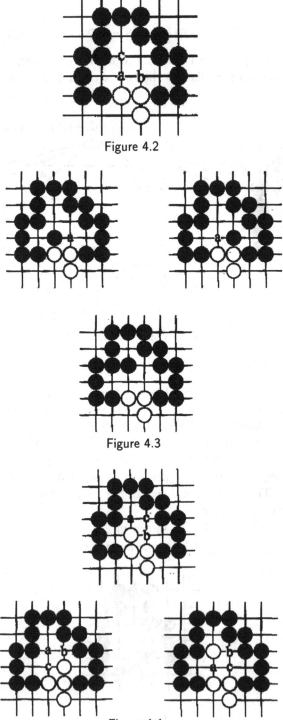

Figure 4.2

Figure 4.3

Figure 4.4

analyses are shown in Figures 4.5 through 4.10:

$$\hat{G}_1^{L_1} = 3 + \{1| \uparrow\},$$
$$\check{G}_1^{L_1} = 2 + \{2||1|0\},$$
$$\hat{G}_1^{L_2} = 4||3|2\tfrac{1}{2},$$
$$\check{G}_1^{L_2} = 4||3|2\tfrac{1}{4},$$
$$G_1^{L_3} = 2 + \tfrac{1}{2}*,$$
$$\hat{G}_1^{R_1} = \check{G}_1^{R_1} = 1 \downarrow,$$
$$\hat{G}_1^{R_2} = 1*,$$
$$\check{G}_1^{R_2} = \{(\tfrac{1}{2}, 1|0)||1\},$$
$$\hat{G}_1^{R_3} = 4||2|1|||1 = 2|1,$$
$$\check{G}_1^{R_3} = 4|\tfrac{1}{4}||1 = \tfrac{1}{2}.$$

Finally, we get the analysis of chilled game G itself as shown in Figures 4.11 and 4.12:

$$\hat{G}_1 = 2 + \{1| \uparrow ||(\downarrow, *)\} \quad \text{and} \quad \check{G}_1 = \{3||2|1\tfrac{1}{4}|||1\tfrac{1}{2}\} = 2|1\tfrac{1}{2}.$$

Thus, Black should play a in Figure 4.2 when Black is the ko-master but play b when White is the ko-master. White should play a or b when Black is the ko-master but play c when White is the ko-master. These are quite interesting results.

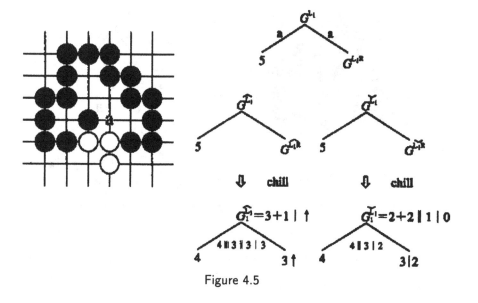

Figure 4.5

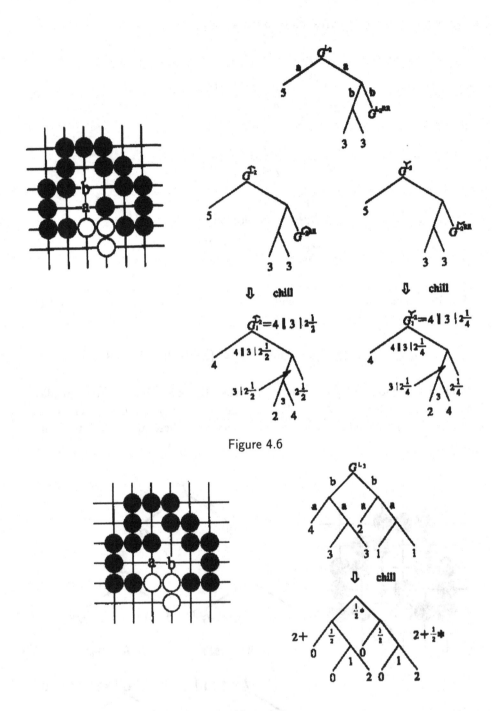

Figure 4.6

Figure 4.7

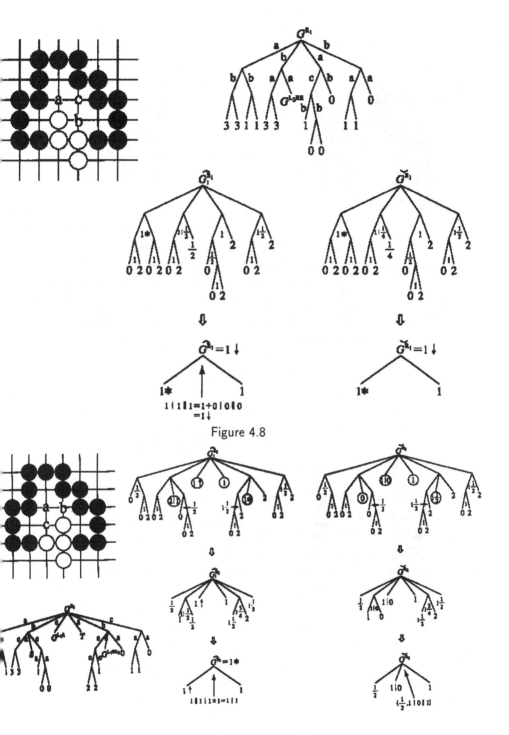

Figure 4.8

Figure 4.9

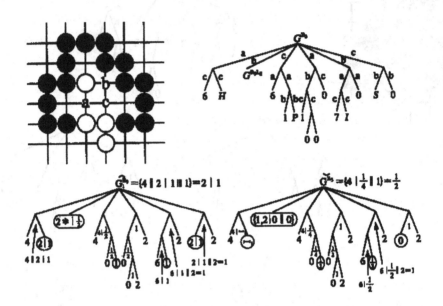

Figure 4.10

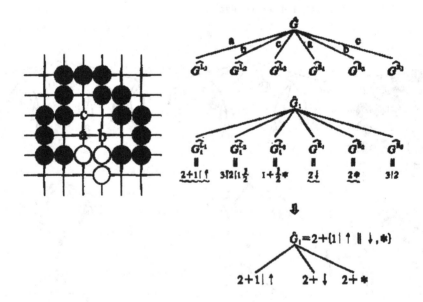

Figure 4.11

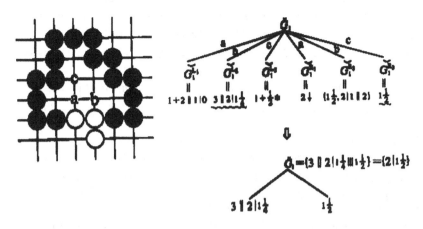

Figure 4.12

Acknowledgements and Conclusion

This study is being conducted under a Waseda University Grant for Special Research Projects (#2000A-005).

The author is very grateful to Prof. Elwyn Berlekamp of UC, Berkeley, for his kind advice and direction. The author is also grateful to the members of Professor Berlekamp's research group, especially David Wolfe, Yonghoan Kim, William Spight, Martin Müller, Kuo-Yuan Kao and William Fraser, for their comments and many suggestions. The author thanks Masahiro Okasaki and Yasuo Kiga for providing the group with a large amount of go literature. We have started to study how to find the best move in a given late-stage endgame of Go and have obtained good results for some patterns including simple kos or simple hidden kos as shown in this article. The study of Go has also helped us to develop mathematical game theory itself. As a way of developing the theory furthur, we are now studying how to find the best move at slightly earlier stages of the endgame and involving more complicated kos. Endgames which include kos give us good examples of loopy games to work with.

References

[1] Elwyn Berlekamp, John Conway and Richard Guy. *Winning Ways*. Academic Press, 1982.

[2] Elwyn Berlekamp. Introductory Overview of Mathematical Go Endgames. *Proceedings of Symposia in Applied Mathematics, 43*. American Mathematics Society, 1991.

[3] Elwyn Berlekamp. The Economist's View of Combinatorial Games. *Games of No Chance*. Cambridge University Press, 1996.

[4] Elwyn Berlekamp and David Wolfe. *Mathematical Go*. A K Peters, 1994.

[5] Charles Bouton. Nim, a Game with a Complete Mathematical Theory. *Annals of Mathematics [Second Series, v. 3]*. 1902.

[6] John Conway. *On Numbers and Games*. Academic Press, 1976.

[7] Kuo-Yuan Kao. *Sums of Hot and Tepid Combinatorial Games*. Ph. D. Thesis, UNC, Charlotte, 1997.

[8] Yasuo Kiga. *On Mathematical Go (Kaiseki no Urodangi)*. 1993 (in Japanese).

[9] Yonghoan Kim. *New Values in Domineering and Loopy Games in Go*. Ph. D. Thesis, UC, Berkeley, 1995.

[10] Martin Müller. *Computer Go as a Sum of Local Games: An Application of Combinatorial Game Theory*. Ph. D. Thesis, ETH Zürich, 1995.

[11] Martin Müller, Elwyn Berlekamp and William Spight. *Generalized Thermography: Algorithms, Implementation, and Application to Go Endgames*. International Computer Science Institute, Berkeley TR-96-030, 1996.

[12] William Spight. Extended Thermography for Multiple Kos in Go. *Computers and Games*. Springer-Verlag, 1999.

[13] Takenobu Takizawa. Mathematical Game Theory and Its Application(1)–(5). *Journal of Liberal Arts Nos. 99–108*. The School of Political Science and Economics, Waseda University, 1996–2000 (in Japanese).

[14] Takenobu Takizawa. Mathematical Game Theory and Its Application to Go Endgames. *IPSJ SIG Notes 99-GI-1*. Information Processing Society of Japan, 1999 (in Japanese).

TAKENOBU TAKIZAWA
takizawa@mn.waseda.ac.jp
SCHOOL OF POLITICAL SCIENCE AND ECONOMICS
WASEDA UNIVERSITY
TOKYO
JAPAN

More Games of No Chance
MSRI Publications
Volume **42**, 2002

Go Endgames Are PSPACE-Hard

DAVID WOLFE

ABSTRACT. In a Go *endgame*, each local area of play has a polynomial
size canonical game tree and is insulated from all other local areas by live
stones. Using techniques from combinatorial game theory, we prove that
the Go endgame is PSPACE-hard.

1. Introduction

Go is an ancient game which has been played for several millennia throughout
Asia. Although *playable* rules are relatively simple (tournament rules can include
many technicalities), the game is strategically very challenging. Go is replacing
Chess as the pure strategy game of choice to serve as a test-bed for artificial
intelligence ideas [9] [7] [6].

Go was proved PSPACE-hard by Lichtenstein and Sipser [8], and was later
proved EXPTIME-complete (Japanese rules) by Robson [12]. More recently,
Crâşmaru and Tromp [5] proved that Go positions called ladders are PSPACE-
complete. Yedwab conjectured that the Go endgame is hard [14]. The *endgame*
occurs when each local area of play has a polynomial size canonical game tree
and is insulated from all other local areas by live stones. A combinatorial game
theorist will recognize this as a *sum* of simple local positions.[1]

Morris succeeded in proving that sums of small local game trees are PSPACE-
complete [11]. Yedwab restricted the games to be of depth 2 [14], and Moews
showed that sums of games with only three branches are NP-hard [10] [1, p. 109].
(Each of Moew's games is of the form $\{a \parallel b \mid c\}$ where a, b and c are integers.)
Since Yedwab's and Moews' Go-like game trees depend upon scores which are
exponential in magnitude (yet polynomial in the number of bits of the scores),
they did not translate to polynomial sized Go positions.

Key words and phrases. Go, PSPACE, endgame, games.

[1]Another possible definition of endgame is when each play gains at most a constant number
of points; we don't address that notion here. Go endgames with ko's might be outside PSPACE.

Berlekamp and Wolfe show how to analyze certain one-point Go endgames [1]. Some of these endgame positions have *values* which are linearly related to *dyadic rationals* of the form $x = \frac{m}{2^n}$. Since these endgame positions are polynomial in size in the number of *bits of* the numerator and denominator of x, their techniques can be combined with those of Yedwab and Moews to prove that the Go endgame is PSPACE-hard.

Robson observed that in Japanese rules (where repetitions of a recent position are forbidden), Go is easily seen to be in EXPTIME. However, he conjectures that according to Chinese rules (where any previous position is forbidden), Go is EXPSPACE-complete [13]. This paper fails to resolve his conjecture, but the techniques may be applicable.

2. Proof Sketch

We'll prove that the Go endgame is PSPACE-hard using a series of reductions, as suggested by Yedwab [14] shown in Figure 1.

3-QBF

⇓ Yedwab

PARTITION GAME

⇓ Yedwab, Moews

SWITCH PICK GAME

⇓

FRACTIONAL SWITCH GAME

⇓ (Yedwab, Moews)

GAME SUM

⇓

GO ENDGAME

Figure 1. Proof outline. Yedwab and Moews reduced something like the SWITCH PICK GAME directly to GAME SUM. The FRACTIONAL SWITCH GAME is introduced here to to keep the GO ENDGAME polynomial in size.

This paper will introduce each of these problems and reductions in sequence.

3. The Artificial Games

We begin reducing from the canonical PSPACE-complete problem Quantified Boolean Formula (QBF) in conjunctive normal form with three literals per clause:

3-QBF

> INSTANCE: A formula of the form
>
> $$\exists x_1 \forall x_2 \exists x_3 \forall x_4 \ldots \exists x_n : C_1 \wedge C_2 \wedge C_3 \wedge \cdots \wedge C_m$$
>
> where each of the clauses C_i is a disjunct of three literals ($l_{i1} \vee l_{i2} \vee l_{i3}$), and each literal is either some variable, say x_k, or its complement, $\overline{x_k}$, for $1 \leq k \leq n$.
>
> QUESTION: Is the formula true?

We reduce 3-QBF to the following partition game:

PARTITION GAME

> INSTANCE: A collection of $2N$ non-negative integers X_i and $\overline{X_i}$, $1 \leq i \leq N$ and a *target* integer T.
>
> QUESTION: Players Left (L) and Right (R) alternate selecting numbers for inclusion in the set S. Left chooses X_1 or $\overline{X_1}$, Right chooses X_2 or $\overline{X_2}$ and so forth. At the end of the game, S will have N elements. Left wins if the elements sum to exactly T.

Lemma 0.1. *The PARTITION GAME is PSPACE-complete.*

Proof. An example of the reduction is shown in Figure 2.

From a 3-CNF formula, F, with n variables and m clauses, construct $n + 12m$ pairs of integers each with $4m$ base b digits, where b is sufficiently large to prevent carries. (A choice of $b = 2n + m + 1$ will work.) The $4m$ digits are allocated as follows: One for each l_{ij} and one for each C_i. Denote by $D(C_i)$ (or $D(l_{ij})$) the value of an integer with the digit position C_i (or l_{ij}) set to 1, all other positions 0. $D(C_i)$ will be a power of b.

As in the figure, the $n + 12m$ pairs of integers are as follows:

(n **variable pairs**): One pair for each variable x_k:

$$\sum_{x_k = l_{ij}} D(l_{ij}) \quad \text{and} \quad \sum_{\overline{x_k} = l_{ij}} D(l_{ij}).$$

($6m$ **clause pairs**): Two pairs for each literal l_{ij}: Left's pair is 0 and $D(C_i) + D(l_{ij})$. Right's pair is 0 and 0.

($6m$ **garbage collection pairs**): Two pairs for each literal l_{ij}: Left's pair is $D(l_{ij})$ and 0. Right's pair is 0 and 0.

The sum T has a 1 in every digit position.

$$(\text{3-QBF}) \quad F = \exists x_1 \forall x_2 \exists x_3 : (x_1 \vee x_2 \vee x_3) \wedge (\overline{x_1} \vee \overline{x_2} \vee x_3)$$

	l_{11}	l_{12}	l_{13}	l_{21}	l_{22}	l_{23}				
	x_1	x_2	x_3	$\overline{x_1}$	$\overline{x_2}$	x_3	C_1	C_2		
$X_1 =$	1	0	0	0	0	0	0	0		
$\overline{X_1} =$	0	0	0	1	0	0	0	0		
$X_2 =$	0	1	0	0	0	0	0	0		
$\overline{X_2} =$	0	0	0	0	1	0	0	0		
$X_3 =$	0	0	1	0	0	1	0	0		
$\overline{X_3} =$	0	0	0	0	0	0	0	0		
$X_5 =$	1	0	0	0	0	0	1	0		$X_4 = \overline{X_4} = \overline{X_5} = 0$
$X_7 =$	0	1	0	0	0	0	1	0		$X_6 = \overline{X_6} = \overline{X_7} = 0$
$X_9 =$	0	0	1	0	0	0	1	0		$X_8 = \overline{X_8} = \overline{X_9} = 0$
$X_{11} =$	0	0	0	1	0	0	0	1		$X_{10} = \overline{X_{10}} = \overline{X_{11}} = 0$
$X_{13} =$	0	0	0	0	1	0	0	1		$X_{12} = \overline{X_{12}} = \overline{X_{13}} = 0$
$X_{15} =$	0	0	0	0	0	1	0	1		$X_{14} = \overline{X_{14}} = \overline{X_{15}} = 0$
$\overline{X_{17}} =$	1	0	0	0	0	0	0	0		$X_{16} = \overline{X_{16}} = \overline{X_{17}} = 0$
$X_{19} =$	0	1	0	0	0	0	0	0		$X_{18} = \overline{X_{18}} = \overline{X_{19}} = 0$
$X_{21} =$	0	0	1	0	0	0	0	0		$X_{20} = \overline{X_{20}} = \overline{X_{21}} = 0$
$X_{23} =$	0	0	0	1	0	0	0	0		$X_{22} = \overline{X_{22}} = \overline{X_{23}} = 0$
$X_{25} =$	0	0	0	0	1	0	0	0		$X_{24} = \overline{X_{24}} = \overline{X_{25}} = 0$
$X_{27} =$	0	0	0	0	0	1	0	0		$X_{26} = \overline{X_{26}} = \overline{X_{27}} = 0$
$T =$	1	1	1	1	1	1	1	1		

Figure 2. A sample reduction of 3-QBF to PARTITION GAME. All integers are in base 9. All omitted X_i and $\overline{X_i}$ are chosen to be 0.

If F is satisfiable, Left should be able to win a "formula game" against Right in which Left and Right alternate selecting truth values for the variables, and Left wins if the formula evaluates to True.

This yields a strategy for Left to win the PARTITION GAME as follows. Left's challenge is to assure that exactly one "1" digit has been selected for each column C_i. First, for each variable x_k which Left selects TRUE Left chooses the integer $\overline{X_k}$. For each variable x_k which Left selects FALSE, Left chooses the integer X_k. Left interprets Right's choices of variable pairs similarly. Left then has full control over the remaining clause pairs and garbage collection pairs.

Now, since F is true, each clause has some true literal l_{ij}. Left selects the integer with the two digits, one digit corresponding to the true variable, and one digit corresponding to C_i. Left uses the garbage collection pairs to include any remaining literal columns.

The converse, that a winning strategy for the PARTITION GAME yields a winning strategy for the 3-QBF *formula game*, is similar.   □

4. Games and Game Sums

For a more complete introduction to combinatorial game theory, refer to [3], [4] or [1, Ch. 3]. This section briefly reviews the definitions and key results needed to get through the PSPACE-hardness reduction.

A *game* $G = \{\mathcal{G}_L \mid \mathcal{G}_R\}$ is defined recursively as a pair of sets of games $\mathcal{G}_L$ and $\mathcal{G}_R$. The two players, named Left and Right, or Black and White (respectively), play alternately. If it is Left's move, she selects an element from $\mathcal{G}_L$ which then becomes the current position. Right, if it is his move, would move to one of $\mathcal{G}_R$. If a player cannot move (because $\mathcal{G}_L$ or $\mathcal{G}_R$ is empty), that player loses.

In a *sum* of games $G + H$, a player can move on either G or H, leaving the other unchanged. Formally,

$$G + H = \{(\mathcal{G}_L + H) \cup (G + \mathcal{H}_L) \mid (\mathcal{G}_R + H) \cup (G + \mathcal{H}_R)\}$$

Here, a game added to a set adds the game to each element of the set. I.e.,

$$\mathcal{G}_L + H = \{G_L + H : G_L \in \mathcal{G}_L\}$$

In order to reduce the number of braces, we often omit them, depending on the | to separate the Left and Right options. Also, we write ‖ as a lower precedence | . So, $A \parallel B \mid C$ means $\{\{A\} \mid \{\{B\} \mid \{C\}\}\}$

The *negative* of the game G, $-G = \{-\mathcal{G}_R \mid -\mathcal{G}_L\}$, reverses the roles of the players.

A game $G = 0$ if the player to move (whether Left or Right) is doomed to lose (under perfect play). $G > 0$ if Left wins whether she moves first or second. $G < 0$ if Right always wins. If the first player to move can force a win, we say that G is incomparable with 0 and write $G <> 0$. Observe that these are the only four possibilities.

To compare two games, $G \geq H$ if and only if $G - H \geq 0$. Under this definition, games form a group with a partial order, where the zero of the group consists of all games $G = 0$.

Certain elements of the group, defined below, are *numbers* which (in finite games) are always dyadic rational integers. Here, n, a, and b are integers.

$$
\begin{aligned}
0 &= \{\mid\} \\
n &= \{n-1 \mid\} & n > 0 \\
n &= \{\mid n+1\} & n < 0 \\
\tfrac{a}{2^b} &= \left\{\tfrac{a-1}{2^b} \mid \tfrac{a+1}{2^b}\right\} & a \text{ odd}, b > 0
\end{aligned}
$$

These games add as expected, so for instance $\frac{1}{2} + \frac{1}{4} = \frac{3}{4}$. The *number avoidance theorem* states that when playing a sum of games $G + x$ where x is a number and G is not, it is best to move on the summand G. I.e., if there exists a winning move from $G + x$, then there exists a winning move to some $G^L + x$, $G^L \in \mathcal{G}_L$.

Three more game values are relevant in this paper:

$$* = \{0 \mid 0\}$$
$$\uparrow = \{0 \mid *\}$$
$$\Uparrow = \uparrow + \uparrow$$

All three are *infinitesimal*, smaller than all positive numbers and larger than all negative numbers. The game $*$ is its own negative and is incomparable with 0. The game $\uparrow$ exceeds 0 but is incomparable with $*$. Lastly, $\Uparrow > *$.

5. Switch Games

A switch $\pm x$ is the game $\{x \mid -x\}$. Our games will typically involve a finale consisting of a collection of switches $\{\pm x_1, \pm x_2, \ldots, \pm x_n\}$, where $x_1 \geq x_2 \geq \cdots \geq x_n \geq 0$. Players alternately choose the largest available switch $\pm x_i$, so that after play the final score will be $x_1 - x_2 + x_3 - x_4 \cdots \pm x_n$. This final score is the *outcome* of the switch game.

Fact 1. A player can successfully choose a switch which is not the largest (and still achieve the outcome) only if the highest occurs with even multiplicity. Similarly, if the switch values are each a multiple of ε, then each time a player bypasses a large switch of odd multiplicity in lieu of a smaller switch, the player will lose at least ε but at most x_1 relative to the outcome.

SWITCH PICK GAME

INSTANCE: A target T and a set X of n pairs of switches,

$$X = \{(\pm X_1, \pm \overline{X_1}), (\pm X_2, \pm \overline{X_2}), \ldots, (\pm X_n, \pm \overline{X_n})\}.$$

All values are integers.

QUESTION: Can Left guarantee a win the following game? Left begins by selecting either $\pm X_1$ or $\pm \overline{X_1}$. Right then selects $\pm X_2$ or $\pm \overline{X_2}$. Players alternate until one from each of the n pairs of switches has been selected. Let Z be this set of n selected switches. Left wins if the switch game Z has outcome T.

Lemma 0.2. *SWITCH PICK GAME is PSPACE-complete.*

Sketch of proof (by example). An example of the reduction is shown in Figure 3. The reader can verify that plays on one game correspond exactly to plays on the other. □

We can normalize all switches by dividing by the smallest power of 2 which exceeds all X_i, $\overline{X_i}$ and T. Letting this power be 2^β we arrive at the following PSPACE-complete variant:

PARTITION GAME

$X_1 = 3$	$\overline{X_1} = 4$
$X_2 = 2$	$\overline{X_2} = 7$
$X_3 = 1$	$\overline{X_3} = 4$
$X_4 = 0$	$\overline{X_4} = 0$
$X_5 = 1$	$\overline{X_5} = 5$

$$T = 14$$

SWITCH PICK GAME

$X_1 = \pm(2^{t+5} + 3)$	$\overline{X_1} = \pm(2^{t+5} + 4)$
$X_2 = \pm(2^{t+4} - 2)$	$\overline{X_2} = \pm(2^{t+4} - 7)$
$X_3 = \pm(2^{t+3} + 1)$	$\overline{X_3} = \pm(2^{t+3} + 4)$
$X_4 = \pm(2^{t+2} - 0)$	$\overline{X_4} = \pm(2^{t+2} - 0)$
$X_5 = \pm(2^{t+1} + 1)$	$\overline{X_5} = \pm(2^{t+1} + 5)$

$$T = 14 + 2^{t+5} - 2^{t+4} + 2^{t+3} - 2^{t+2} + 2^{t+1}$$

Figure 3. Reducing PARTITION to SWITCH PICK GAME. The quantity t is chosen to be large enough to dictate the order of play in the selected switch game, e.g., so 2^t exceeds all the PARTITION GAME's X_i and $\overline{X_i}$. Here, $t = 3$ suffices.

FRACTIONAL SWITCH GAME

INSTANCE: The same input as SWITCH PICK GAME, but all values X_i, $\overline{X_i}$, T are dyadic rationals less than 1 and all are an even multiple of $2^{-\beta}$.

QUESTION: Can Left win this game as in the SWITCH PICK GAME?

We will now show that the following problem, GAME-SUM, is PSPACE-hard by reduction from FRACTIONAL SWITCH GAME. The quantity $k = 6\beta + 2$ is chosen to be sufficiently large to admit construction of the game sum on a Go board.

GAME-SUM

INSTANCE: A sum of games S each of the form $a \parallel b \mid c$ or $a \mid b \parallel c$ for dyadic rationals $a \geq b \geq c$. Each of these numbers has a polynomial number of bits, and has an integer part which is polynomial in magnitude. Furthermore $a - b > k$ and $b - c > k$.

QUESTION: Can Left force a win moving first on S?

The constraints on a, b and c will make translation to a Go board possible.

For a given instance of FRACTIONAL SWITCH GAME (n, X and T), we will construct a game sum with several components:

$$S = \sum G_i + \sum \overline{G_i} + \sum H_i + I + \overline{I} + \Uparrow$$

The *temperatures*[2] of all components will be sufficiently far apart to almost assure that players must play on the hottest available summand to have a hope of winning. We'd like to force the order of play to be:

---

[2]The terms "temperature" and "hot" are technical concepts of combinatorial game theory. Their inclusion in this paper will aid the intuition of the combinatorial game theorist. Other readers can safely ignore the terms.

$$G_1 \text{ and } \overline{G_1} \text{ (in either order)}$$
$$H_1$$
$$G_2 \text{ and } \overline{G_2}$$
$$H_2$$
$$\vdots$$
$$G_{n-1} \text{ and } \overline{G_{n-1}}$$
$$H_{n-1}$$
$$G_n \text{ and } \overline{G_n}$$
$$I$$
$$\overline{I}$$

To achieve this, G_i and $\overline{G_i}$ will have equal temperatures $t_i = (2n - 2i + 4)k$ which decrease with increasing i. H_i will have temperature $t_i^- = t_i - k$. In truth, the first player will be able to play H_1 before G_1 or $\overline{G_1}$, but we'll see that this will be no better than playing according to the agenda.

If i is odd,

$$G_i = t_i \; \| \; -t_i + X_i + k \; | \; -t_i - X_i - k$$
$$\overline{G_i} = t_i \; \| \; -t_i + \overline{X_i} + k \; | \; -t_i - \overline{X_i} - k$$
$$H_i = 0 \; \| \; -2t_i^- \; | \; -2t_i^-$$
$$\quad\; = 0 \; \| \; -2t_i^- *$$

If i is even,

$$G_i = t_i + X_i + k \; | \; t_i - X_i - k \; \| \; -t_i$$
$$\overline{G_i} = t_i + \overline{X_i} + k \; | \; t_i - \overline{X_i} - k \; \| \; -t_i$$
$$H_i = 2t_i^- * \; \| \; 0 \quad (\text{except omit } H_n)$$

Left chooses to play on either G_1 or $\overline{G_1}$ and Right is compelled to play on the other. Thus, Left decides whether $\pm X_1$ or $\pm \overline{X_1}$ is included in the switch game Z. Left is then compelled to play H_1 to 0. If Left were to play on H_1 before G_1 or $\overline{G_1}$, this is tantamount to giving Right the option as to whether $\pm X_1$ or $\pm \overline{X_1}$ is included in Z. Play on a later game such as G_2 or one of the $\pm X_i$ is too costly as a consequence of Fact 1. The amount gained by such a maneuver is at most the largest X_1, and this is less than the amount lost which is at least k.

Next, Right selects whether $\pm X_2$ or $\pm \overline{X_2}$ is included and Left and Right continue alternating until the last two plays of this phase on G_n and $\overline{G_n}$ (since we omitted H_n). It is now Right's turn, who is free to choose between the games:

$$I = k \; | \; -T \; \| \; -2k$$
$$\overline{I} = T + 2k \; \| \; *$$

and Left will reply on the other. If the switch game Z is lost by Left because the alternating sum is less than T, Right will choose to play $\overline{I}$, Left will reply on I, and Right will move I to $-T$, which is sufficient to win out over the alternating sum. If, however, Z is lost by Left because the alternating sum exceeds T, Right will choose to play I to $-2k$, Left will move $\bar{I}$ to $T + 2k$, and Right will be the first to play on the alternating sum.

At this point, the players will play the switch game $\{(\pm(k + X_i), \pm(k + \overline{X_i}))\}$. Since there are an even number of terms, the alternating sum will have the same outcome as $\{(\pm X_i, \pm \overline{X_i})\}$; the k's will cancel. If this outcome is exactly T, then however Right played I and $\bar{I}$, Left can play the switches until the whole total is 0 or $*$. The infinitesimal $\Uparrow$ included in S is then sufficiently large to assure Left's win. Thus, Left can win if and only if she can arrange that the alternating sum adds to exactly the number T.

6. Switch games in Go

The following *warming* operator is a special case of the Norton product of two games "$g.h$" [3, p. 246], or of the overheating operator defined in [2], "$\int_s^t g$". If $g = \{g^L \mid g^R\}$, define

$$\int g \overset{\text{def}}{=} \int_{1*}^1 g = g.1* = \begin{cases} g & \text{if } G \text{ is an even integer,} \\ g + * & \text{if } G \text{ is an odd integer,} \\ \{1 + \int g^L \mid -1 + \int g^R\} & \text{otherwise.} \end{cases}$$

As a consequence of being a Norton multiple, the warming operator has the following properties [3, p. 246]:

1. linearity: $\int g + \int h = \int(g + h)$,
2. order preserving: If $g \geq h$ then $\int g \geq \int h$.

The following *blocked corridor* in Go has value $n - 2 + \int 2^{1-n}$, where n is the number of empty nodes in the corridor [1]. (In this example, $n = 5$.)

$$= 3 + \int \frac{1}{16} = \{4 \;|||| \; 3 \; ||| \; 2 \; || \; 1 \mid *\}$$

Since the warming operator is linear and order preserving, it suffices to convert the sum of abstract games to Go positions of value $\int\{a \parallel b \mid c\}$, where $b - a \geq k$, $c - b > k$ and a, b and c are multiples of $\frac{1}{2^\beta}$. I'll describe the conversion by example. We'll first analyze the following specific position, and then argue that the position can be augmented to achieve any $\int\{a \parallel b \mid c\}$.

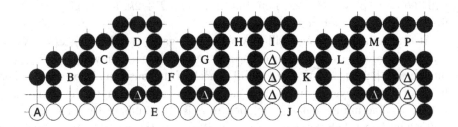

The white group including stone A is assumed to be alive, and all black stones are alive. Black has 2 sure points in region P, and there is one zero point play (or *dame*, "dah-meh") above A. Together these are worth $2*$. In addition, if Black plays at E, Black captures 56 points to the right of E. The blocked corridor at B is worth $\int \frac{1}{2}$, C is worth $1 \int \frac{1}{4}$ (i.e., $1 + \int \frac{1}{4}$) and area D is 3 points of territory. These total to $60 \int \frac{3}{4}$. If White plays at E and Black replies at J, Black nets $34 * \int \frac{11}{8}$. The $*$ represents the zero point play (or *dame*) at I. Lastly, if White plays at E and J, the resulting position is worth $12 * \int \frac{17}{8}$. Thus the original position is worth

$$2* \; + \; \left\{ 60 \int \frac{3}{4} \;\middle|\middle|\; 35 * \int \frac{11}{8} \;\middle|\; 12 * \int \frac{17}{8} \right\}$$

which, by applying the definition of $\int$ is

$$= 2* + * + \int \left\{ 59\frac{3}{4} \;\middle|\middle|\; 35\frac{3}{8} \;\middle|\; 16\frac{1}{8} \right\} = \int \left\{ 61\frac{3}{4} \;\middle|\middle|\; 37\frac{3}{8} \;\middle|\; 18\frac{1}{8} \right\}$$

An informal recipe should convince the reader that there are enough degrees of freedom to generate any Go position $\int\{a \parallel b \mid c\}$ (where a, b and c are constrained as above):

1. A group (A) invading β corridors of increasing length. (In the example, $\beta = 3$ and the corridors are marked B, C and D.) The binary expansion of the fractional part $a - \lfloor a \rfloor$ dictates which of these β corridors are blocked. (Here, $a - \lfloor a \rfloor = .110$, so only the third corridor is blocked.)

2. A second group invading β corridors which threatens to connect to A. These corridors are blocked according to the quantity $b - a - \lfloor b - a \rfloor$. The third group's corridors should account for $c - b - a - \lfloor c - b - a \rfloor$.)

3. Adjustments to the integer differences $\lfloor b-a \rfloor$ and $\lfloor c-b \rfloor$ are made by extending the White stones marked $\triangle$. Each additional stone adds two points; an empty node at, say, I, adds one.

4. Lastly, shift the value of the Go position by any integer by adding territory to Black or White (area P). Include a dame (say, at the point above A) as needed to adjust by $*$.

The construction of one switch $\int\{a \parallel b \mid c\}$ requires a number of White stones which is linear in β. The choice of $k = 6\beta + 2$ suffices. Using the fact that

$$\text{(board diagram)} = \int \uparrow = \int \{0 \mid *\},$$

the last games $\int \bar{I} = \int \{T + 2k \parallel *\}$ and $\int \Uparrow$ are also constructible on a Go board.

References

[1] Elwyn Berlekamp and David Wolfe. *Mathematical Go: Chilling Gets the Last Point.* A K Peters, Ltd., Wellesley, Massachusetts, 1994.

[2] Elwyn R. Berlekamp. Blockbusting and domineering. *Journal of Combinatorial Theory*, 49(1):67–116, September 1988.

[3] Elwyn R. Berlekamp, John H. Conway, and Richard K. Guy. *Winning Ways.* Academic Press, New York, 1982.

[4] John H. Conway. *On Numbers and Games.* Academic Press, London/New York, 1976.

[5] Marcel Crâşmaru and John Tromp. Ladders are pspace-complete. In T. A. Marsland and I. Frank, editors, *Computers and Games: Second International Conference, CG 2000, Hamamatsu, Japan, October 2000*, pages 241–249. Springer, 2001.

[6] George Johnson. To test a powerful computer, play an ancient game. *New York Times*, July 29, 1997.

[7] Anders Kierulf. *Smart Game Board: a Workbench for Game-Playing Programs, with Go and Othello as Case Studies.* PhD thesis, Swiss Federal Institute of Technology (ETH) Zürich, 1990. Nr. 9135.

[8] David Lichtenstein and Michael Sipser. Go is polynomial-space hard. *Journal of the Association for Computing Machinery*, 27(2):393–401, April 1980.

[9] J. McCarthy. Chess as the drosophila of AI. In T. A. Marsland and J. Schaeffer, editors, *Computers, Chess, and Cognition*, pages 227–237. Springer Verlag, New York, 1990. Other games replacing chess in AI.

[10] David Moews. *On Some Combinatorial Games Related to Go.* PhD thesis, University of California, Berkeley, 1993.

[11] F. L. Morris. Playing disjunctive sums is polynomial space complete. *International Journal of Game Theory*, 10:195–205, 1981.

[12] J. M. Robson. The complexity of Go. In *Information Processing; proceedings of IFIP Congress*, pages 413–417, 1983.

[13] J. M. Robson. Combinatorial games with exponential space complete decision problems. In *Lecture Notes in Computer Science. Mathematical Foundations of Computer Science 1984*, pages 498–506. Springer-Verlag, 1984.

[14] Laura Yedwab. On playing well in a sum of games. Master's thesis, M.I.T., August 1985. MIT/LCS/TR-348.

DAVID WOLFE
DEPARTMENT OF MATHEMATICS AND COMPUTER SCIENCE
GUSTAVUS ADOLPHUS COLLEGE
800 WEST COLLEGE AVENUE
SAINT PETER, MN 56082
UNITED STATES
  wolfe@gustavus.edu
  http://www.gustavus.edu/~wolfe

More Games of No Chance
MSRI Publications
Volume **42**, 2002

Global Threats in Combinatorial Games: A Computation Model with Applications to Chess Endgames

FABIAN MÄSER

ABSTRACT. The end of play in combinatorial games is determined by the normal termination rule: A player unable to move loses. We examine combinatorial games that contain *global threats*. In sums of such games, a move in a component game can lead to an immediate overall win in the sum of all component games. We show how to model global threats in Combinatorial Game Theory with the help of infinite loopy games. Further, we present an algorithm that avoids computing with infinite game values by cutting off branches of the game tree that lead to global wins. We apply this algorithm to combinatorial chess endgames as introduced by Elkies [4] where this approach allows to deal with positions that contain entailing moves such as captures and threats to capture. As a result, we present a calculator that computes combinatorial values of certain pawn positions which allow the application of Combinatorial Game Theory.

1. Global Wins and Global Threats

Combinatorial game theory (CGT) applies the divide and conquer paradigm to game analysis and game tree search. We decompose a game into independent components (local games) and compute its value as the sum of all local games. The end of play in a sum of combinatorial games is determined by the normal termination rule: A player unable to move loses. Thus, in a sum of games, no single move or game can be decisive by itself. We investigate a class of games where a move in a local game may lead to an overall win in the sum of all local games. We call such a move *globally winning*.

Examples of globally winning moves are moves that capture a vital opponent piece such as *checkmate*,[1] moves that promote a piece to a much more powerful one like promoting a king in *checkers*, or moves that "escape" in games where one side has to try to catch the other side's pieces like in the game *Fox and Geese*

---

[1] Although checkmate does not actually capture the enemy king, it creates the unstoppable threat to do so.

(*Winning Ways* [2], chapter 20). Figure 1 shows a Fox and Geese position where the fox escapes with his last move from e5 to d4 and obtains "an infinite number of free moves". If this game were a component of a sum game S, the fox side would never lose in S.

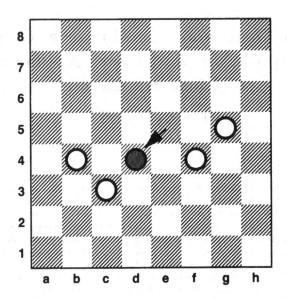

Figure 1. A position in the game *Fox and Geese*: the last move by the (dark) fox *escapes* the geese that are only allowed to move upwards.

We are mainly interested in games where none of the players can win by executing a global threat if the opponent defends optimally. Such games are finally decided by normal termination and have finite combinatorial values. However, when we search the tree of all moves in order to compute a game's value, we also have to execute and undo globally winning moves. In section 2, with the help of infinite games, we model the situation that arises after one of the players has executed a globally winning move. A global win is equivalent to having an infinite number of moves available. In section 3, we present an algorithm that computes combinatorial values of games with global threats. In order to avoid computing with infinite game values, the algorithm cuts of the game tree just before a globally winning move is executed. Section 4 applies these techniques to king and pawn endgames in chess.

Regarding chess endgames, Elkies [4] writes: "The analysis of such positions is complicated by the possibility of pawn trades which involve entailing moves: an attacked pawn must in general be immediately defended, and a pawn capture parried at once with a recapture. Still we can assign standard CGT values to many positions ... in which each entailing line is dominated by a non-entailing one." This paper introduces an algorithm that solves this problem in general.

2. A CGT Model Based on Loopy Games

A natural way to model global wins in combinatorial game theory is to define values G_{lwin} and G_{rwin}. The game G_{lwin} stands for the situation that *Left* has executed a globally winning move. It must be greater than any finite game G. Analogously, G_{rwin} stands for the situation that *Right* has executed a globally winning move and has to be smaller than any finite game G.

$$\forall \text{ finite games } G : G_{rwin} < G < G_{lwin} \qquad (2\text{--}1)$$

The simplest games (they consist of only one position) that satisfy (2–1) are the loopy games $on = \{on \mid \}$ and $off = \{ \mid off\}$ (*Winning Ways* [2], chapter 11). The game *on* is greater than any *ender*[2] G. *Left* wins any difference (*on* - G) by just playing in *on* until *Right* runs out of moves in $(-G)$.

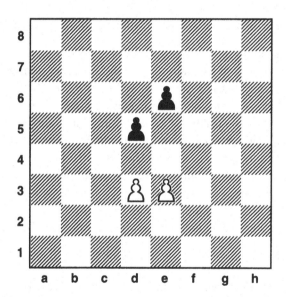

Figure 2. Global threats in a chess position: Both players have to prevent the opponent from promoting a pawn to a queen. Play will end when the pawns are blocked and neither player has any moves left.

Figure 2 shows a chess example. In the context of king and pawn endgames, we consider promoting a pawn to be globally winning.[3] For instance, White's move e3-e4 leads to a symmetrical position where both players have the choice either to capture or to push the more advanced pawn. We compute its value as $\{*, \{on \mid *\} \mid *, \{* \mid off\}\}$.

---

[2]a game in which no player can play an infinite sequence of moves.

[3]This is true for a majority of pawn endgames, and we limit our attention to these.

As the values *on* and *off* only appear as *threats*, the resulting game value of Figure 2 is finite. None of the players is able to force the promotion of a pawn if his opponent plays correctly. Here is a short analysis of the game position: White's move d3-d4 and Black's move e6-e5 both lead to zero positions. It turns out that both other moves are reversible (White's e3-e4 and Black's d5-d4), and thus the game value is $G = \{0 \mid 0\} = *$.

If both players have the possibility to force a global win in different subgames, the result of the sum is a "draw by repetition". In the sum $(on + off)$, both players will always have moves available, therefore none will lose:

$$on + off = \{on + off \mid on + off\} = dud$$

The result *dud* ("deathless universal draw" [2]) offers a pass move for both players, so any sum S containing at least one component game of value *dud* will never be brought to an end $(dud + G = dud)$.

3. An Algorithm Based on Cutoffs in the Game Tree

While the model presented in section 2 works fine in theory, it is not so easy to implement in practice. We usually map finite combinatorial games "one to one" to data structures by their inductive definition $G = \{G^L \mid G^R\}$ with the basis $0 = \{\mid\}$. Following a path of left and right options, we are sure that finally the zero game will be reached. Almost all algorithms that work on combinatorial games are based on their inductive nature. Obviously, loopy games do not fit into this model. Either, we must extend the model to handle loopy games, or we must somehow avoid loopy games in our computations.

In this section, we pursue the second way. We present a computation model for local games with global threats that cuts off branches of the game tree that lead to global wins. The model is based on the following two lemmas:

(i) If a player has the chance to execute a globally winning move, he will always do so.

(ii) Any move to a position which offers the opponent a globally winning move is "bad". Such a move need not be considered when evaluating a player's options.

Both lemmas follow directly from the rules for *simplifying games* (see [3] or [2]). The first one is easily deduced: As for any game G, the equation $off \leq G \leq on$ applies, a move to a globally winning position always *dominates* any other possible move. The second one is deduced from the rule of replacing reversible moves. If in the game $G = \{G^L \mid G^R\}$, *Left* plays a move to G^{Li} which contains a right move to *off*, then *Left's* move to G^{Li} is *reversible* $(G \geq off)$ and is replaced by all left options of *off*. As *off* has no left options, *Left's* move to G^{Li} is simply omitted. The same holds of course for a right move to a game G^{Rj} that offers a left move to *on*.

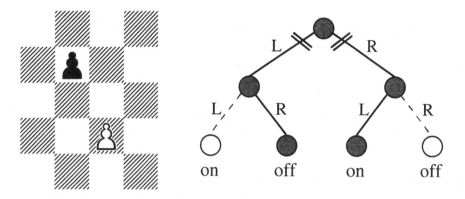

Figure 3. A game of value zero and its game tree. Each player's move leads to an immediate opponent win. This is equivalent to having no moves at all. $G = \{\{on \mid off\} \mid \{on \mid off\}\} = \{\mid\} = 0$

The chess position on the left side of Figure 3 illustrates this. Both players have exactly one move which leads to a position where the two pawns attack each other. The player who captures his opponent's pawn will go on to promote his pawn to a queen. The value of this position is computed as $G = \{\{on \mid off\} \mid \{on \mid off\}\}$ which simplifies to $\{\mid\} = 0$.

Applying the model presented in section 2, we compute the value of G by producing its game tree up to the terminal positions *on* and *off*. Instead, based on the second lemma, we can immediately cut off both players' moves as we already know that they will be reversible (see the right side of Figure 3). This leads directly to the same value $G = \{\mid\} = 0$. There are two evident advantages of this approach. First, we avoid most calculations with loopy games. In this simple example, we do not have to deal with loopy games at all. Second, thanks to the cutoffs, we minimize the number of nodes to be searched in the game tree.

Now we are ready to formulate an algorithm for evaluating local games with global threats. We consider both players' options in each position. Additionally, however, we make use of the information who played the last move. This allows us to cut off moves that lead to global wins for the opponent. It might seem unusual to make use of to-play-information in combinatorial game tree search, but this also occurs implicitly in conventional CGT. The same rule of replacing reversible options that allows us to cut off the game tree is based on "good replies" to an opponent's move, thus also makes use of to-play-information.

3.1. Result Types. In contrast to finite combinatorial games, the value of a game that contains global threats might be *on* or *off* i.e. a forced global win for one of the players. This is the case if a player cannot prevent his opponent from finally playing a globally winning move no matter how he defends. In order to separate finite combinatorial values from global wins, we distinguish the following *result types* of local games:

- Type *win*: If the players move alternately including the right to pass, *Left* will win by executing a globally winning move no matter how *Right* defends and no matter who starts.
- Type *loss*: If the players move alternately including the right to pass, *Right* will win by executing a globally winning move no matter how *Left* defends and no matter who starts.
- Type CGT: None of the players can force a win by global threat. In this case, a combinatorial value $G = \{G^L \mid G^R\}$ can be computed for the actual game position.

3.2. Game Tree Search. In order to evaluate a game, we produce its tree starting from the current position. In every position, we recursively evaluate both players' options. As the possible result types are ordered ($win > CGT > loss$ from *Left's* point of view and $loss > CGT > win$ from *Right's* point of view), we can determine the best result types both players can get if they have the move. If a player's best result type is CGT, we also compute his combinatorial game options (G^L respectively G^R). Combined with the information on who is to play, we compute the result type of the actual position as shown in Figure 4. In case the actual result type is CGT, we also compute the combinatorial game value of the actual position.

L \ R	loss	CGT	win
loss	loss	CGT_1	CGT_2
CGT	loss / CGT	CGT_3	CGT_4
win	loss / win	CGT / win	win

Figure 4. Result Types of *Global Threats Evaluation*: the table indicates the result type of a local game depending on the best result types of *Left's* (rows) and *Right's* (columns) options and on the right to make the next move. The split entries show the result types for *Left* to play (lower left) and *Right* to play (upper right).

In the four cases labeled CGT, we can compute a finite combinatorial value for the actual game position.

(i) All *left* options lead to games of type *loss* while *Right* has at least one move that leads to a finite combinatorial game. As *Left* has no good moves, the value of the actual position is $G = \{ \mid G^R\}$.

(ii) Neither player has any good moves. The position is a *mutual Zugzwang*. $G = \{ \mid \} = 0$. We have already seen an example of such a situation in Figure 3.

(iii) Both players' best options all lead to finite combinatorial values. The actual game value is computed as $G = \{G^L \mid G^R\}$.

(iv) In contrast to the first case, *Right* has no good move while *Left* has at least one CGT option. $G = \{G^L \mid \}$.

If both players' best result types are *win* (resp. *loss*), the result type of the actual game position is of course also *win* (resp. *loss*). In the remaining three cases, the result type of the actual position will depend on who has the right to move. If the player to move can move to a winning position, he will of course do so and the result type is determined as a win in his favor. Of special interest are the positions where the player to move has one or more moves that lead to games of type *CGT* while his opponent would be winning if he was to play. These are the only cases where the loopy games *on* and *off* occur in our combinatorial game values which are either $\{on \mid G^R\}$ or $\{G^L \mid off\}$). Fortunately, *on* and *off* only appear as *threats*. For example, *Left* plays a move to a position of value $\{on \mid G^R\}$ when *Right* immediately has to move to one of the options in G^R as *Left threatened* to move to *on*.

3.3. Implementation. The function *GTSearch* searches the game tree depth first and computes result types and combinatorial game values of local games that contain global threats. Its specification is

- **in**:

    - *toplay*: the player (constants *kWhite*, *kBlack*, *kNoPlayer*) who has the move. In the starting position, *kNoPlayer* is passed. After at least one move is played, the right to move is determined.

- **out**:

    - *return value*: the result type (constants *kWin*, *kLoss* or *kCGT*) of the current position.
    - *value*: the combinatorial game value of the current position. This value is "valid" only in case *kCGT* is returned.

The algorithm performs the following steps (numbers refer to comments in the code):

(i) Check termination:
determine if the actual position is a global win for one of the players. If so, we are finished and return *kWin* resp. *kLoss*.

(ii) Recursively evaluate *Left's* options:
We store the best result type ($kWin > kCGT > kLoss$) achieved so far in the variable *bestL*. For every option of type *kCGT*, we include its combinatorial

value in the set G^L. If we find an option leading to result type $kWin$, we can skip the remaining options (cutoff!).

(iii) Recursively evaluate *Right's* options:

As in step 2, we compute the values $bestR$ and G^R.

(iv) Compose the result:

According to the table in Figure 4, we compute the result type of the actual position. In case the result type is $kCGT$, we also compute the combinatorial game value $\{G^L \mid G^R\}$ and return it in the out-parameter *value*.

```
function GTSearch(toplay: TPlayer; var value: TGameValue): TResultType;
begin
  if GlobalWin(kLeft) then return kWin; endif; /* 1 */
  if GlobalWin(kRight) then return kLoss; endif;
  G^L ← { }; bestL ← kLoss;
  forall left moves m do /* 2 */
    ExecMove(m);
    res ← GTSearch(kRight, val);
    UndoLastMove();
    if res = kWin then bestL ← kWin; break endif; /* cutoff! */
    if res = kCGT then bestL ← kCGT; G^L ← G^L∪ val endif;
  endfor;
  G^R ← { }; bestR ← kWin;
  forall right moves m do /* 3 */
    ExecMove(m);
    res ← GTSearch(kLeft, val);
    UndoLastMove();
    if res = kLoss then bestR ← kLoss; break endif; /* cutoff! */
    if res = kCGT then bestR ← kCGT; G^R ← G^R∪ val endif;
  endfor;
  return ComposeResult(bestL, bestR, value, G^L, G^R, toplay); /* 4 */
end GTSearch;
```

4. Application to Chess Endgames

Elkies [4] has shown that certain types of chess endgames (mostly king and pawn endgames) can be analyzed using Combinatorial Game Theory. If both players' kings (or any other remaining pieces) are bound by *mutual Zugzwang* (mZZ), the remaining pawn moves (*tempi* in the chess literature) decide who will have to give way. Thus, when we identify a mZZ, we try to decompose the position and calculate local values for the independent pawn chunks. Figure 5 shows a position from an actual game (Sveda - Sika, Brno 1929) which has been solved by Elkies. Its value is ↑ (queenside, a and b files) plus 0 (center, a mZZ involving both kings) plus ↓↓ * which adds up to ↓ *, a first player win.

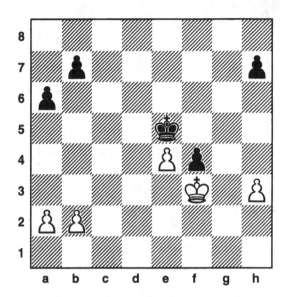

Figure 5. Mutual Zugzwang: Sveda - Sika, Brno 1929 The player who first has to move his king loses. Therefore the player who makes the last pawn move in the sum game of the queenside and kingside pawns will win.

Position with pawns on one file only, like the kingside in Figure 5, are easy to analyze. The pawns will get blocked in any possible line of play. Because of the possibility of moving a pawn by two squares from its original square they are not completely trivial. Positions with two or more pawns on each side (for example the queenside in Figure 5) are more complex. The key to the analysis of such structures are the *global threats* of promoting pawns that reach their final rank. With the computation model presented in section 2, we can also handle positions like the one shown in Figure 6.

The kingside with king and pawn each is "the same" mutual Zugzwang that we have already seen in the Sveda-Sika game. The pawn structure on the queenside looks like it will block after a few moves, but in fact thanks to his far advanced pawns white to move can force a global win by sacrificing two pawns in order to promote the third one. After 1.b5–b6 a7×b6 (c7×b6 2.a5–a6 etc.) 2.c5–c6 b7×c6 3.a5–a6 the a-pawn is unstoppable. In the chess literature, this maneuver is known as a *breakthrough*. Black to play, on the other hand, cannot do the same, as white would be much faster promoting one of his own pawns. His only move that does not allow white to win by global threat is 1... b7-b6 leading to a value $G_1^L = \{0, \{on \mid 0\} \mid off\}$.

We conclude that captures and attacks are often *entailing moves* because they usually lead to global threats in form of the promotion of a pawn. In

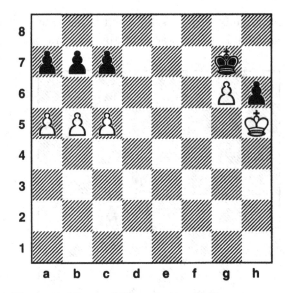

Figure 6. The breakthrough: White to play sacrifices two pawns in order to promote the third one. Black to play loses by Zugzwang.

many pawn structures, however, especially if both sides have an equal number of pawns, none of the players can force a win by promoting a pawn, and we can calculate a combinatorial game value for the given position.

4.1. Implementation. Based on the algorithm presented in section 3, we have implemented a pawn structure calculator within the *Game Bench* [6] project. The Game Bench is an application framework for programs that implement combinatorial games. One of its main goals is to separate algorithms and data structures common to all combinatorial games from the details of specific games and make them available to the game programmer.

The Game Bench is written in Java which makes it portable to almost all of today's computer platforms. Its variety of game independent support makes the Game Bench well suited for "fast prototyping". On the other hand, thanks to "just in time compilation" of Java bytecode, it allows serious game analysis and time critical calculations of combinatorial games as well. Furthermore, on Unix and Windows platforms, David Wolfe's *Gamesman's Toolkit* [7] is used in the form of a dynamic library of efficient C-functions. This library performs all basic CGT computations which are the most time critical part of algorithms like combinatorial game tree search.

4.2. Results and Conclusions. Applying combinatorial game theory to king and pawn chess endgames involves the following three steps:

(i) Detection of *Zugzwang* and identification of the component games.
(ii) Evaluation of the local component games.
(iii) Combination of the local results.

Step one is chess specific, whereas steps two and three are independent of the game we want to analyze. Although we can detect Zugzwang positions involving the kings (and possibly other pieces) automatically, e.g. with the help of a local minimax search, the conclusion that such a Zugzwang will be constant is of a heuristic nature.

Let's again look at the Sveda-Sika game (Figure 5): If White starts, play might continue 1.h3–h4 a6–a5 2.h4–h5 a5–a4 3.h5–h6. Black could now consider to abandon his f4-pawn and instead attack the white pawn on h6. After the moves 3...♚e5–f6 4.♚f3×f4 ♚f6–g6 5.♚f4–e5 ♚g6×h6, material is balanced but white has an easy win with 6.♚e5–f6 as his e-pawn is unstoppable. Thus, our initial assumption that the side to move its king first loses, was correct. But in other cases, breaking out of the Zugzwang might upset the combinatorial evaluation of the position.

Thus, we see the main application of a pawn structure calculator as a tool to support human chess players when analyzing and explaining certain chess endgames. Positions in which the pawns fight for extra tempi are only vaguely described in the chess literature, and a verification of the results of human analysis by brute force minimax search is still beyond the scope of today's leading chess software if the positions are complex enough.

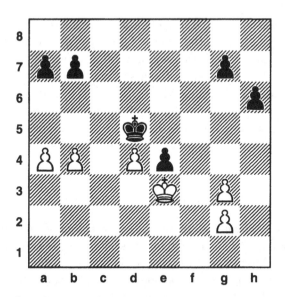

Figure 7. A first player win: Popov - Dankov, Albena 1978. The queenside and center are both of value 0. On the kingside, the first player to move forces his opponent into a zugzwang position. The kingside and therefore the whole game has a value of $G = \{0 \mid -1\}$.

Figure 7 shows a position from the game Popov vs. Dankov (Albena 1978). The relative position of the two kings is the same as in the Sveda-Sika game.

The player to move his king loses his central pawn. Thus, the game is decided by the subgames on the queenside and kingside.

- The queenside is a game of value $G_{QS} = 0$. With the white pawns advanced to the fourth rank, Black gets no advantage from the double step option. The player to move gets blocked immediately.
- The kingside is a bit more complex. Black to play gains a considerable advantage with 1...h6-h5. In fact, due to White's option to sacrifice a pawn for a tempo with g3-g4, it "only" leads to a value of -1. White to play has only one move. Thanks to the possibility of sacrificing one of the doubled pawns for a tempo, it leads to a position of value 0. The main line runs 1.g4 g6 2.g5 h×g5 3.g4 and a zero position with no moves for either side is reached. The value of the kingside is $G_{KS} = \{0 \mid -1\}$.

Thus, the sum $G_{QS} + G_{KS} = \{0 \mid -1\}$ is a first player win. Awerbach [1] assesses this position correctly, but does not offer other calculation methods than brute force search. He write (in German): "With pawns on one or two files, computing spare tempi is not too difficult..." Conventional chess programs, on the other hand, have more problems evaluating this position. The following results are computed on a PC (466 MHz Intel Celeron, 128 MByte Ram) running Linux.

- The CGT algorithm presented in section 3 requires a total of less than 1000 evaluated nodes to compute the values of the kingside and the queenside.
- Combinatorial evaluation of the combined (kingside and queenside) pawn structure yields the same result of course, but takes much longer. Almost $200,000$ nodes need to be evaluated.
- In order to illustrate the complexity of a full-width alpha-beta search, we ran Crafty [5] on the game position with White to move. Only after evaluating more than $2 \cdot 10^9$ nodes, the program indicated 1.g3–g4 leading to a white advantage.

5. Summary and Outlook

Games with global threats are an interesting extension of conventional combinatorial games. They model entailing moves such as captures in king and pawn chess endgames. As we can see in Figure 6, the heuristic that a capture should be answered with a recapture fails to produce exact results. We can represent global threats with infinite, loopy games. On the other hand, we propose an algorithm for game tree search that avoids dealing with infinite game values by cutting off branches of the game tree that lead to global wins.

Certain chess endgames allow the application of combinatorial game theory [4]. We decompose the pawn structure into independent local games, calculate their values and finally compute the value of the sum by rules of CGT. However, promoting a pawn to a queen in a local game results in a global win in the sum of all component games. The analysis of local pawn structures becomes very complex if the number of pawns increases. A calculator that computes values of such pawn structures automatically is a useful tool for the game analyst, especially as most game positions are still too complex to be searched at full width by conventional chess programs.

The analysis of pawn structures presented in this paper has potential applications beyond CGT. We have used divide and conquer in combination with CGT to combine the results of local games. But the results computed by the *GT-Search* algorithm can be used in more general settings as well. As an example, the information that a cluster of white and black pawns can generate a passed pawn for one of the players may be used in a heuristic evaluation function that rates pawn structures.

References

[1] J. Awerbach. *Bauernendspiele*. Sportverlag, Berlin, 1983.

[2] E. Berlekamp, J. H. Conway, and R. Guy. *Winning Ways for Your Mathematical Plays*. Academic Press, New York, NY, USA, 1982.

[3] J. H. Conway. *On Numbers and Games*. A K Peters, 2001.

[4] N. D. Elkies. On Numbers and Endgames: Combinatorial Game Theory in Chess Endgames. In R. J. Nowakowski, editor, *Games of No Chance*, pages 135–150. Cambridge University Press, New York, 1996.

[5] R. Hyatt. Crafty v17.9, a very strong freeware chess program, rated 2499 on the Elo scale by the Swedish Chess Computer Association in August 2000.

[6] F. Mäser. *Divide and Conquer in Game Tree Search: Algorithms, Software and Case Studies*. PhD thesis, ETH Zürich, 2001.

[7] D. Wolfe. The Gamesman's Toolkit. In R. J. Nowakowski, editor, *Games of No Chance*, pages 93–98. Cambridge University Press, New York, 1996.

FABIAN MÄSER
ERGON INFORMATIK AG
BÄCHTOLDSTRASSE 4
8044 ZÜRICH
SWITZERLAND
fabian.maeser@ergon.ch

More Games of No Chance
MSRI Publications
Volume **42**, 2002

The Game of Hex: The Hierarchical Approach

ABSTRACT. Hex is a beautiful and mind-challenging game with simple rules
and a strategic complexity comparable to that of Chess and Go. Hex positions do not tend to decompose into sums of independent positions. Nevertheless, we demonstrate how to reduce evaluation of Hex positions to an
analysis of a hierarchy of simpler positions. We explain how this approach
is implemented in Hexy, the strongest Hex-playing computer program, and
Gold Medalist of the 5th Computer Olympiad in London, August 2000.

1. Introduction

The rules of Hex are extremely simple. Nevertheless, Hex requires both deep
strategic understanding and sharp tactical skills. The massive game-tree search
techniques developed over the last 30–40 years mostly for Chess (Adelson-Velsky,
Arlazarov, and Donskoy 1988; Marsland 1986), and successfully used for Checkers (Schaeffer et al. 1996), and a number of other games, become less useful
for games with large branching factors like Hex and Go. For a classic 11 × 11
Hex board the average number of legal moves is about 100 (compare with 40 for
Chess and 8 for Checkers).

Combinatorial (additive) Game Theory provides very powerful tools for analysis of sums of large numbers of relatively simple games (Conway 1976; Berlekamp,
Conway, and Guy 1982; Nowakowski 1996), and can be also very useful in situations, when complex positions can be decomposed into sums of simpler ones.
This method is particularly useful for an analysis of Go endgames (Berlekamp
and Wolfe 1994; Müller 1999).

Hex positions do not tend to decompose into these types of sums. Nevertheless, many Hex positions can be considered as combinations of simpler subgames.
We concentrate on the hierarchy of these subgames and define a set of deduction
rules, which allow to calculate values of complex subgames recursively, starting
from the simplest ones. Integrating the information about subgames of this hierarchy, we build a far-sighted evaluation function, foreseeing the potential of Hex
positions many moves ahead.

In Section 2 we introduce the game of Hex and its history. In Section 3 we
discuss the concept of virtual connections. In Section 4 we introduce the AND

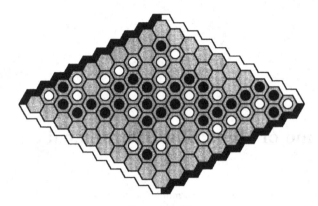

Figure 1. The chain of black pieces connects black boundaries. Black has won the game.

and OR deduction rules. In Section 5 we show how to recursively calculate the hierarchy of virtual connections. In Section 6 we present an electrical resistor circuits model, which allows us to combine information about the hierarchy of virtual connections into a global evaluation function. In Section 7 we explain how this approach is implemented in Hexy, the strongest Hex-playing computer program, and the Gold Medallist of the 5th Computer Olympiad in London, August 2000. A Windows version of the program is publicly available at http://home.earthlink.net/~vanshel.

The major ideas of this work were presented on the MSRI Combinatorial Game Theory Workshop in Berkeley, July 2000 and on the 17th National Conference on Artificial Intelligence in Austin, July-August 2000 (Anshelevich 2000a).

2. Hex and Its History

The game of Hex was introduced to the general public in Scientific American by Martin Gardner (Gardner 1959). Hex is a two-player game played on a rhombic board with hexagonal cells (see Figure 1). The classic board is 11×11, but it can be any size. The 10×10, 14×14 and even 19×19 board sizes are also popular. The players, Black and White, take turns placing pieces of their color on empty cells of the board. Black's objective is to connect the two opposite black sides of the board with a chain of black pieces. White's objective is to connect the two opposite white sides of the board with a chain of white pieces (see Figure 1). The player moving first has a big advantage in Hex. In order to equalize chances, players often employ a "swap" rule, where the second player has the option of taking the first player's opening move.

Despite the simplicity of the rules, the game's strategic and tactical ideas are rich and subtle. An introduction to Hex strategy and tactics can be found in the book written by Cameron Browne (Browne 2000).

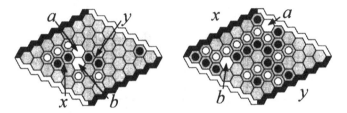

Figure 2. Groups of black pieces, x and y, form two-bridges. In the position on the right, those groups are connected to the black boundaries.

Hex was invented by a Danish poet and mathematician Piet Hein in 1942 at the Niels Bohr Institute for Theoretical Physics, and became popular under the name of Polygon. It was rediscovered in 1948 by John Nash, when he was a graduate student at Princeton (Gardner 1959). Parker Brothers marketed a version of the game in 1952 under the name Hex.

The game of Hex can never end in a draw. This follows from the fact that if all cells of the board are occupied then a winning chain for Black or White must necessarily exist. While this two-dimensional topological fact may seem obvious, it is not at all trivial. In fact, David Gale demonstrated that this result is equivalent to the Brouwer fixed-point theorem for 2-dimensional squares (Gale 1979). It follows that there exists a winning strategy either for the first or second player. Using a "strategy stealing" argument (Berlekamp, Conway, and Guy 1982), John Nash showed that a winning strategy exists for the first player. However, this is only a proof of existence, and the winning strategy is not known for boards larger than 7×7.

S. Even and R. E. Tarjan (Even and Tarjan 1976) showed that the problem of determining which player has a winning strategy in a generalization of Hex, called the Shannon switching game on vertices, is PSPACE complete. A couple of years later S. Reisch (Reisch 1981) proved this for Hex itself.

A Hex-playing machine was built by Claude Shannon and E. F. Moore (Shannon 1953). Shannon associated a two-dimensional electrical charge distribution with any given Hex position. His machine made decisions based on properties of the corresponding potential field. We gratefully acknowledge that our work is greatly inspired by the beauty of the Shannon's original idea.

3. Virtual Connections and Semi-Connections

In this and the two following sections we characterize Hex positions from Black's point of view. White's point of view can be considered in a similar way.

Consider the four polygonal boundary bands as additional cells (see Figure 1). We assume that black boundary cells are permanently occupied by black pieces, and white boundary cells are permanently occupied by white pieces.

Consider the two positions in Figure 2. In both positions White cannot prevent Black from connecting the two groups of connected black pieces, x and y,

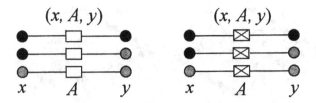

Figure 3. Diagrams of virtual connections (on the left) and virtual semi-connections (on the right): black-black, black-empty, and empty-empty.

even if White moves first, because there are two empty cells a and b adjacent to both x and y. If White occupies one of those empty cells, then Black can move to the other. Note that the black connection between groups x and y is secured as long as two cells a and b stay empty. Black can postpone moving to either a or b and can use his precious moves for other purposes. In this type of situation we say that the groups of black pieces x and y form a *two-bridge*. In a battle, where Black tries to connect groups x and y, and White tries to prevent it, the result of this battle is predictable two moves ahead. This provides an important advantage to Black. In the position on the left this advantage is local. In the position on the right this advantage is decisive, and White should resign immediately.

The following definitions generalize the two-bridge concept. First we need to clarify some terms. We say that a cell is *black* if and only if it is occupied by a black piece, and we refer to a group of connected black cells as a single black cell.

Definition. Let x and y be two different cells, and A be a set of empty cells of a position. We assume that $x \notin A$ and $y \notin A$. The triplet (x, A, y) defines a *subgame*, where Black tries to connect cells x and y with a chain of black pieces, White tries to prevent it, and both players can put their pieces only on cells in A. We say that x and y are *ends* of the subgame, and A is its *carrier*.

Definitions. A subgame is a *virtual connection* if and only if Black has a winning strategy even if White moves first.

A subgame is a *virtual semi-connection* if and only if Black has a winning strategy if he moves first, and does not have one if he moves second.

We represent virtual connections and semi-connections with diagrams as in Figure 3.

In practice, it is more convenient to use the following recursive definitions.

Definitions. A subgame is a virtual connection if and only if for every White's move there exists a Black's move such that the resulting subgame is a virtual connection.

A subgame is a virtual semi-connection if and only if it is not a virtual connection, and there exists a Black's move such that the resulting subgame is a virtual connection.

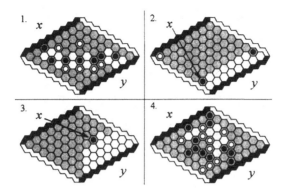

Figure 4. Black cells x and y form virtual connections. In each diagram the cell y is formed by the black pieces connected to the bottom right black boundary. The cells of their carriers are marked white. 1: A chain of two-bridges; $depth = 12$. 2: A ladder; $depth = 14$. 3: An edge connection from the fourth row; $depth = 10$. 4: this virtual connection will be analyzed in the next section; $depth = 6$.

Assume that in a given position with a virtual connection, White moves first. The number of moves, which must be made in order for Black to win this subgame, under the condition that Black does his best to minimize this number, and White does his best to maximize it, characterizes the *depth of the virtual connection*. In other words, the depth of virtual connection is a depth of a game-tree search required to discover this virtual connection. Thus, virtual connections with the depth d contain information about development of Hex position d moves ahead.

We make several remarks:

- Pairs of neighboring cells form virtual connections with empty carriers. The depths of these virtual connections are equal to zero.
- Two-bridges form virtual connections with a depth of two.
- The ends x and y can form virtual connections with several different carriers. The virtual connection (x, A, y) is *minimal* if and only if there does not exist a virtual connection (x, B, y) such that $B \subset A$ and $B \neq A$. We will be primarily interested in minimal virtual connections.
- A special role is played by a *winning virtual connection* formed by the additional boundary cells. If it exists, then there exists a global winning strategy for Black, even if White moves first.

In Figures 4 and 5 you can see samples of virtual connections and virtual semi-connections.

4. Deduction Rules

In this section we define two binary operations, conjunction ($\wedge$) and disjunction ($\vee$), on the set of subgames belonging to the same position. These

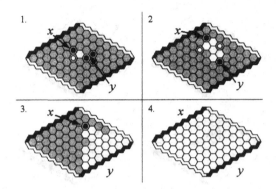

Figure 5. Black cells x and y form virtual semi-connections. The cells of their carriers are marked white. Diagram 4 shows the initial position. According to the Nash theorem mentioned in Section 2, the initial position is a virtual semi-connection.

operations will allow us to build complex virtual connections starting from the simplest ones.

Definition. Let two subgames $G = (x, A, u)$ and $H = (u, B, y)$ with common end u and different ends $x \neq y$ belong to the same position, and $x \notin B$, $y \notin A$.

If common end u is black, then conjunction of these subgames is the subgame $G \wedge H = (x, A \cup B, y)$.

If common end u is empty, then conjunction of these subgames is the subgame $G \wedge H = (x, A \cup u \cup B, y)$.

Definition. Let two subgames $G = (x, A, y)$ and $H = (x, B, y)$ with common ends x and y belong to the same position. Then disjunction of these subgames is the subgame $G \vee H = (x, A \cup B, y)$.

Theorem 1 (The AND deduction rule). *Let two subgames $G = (x, A, u)$ and $H = (u, B, y)$ with common end u and different ends $x \neq y$ belong to the same position, and $x \notin B$, $y \notin A$. If both subgames G and H are virtual connections and $A \cap B = \varnothing$, then*

(a) *$G \cap H$ is a virtual connection if u is black, and*

(b) *$G \cap H$ is a virtual semi-connection if u is empty.*

Proof. If cell u is empty, then Black can occupy this cell, and this reverts to case (a). Since $A \cap B = \varnothing$, White cannot attack both virtual connections simultaneously. Suppose that White occupies a cell $a \in A$. Since the subgame $G = (x, A, u)$ is a virtual connection, then there exists a cell $b \in A$ where Black can play to create a new virtual connection (x, A', u). The new carrier A' is obtained from A by removing two cells a and b. (The new virtual connection belongs to a position different than the original one). In short, if White occupies a cell from A, then Black can restore the first virtual connection by moving to

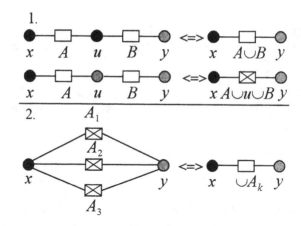

Figure 6. 1: The AND deduction rule. 2: The OR deduction rule.

an appropriate cell of A. The same is true for B, and thus the result follows by induction. $\qquad\square$

Diagram 1 in Figure 6 shows a graphical representation of this deduction rule.

Theorem 2 (The OR deduction rule). *Let subgames* $G_k = (x, A_k, y)$ ($k = 1, 2, \ldots, n$, *for* $n > 1$) *with common ends* x *and* y *belong to the same position. If all games* G_k *are virtual semi-connections and*

$$\bigcap_{k=1}^{n} A_k = \varnothing,$$

then $G = \bigvee G_k$ *is a virtual connection.*

Proof. If White occupies a cell $a \in A_i$, there exists a different carrier A_j such that $a \notin A_j$. Therefore, Black can move to A_j and convert the virtual semi-connection G_j to a virtual connection. $\qquad\square$

Diagram 2 in Figure 6 graphically represents this deduction rule (for $n = 3$).

Theorem 3 (The OR decomposition). *Let a subgame* $G = (x, A, y)$ *be a minimal virtual connection, with* $A \neq \varnothing$. *There exist virtual semi-connections* $G_k = (x, A_k, y)$ ($k = 1, 2, \ldots, n$, *for* $n > 1$) *such that*

$$\bigcap_{k=1}^{n} A_k = \varnothing$$

and $G = \bigvee G_k$.

Proof. Since G is a minimal virtual connection, then for every White's move $a \in A$, the game $G_a = (x, A - a, y)$ is a virtual semi-connection. Besides, $G = \bigvee G_a$ and

$$\bigcap_{a} A - a = \varnothing.$$
$\qquad\square$

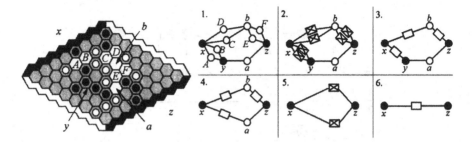

Figure 7. Diagram 1 represents the subgame on the board. Diagram 3 is obtained from Diagram 1 by applying the AND deduction rule six times and then the OR deduction rule three times. Diagram 4 results from the AND deduction rule. The winning virtual connection in Diagram 6 follows from application of the AND deduction rule 2 times and final application of the OR deduction rule.

The last theorem means that the OR deduction rule provides a universal way of building virtual connections from virtual semi-connections. On the other hand, there exist virtual semi-connections, which cannot be obtained from virtual connections by applying the AND deduction rule. An example will be given in the next section.

5. Hierarchy of Virtual Connections

Figure 7 demonstrates how the AND and OR deduction rules can be used for proving virtual connections. Diagram 1 in Figure 7 represents the position on the board. The sequence of transformations in diagrams 2 through 6 graphically demonstrates the application of the AND and OR deduction rules, and proves that Black has a winning position, even if White moves first.

The H-process. Consider the simplest virtual connections, namely pairs of neighboring cells, as the first generation of virtual connections. Applying the AND deduction rule to the appropriate groups of the first generation of virtual connections we build the second generation of virtual connections and semi-connections. Then we apply the AND and OR deduction rules to both the first and the second generations of virtual connections and semi-connections to build the third generation of virtual connections and semi-connections, etc. This process stops when no new virtual connections are produced.

In general, this process can start from any initial set of virtual connections and semi-connections.

This iterative process can build all of the virtual connections shown in Figures 2 and 4. A formal proof for the subgame on Diagram 2 in Figure 4 is provided in Appendix as an example.

Is the set of the AND and OR deduction rules complete, i.e. can this process build all virtual connections? The answer is negative. The diagram in Figure 8

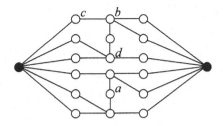

Figure 8. The two black cells form a virtual connection, which cannot be built using the AND and OR deduction rules.

represents a counter-example of a virtual connection that cannot be built by the H-process.

It is easy to check that this subgame is a virtual connection. Indeed, if White plays at a, Black can reply with b, forcing White to occupy c. Then Black plays d securing the win. This virtual connection is a disjunction of two equivalent virtual semi-connections with disjoint carriers. A computer program was used to verify that no combination of the AND and OR deduction rules can establish neither these virtual semi-connections nor the overall virtual connection.

6. Electrical Resistor Circuits

The H-process introduced in the previous section is useful in two ways. First, in some positions it can reach the ultimate objective by building a winning virtual connection between either black or white boundaries. Second, even if it is impossible due to incompleteness of the AND and OR deduction rules and the limited computing resources, the information about connectivity of subgames is useful for the evaluation of the entire position.

In this section we introduce a family of evaluation functions based on an *electrical resistor circuit* representation of Hex positions. One can think of an electrical circuit as a graph. Edges of the graph play a role of electrical links (resistors). The resistance of each electrical link is equal to the length of the corresponding edge of the graph. Yet, we consider that the "electrical circuit" language better suits our needs.

With every Hex position we associate two electrical circuits. The first one characterizes the position from Black's point of view (Black's circuit), and the second one from White's point of view (White's circuit). To every cell c of the board we assign a resistance r in the following way:

$$r_B(c) = \begin{cases} 1, & \text{if } c \text{ is empty,} \\ 0, & \text{if } c \text{ is occupied by a black piece,} \\ +\infty, & \text{if } c \text{ is occupied by a white piece,} \end{cases}$$

for Black's circuit, and

$$r_W(c) = \begin{cases} 1, & \text{if } c \text{ is empty,} \\ 0, & \text{if } c \text{ is occupied by a white piece,} \\ +\infty, & \text{if } c \text{ is occupied by a black piece,} \end{cases}$$

for White's circuit. For each pair of neighboring cells, (c_1, c_2), we associate an electrical link with resistance

$$r_B(c_1, c_2) = r_B(c_1) + r_B(c_2), \quad \text{for Black's circuit,}$$

$$r_W(c_1, c_2) = r_W(c_1) + r_W(c_2), \quad \text{for White's circuit.}$$

These circuits take into account only virtual connections between neighboring cells, and describe the microstructure of Hex position.

We are now going to enhance these circuits by including information about more complex known virtual connections. We focus on Black's circuits only. White's circuits can be dealt with in a similar way.

A seemingly natural way of doing this is to add an additional electrical link between cells x and y to Black's circuit if x and y form a virtual connection. Then all virtual connections would be treated as neighboring cells. However, virtual connections between nearest neighbors are stronger than other virtual connections, so our circuit should reflect this. Instead of connecting black cells x and y with a shortcut, we add other links to Black's circuit in the following way. If an empty cell c is a neighbor of one of the ends of this virtual connection, say x, then we also treat this cell c as a neighbor of the other end y. This means that we connect cells c and y with an additional electrical link in the same way as actual neighbors.

Let R_B and R_W be distances between black boundaries in Black's circuit and between white boundaries in White's circuit, correspondingly. We define an evaluation function:

$$E = \log(R_B/R_W).$$

One of the reasonable distance metrics is the length of the shortest path on the graph, connecting boundaries. We can also measure distances in a different way. Apply an electrical voltage to the opposite boundaries of the board and measure the total resistance between them, R_B for Black's circuit, and R_W for White's circuit (see Figure 9).

We prefer this way for measuring distances, because according to the Kirchhoff electrical current laws, the total resistance takes into account not only the length of the shortest path, but also all other paths connecting the boundaries, their lengths, and their intersections.

Virtual connections with the depth d contain information about development of Hex position d moves ahead. Thus, we can expect that by including electrical links, which correspond to virtual connections with depth less or equal than d, we obtain an evaluation function with foreseeing abilities up to d moves ahead.

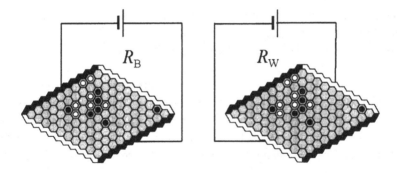

Figure 9. Black's and White's circuits.

7. Hexy Plays Hex

Hexy is a Hex-playing computer program, which utilizes the ideas presented in this paper. It runs on a standard PC with Windows, and can be downloaded from the website http://home.earthlink.net/~vanshel.

Hexy uses a selective alpha-beta search algorithm, with the evaluation functions described in the previous section. For every node to be evaluated, Hexy calculates the hierarchy of virtual connections for both Black's and White's circuits using the H-process described in Section 5. Then Hexy calculates distances R_B and R_W between Black and White boundaries, correspondingly. For calculation of the shortest path, a version of Dijkstra algorithm is applied. For calculation of the total electrical resistance between boundaries, Hexy solves the Kirchhoff system of linear equations using a method of iterations (see (Strang 1976), for example).

In practice, Hexy does not start the H-process from pairs of neighboring cells, but looks for changes in the hierarchy of virtual connections, caused by an additional piece, placed on the board. Besides, the program keeps track of only minimal available virtual connections and semi-connections.

The program has two important thresholds, D and N. The parameter D is the depth of the game-tree search. The second parameter, N, sets the limit to the number of different minimal virtual connections with the same ends, built by the program. This threshold indirectly controls the total number of calculated minimal virtual connections. The larger N, the more minimal virtual connections the H-process builds for every node of the game-tree. However, we do not put any limits on the number of iterations of the algorithm, or the total number of virtual connections, or their depths. The process stops when the next iteration of the algorithm does not produce new virtual connections.

There is an obvious trade-off between parameters D and N, and finding an optimum is an important task. Since our major objective has been a creation of Hex-playing program, which can provide fun for Hex fans, we confine ourselves by a condition, that Hexy should be able to complete a game on the 10×10 board for less then 8 minutes on a standard PC with 300 MHz processor and

32 MB RAM. Thus, we try to find optimal (in terms of playing strength) values of these thresholds, satisfying the above condition. Experiments show that the dependence of Hexy's strength on the parameter N, which controls the number and the depth of virtual connections, is much more dramatic than its dependence on the depth D of game-tree search. The best results determined experimentally, are obtained with values of $D = 3$ and $N = 20$ (for a 10×10 board). This version of Hexy (called Advanced level) performs a very shallow game-tree search (200–500 nodes per move), but routinely detects virtual connections with depth 20 or more. It means that this version of Hexy routinely foresees some lines of play 20 or more moves ahead.

Hexy demonstrates a clear superiority over all known Hex-playing computer programs. This program won a Hex tournament of the 5th Computer Olympiad in London on August 2000, with the perfect score (Anshelevich 2000b).

Hexy was also tested against human players on the popular game website Playsite (http://www.playsite.com/games/board/hex). Hexy cannot compete with the best human players. Nevertheless, after more than 100 games, the program achieved a rating within the highest Playsite red rating range.

8. Conclusion

In this paper we have described a hierarchical approach to the game of Hex, and explained how this approach is implemented in Hexy - a Hex-playing computer program. Hexy does not perform massive game-tree search. Instead, this program spends most computational resources on deep analysis of a relatively small number of Hex positions.

We have concentrated on a hierarchy of positive subgames of Hex positions, called virtual connections, and have defined the AND and OR deduction rules, which allow to build complex virtual connections recursively, starting from the simplest ones. Integrating the information about virtual connections of this hierarchy, we have built a far-sighted evaluation function, foreseeing the potential of Hex positions many moves ahead.

The process of building virtual connections, the H-process, has its own cost. Nevertheless, the resulting foreseeing abilities of the evaluation function greatly outweigh its computational cost. This approach is much more efficient than brute-force search, and can be considered as both alternative and complimentary to the alpha-beta game-tree search.

Acknowledgements

I would like to express my gratitude to the organizers and the participants of the Combinatorial Game Theory Workshop in MSRI, Berkeley, July 2000, for the exciting program and fruitful discussions.

Appendix

In this Appendix we show how to prove that the ladder in Figure 10 is a virtual connection using the AND and OR deduction rules. We use abbreviation VC for virtual connections, VSC for virtual semi-connections, brackets [] for carriers, and parentheses () for subgame triplets.

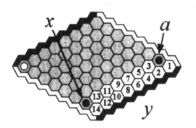

Figure 10. Black cells x and y form a ladder. The cells of the carrier are enumerated.

Examples:

$[a, b]$ is a carrier consisting of t wo cells a and b.

$[]$ is an empty carrier.

$(x, [a, b, c, d], y)$ is a subgame with ends x and y and carrier $[a, b, c, d]$.

$(x, [], y)$ is a subgame with ends x and y and an empty carrier.

The following sequence of deductions proves that the subgame in Figure 10 is a virtual connection.

$(3, [], a)$ is VC, $(a, [], 1)$ is VC. Apply AND: $(3, [], 1)$ is VC.

$(3, [], 1)$ is VC, $(1, [], y)$ is VC. Apply AND: $(3, [1], y)$ is VSC.
$(3, [], 2)$ is VC, $(2, [], y)$ is VC. Apply AND: $(3, [2], y)$ is VSC.
$(3, [1], y)$ is VSC, $(3, [2], y)$ is VSC. Apply OR: $(3, [1, 2], y)$ is VC.

$(5, [], 3)$ is VC, $(3, [1, 2], y)$ is VC. Apply AND: $(5, [1, 2, 3], y)$ is VSC.
$(5, [], 4)$ is VC, $(4, [], y)$ is VC. Apply AND: $(5, [4], y)$ is VSC.
$(5, [1, 2, 3], y)$ is VSC, $(5, [4], y)$ is VSC. Apply OR: $(5, [1, 2, 3, 4], y)$ is VC.

$(7, [], 5)$ is VC, $(5, [1, 2, 3, 4,], y)$ is VC. Apply AND: $(7, [1, 2, 3, 4, 5], y)$ is VSC.
$(7, [], 6)$ is VC, $(6, [], y)$ is VC. Apply AND: $(7, [6], y)$ is VSC.
$(7, [1, 2, 3, 4, 5], y)$ is VSC, $(7, [6], y)$ is VSC. Apply OR: $(7, [1, 2, 3, 4, 5, 6], y)$ is VC.

$(9, [], 7)$ is VC, $(7, [1, 2, 3, 4, 5, 6], y)$ is VC. Apply AND: $(9, [1, 2, 3, 4, 5, 6, 7], y)$ is VSC.
$(9, [], 8)$ is VC, $(8, [], y)$ is VC. Apply AND: $(9, [8], y)$ is VSC.
$(9, [1, 2, 3, 4, 5, 6, 7], y)$ is VSC, $(9, [8], y)$ is VSC. Apply OR:
    $(9, [1, 2, 3, 4, 5, 6, 7, 8], y)$ is VC.

$(11, [], 9)$ is VC, $(9, [1, 2, 3, 4, 5, 6, 7, 8], y)$ is VC. Apply AND:
  $(11, [1, 2, 3, 4, 5, 6, 7, 8, 9], y)$ is VSC.
$(11, [], 10)$ is VC, $(10, [], y)$ is VC. Apply AND: $(11, [10], y)$ is VSC.
$(11, [1, 2, 3, 4, 5, 6, 7, 8, 9], y)$ is VSC, $(11, [10], y)$ is VSC. Apply OR:
  $(11, [1, 2, 3, 4, 5, 6, 7, 8, 9, 10], y)$ is VC.

$(13, [], 11)$ is VC, $(11, [1, 2, 3, 4, 5, 6, 7, 8, 9, 10], y)$ is VC. Apply AND:
  $(13, [1, 2, 3, 4, 5, 6, 7, 8, 9, 10, 11], y)$ is VSC.
$(13, [], 12)$ is VC, $(12, [], y)$ is VC. Apply AND: $(13, [12], y)$ is VSC.
$(13, [1, 2, 3, 4, 5, 6, 7, 8, 9, 10, 11], y)$ is VSC, $(13, [12], y)$ is VSC. Apply OR:
  $(13, [1, 2, 3, 4, 5, 6, 7, 8, 9, 10, 11, 12], y)$ is VC.

$(x, [], 13)$ is VC, $(13, [1, 2, 3, 4, 5, 6, 7, 8, 9, 10, 11, 12], y)$ is VC. Apply AND:
  $(x, [1, 2, 3, 4, 5, 6, 7, 8, 9, 10, 11, 12, 13], y)$ is VSC.
$(x, [], 14)$ is VC, $(14, [], y)$ is VC. Apply AND: $(x, [14], y)$ is VSC.
$(x, [1, 2, 3, 4, 5, 6, 7, 8, 9, 10, 11, 11, 12, 13], y)$ is VSC, $(x, [14], y)$ is VSC. Apply
  OR: $(x, [1, 2, 3, 4, 5, 6, 7, 8, 9, 10, 11, 12, 13, 14], y)$ is VC.

References

Adelson-Velsky, G.; Arlazarov, V.; and Donskoy, M. 1988. Algorithms for Games. Springer-Verlag.

Anshelevich, V. V. 2000a. The Game of Hex: An Automatic Theorem Proving Approach to Game Programming. Proceedings of the Seventeenth National Conference on Artificial Intelligence (AAAI-2000), 189–194, AAAI Press, Menlo Park, CA.

Anshelevich, V. V. 2000b. Hexy Wins Hex Tournament. The ICGA Journal, 23(3):181–184.

Berlekamp, E. R.; Conway, J. H.; and Guy, R. K. 1982. Winning Ways for Your Mathematical Plays. New York: Academic Press.

Berlekamp, E.R.; and Wolfe, D. 1994. Mathematical Go: Chilling Gets the Last Point. A. K. Peters, Wellesley.

Browne, C. 2000. Hex Strategy: Making the Right Connections. A. K. Peters, Natick, MA.

Conway, J. H. 1976. On Numbers and Games. London. Academic Press.

Even, S.; and Tarjan, R. E. 1976. A Combinatorial Problem Which Is Complete in Polynomial Space. Journal of the Association for Computing Machinery 23(4): 710–719.

Gale, D. 1979. The Game of Hex and the Brouwer Fixed-Point Theorem. American Mathematical Monthly 86: 818–827.

Gardner, M. 1959. The Scientific American Book of Mathematical Puzzles and Diversions. New York: Simon and Schuster.

Marsland, T. A. 1986. A Review of Game-Tree Pruning. Journal of the International Computer Chess Association 9(1): 3–19.

Müller, M. 1999. Decomposition search: A combinatorial games approach to game tree search, with applications to solving Go endgames. In IJCAI-99, 1: 578–583.

Nowakowski, R. (editor) 1996. Games of No Chance. MSRI Publications 29. New York. Cambridge University Press.

Reisch, S. 1981. Hex ist PSPACE-vollständig. Acta Informatica 15: 167–191.

Schaeffer, J.; Lake, R.; Lu, P.; and Bryant M. 1996. Chinook: The World man-Machine Checkers Champion. AI Magazine 17(1): 21–29.

Shannon, C. E. 1953. Computers and Automata. Proceedings of Institute of Radio Engineers 41: 1234–1241.

Strang, G. 1976. Linear Algebra and Its Applications. Academic Press.

VADIM V. ANSHELEVICH
vanshel@earthlink.net

More Games of No Chance
MSRI Publications
Volume **42**, 2002

Hypercube Tic-Tac-Toe

SOLOMON W. GOLOMB AND ALFRED W. HALES

ABSTRACT. We study the analogue of tic-tac-toe played on a k-dimensional hypercube of side n. The game is either a first-player win or a draw. We are primarily concerned with the relationships between n and k (regions in n-k space) that correspond to wins or draws of certain types. For example, for each given value of k, we believe there is a critical value n_d of n below which the first player can force a win, while at or above this critical value, the second player can obtain a draw. The larger the value of n for a given k, the easier it becomes for the second player to draw. We also consider other "critical values" of n for each given k separating distinct behaviors. Finally, we discuss and prove results about the misère form of the game.

1. Introduction

Hypercube tic-tac-toe is a two-person game played on an n^k "board" (i.e. a k-dimensional hypercube of side n). (The familiar 3×3 game has $k = 2$ and $n = 3$. Several editions of the 4^3 game, $k = 3$ and $n = 4$, are commercially available.) In all these games the two players take turns. Each player claims a single one of the n^k cells with his/her *symbol* (traditionally O's and X's, or "noughts and crosses", as the game is known in the UK), and the first player to complete a "path" of length n (in any straight line, including any type of diagonal) is the winner. If all n^k cells are filled (with the two kinds of symbols) but no solid-symbol path has been completed, the game is declared a draw.

Since the first move cannot be a disadvantage, with best play the first player should never lose. Hence, in the ideal world, the first player seeks a win, while the second player tries to draw. For each given value of k, we believe there is a critical value n_d of n below which the first player can force a win, while at or above this critical value, the second player can obtain a draw. This exact value of n is exceedingly difficult to determine as a function of k. (The larger the value of n for a given k, the easier it becomes for the second player to draw.)

There are several other "critical values" of n for each given k. The smallest of these is the value of n below which the first player *must* win, no matter how well or poorly the two players play. Thus, for all $k > 1$, the 2^k board is a win for the first player on his/her second move, independent of the actual sequence of moves. Another critical value, $n_p \geq n_d$, is the value at or above which the second player can force a draw by a "pairing strategy".

There are exactly $((n+2)^k - n^k)/2$ possible winning paths on the n^k board. If it is possible to "dedicate" two cells of the hypercube exclusively to each path, the second player can draw by occupying the second dedicated cell whenever the first player occupies the first dedicated cell on a single path. A *necessary* condition for a pairing strategy to exist is that the number of cells must be at least twice the number of paths, i.e.,

$$n^k \geq (n+2)^k - n^k,$$

which is easily seen to be equivalent to

$$n \geq \frac{2}{2^{1/k} - 1}.$$

The *Hales–Jewett Conjecture* is that for every k, when $n \geq \frac{2}{2^{1/k}-1}$, i.e. for all $n \geq \left\lceil \frac{2}{2^{1/k}-1} \right\rceil$, the second player can force a draw. A stronger conjecture would be that for each k, for all $n \geq n_k = \left\lceil \frac{2}{2^{1/k}-1} \right\rceil$, a draw for the second player by "pairing strategy" can be found.

It is very tempting to conjecture that

$$n_k = \left\lceil \frac{2}{2^{1/k} - 1} \right\rceil = \left\lfloor \frac{2k}{\ln 2} \right\rfloor$$

for all integers $k \geq 1$. Somewhat surprisingly, this conjecture is false. Even more surprisingly, the first failure of this conjecture occurs at $k = 6,847,196,937$ dimensions, where a "board" of side $\left\lfloor \frac{2k}{\ln 2} \right\rfloor = 19{,}756{,}834{,}129$ is too small (by just a little) to allow a pairing strategy. The next failure occurs at $k = 27,637,329,632$ dimensions, where a "board" of side $n = \left\lfloor \frac{2k}{\ln 2} \right\rfloor = 79{,}744{,}476{,}806$ is too small (by just a little) to allow a pairing strategy.

Are there infinitely many (albeit incredibly sparse) exceptional values of k? Can an explicit pairing strategy be exhibited for specific pairs or classes of pairs, of k and n_k? Are there values of k such that no pairing strategy exists when $n = n_k$? These questions and others will be explored.

In the (n, k)-plane, "phase changes" occur from "forced win for first player", to "win by strategy for first player", to "draw by strategy" for second player, to "draw by pairing strategy" for second player. It should be easier to describe these *regions* in (n, k) "phase space" than to calculate the locations of their precise boundaries.

What we have just discussed is the normal form of the game. In the *misère* form, the first player to form a straight path of length n is the *loser*. We will consider the misère form later on, but unless otherwise specified we will always be talking about the normal form.

2. An Elementary Result

Theorem 1. *The number of winning paths on the n^k hypercube is*

$$\tfrac{1}{2}\big((n+2)^k - n^k\big).$$

Proof A (Geometric/Intuitive): Embed the n^k hypercube in an $(n+2)^k$ hypercube which extends one unit farther in each direction in each of the k dimensions than the original hypercube (see Figure 1). Then each winning path in the n^k hypercube terminates in exactly t wo "border cells" of the enlarged hypercube, and these t wo border cells are unique to that path. Moreover, every border cell is at the end of a path, so that the $(n+2)^k - n^k$ border cells are in t wo-to-one correspondence with the winning paths. □

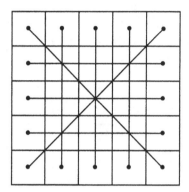

Figure 1. The familiar $n = 3, k = 2$ tic-tac-toe board is embedded in an $n = 5, k = 2$ board. Each of the 8 winning paths terminates in exactly 2 border cells of the 5×5 board: $\frac{1}{2}(5^2 - 3^2) = 8$.

Proof B (Algebraic/Rigorous): Represent each cell of the n^k hypercube by its coordinate k-vector $\alpha = (a_1, a_2, \ldots, a_k)$, where $1 \le a_i \le n$ for each i, $1 \le i \le k$. A winning path P is an ordered sequence of n such vectors, $P = \{\alpha_1, \alpha_2, \ldots, \alpha_n\}$, in which the ith component, for each i, either runs from 1 up to n, or from n down to 1, or remains constant at any one of the n values, except that we do not allow all k components to remain constant (since all n vectors in P would then degenerate to the same cell, and we would not have a path). Thus the number of allowed sequences which represent paths is $(n+2)^k - n^k$. However, the path $P = \{\alpha_1, \alpha_2, \ldots, \alpha_n\}$ is the same as the path $P' = \{\alpha_n, \alpha_{n-1}, \ldots, \alpha_1\}$, so there are only $\frac{1}{2}((n+2)^k - n^k)$ *unoriented* paths. □

3. Regions in "$n - k$ Space"

We first observe that having the first move cannot be a disadvantage; so the first player looks for a winning strategy, while the second player looks for a drawing strategy. (This assumes intelligent players of comparable skill.)

If $k \geq 2$ and $n = 2$, the first player must complete a winning path on her second move, independently of how well or poorly she plays. More generally, when n is "small" compared to k, it is impossible to assign all n^k cells to t wo players without at least one winning path having been created for one of the t wo players. For $k = 1$, we see that $n = 1$ is "small enough" but $n = 2$ is not. For $k = 2$, we have $n = 2$ is "small enough" (as mentioned more generally) but $n = 3$ is not (i.e. *ordinary* tic-tac-toe *can* result in a draw, and in fact always will with best play by both players).

It is also known that for $k = 3$, no draw is possible on the 3^3 "board", but that draws *are* possible on the 4^3 "board" (which has been available commercially from several manufacturers). The "obvious" conjecture that the critical value for all k is $n = k$ (since it is true for $k = 1, 2, 3$) was first disproved some forty years ago (by A. W. Hales), as follows:

Form the 4^4 hypercube as the tensor product of the following t wo 4^2 "boards", where we represent the cells of the t wo players by $+$ (for $+1$) and $-$ (for -1).

+	+	−	−
+	−	+	−
−	+	−	+
−	−	+	+

$\ast$

+	−	+	+
+	+	+	−
+	+	−	+
−	+	+	+

Note that the *left* factor has 2 plusses and 2 minuses on each tic-tac-toe path (an even number, but not 0 or 4, of each; while the *right* factor has 3 plusses and 1 minus on each tic-tac-toe path. Each "winning path" on the resulting 4^4 hypercube is either a constant from one factor times a path from the other factor (and therefore not "four identical symbols"), or the term-by-term products of a path from the first factor and a path from the second factor; but clearly such a term-by-term product path will have an *odd* number of minuses, and therefore cannot have all four cells the same.

Since the left factor has equally many $+$'s and $-$'s, this will also be true of the tensor product. Thus, the 4^4 draw could occur in an actual game, especially if the t wo players cooperated to achieve it. □

In general, this suggests that the critical n for each k (for "no draw is possible on the n^k board) satisfies $n \leq k$, and this n is monotonically non-decreasing as k increases. (The monotonic property is easily proved. The precise expression of this critical n as a function of k is not known.)

The principal regions in $n - k$ space are the following:

1. The first player *must* win (as when $n = 2$ for $k \geq 2$).

2. Since no draw is possible, the first player should have a relatively easy win (as when $n = k = 3$, where playing in the center on the first move is already devastating).

3. Although draws are possible, there is a win for the first player with best play. (It is known [1] that exhaustive computer searching has shown that the 4^3 "board" is in this category.)

4. Although there is no trivial drawing strategy for the second player (as in region 5, below), the second player can always draw with best play. (This is the case for the familiar 3^2 board. While mathematicians will consider the drawing strategy "trivial" because it is so easily learned, it does not meet our definition of "trivial" given in Region 5; nor does it meet the layman's notion of "trivial" since this game is still widely played.)

5. The second player has a "trivial" draw by a *pairing strategy*. In a pairing strategy, two of the n^k cells are explicitly dedicated to each of the $\frac{1}{2}((n + 2)^k - n^k)$ winning paths. (There may be some undedicated cells left over.) Whenever the first player claims one dedicated cell, the second player then immediately claims the other cell dedicated to the same path, if he hasn't already claimed it. (If he already has, he is free to claim *any* unclaimed cell.) Clearly, the first player can never complete a winning path if the second player is able to follow this strategy.

When $k = 1$, the line of length $n = 2$ *forces* the second player to draw by an automatic pairing of the only two "cells" of the "board".

When $k = 2$, the smallest board with a pairing strategy has $n = 5$, as shown in Figure 2.

The second player can even give the first player a "handicap" of the center square, as well as the first move, and still draw by the pairing shown in Figure 2.

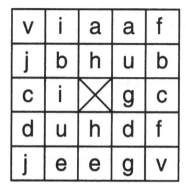

v	i	a	a	f
j	b	h	u	b
c	i	✕	g	c
d	u	h	d	f
j	e	e	g	v

Figure 2. A pairing strategy for the 5 × 5 board. Two each of a through e are dedicated to the rows, two each of f through j are dedicated to the columns, and two each of u and v to the diagonals. The center cell is left undesignated.

A more suggestive way to indicate the same pairing strategy is shown in Figure 3.

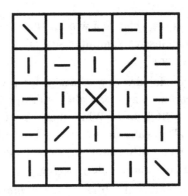

Figure 3. The pairing strategy for the 5 × 5 board, shown with horizontal, vertical, and diagonal strokes to indicate the type of "winning paths" to which the cells are dedicated. Note the symmetry of this pattern relative to each of the two diagonals, as well as under 180° rotation.

On the 6 × 6 board, there are $6^2 = 36$ cells and $\frac{1}{2}(8^2 - 6^2) = 14$ paths. If we dedicate *all six* cells on each diagonal to that diagonal, we have $36 - 12 = 24$ remaining cells, to assign to $14 - 2 = 12$ remaining paths. This can be done as shown in Figure 4, which has the full D_4 symmetry of the square board.

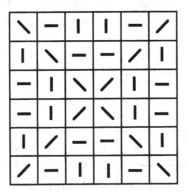

Figure 4. Representation of a pairing strategy on the 6 × 6 board. The horizontal and vertical midlines can be regarded as "reflectors" for this pattern. (So too can the diagonals.)

For $k = 3$, the smallest "board" with a pairing strategy is the 8^3, which has 512 cells and $\frac{1}{2}(10^3 - 8^3) = 244$ paths. However, the four "body diagonals" have 8 cells each, and if we dedicate *all* of these to their respective (non-overlapping) body diagonals, we are left with $512 - 32 = 480$ *cells*, and $244 - 4 = 240$ *paths*, i.e., exactly two cells available per path. If we divide the 8^3 "board" into octants, each 4^3, by the three mid-planes, we can assign "strokes" to the sixty available cells in the first octant (the other 4 cells were on a body diagonal), and then use the three mid-planes as mirrors to assign "strokes" (as in Figures 5 and 6) to all the remaining cells in the other octants, to end up with two dedicated cells per winning path (having treated the body-diagonal paths separately.)

To show the formation of the assignment, for pairing strategy purpose, to the $4 \times 4 \times 4$ "first octant" of the $8 \times 8 \times 8$ board, we first show, in Figure 5, the dedication of cells to ranks, files, and (vertical) columns.

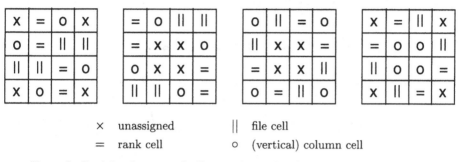

×	unassigned	‖	file cell
=	rank cell	o	(vertical) column cell

Figure 5. Partial assignment of cells to paths, in the "first octant" of the $8 \times 8 \times 8$ game.

We complete the assignment with three orientations of diagonals (including body diagonals, to which *all* their cells are dedicated) in Figure 6.

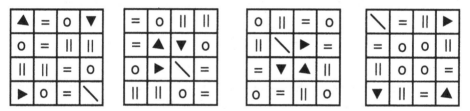

Figure 6. Assignment of cells to "winning paths", in the "first octant" of the $8 \times 8 \times 8$ game. The arrows indicate the type of diagonals to which these cells are dedicated. These four 4×4 patterns are to be stacked with the left-most on top. Then the upper left corner of that top layer is at a corner of the $8 \times 8 \times 8$ board, and that cell is dedicated to the corresponding body diagonal.

Several of these examples of pairing strategies were shown in [4].

In general, a *necessary* condition for a pairing strategy to exist is that the number of cells must at least be equal to twice the number of winning paths, that is,

$$n^k \geq (n+2)^k - n^k, \quad \text{or, equivalently,} \quad n \geq \frac{2}{2^{1/k} - 1}.$$

Accordingly, let us *define* $n_k = \left\lceil \frac{2}{2^{1/k} - 1} \right\rceil$.

Thus, $n_1 = 2$, $n_2 = 5$, $n_3 = 8$, and we have seen that pairing strategies really do exist for these values of n_k. At the present time (July, 2000), successful pairing strategies have also been reported for the next two cases: $n_4 = 11$ and $n_5 = 14$. Whether a pairing strategy exists for every $(n_k)^k$ hypercube is *not* known, but it *has* been shown [3] that a somewhat more elaborate drawing strategy exists for the second player for these cases, at least for all large k (specifically, for $k \geq 100$), proving a conjecture in [2] for these values.

In the next section, we will examine the validity of replacing $n_k = \left\lceil \frac{2}{2^{1/k}-1} \right\rceil$, the round-up of an *exponential* expression in k, by the much simpler $n_k = \left\lfloor \frac{2k}{\ln 2} \right\rfloor$, the round-down of a *linear* expression in k.

4. The Linearized Approximation to n_k

We have $n_k = \left\lceil \frac{2}{2^{1/k}-1} \right\rceil$, where, letting $a = 2^{1/k}$, we have

$$\frac{1}{2^{1/k}-1} = \frac{a^k - 1}{a-1} = 1 + a + a^2 + \cdots + a^{k-1} \approx \int_0^k a^t dt = \int_0^k e^{t \ln a} dt$$

$$= \frac{1}{\ln a} \cdot a^t \Big|_{t=0}^k = \frac{a^k - 1}{\ln a} = \frac{k}{\ln 2}(2-1) = \frac{k}{\ln 2},$$

and since taking the upper limit of integration to be k (rather than, say, $k-1$), this suggests that $\frac{1}{2^{1/k}-1}$ has been rounded upward to $\frac{k}{\ln 2}$, giving some heuristic motivation to believing the "identity" $n_k = \left\lceil \frac{2}{2^{1/k}-1} \right\rceil \overset{?}{=} \left\lfloor \frac{2k}{\ln 2} \right\rfloor$.

Alternatively,

$$2 > \frac{(n+2)^k}{n^k} = \left(1 + \frac{2}{n}\right)^k \approx e^{2k/n},$$

from which $n_k \approx \frac{2k}{\ln 2}$.

This belief is easily strengthened by routine computer verification for the first 10^j values of k, for each $j = 1, 2, 3, 4, 5, 6, 7, 8, 9$. (Multiple precision is certainly required long before reaching $k = 10^9$.) Surely this constitutes "proof beyond a reasonable doubt", and would almost certainly convince not only a jury, but an engineer, a statistician, even a physicist, but (we hope) not a true mathematician. Because this purported "identity" is not always true!

The first failure occurs at $k = 6,847,196,937$. That is, if one is playing hypercube tic-tac-toe in $k = 6,847,196,937$ dimensions on a board which is $n = \left\lfloor \frac{2k}{\ln 2} \right\rfloor = 19,756,834,129$ on a side, the number of cells is slightly less than twice the number of winning paths, so no true pairing strategy can possibly exist! (If the two players each make 10^9 moves per second, how many eons will it take to claim all n^k cells?)

And, *horribile dictu*, this first failure is not the last! It is, to be sure, rather isolated, but the *second* failure occurs at $k = 27,637,329,632$, where the value $n = \left\lfloor \frac{2k}{\ln 2} \right\rfloor = 79,744,476,806$ again fails to allow twice as many cells as paths on the n^k "board". More careful power series analysis shows that the difference $\frac{2k}{\ln 2} - \frac{2}{2^{1/k}-1}$ equals $1 - \varepsilon$, where $0 < \varepsilon < \frac{\ln 2}{6k}$. Using results from the theory of diophantine approximation, "failure" can only occur when k is the denominator of a continued fraction convergent for $\frac{2}{\ln 2}$, which greatly facilitates computation.

Worst of all, we *believe* (though it is not yet proved) that the set of k's for which

$$\left\lfloor \frac{2k}{\ln 2} \right\rfloor \neq \left\lceil \frac{2}{2^{1/k} - 1} \right\rceil$$

is an *infinite* subsequence of the positive integers! This is related to the "Markov constant" $M\left(\frac{2}{\ln 2}\right)$. (Fortunately the chance of landing on one of these deadly values of k "at random" is not very great.)

A conjecture about all positive integers k that fails for the first time at $k = 6{,}847{,}196{,}937$ is impressive, but not record-setting. However, the number of cells in the n^k hypercube for this value of k and $n = \left\lfloor \frac{2k}{\ln 2} \right\rfloor$ may be one of the larger integers that has occurred "naturally".

5. Hypercube Tic-Tac-Toe and Combinatorial Phase Space

The five regions described in Section 2 above in $n-k$ space partition the lattice points in one quadrant of the Euclidean plane into five "connected" regions. (If we use cells of quadrille paper rather than points for each pair (k, n), these regions are more likely to be connected and simply connected.) The hard problem is to find the precise boundaries of these regions — i.e. to locate exactly where the "phase transitions" occur, between the different "states" in game space. What is undoubtedly easier, and probably more "useful", is to obtain qualitative results on the shapes of these regions and their boundaries, and to get fairly good inequalities of the sort: "if $c_1 k < n < c_2 k$, then (k, n) is in region j", for each j from 1 to 5.

The connectedness of Region 5 in n-k phase space is actually provable. The first part is that a pairing pattern on n^{k+1} obviously imposes a pairing pattern on n^k. It is also true that a pairing pattern on n^k extends to a pairing pattern on $(n + 1)^k$, but here one must be careful. Instead of extending at the edges, it is easier to extend from the middle.

Assume that the successful pairing designations have already been made on the n^k "board". We now insert k mid-hyperplanes into the n^k configuration. If n is odd, all split cells are "replicated" (i.e. their designations as rank cells, body diagonal cells, etc. are inherited by each offspring cell). If n is even, use the mid-hyperplanes as one-way mirrors to generate a duplication of one of the adjacent layers. The crucial point is that by adjoining the new layers "centrally", all diagonals remain diagonals.

By centrally enlarging the "board", all new paths are blocked if all old paths were blocked; all new paths have at least two dedicated cells if all old paths had at least two dedicated cells.

In Figure 7, we see this "central enlargement" illustrated to go from 3×3 to 4×4, and from 4×4 to 5×5.

DIMENSION

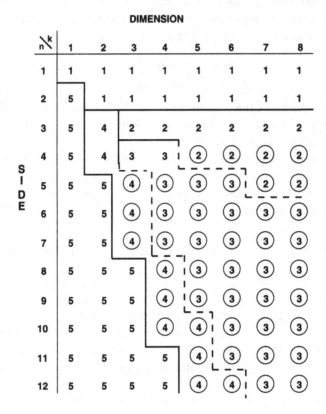

n\k	1	2	3	4	5	6	7	8
1	1	1	1	1	1	1	1	1
2	5	1	1	1	1	1	1	1
3	5	4	2	2	2	2	2	2
4	5	4	3	3	(2)	(2)	(2)	(2)
5	5	5	(4)	(3)	(3)	(3)	(2)	(2)
6	5	5	(4)	(3)	(3)	(3)	(3)	(3)
7	5	5	(4)	(3)	(3)	(3)	(3)	(3)
8	5	5	5	(4)	(3)	(3)	(3)	(3)
9	5	5	5	(4)	(3)	(3)	(3)	(3)
10	5	5	5	(4)	(4)	(3)	(3)	(3)
11	5	5	5	5	(4)	(3)	(3)	(3)
12	5	5	5	5	(4)	(4)	(3)	(3)

SIDE

Table 1. Regions in $n - k$ phase space. (Dotted boundaries and circled numbers are uncertain.)

Figure 7. Central enlargements of even and odd boards.

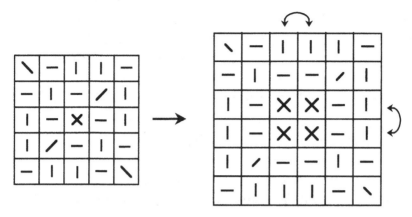

Figure 8. Extending a "pairing strategy" from 5^2 to 6^2.

In Figure 8, an actual pairing strategy is extended from the 5×5 board to the 6×6 board by this method.

Note that we have shown more than just connectedness. We have shown that Region 5 is "row-column convex" — i.e. that any horizontal or vertical line joining t wo points of the region lies entirely in the region. This implies that the region is simply connected.

A similar argument shows that Region 2 is also "row-column convex".

6. Mis`ereHypercube Tic-Tac-Toe

There is one sweepingly general result.

Theorem 2. *The first player can achieve at least a draw on the n^k board whenever $n > 1$ is odd (in the mis`er case).*

Proof. When n is odd, the n^k "board" has a central cell. The first player should start by claiming this central cell, and thereafter playing diametrically opposite every subsequent move of her opponent. It is clear that the first player will never be completing a path that includes the central cell; and any other path completed by the first player will be a mirror image of a path already completed by the second player.   □

Corollary. *For (n, k) in "Region 2", i.e. where the n^k hypercube is too small to fail to have a path when all filled in, if $n > 1$ is odd, the first player wins the mis`er game using the strategy in the proof of Theorem 2.*

Example. On the $3 \times 3 \times 3$ "board", one would naively expect that the *worst* move for the first player (under mis`ererules) would be to claim the central cube, since this is on the most paths (thirteen paths, versus seven paths for a corner cell, five paths for a mid-face cell, and only four paths for an edge cell). Yet, by the Corollary, this is a winning move for the first player. (A computer program

e	f	g
l	m	h
k	j	i

a	b	c
d	⊗	d
c	b	a

i	j	k
h	m	l
g	f	e

Figure 9. First player winning strategy for $3 \times 3 \times 3$ Misère Tic-Tac-Toe.

could determine whether or not it is the only winning first move; but in any case, it is the first move with the simplest winning strategy.)

7. Tic-Tac-Toe on Projective Planes

The n^2 board resembles the structure of a finite affine geometry. It can be extended to the finite projective plane of order n, with a total of $n^2 + n + 1$ points on $n^2 + n + 1$ lines, each line containing $n + 1$ points. Tic-Tac-Toe generalizes to these projective "boards" in the obvious way: the two players take turns claiming *points*, and the first player to complete a *line* is the winner. This game is an easy win (for the first player) on the 7-point plane, and a fairly easy draw (for the second player) on the 13-point plane. (On the 7-point plane, no draw is possible.) Note that in projective Tic-Tac-Toe, if one player completes a path, the other cannot possibly, even if given all the remaining "points".

8. On the Boundary Between Regions 3 and 4

We call the n^k-game a *win* if the first player wins, given best play by both sides; otherwise, we call it a *draw*.

We offer the following three hypotheses, all assuming $n_1 \geq 2$ and $k_1 \geq 2$.

Hypothesis 1. If the $n_1^{k_1}$ game is a draw, then the $n_1^{k_1-1}$ game is a draw. ("row convexity")

Hypothesis 2. If the $n_1^{k_1}$ game is a draw, then the $(n_1 + 1)^{k_1}$ game is a draw ("column convexity").

Hypothesis 3. If the $n_1^{k_1}$ game is a draw, then the n^k game is a draw for all $k \leq k_1$ and all $n \geq n_1$.

Clearly, Hypothesis 3 is true if and only if Hypotheses 1 and 2 are both true. In this case, the union of regions 1, 2 and 3, and the union of regions 4 and 5, are both connected and simply connected.

The next result is not quite so obvious.

Theorem 3. *If, for specific k_1 and even n_1, Hypothesis 1 is true, then Hypothesis 2 is true.*

Proof. Interpreting geometrically the binomial expansion

$$(n_1 + 1)^{k_1} = n_1^{k_1} + k_1 n_1^{k_1 - 1} + \binom{k_1}{2} n_1^{k_1 - 2} + \cdots + k_1 n_1 + 1,$$

we see that the $(n_1 + 1)^{k_1}$ hypercube can be decomposed, relative to its k_1 mid-hyperplanes (which are unique because $n_1 + 1$ is odd) into $2^{k_1} - 1$ hypercubes of side n_1 and dimension j, $1 \leq j \leq k_1$, plus the unique central cell. For example, the single $n_1^{k_1}$ hypercube results from squeezing back together the 2^{k_1} pieces which are left after all the mid-hyperplanes have been removed; the k_1 hypercubes of size $n_1^{k_1 - 1}$ result from squeezing back together the cells that are, respectively, in each of the k_1 hyperplanes but not in t wo or more; etc. Such a decomposition of the 5^3 hypercube ($n_1 = 4$, $k_1 = 3$) is shown in Figure 10. (This figure is used only to indicate how the decomposition works, and *not* to suggest that the 4^3-game is a draw, which in fact it is not.)

```
A A C A A     A A C A A     D D G D D     A A C A A     A A C A A
A A C A A     A A C A A     D D G D D     A A C A A     A A C A A
B B E B B     B B E B B     F F X F F     B B E B B     B B E B B
A A C A A     A A C A A     D D G D D     A A C A A     A A C A A
A A C A A     A A C A A     D D G D D     A A C A A     A A C A A
```

Figure 10. The 5^3 (hyper)cube decomposed into: one 4^3 (hyper)cube (the A-cells); three 4^2 hypercubes (i.e., squares), indicated by the letters B,C, and D, respectively; three 4^1 hypercubes (i.e., lines), indicated by the letters E, F, and G, respectively; and the central cell, X.

Using Hypothesis 1, the second player draws by always replying to the first player in the same sub-hypercube where the first player has just moved, and using the drawing strategy for that sub-hypercube. (If the first player ever occupies the central cell, the second player then gets a "free move", which may lead to additional free moves later on, as in the proof that Region 5 is "row-column convex".) Because $n_1 + 1$ is odd, every path in the $(n_1 + 1)^{k_1}$ hypercube requires a winning path in one of the $2^k - 1$ sub-hypercubes, and by the strategy just described, no such path will ever be completed by the first player. $\square$

Notes. 1. It appears that Hypothesis 1 may be easier to prove than the n_1-is-odd case of Hypothesis 2.

2. Analogous to the theorem just proved, it is possible to show that "$n_1^{k_1}$ is a draw" implies "$(n_1 + 2)^{k_1}$ is a draw" whether n_1 is even or odd. (This greatly reduces the uncertaint y in the shape of the boundary between Region 3 and Region 4.)

3. We can prove the assertion in Note 2 as follows: 1) We embed the $n_1^{k_1}$ hypercube in an $(n_1 + 2)^{k_1}$ hypercube, as in the geometric approach to counting winning paths. We then deal with the $2k_1$ added hypersurfaces in much the same

way as with the k_1 mid-hyperplanes in the previous theorem. (This time, the sub-hypercubes of dimensions $j = 1, 2, \ldots, k_1$, are still geometrically connected, and don't need to be squeezed back together.) There are now 2^{k_1} zero-dimensional vertices, so if the first player ever claims one of them, the second player can claim the diametrically opposite one (just to give a rule for dealing with free moves).

4. In Table 1, Region 5 is "row-column convex" and progagates mostly vertically. In a strongly analogous sense, Region 2 is also "row-column convex" and progagates mostly horizontally.

9. Additional Notes

(i) On the 3^3 board, the central cell is so powerful that if the first player is forbidden to occupy it on his initial move, the second player wins by occupying it in reply (in the normal game). The entire game tree is easily searched "by hand".

(ii) A further generalization of the n^k game with n in a row is to the n^k game with r in a row. That is, the first player to claim r consecutive cells along any straight path is the winner (or, in the misère version, the loser). Games that are dull draws for given n and k may become interesting when it is merely required to get r-in-a-row (for some $r < n$). This generalization provides a common framework for Tic-Tac-Toe and Go Moku.

References

[1] Patashnik, O., "Qubic: $4 \times 4 \times 4$ Tic-Tac-Toe", *Mathematics Magazine*, vol. 53, no. 4, Sept. 1980, 202–216.

[2] Hales, A. W., and Jewett, R. I., "Regularity and Positional Games," *Transactions of the American Mathematical Society*, vol. 106, no. 2, Feb. 1963, 222–229.

[3] Beck, J., "On Positional Games", *Journal of Combinatorial Theory (Series A)*, vol. 30, no. 2, March, 1981, 117–133.

[4] Berlekamp, E. R., Conway, J. H., and Guy, R. K., *Winning Ways for Your Mathematical Plays*, Academic Press, 1982; see especially Vol. II, pp. 667–679.

[5] Beck, J., "Games, Randomness and Algorithms," in *The Mathematics of Paul Erdős, I*, R. L. Graham and J. Nešetřil, eds., Springer-Verlag, 1997.

Supplemental Annotated Bibliography

1. Reference [2] first appeared, with the same title and authors, as Jet Propulsion Laboratory Report no. 32–134, January 31, 1962.

2. Several results presented at the January, 1972, MAA meeting in Las Vegas, Nevada (including the pairing pattern on the 8^3, and how to enlarge a pairing from n^k to $(n+1)^k$) were contained in a letter dated April 6, 1970, from S. W. Golomb to J. L. Selfridge. Some of this material was included in [4].

3. Three of Martin Gardner's *Mathematical Games* columns in *Scientific American* dealing with aspects of generalized Tic-Tac-Toe were:

   a. March, 1957, included in *The Scientific American Book of Mathematical Puzzles and Diversions* (Simon and Schuster, 1959) is rather elementary, but does describe winning paths on the 4^4 game (without the proof that draws are possible).

   b. August, 1971, included in *Wheels, Life, and Other Mathematical Amusements* (Freeman, 1983), Chapter 9, describes pairing strategies on n^2 boards for all $n \geq 5$.

   c. April, 1979, included in *Fractals, Hypercards, and More* (Freeman, 1992) deals with the generalization of Tic-Tac-Toe to polyominoes.

4. Reference [5], József Beck's chapter "Games, Randomness and Algorithms", in *The Mathematics of Paul Erdős, I*, edited by R. L. Graham and J. Nešetřil, Springer-Verlag, 1997, has what is probably the most up-to-date published results on n^k-hypercube Tic-Tac-Toe. One example is that the second player can draw (not necessarily by a pairing strategy) provided that $n > (\log_2 3 + \varepsilon)k$ and $n > n_0(\varepsilon)$, which is asymptotically better than the $n > \frac{2k}{\ln 2}$ conjecture first proposed in Reference [2]. (The underlying method for this improved result is attributed to Erdős and Selfridge.)

   In recent, as yet unpublished work, Beck has shown that the second player can force a draw if $k = O(n^2/\log n)$, improving his previous result of $k = O(n^{3/2})$. This is close to best possible, in a sense, since it is known that if $k = \Omega(n^2)$ then the first player has a "weak win", i.e. can occupy n-in-a-line, though not necessarily first.

   See, also, P. Erdős and J. L. Selfridge, "On a combinatorial game", *Journal of Combinatorial Theory, Series A*, vol. 14, no. 3, May 1973, 298–301.

5. Another reference is: J. L. Paul, *Tic-Tac-Toe in n-dimensions*, Mathematics Magazine, vol. 51, no. 1, Jan. 1978, 45–49.

6. A more elementary reference is: Mercer, G. B., and Kolb, J. R., "Three-dimensional ticktacktoe", *Mathematics Teacher*, vol. 64, no. 2, February, 1971, 119–122. (They show that the number of winning paths on the $n \times n \times n$ cube is $3n^2 + 6n + 4$, without any of the more general insights contained in Theorem 1.)

Historical Note: Except for the "conjecture" on n_k with its counterexamples (from 1999) in Section 4, the even more recent work described in Section 8, and the results cited in references [1], [3], [4], and [5], the work in this paper predates 1972. Some results already appeared in [2]. Other results were presented at the January, 1972, meeting of the MAA in Las Vegas, Nevada (in a session titled

"Players, Probabilities, and Profits", on the morning of 21 January) by S. W. Golomb, in a talk titled "Games Mathematicians Play".

SOLOMON W. GOLOMB
UNIVERSITY PROFESSOR
UNIVERSITY OF SOUTHERN CALIFORNIA, EEB-504A
3740 MCCLINTOCK AVENUE
LOS ANGELES, CA 90089-2565
UNITED STATES
  c/o milly@usc.edu

ALFRED W. HALES
CENTER FOR COMMUNICATIONS RESEARCH
4320 WESTERRA COURT
SAN DIEGO, CA 92121
UNITED STATES
  hales@ccrwest.org

More Games of No Chance
MSRI Publications
Volume **42**, 2002

Transfinite Chomp

SCOTT HUDDLESTON AND JERRY SHURMAN

ABSTRACT. Chomp is a Nim-like combinatorial game played in $\mathbb{N}^d$ or some finite subset. This paper generalizes Chomp to transfinite ordinal space Ω^d. Transfinite Chomp exhibits regularities and closure properties not present in the smaller game. A fundamental property of transfinite Chomp is the existence of certain initial winning positions, including rectangular positions $2 \times \omega$, $3 \times \omega^\omega$, $2 \times 2 \times \omega^3$, and $2 \times 2 \times \omega \times \omega$. Many open questions remain for both transfinite and finite Chomp.

Introduction and Notation

In the game of Chomp, cookies are laid out at the lattice points $\mathbb{N}^d$ where $\mathbb{N}$ denotes the natural numbers and play is in $d \in \mathbb{Z}^+$ dimensions. The cookie at the origin is poisonous. Two players alternate biting into the configuration, each bite eating the cookies in an infinite box from some lattice point outward in all directions, until one player loses by eating the poison cookie. The game can start from a position with finitely many bites already taken from $\mathbb{N}^d$ rather than from all of $\mathbb{N}^d$. Chomp was invented by David Gale in [Ga74] and christened by Martin Gardner. When Chomp begins from a finite rectangle it is isomophic to an earlier game, Divisors, due to Schuh [Sch52]. See also [BCG82b], pp.598–606.

This paper considers Chomp on Ω^d where Ω denotes the ordinals, a subject the first author began studying in the early 1990s. This transfinite version of Chomp has been mentioned in *Mathematical Intelligencer* columns [Ga93; Ga96]; these are anthologized in [Ga98].

Identifying each ordinal a with the set $\Omega_a = \{x \in \Omega : x < a\}$, the sets

$$a, \qquad \bar{a} = \{x \in \Omega : x \geq a\} \quad \text{for } a \in \Omega,$$
$$a \times b, \qquad \overline{(a,b)} = \bar{a} \times \bar{b} \qquad \text{for } (a,b) \in \Omega^2$$

are the boxes at the origin and the Chomp bites in one and two dimensions. Similarly in d dimensions, the Chomp boxes and bites are the corresponding

---

1991 *Mathematics Subject Classification.* 91A46, 91A44, 03E10, 03D60.

$v_1 \times \cdots \times v_d$ and $\overline{v} = \overline{v}_1 \times \cdots \times \overline{v}_d$ for $v \in \Omega^d$. Every Chomp position X is a finite union of boxes, and conversely. As we will see, every Chomp game must terminate after finitely many bites.

Working in the ordinals gives a more satisfyingly complete picture of Chomp. For example, a position of two equal-height columns, $X = 2 \times h$, is an N-position (next player wins) for all $h \in \mathbb{Z}^+$, but it is a P-position (previous player wins) for $h = \omega$, and then it is an N-position again for all $h > \omega$. For three equal-height columns, $3 \times h$ is a P-position if and only if $h = \omega^\omega$. Another result is that the six-dimensional position $\omega \times \omega \times \omega \times 2 \times 2 \times 2$—i.e., the cartesian product of $\mathbb{N}^3$ with a $2 \times 2 \times 2$ cube—is a P-position.

The join operator on Chomp positions is denoted "+," the difference operator "−," and "$\subset$" denotes proper subposition. Every Chomp bite is a difference operation

$$X \mapsto X - \overline{v} \qquad \text{for some } v \in \Omega^d,$$

valid only for positions X such that $X - \overline{v} \subset X$. Since parts of the analysis use the last direction for special purposes, vectors $v \in \Omega^k$ are often written $v = (u, k)$ with $u \in \Omega^{d-1}$ and $k \in \Omega$.

The minimal excluded element operator is denoted mex. Thus for any proper subset $S \subset \Omega$,

$$\mathrm{mex}(S) = a, \text{ where } a \notin S \text{ and } x \in S \text{ for all } x < a.$$

1. The Ordinals, Very Briefly

This section gives the bare basics on ordinals needed to read the paper.

The ordinal numbers Ω with ordinal addition $\uplus$ and ordinal multiplication $\star$ extend the natural numbers $(\mathbb{N}, +, \cdot)$ to the infinite. The operations are associative and satisfy the distributive property, but they are not commutative. Every ordinal x has an ordinal successor $x \uplus 1$ and the supremum of every set of ordinals is an ordinal. The finite ordinals are just $\mathbb{N}$. The first infinite ordinal is $\omega = \sup(\mathbb{N})$. Infinite ordinals include ω, $\omega \uplus 1$, $\omega \uplus 2$, ..., $\omega \star 2$, $\omega \star 2 \uplus 1$, ..., $\omega \star 3$, ..., ω^2, $\omega^2 \uplus 1$, ..., $\omega^2 \star 2$, ..., $\omega^2 \star 3 \uplus \omega \star 5 \uplus 19$, ..., ω^3, ..., ω^4, ..., ω^ω, ...

The ordinals are totally ordered and they are well founded, meaning every nonempty subset contains a minimal element and so the mex operator makes sense.

Every nonzero ordinal x can be written uniquely as

$$x = \hat{c} \uplus \omega^{e_0} \star c_0$$

with e_0 an ordinal, $0 < c_0 < \omega$, and $\hat{c}$ an ordinal multiple (possibly 0) of $\omega^{e_0 \uplus 1}$. Recursively expanding $\hat{c}$ as long as it is nonzero gives a unique expression

$$x = \omega^{e_k} \star c_k \uplus \omega^{e_{k-1}} \star c_{k-1} \uplus \cdots \uplus \omega^{e_1} \star c_1 \uplus \omega^{e_0} \star c_0$$

for some finite k, with $e_k > e_{k-1} > \cdots > e_1 > e_0$ a descending chain of ordinals, and $0 < c_i < w$ for $i \in \{0, \ldots, k\}$. This form for an ordinal number is its base ω expansion, also known as its Cantor normal form.

Instead of ordinal addition and multiplication, we nearly always use commutative operators called *natural* addition and multiplication, denoted by ordinary "+" and "·". Natural addition of two ordinals written in base ω simply adds coefficients of equal powers of ω, where missing terms are taken to have coefficient 0. Ordinal addition satisfies

$$\omega^e \uplus y = \begin{cases} \omega^e + y & \text{if } y < \omega^{e+1} \\ y & \text{if } y \geq \omega^{e+1}. \end{cases}$$

This property and associativity completely define ordinal addition. Though we seldom use ordinal addition, explicitly noting when we do so, we do make a point of arranging natural operations to agree with the ordinal operations, e.g., writing $\omega + 1$ (which equals $\omega \uplus 1$) rather than $1 + \omega$ (which does not equal $1 \uplus \omega = \omega$), and writing $\omega \cdot 2$ (which equals $\omega \star 2$) rather than $2 \cdot \omega$ (which does not equal $2 \star \omega = \omega$).

2. Size

Every Chomp position X has an ordinal size, denoted $\text{size}(X)$.

To compute size, start by expressing X as a finite overlapping sum of boxes at the origin. Each side of each box is uniquely expressible as a finite sum of powers of ω (including $1 = \omega^0$), e.g., $\omega \cdot 2 + 3 = \omega + \omega + 1 + 1 + 1$. (Here and elsewhere we use so-called "natural addition" or "polynomial addition" of ordinals, as compared to concatenating order types, or "ordinal addition.") This decomposition of the sides induces a unique decomposition of each box as a finite disjoint sum of translated boxes all of whose sides are powers of ω. Construct the finite set S of translated boxes that decompose the boxes of X, and then remove any box contained in some other box of S, creating a new set S'. Then $\text{size}(X)$ is just the sum of the sizes of the elements of S', with the size of each box being the product of the lengths of its sides.

For example, when X is finite, $\text{size}(X)$ is just the number of points in X. In this case, S consists of distinct unit boxes, eliminating repeats since it is a set rather than a multiset. And S' is simply S again since there are no inclusions of distinct unit boxes.

As another example, consider the position $X = (\omega + 1) \times 2 + (\omega \cdot 2) \times 1$. Decomposing the summands yields (using "@" to specify translation)

$$S_{(\omega+1)\times 2} = \{(\omega \times 1)@(0,0), (\omega \times 1)@(0,1), (1 \times 1)@(\omega,0), (1 \times 1)@(\omega,1)\},$$
$$S_{(w\cdot 2)\times 1} = \{(\omega \times 1)@(0,0), (\omega \times 1)@(\omega,0)\},$$

and

$$S' = S_{(\omega+1)\times 2} \cup S_{(w\cdot 2)\times 1} - \{(1 \times 1)@(\omega, 0)\},$$

because the subtrahend is a subset of $(\omega \times 1)@(\omega, 0)$ in $S_{(w\cdot 2)\times 1}$. Thus $\text{size}(X) = \text{size}(S') = \omega \cdot 3 + 1$. (See Figure 1, where the including box is shaded.)

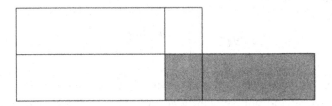

Figure 1. A Chomp position of size $\omega \cdot 3 + 1$.

As a third example, let $X = 1 \times \omega \times \omega^2 + \omega \times \omega^2 \times 1 + \omega^2 \times 1 \times \omega$. (See Figure 2.) Then $\text{size}(X) = \omega^3 \cdot 3$. Here S' contains just three terms, the summands of X.

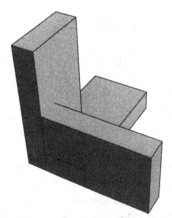

Figure 2. A Chomp position of size $\omega^3 \cdot 3$.

This example illustrates the difficulty of trying to compute size from a fully disjoint decomposition: there are finite fully disjoint decompositions of X, but they all have component sums exceeding $\text{size}(X)$; and there are fully disjoint decompositions whose component sums come to $\text{size}(X)$, but they are all infinite. In our construction, the finite decomposition used to compute $\text{size}(X)$ is semi-disjoint, meaning whenever t wo elements of S' have nonempty intersection, then the size of the intersection is at least a factor of ω less than the size of either intersectand.

If Y is reachable from X by one bite, then $\text{size}(Y) < \text{size}(X)$. To see this, note that S'_Y changes one or more components in S'_X, either removing them or replacing them by components with smaller total sum. It follows that Chomp terminates in finitely many moves.

3. Grundy Values

The Grundy value function on Chomp positions,

$$G : \{\text{Chomp positions in } \Omega^d\} \longrightarrow \Omega,$$

is

$$G(X) = \text{mex}\{G(Y) : Y \text{ can be reached from } X \text{ in one bite}\}.$$

In particular, the poison cookie has Grundy value 1, and a column of single cookies has Grundy value equal to its height. Only the empty position has Grundy value 0 since it is reachable from any other position.

A Chomp position X is a P-position if and only if $G(X) = 1$. The case $X = \varnothing$ is clear. As for nonempty X, observe that if $G(X) = 1$ then every nonempty Y left by a bite into X satisfies $G(Y) > 1$, so some second bite into Y gives a nonempty position X' with $G(X') = 1$ again. (In general, any Grundy value-increasing Chomp bite is *reversible* in this fashion—a fact we will exploit several times.) So if $G(X) = 1$, the previous player wins by reversing bites until the next player is left with the poison cookie; and if $G(X) > 1$, the next player wins by biting X down to Y with $G(Y) = 1$.

A simple upper bound on Grundy values is clear: since Chomp bites decrease position size, it follows by induction that $G(X) \leq \text{size}(X)$ for all positions X.

Though it only seems to matter whether or not Grundy values are 1, knowing them in general will let us construct P-positions and execute winning strategies.

4. Other Termination Criteria

The astute reader may reasonably object that P-positions should have Grundy value 0.

In fact, Chomp can be defined with different termination conditions, leading to definitions of Grundy value different from the unrestricted definition used so far here. (We will use a certain restricted Grundy value later in this paper.)

The poison cookie description of Chomp can be viewed in two ways. First, the poison cookie specifies 1-restricted Chomp, i.e., the bite $\overline{(0,\ldots,0)}$ is forbidden and the last bite wins. The restricted Grundy values for this game are smaller by 1 than unrestricted Grundy values in the finite case (so now a P-position is specified by Grundy value 0 as it should be), but they catch up at $G = \omega$ and are equal from then on. In a larger theory of Grundy values, the poison cookie bite has the special value "loony" as defined in [BCG82a], Chapter 12.

Second, the poison cookie describes unrestricted misère Chomp, i.e., all bites are allowed but the last bite loses. In this context, an unrestricted Grundy value of 1 is the natural P-position criterion. Unrestricted Chomp Chomp is "tame" in the sense of [BCG82a], Chapter 13, so its misere analysis is tractable.

One can restrict Chomp more generally, disallowing a set of moves, and one can play a misere version of restricted Chomp. Even 1-restricted Chomp is not tame, however (e.g., the 1-restricted misere position $(3 \times 1) + (1 \times 3)$, like the misere Nim sum $2 + 2$, does not reduce to a Nim heap), and its misere analysis already requires general misere game theory.

For an isolated game, 1-restricted Chomp and unrestricted misere Chomp are the same. But a sum of Chomp positions played unrestricted misere is not equivalent to the same sum played 1-restricted: in the misere sum, only the last poison cookie is fatal. The 1-restricted sum is a P-position if and only if the 1-restricted Grundy values of the components have Nim sum 0. The unrestricted misere sum is equivalent to misere Nim, where the sum is a P-position if and only if the Nim sum of the unrestricted Grundy values equals

$$\begin{cases} 1 & \text{if every component has Grundy value 0 or 1,} \\ 0 & \text{if any component has Grundy value 2 or more.} \end{cases}$$

5. The Fundamental Theorem

The fundamental theorem of Chomp requires a construction. Let $d > 1$ be a finite ordinal, let A be a d-dimensional Chomp position, and let B (standing for "base") be a nonnull $(d-1)$-dimensional Chomp position. For any ordinal h, let $E(A, B, h)$ denote the d-dimensional Chomp position consisting of A and an infinite column over B in the last direction, the whole thing then truncated in the last direction at height h. That is,

$$E(A, B, h) = A + (B \times \Omega) - \overline{(0, \dots, 0, h)}.$$

Call this the *extension of A by B to height h*. (See Figure 3 for two examples with the same A and B; in the second example the truncation eats into A.)

Theorem 5.1 (Fundamental Theorem of Chomp). *Suppose d, A, and B are given, where $d > 1$ is a finite ordinal, A is a d-dimensional Chomp position, and B is a nonnull $(d-1)$-dimensional Chomp position. Then there is a unique ordinal h such that $E(A, B, h)$, the extension of A by B to height h, is a P-position, meaning the second player wins.*

Once one realizes this, the proof is almost self-evident: extend A by B to the first—and only—height that doesn't give an N-position. Here is the formal argument:

Uniqueness is easy, granting existence. Let h be the least ordinal such that $E(A, B, h)$ is a P-position, and consider any greater ordinal, $k > h$. Since $E(A, B, k)$ reaches $E(A, B, h)$ in one bite, it is an N-position.

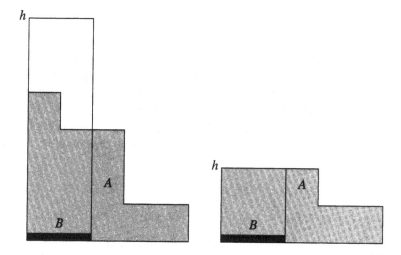

Figure 3. Extensions of A by B to height h.

In showing that h exists we further show how (in principle) to find it. We will construct a height function H on suitable pairs A, B of Chomp positions, such that

$$H(A, B) = h \quad \text{if} \quad E(A, B, h) \text{ is a P-position.}$$

The construction is by an outer induction on B and an inner induction on A.

The outer basis case is the poison cookie $B = 1^{d-1} \subset \Omega^{d-1}$, denoted B_0.

The inner basis case is $A = \varnothing$. The extension of $\varnothing$ by B_0 to height 1, $E(\varnothing, B_0, 1)$, gives the d-dimensional poison cookie, a P-position. Thus we have $H(\varnothing, B_0) = 1$.

For the inner induction step, take nonnull A and assume that for each $A' \subset A$, some extension $E(A', B_0, H(A', B_0))$ is a P-position. Let h be the minimal nonzero excluded member from the set of such prior P-position heights $H(A', B)$ as A' ranges over the positions reachable from A by one bite, said bite not eating into the column over B_0. Thus $h = \mathrm{mex}(M_0)$ where

$$M_0 = \{0\} \cup \{h : h = H(A - \bar{v}, B_0),\ A - \bar{v} \subset A,\ v = (u, k),\ B_0 - \bar{u} = B_0,\ h > k\}.$$

Then $E(A, B_0, h)$ is a P-position. For if it were an N-position, then some bite would take it to a P-position, $E(A, B_0, h) - \bar{v}$ for $v = (u, k)$ with $k < h$. There are t wo cases:

(i) If $B_0 - \bar{u} = B_0$ then the bite eats into $E(A, B_0, h)$ without truncating the column over B_0, leaving $E(A - \bar{v}, B_0, h)$ with the bite $\bar{v}$ as in the definition of M_0. This is an N-position since $h \notin M_0$ means $h \neq H(A - \bar{v}, B_0)$.

(ii) If $B_0 - \bar{u} = \varnothing$ then the bite truncates $E(A, B_0, h)$ in the last direction, leaving $E(A, B_0, k)$ with $k < h$; we may assume $k > 0$ else the game was just lost. This is an N-position since $k = H(A - \bar{v'}, B_0)$ for some bite $\bar{v'}$

as in the definition of M_0, and said bite takes $E(A, B_0, k)$ to the P-position $E(A - \overline{v'}, B_0, k)$.

So $H(A, B_0)$ is defined by for all A. Note how the first case relies on h being excluded from M_0, while the second relies on all $k < h$ belonging to M_0.

Returning to the outer induction, suppose that for some B, the height function $H(A', B')$ is defined for all pairs A', B' satisfying either of (i) $\varnothing \subset B' \subset B$, or (ii) $B' = B$ and $A' \subset A$. To define $H(A, B)$, construct a sequence of sets M_i, none of whose elements can serve as $H(A, B)$, and show that $\mathrm{mex}(M_\omega)$ does so. So, starting from an M_0 as above,

$$M_0 = \{0\} \cup \{h : h = H(A - \overline{v}, B), \ A - \overline{v} \subset A, \ v = (u, k), \ B - \overline{u} = B, \ h > k\},$$

adjoin for each succeeding ordinal $i + 1$ the heights of certain P-positions,

$$M_{i+1} = \left\{ \begin{array}{c} h : h = H(A + E(A, B, k) - \overline{v}, B - \overline{u}), \\ v = (u, k), \ \varnothing \subset B - \overline{u} \subset B, \ k \in M_i, \ h > k \end{array} \right\},$$

and let

$$M_\omega = \bigcup_{i < \omega} M_i.$$

Let $h = \mathrm{mex}(M_\omega)$. Then $E(A, B, h)$ is a P-position. For as before, if it were an N-position, then some bite would leave a P-position, $E(A, B, h) - \overline{v}$ for $v = (u, k)$ with $k < h$. This time there are three cases:

(i) If $B - \overline{u} = B$ then the bite eats into $E(A, B, h)$ without truncating the column over B, leaving $E(A - \overline{v}, B, h)$ with the bite $\overline{v}$ as in the definition of M_0. This is an N-position as before.

(ii) If $\varnothing \subset B - \overline{u} \subset B$ then the bite eats into part of the column over B, leaving $A + E(A, B, k) - \overline{v}$. Since $k < h$, we have $k \in M_i$ for some i. Let $h' = H(A + E(A, B, k) - \overline{v}, B - \overline{u})$; either $h' > k$, so $h' \in M_{i+1}$ and thus $h \neq h'$, or $h' \leq k < h$ and again $h \neq h'$; in either case, the bite has left an N-position.

(iii) If $B - \overline{u} = \varnothing$ then the bite truncates $E(A, B, h)$ in the last direction, leaving $E(A, B, k)$ for some $k < h$; we may assume $k > 0$ else N has just lost. If $k \in M_0$ then a bite as in the definition of M_0 leaves a P-position; if $k \in M_{i+1}$ for some i then a bite as in the definition of M_{i+1} leaves a P-position.

This completes the proof. This outer inductive step actually covers the basis case as well: when $B = B_0$, the construction of M_ω simply gives M_0 since all other M_i are empty.

Corollary 5.2 (Size Lemma). *For any nonempty d-dimensional Chomp position A, let $A - (B_0 \times \Omega)$ denote all of A except its intersection with the tower over the $(d-1)$-dimensional poison cookie B_0. Then*

$$H(A, B_0) \leq 1 + \mathrm{size}(A - (B_0 \times \Omega)).$$

In particular, if $A - (B_0 \times \Omega)$ is finite then so is $H(A, B_0)$.

This follows from the proof of the Fundamental Theorem since $H(A, B_0)$ is the minimal excluded element from the set M_0, and every element of M_0 is less than size$(A - (B_0 \times \Omega))$ by induction.

6. Two Constructions

Along with extending Chomp positions to P-positions, The Fundamental Theorem can be used to find Grundy values and to construct extensions with arbitrary Grundy values.

To find the Grundy value of an arbitrary position, let X be d-dimensional and raise it to height 1 in the $(d+1)$st dimension, creating the position $A = X \times 1$; note that A is essentially the same thing as X, i.e., the bites out of A and X correspond perfectly and so $G(A) = G(X)$. Apply the Fundamental Theorem to A and $B = B_0$, the poison cookie in $d+1$ dimensions; this creates a P-position Y from the original X by adding an orthogonal column of single cookies of some height h,

$$Y = (X \times 1) + (1^d \times h).$$

(See Figure 4.) In fact, the column has height $h = G(X)$, because in that case the previous player wins by a pairing strategy: if the next bite is from $X \times 1$ leaving a Grundy value smaller than $G(X)$, bite the orthogonal column down to the same value; if the next bite is from $X \times 1$ and reversible, i.e., it leaves a larger Grundy value, then bite into what's left of $X \times 1$ restoring the Grundy value to $G(X)$; if the next bite is from the orthogonal column, reducing its Grundy value below $G(X)$, then bite into $X \times 1$ reducing its Grundy value by the same amount. So we have constructed a column of the desired height $G(X)$. This sort of pairing strategy will be used frequently throughout the paper.

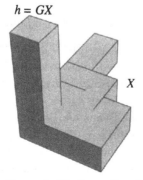

Figure 4. Finding a Grundy value.

To extend a position and get an arbitary Grundy value g, start with X as before and carry out a similar construction, only the column of single cookies extends to height g in a prepended zeroth dimension:

$$A = (1 \times X) + (g \times 1^d).$$

Again apply the Fundamental Theorem to A and any nonempty $1 \times B$ in dimensions $0, \ldots, d$, extending A in the last direction to get a P-position,

$$Y = E((1 \times X) + (g \times 1^d), 1 \times B, h).$$

(See Figure 5, where B is taken to be B_0, the poison cookie.) Again the win is by a pairing strategy, showing that the d-dimensional extension of X,

$$E(X, B, h),$$

has the same Grundy value as the orthogonal column, i.e., the desired value g.

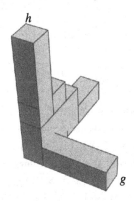

Figure 5. Constructing a position with Grundy value g.

Proposition 6.1 (Beanstalk Lemma). *Let A be any finite d-dimensional Chomp position, and let h be any infinite ordinal. Then the Chomp position*

$$A + (1^{d-1} \times h),$$

obtained by adding a tower of height h to A, has infinite Grundy value.

To see this, note that for every finite ordinal g, the second construction just given—adding a tower to A over B_0 to obtain a position with Grundy value g—adds a tower of finite height by the Size Lemma. Adding an infinite tower to A thus gives an infinite Grundy value.

A more general statement of the Beanstalk Lemma is that for any ordinal h, $G(A + (1^{d-1} \times h))$, where A is finite, has the same highest term as h. That is, if $h = \omega^i \cdot a_i + \cdots$ with $0 < a_i < \omega$ then also $G(A + (1^{d-1} \times h)) = \omega^i \cdot a_i + \cdots$. This phenomenon of the dominant term of size determining the dominant term of Grundy value is called *stratification*, and it is ubiquitous in transfinite Chomp analysis. By contrast, the dominant term of size actually equalling the dominant term of Grundy value is particular to this case, an artifact of a single column's Grundy value being its height.

7. P-Ordered Positions

Define a Chomp position to be *P-ordered* if its P-subpositions are totally ordered by inclusion. Every non-P-ordered position contains a minimal non-P-ordered position $(3 \times 1) + (2 \times 2) + (1 \times 3)$ or $(2 \times 1 \times 1) + (1 \times 2 \times 1) + (1 \times 1 \times 2)$, so every P-position must be, up to congruence, a subposition of $2 \times \Omega$ or $(1 \times \Omega) + (\Omega \times 1)$. Thus the complete list of P-ordered P-positions, up to rotation and inclusion, is

$$2 \times \omega, \{(1 \times (i+1)) + (2 \times i) : 0 \leq i < \omega\}, \{(1 \times a) + (a \times 1) : 0 < a\}.$$

Theorem 7.1. *If X is any Chomp position and P is any P-ordered P-position, then $G(X \times P) = G(X)$. In particular, if X is a P-position, then so is $X \times P$.*

To prove this, let $g = G(X)$, construct the product $X \times P$, and prepend to this main body a column of g single cookies in the zeroth direction, constructing the position

$$Y = (1 \times X \times P) + (g \times 1 \times 1).$$

(These triple products and others to follow refer to $\Omega \times \Omega^d \times \Omega^e$ where $X \subset \Omega^d$ and $P \subset \Omega^e$; thus "1" often will mean 1^d or 1^e.) The idea is to show that Y is a P-position by exhibiting a winning strategy.

Let $Y_0 = Y$. For $i \geq 0$, let Y_{2i+1} be the result of an arbitrary bite applied to Y_{2i} and let Y_{2i+2} be the result of a to-be-specified bite applied to Y_{2i+1}. The specified bite will maintain two invariants of the even positions Y_{2i}:

(I1) Vertical sections of the main body are P-positions, i.e., for all $z \in \Omega$, $Y_{2i} \cap (1 \times \{z\} \times P)$ either is empty or is a P-position. (Strictly speaking, it needs to be translated to the origin — we're being a bit casual to avoid even more notation.)

(I2) What's left of the prepended column retains the same Grundy value as what's left of X, i.e., $G(Y_{2i} \cap (g \times 1 \times 1)) = G(Y_{2i} \cap (1 \times X \times 1))$.

The invariants clearly hold for Y_0.

Bites into Y_{2i} fall under three cases:

(i) $\overline{(0, x, p)}$ with $p > 0$. This bite preserves invariant (I2). Answer it with $\overline{(0, x, p')}$, where p' is the bite into P that restores the section $Y_{2i+1} \cap (1 \times \{x\} \times P)$ to a P-position $Y_{2i+2} \cap (1 \times \{x\} \times P)$. The bite $\overline{(0, x, p)}$ may have left other sections $Y_{2i+1} \cap (1 \times \{\xi\} \times P)$ not P-positions for $\xi \in \overline{x}$; but the answering bite restores all such sections in Y_{2i+2} to P-positions, thus restoring invariant (I1) while preserving (I2). This property depends on P being P-ordered.

(ii) $\overline{(0, x, 0)}$ with $x \neq 0$, i.e., a truncation in the X-direction. This bite preserves invariant (I1). Let $g' = G(Y_{2i} \cap (1 \times X \times 1))$, and $g'' = G(Y_{2i+1} \cap (1 \times X \times 1))$; note $g' \neq g''$. Also note that $G(Y_{2i} \cap (g \times 1 \times 1)) = g'$ by invariant (I2). If $g'' < g'$, answer the bite by truncating the orthogonal column with $\overline{(g'', 0, 0)}$, restoring (I2). If $g'' > g'$, answer the bite with $\overline{(0, x', 0)}$, reversing the bite into X to make $G(Y_{2i+2} \cap (1 \times X \times 1)) = g'$, also restoring (I2). Either of these answering bites also preserves (I1).

(iii) $\overline{(h, 0, 0)}$, i.e., a bite into the orthogonal column. This bite preserves (I1). We had $G(Y_{2i} \cap (1 \times X \times 1)) > h$ before this bite, so answer it with any bite $\overline{(0, x, 0)}$ that makes $G(Y_{2i+2}) \cap (1 \times X \times 1) = h$, restoring (I2) while preserving (I1).

This completes the proof.

Applying the theorem twice, if X is any P-position and P_1, P_2 are P-ordered P-positions, then $X \times P_1 \times P_2$ is a P-position. In particular, letting $X = P_1 = P_2 = \omega \times 2$ and then permuting axes, $\omega \times \omega \times \omega \times 2 \times 2 \times 2$ is a P-position. The winning strategy is

- Answer $\overline{(r, t, i+1, s, u, 0)}$ with $\overline{(r, t, i, s, u, 1)}$ and vice versa.
- Answer $\overline{(r, i+1, 0, s, 0, 0)}$ with $\overline{(r, i, 0, s, 1, 0)}$ and vice versa.
- Answer $\overline{(i+1, 0, 0, 0, 0, 0)}$ with $\overline{(i, 0, 0, 1, 0, 0)}$ and vice versa.

8. Side-Top Positions

Consider two 2-dimensional Chomp positions S and T (for "side" and "top"), with S finite and T at most two cookies wide. Let $S@(2, 0)$ denote the translation of S rightward by 2 and let $T@(0, \omega)$ denote the translation of T upward by ω. Construct the Chomp position

$$U = (2 \times \omega) + S@(2, 0) + T@(0, \omega),$$

i.e., a 2-wide, ω-high column with finitely much added to its side and any amount to its top. (See Figure 6.) This section gives the criterion for U to be a P-position.

Doing so requires a certain notion of restricted Grundy value, cf. the section earlier in this paper of other termination criteria. Let $\mathcal{X}$ denote a collection of Chomp positions. The $\mathcal{X}$-restricted Grundy value function

$$G_{\mathcal{X}} : \mathcal{X} \longrightarrow \Omega$$

is defined as

$$G_{\mathcal{X}}(X) = \text{mex}\{G_{\mathcal{X}}(Y) : Y \in \mathcal{X}, Y \text{ can be reached from } X \text{ in one bite}\}.$$

Let $\square S$ denote $S@(2, 0)$ left-extended two units to the axis, i.e.,

$$\square S = (2 \times s) + S@(2, 0),$$

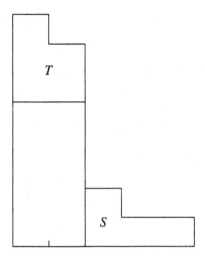

Figure 6. A side-top position.

where s is the height of S. Let $\mathcal{S}$ denote the set of all side positions $\tilde{S}$ such that $H(\square\tilde{S}, 2)$ is infinite.

Theorem 8.1 (Side-Top Theorem). *Let a side-top position $U = (2 \times \omega) + S@(2,0) + T@(0,\omega)$ be given. If $H(\square S, 2)$ is finite then U is an N-position. If $H(\square S, 2)$ is infinite then*

$$U \text{ is a P-position} \quad \Longleftrightarrow \quad G_{\mathcal{S}}(S) = G(T).$$

The first statement is clear, for if $H(\square S, 2)$ is finite then the bite truncating U to that height in the second direction wins. For the rest of the proof, $H(\square S, 2)$ is infinite.

To show "$\Longleftarrow$" of the second statement, assume $G_{\mathcal{S}}(S) = G(T)$. Take any bite $\overline{(a, b)}$, other than the complete bite $a = b = 0$ of course.

If b is infinite then the bite takes $T@(0, \omega)$ down to some $\tilde{T}@(0, \omega)$ and leaves the rest of U intact. Let $g = G(\tilde{T}) \neq G(T)$. If $g > G(T)$ then a further bite into $\tilde{T}@(0, \omega)$ reverses the first one, while if $g < G(T) = G_{\mathcal{S}}(S)$ then a bite into $S@(2, 0)$ takes it down to some $\tilde{S}@(2, 0)$ with $G(\tilde{S}) = g$ and $H(\square\tilde{S}, 2)$ infinite.

Now we consider b finite. If $a = 0$ then the bite leaves a finite position, thus leaving a losing position since $H(\square S, 2)$ is infinite. If $a = 1$ then the bite leaves a position with one infinite column plus a finite part. The Beanstalk Lemma from earlier shows that this has infinite Grundy value, meaning its Grundy value isn't 1, so it is a losing position.

Finally, if $a \geq 2$ then player N has bitten into $S@(2, 0)$, leaving position $(2 \times \omega) + \tilde{S}@(2, 0) + T@(0, \omega)$ for some finite $\tilde{S}$. Either $H(\square\tilde{S}, 2)$ is finite or it is infinite. If it is finite then the bite truncating the second direction to that height leaves a P-position. If it is infinite then $\tilde{S} \in \mathcal{S}$. Let $g = G(\tilde{S})$, so that $g \neq G_{\mathcal{S}}(S) = G(T)$; if $g < G(T)$ then bite into $T@(0, \omega)$, getting $\tilde{T}@(0, \omega)$ with

$G(\tilde{T}) = g$; if $g > G_S(S)$ then bite into $\tilde{S}@(2,0)$ to restore Grundy value $G_S(S)$. This covers all cases.

For "$\Longrightarrow$" of the second statement, let $G_S(S) \neq G(T)$. If $G_S(S)$ is larger, bite into $S@(2,0)$ to produce a P-position as just shown; if $G(T)$ is larger, bite into $T@(0,\omega)$ similarly.

9. Two-wide Chomp

A two-wide Chomp position takes the form $X = (1 \times h) + (2 \times k)$ with $h \geq k$. Thus

$$h = \omega^i \cdot u + a, \quad k = \omega^i \cdot v + b,$$

where $u > 0$, $u \geq v$, $a \geq b$ when $u = v$, $a < \omega^i$, $b < \omega^i$, and when $i < \omega$ then also $u < \omega^i$. This section gives the Grundy values $G(X)$, renotated for convenience

$$G(\omega^i \cdot u + a, \omega^i \cdot v + b).$$

When $i = 0$, the position is just two columns of finite heights $u \geq v$. Computing from first principles gives the following Grundy values shown in Table 1. Reading down the diagonals shows what's going on. For $u - v$ even, the Grundy

$G(u,v)$	$v = 0$	1	2	3	4	5	6	$\cdots$
$u = 0$	0							
1	1	2						
2	2	1	3					
3	3	4	1	5				
4	4	3	5	1	6			
5	5	6	4	7	1	8		
6	6	5	7	4	8	1	9	
7	7	8	6	9	4	10	1	$\cdots$
8	8	7	9	6	10	4	11	$\cdots$
9	9	10	8	11	7	12	4	$\cdots$
10	10	9	11	8	12	7	13	$\cdots$
11	11	12	10	13	9	14	7	$\cdots$

Table 1. Two-wide Grundy values

values skip every third value; for $u - v$ odd, they iterate at every second step for a while and then stabilize. The formula is

$$(9\text{-}1) \qquad G(u,v) = \begin{cases} u - v + \lfloor \frac{3v+1}{2} \rfloor & \text{if } u - v \text{ is even,} \\ \min\{u - v + \lfloor \frac{v}{2} \rfloor, \frac{3(u-v)-1}{2}\} & \text{if } u - v \text{ is odd.} \end{cases}$$

The next case is $i = 1$, i.e., the column heights are $\omega \cdot u + a$, $\omega \cdot v + b$, with all of u, v, a, b finite and $u > 0$, $u \geq v$, $a \geq b$ when $u = v$.

Consider the subcase $u = v$, $a = b = 0$. Biting only into the right column gives a Chomp position whose left column has greater order of magnitiude than its right; by an extension of the Beanstalk Lemma such a position has infinite Grundy value. On the other hand, biting into both columns and thus truncating the position at a lower height gives (by induction) a finite Grundy value. Thus, $G(\omega^i \cdot u, \omega^i \cdot u)$ is the mex of the set of Grundy values of smaller positions with two columns of equal height h. These run through

$$\mathbb{N} \setminus \{3k + 1\} \quad \text{as } 0 \le h < \omega,$$
$$\{3k + 1\} \setminus \{9k + 4\} \quad \text{as } \omega \le h < \omega \cdot 2,$$
$$\{9k + 4\} \setminus \{27k + 13\} \quad \text{as } \omega \cdot 2 \le h < \omega \cdot 3,$$

$$\vdots$$

$$\left\{ 3^{u-1}k + \frac{3^{u-1} - 1}{2} \right\} \setminus \left\{ 3^u k + \frac{3^u - 1}{2} \right\} \quad \text{as } \omega \cdot (u - 1) \le h < \omega \cdot u.$$

And thus $G(\omega, \omega) = 1$, $G(\omega \cdot 2, \omega \cdot 2) = 4$, $G(\omega \cdot 3, \omega \cdot 3) = 13$, and in general $G(\omega \cdot u, \omega \cdot u) = (3^u - 1)/2$. Each time the column heights reach a new multiple of ω, the Grundy values have filled up the naturals minus an arithmetic progression, with the complementary progression—an iterate of the initial $\{3k+1\}$—becoming sparser each time.

Continuing to assume $i = 1$ and $u = v$ but now allowing a and b nonzero gives the position $X = (2 \times \omega^i \cdot u) + ((1 \times a) + (2 \times b))@(0, \omega^i \cdot u)$. By the Beanstalk Lemma, the only new bites giving positions of small enough Grundy value to worry about are the bites into the finite top part. These bites give Grundy values from the beginning of the omitted arithmetic progression $\{3^u k + (3^u - 1)/2\}$, thus

$$(9\text{--}2) \qquad G(\omega \cdot u + a, \omega \cdot u + b) = 3^u G(a, b) + \frac{3^u - 1}{2}.$$

We already know the right side here, thanks to (1).

Still keeping $i = 1$ but now taking $u > v$, the formula is

$$(9\text{--}3) \qquad G(\omega \cdot u + a, \omega \cdot v + b) = \omega \cdot (u - v) + (a \oplus b) \quad \text{when } u > v,$$

where "$\oplus$" denotes Nim sum. This is easy to see when $b = 0$ and the position is $X = (2 \times \omega \cdot v) + (1 \times \omega \cdot (u - v) + a)@(0, \omega \cdot v)$: bites leaving two equal height columns give all finite Grundy values except an arithmetic progression; bites into the top part give $\omega \cdot (u - v) + a$ more Grundy values, filling in the missing progression and then all other values less than $\omega \cdot (u - v) + a$; and bites into the right column give larger Grundy values.

To prove (3) when $b > 0$, first note the recursive formula

$$G(\omega \cdot u + a, \omega \cdot v + b) = G(\omega \cdot (u - v) + a, b),$$

which follows from observing that taking base $B = 2 \times \omega \cdot v$ and top $T = (1 \times \omega \cdot (u-v)+a)+(1 \times b)@(1,0)$, the orthogonal sum of positions $B+T@(0,\omega\cdot v)$ and T (cf. various prior constructions) is a P-position. The pairing strategy is clear: match bites into either top part, and if $B + T@(0, \omega \cdot v)$ is bitten down to two columns of equal height less than $\omega \cdot v$, giving a position of finite Grundy value, the Beanstalk Lemma says that a matching bite into T gives the same Grundy value.

With the recursive formula established it suffices to prove (3) when the position is T. To do so, take the orthogonal sum of T with a single column of height $\omega \cdot (u - v) + (a \oplus b)$. This time the pairing strategy to win is: the Nim sum allows matching bites into the finite parts; the simple upper bound $G \leq$ size from Section 2 shows that other bites into T leave positions of Grundy value less than $\omega \cdot (u - v)$, matched by bites into the column; and the Beanstalk Lemma shows that $G(T) \geq \omega \cdot (u - v)$, so bites deeper than $a \oplus b$ into the column can be matched by bites into T.

The analysis for $2 < i < \omega$ is similar to $i = 1$. When $u = v$ the formula is

$$(9\text{-}4) \qquad G(\omega^i \cdot u + a, \omega^i \cdot u + b) = \omega^{i-1} \cdot u + G(a, b).$$

This is easy to establish first when $a = b = 0$, and then in general. When $u > v$, the formula is

$$(9\text{-}5) \qquad G(\omega^i \cdot u + a, \omega^i \cdot v + b) = \omega^i \cdot (u - v) + (a \oplus b).$$

Again this follows from the recursive formula

$$G(\omega^i \cdot u + a, \omega^i \cdot v + b) = G(\omega^i \cdot (u - v) + a, b),$$

which is established as above.

Finally, for $i \geq \omega$, the formula is

$$(9\text{-}6) \quad G((\omega^\omega)^j \cdot u + a, (\omega^\omega)^j \cdot v + b) = (\omega^\omega)^j \cdot G(u, v) + \begin{cases} G(a, b) & \text{if } u = v, \\ a \oplus b & \text{if } u > v. \end{cases}$$

Here $j > 0$, $0 < u < \omega^\omega$, $u \geq v$, $a < (\omega^\omega)^j$, $b < (\omega^\omega)^j$.

To see this, first let $j = 1$ and $a = b = 0$. Build up the formulas $G(\omega^\omega, \omega^\omega) = \omega^\omega \cdot 2 = \omega^\omega \cdot G(1, 1)$, $G(\omega^\omega \cdot 2, \omega^\omega) = \omega^\omega \cdot 1 = \omega^\omega \cdot G(2, 1)$, $G(\omega^\omega \cdot 2, \omega^\omega \cdot 2) = \omega^\omega \cdot 3 = \omega^\omega \cdot G(2, 2)$, ..., and in general,

$$G(\omega^\omega \cdot u, \omega^\omega \cdot v) = \omega^\omega \cdot G(u, v) \text{ for finite } u \geq v$$

by considering the bites into these configurations.

Similarly, $G(\omega^\omega \cdot \omega, \omega^\omega \cdot \omega) = \omega^\omega$: bites truncating both columns give positions with Grundy values $\omega^\omega \cdot (\mathbb{N} \setminus \{3k + 1\})$, and bites truncating the second column give positions with Grundy value on the order of $\omega^\omega \cdot \omega$. This continues on to $G(\omega^\omega \cdot \omega \cdot 2, \omega^\omega \cdot \omega \cdot 2) = \omega^\omega \cdot 4$, $G(\omega^\omega \cdot \omega \cdot 3, \omega^\omega \cdot \omega \cdot 3) = \omega^\omega \cdot 13$, ..., until the gaps in $\mathbb{N}$ are filled in and $G(\omega^\omega \cdot \omega^2, \omega^\omega \cdot \omega^2) = \omega^\omega \cdot \omega$.

Essentially the same argument now gives $G(\omega^\omega \cdot \omega^3, \omega^\omega \cdot \omega^3) = \omega^\omega \cdot \omega^2$, $G(\omega^\omega \cdot \omega^4, \omega^\omega \cdot \omega^4) = \omega^\omega \cdot \omega^3$, ..., until catching up at $G((\omega^\omega)^2, (\omega^\omega)^2) = (\omega^\omega)^2$. From here the whole argument repeats to establish the formula for $j = 2$, $a = b = 0$, and similarly for all ordinals j.

The term in (9–6) when $(a, b) \neq (0, 0)$ is argued exactly as in earlier cases. Formulas (9–1)–(9–6) cover all cases.

10. Three-Wide Chomp

Let A be a three-wide Chomp position,

$$A = (1 \times u) + (2 \times v) + (3 \times x), \qquad u \geq v \geq x.$$

This section gives the conditions for A to be a P-position.

First we dispense with some small positions and some large ones: when the right column has finite height, i.e., $x < \omega$, the Beanstalk Lemma reduces the case $v < \omega$ to finite calculation, and the Side-Top Theorem covers the case $v \geq \omega$. On the other hand, the discussion below will show that the tall box $A = 3 \times \omega^\omega$ is a P-position, so any superposition of this box, i.e., any other position with $x \geq \omega^\omega$, is an N-position.

This leaves the case $\omega \leq x < \omega^\omega$, where the analysis is detailed. The first step is to decompose A into components, two bottoms (denoted with a subscript "b") and three tops (subscript "t"). For some unique i with $1 \leq i < \omega$, we have $\omega^i \leq x < \omega^{i+1}$, so write

$$
\begin{aligned}
u &= \omega^{i+1} \cdot u_{i+1} + \omega^i \cdot u_i + a & (u_i < \omega, a < \omega^i), \\
v &= \omega^{i+1} \cdot v_{i+1} + \omega^i \cdot v_i + b & (v_i < \omega, b < \omega^i), \\
x &= \phantom{\omega^{i+1} \cdot v_{i+1} +} \omega^i \cdot x_i + c & (0 < x_i < \omega, c < \omega^i).
\end{aligned}
$$

The precise decomposition of A depends on the nature of u, v, and x. Specifically,

(i) If $u_{i+1} = v_{i+1} = 0$ and $u_i = v_i = x_i$ then let

$$U_b = 2 \times \omega^i \cdot u_i, \quad X_b = (1 \times \omega^i \cdot x_i)@(2, 0),$$
$$U_t = ((1 \times a) + (2 \times b) + (3 \times c))@(0, \omega^i \cdot u_i), \quad V_t = \varnothing, \quad X_t = \varnothing.$$

(ii) If $v_{i+1} = 0$ and $v_i = x_i$ and either $u_{i+1} > 0$ or $u_i > v_i$ then let

$$U_b = (1 \times (\omega^{i+1} \cdot u_{i+1} + \omega^i \cdot u_i)) + (2 \times \omega^i \cdot v_i),$$
$$X_b = (1 \times \omega^i \cdot x_i)@(2, 0), \quad U_t = (1 \times a)@(0, \omega^{i+1} \cdot u_{i+1} + \omega^i \cdot u_i),$$
$$V_t = ((1 \times b) + (2 \times c))@(1, \omega^i \cdot v_i), \quad X_t = \varnothing.$$

(iii) If $u_{i+1} = v_{i+1} > 0$ and $u_i = v_i$ then let

$$U_b = 2 \times \omega^{i+1} \cdot u_{i+1}, \quad X_b = (1 \times \omega^i \cdot x_i)@(2,0),$$
$$U_t = ((1 \times (\omega^i \cdot u_i + a)) + (2 \times (\omega^i \cdot v_i + b)))@(0, \omega^{i+1} \cdot u_{i+1}),$$
$$V_t = \varnothing, \quad X_t = (1 \times c)@(2, \omega^i \cdot x_i).$$

(iv) Otherwise let

$$U_b = (1 \times (\omega^{i+1} \cdot u_{i+1} + \omega^i \cdot u_i)) + (2 \times (\omega^{i+1} \cdot v_{i+1} + \omega^i \cdot v_i)),$$
$$X_b = (1 \times \omega^i \cdot x_i)@(2,0), \quad U_t = (1 \times a)@(0, \omega^{i+1} \cdot u_{i+1} + \omega^i \cdot u_i),$$
$$V_t = (1 \times b)@(1, \omega^{i+1} \cdot v_{i+1} + \omega^i \cdot v_i), \quad X_t = (1 \times c)@(2, \omega^i \cdot x_i).$$

(See Figures 7a and 7b. Note that 7b only illustrates one instance of the fourth, general case.)

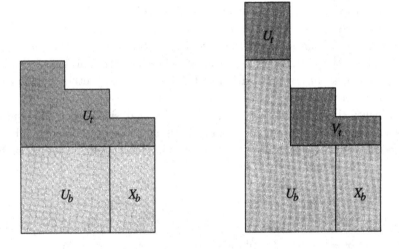

Figure 7a. Three-wide positions: first and second cases.

The criterion to be established is that in these cases the three-wide Chomp position A is a P-position if and only if

$$(10\text{-}1) \qquad G(U_b) = G(X_b) \quad \text{and} \quad G(U_t) \oplus G(V_t) \oplus G(X_t) = 0.$$

This reduces the three-wide question to questions about one-wide and two-wide subpositions: even in the case with a three-wide top subposition, the bottom subpositions already fail the test.

To show the criterion it suffices to show that if A is a position that satisfies (7) then every bite into A gives a position that doesn't; and if A is a position not satisfying (7) then some bite into A gives a position that does.

So assume first that A satisfies (7). Bite into A, obtaining a new position with components U_b', X_b', etc. If the bite doesn't touch any bottom component then it changes exactly one top component, violating $G(U_t') \oplus G(V_t') \oplus G(X_t') = 0$ in (7).

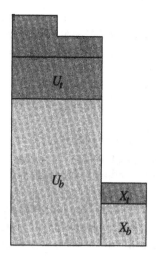 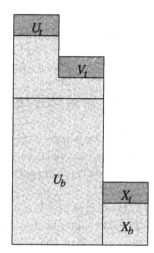

Figure 7b. Three-wide positions: third and fourth cases.

If the bite reduces U_b to U_b' but doesn't touch X_b then $G(U_b') \neq G(U_b) = G(X_b) = G(X_b')$, again violating (7). If the bite reduces X_b to X_b' but doesn't touch U_b then possibly $U_b' \supset U_b$ and $G(U_b') > G(U_b)$ in the new decomposition, but in any case $G(U_b') \geq G(U_b) = G(X_b) > G(X_b')$. Finally, if the bite touches both bottom components then it is either $\overline{(1,z)}$ for some z, in which case $G(U_b') > G(U_b)$ by two-wide results and $G(X_b) > G(X_b')$; or the bite is $\overline{(0,z)}$ for some z, in which case $U_b' = 2 \times h$ and $X_b' = (1 \times h)@(2,0)$ for some h, so that $G(U_b') \neq G(X_b')$.

For the other half of the argument, assume that position A, with components U_b, X_b, etc., doesn't satisfy (7). We seek a bite into A, giving a new position with components U_b', X_b', etc., that does.

If $G(U_b) = G(X_b)$ then $G(U_t) \oplus G(V_t) \oplus G(X_t) \neq 0$ and the disjoint grouping strategy (cf. the pairing strategy used earlier) provides a bite into exactly one top component restoring the second condition in (7).

If $G(U_b) < G(X_b)$ then the answering bite to restore (7) is some $\overline{(2,z)}$ that touches X_b. All such bites satisfy $U_b' \supseteq U_b$ (proper containment only when the decomposition index i is altered by the bite), $G(U_b') \geq G(U_b)$, and $G(X_b') < G(X_b)$. The needed bite has $z = \omega^j \cdot z_j + c'$ such that $\omega^j \cdot z_j = G(U_b')$ and $c' = G(U_t') \oplus G(V_t')$.

If $G(U_b) > G(X_b)$, then the answering bite is some $\overline{(0,z)}$ touching U_b but not X_b. All such bites give $G(U_b') < G(U_b)$ in this decomposition. To make the top Grundy values sum to zero, use the fact that for any component U_b arising in our decomposition, any bite $\overline{(1,y)}$ in U_b increases the Grundy value, i.e., $G(U_b - \overline{(1,y)}) > G(U_b)$; it follows by the definition of Grundy value as a mex that $G(X_b)$ takes the form $G(U_b - \overline{(0,z')})$ for some z'. A larger bite height $z > z'$ will leave bottom piece $U_b' = U_b - \overline{(0,z')}$ with the right Grundy value and also leave a two-wide top piece U_t' with $G(U_t') = G(X_t')$.

To complete the discussion of three-wide Chomp, we show that the tall box $3 \times \omega^\omega$ is a P-position by itemizing how every bite leaves an N-position. That is, every bite into $3 \times \omega^\omega$ can be answered with another bite leaving a smaller P-position described above. Bites take the form $\overline{(e, z)}$ with $0 \leq e < 3$ and $0 \leq z < \omega^\omega$. One can check that the P-positions enumerated don't include any position $3 \times \omega^\omega - \overline{(e, z)}$. Responses to the bite are analyzed by the cases $e = 2, 1, 0$.

If $e = 2$, the bite truncates the third column. Let $u = H(3 \times z, 2)$. If $u < \omega^\omega$ then by definition the answering bite is $\overline{(0, u)}$. To show $u < \omega^\omega$ we may assume $u \geq \omega$, else there is nothing to prove. The argument has two cases, depending on z.

When $z < \omega$, the position $3 \times \omega^\omega - \overline{(2, z)}$ is a Side-Top position and the Side-Top Theorem tells us how to proceed: recalling the notion of restricted Grundy value G_S, bite the top part of the position down to $2 \times a$ such that $G(2 \times a) = G_S(1 \times z) \leq G(1 \times z) = z < \omega$. From the results on two-wide Chomp we know that $G(2 \times a) < \omega$ if and only if $a < \omega^2$; since we are biting into $2 \times \omega^\omega$, there is plenty of room to bite the top part down to $2 \times a$.

When $z \geq \omega$, write, using the terminology of the three-wide algorithm, $z = x = \omega^i \cdot x_i + c$ with $0 < x_i < \omega$ and $c < \omega^i$. Now the answering bite to $\overline{(2, x)}$ is $\overline{(0, u)}$, where $u = \omega^{i+1} \cdot x_i + a$ is found from the algorithm with $G(2 \times a) = c$. From the two-wide results, a is unique and satisfies $a < \omega^{i+1}$. This completes the case $e = 2$.

If $e = 1$, the bite is $\overline{(1, z)}$, truncating the last two columns at the same height. Let $u = H(3 \times z, 1)$. Again the answering bite is clearly $\overline{(0, u)}$ once we know that $u < \omega^\omega$, and again showing this breaks into two cases depending on z. When $z < \omega$, the Beanstalk Lemma says that $u < \omega$ as well. When $z \geq \omega$, write $z = x = \omega^i \cdot x_i + c$ with $0 < x_i < \omega$ and $c < \omega^i$, and u works out to $u = \omega^i \cdot 2x_i + G(2 \times c)$ — to see this, check that the position $1 \times u + 3 \times x$ is identified by the three-wide algorithm as a P-position. Note $G(2 \times c) < \omega^i$ by two-wide results. This completes the case $e = 1$.

If $e = 0$, the bite is $\overline{(0, z)}$, leaving a rectangle $3 \times z$ with $z < \omega^\omega$. When $z < \omega$, the rectangle is finite. A result on finite Chomp says that the rectangle must be an N-position, so some finite bite into it leaves a P-position. The bite takes the form $\overline{(1, v)}$ or $\overline{(2, x)}$ and is believed to be unique.

When $\omega \leq z < \omega^2$, write $z = \omega \cdot (n + 1) + c$ with $0 \leq n < \omega$ and $c < \omega$. Answer the bite $\overline{(0, z)}$ with $\overline{(2, x)}$ for the unique $x < \omega$ that satisfies $G_S(x) = G(2 \times (\omega \cdot n + c))$, where again G_S is restricted Grundy value as in the Side-Top Theorem and x is found from that theorem.

When $z \geq \omega^2$, write $z = \omega^{i+1} \cdot z_{i+1} + c$ with $0 < z_{i+1} < \omega$ and $c < \omega^{i+1}$. Answer $\overline{(0, z)}$ with $\overline{(2, x)}$ where $x = \omega^i \cdot z_{i+1} + G(2 \times c)$ is found from the three-wide P-position algorithm; again $G(2 \times c) < \omega^i$ by two-wide results. In this case, when z_{i+1} is even there is sometimes also a winning bite $\overline{(1, v)}$; finding it is an exercise for the interested reader. This completes the case $e = 0$.

11. A Three-Dimensional Example

This section will present the three-dimensional Chomp P-positions with 2-by-2 base, including $2 \times 2 \times \omega^3$.

We need a variant of the Side-Top Theorem. Consider a pair of 3-dimensional Chomp positions F and T (for "front" and "top"), with F a finite subset of $2 \times \Omega \times \Omega$, and T a subset of $1 \times 2 \times \Omega$. Let $B = 1 \times 2 \times \omega$ be the one-deep, two-wide infinite brick of height ω. Let $F@(1,0,0)$ denote the translation of F one unit forward, meaning perpendicular to the wide dimension of the brick, and let $T@(0,0,\omega)$ denote the translation of T upward by ω. Construct the Chomp position

$$U = B + F@(1,0,0) + T@(0,0,\omega),$$

the brick with finitely much added to its front and any amount to its top. (See Figure 8.)

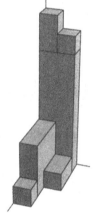

Figure 9. A front-top position.

Let $\square F$ denote $F@(1,0,0)$ back-extended one unit to the wall, i.e., $\square F = F \cap (1 \times 2 \times \Omega) + F@(1,0,0)$. Let $\mathcal{F}$ denote the set of all front positions $\tilde{F}$ such that $H(\square \tilde{F}, 1 \times 2)$ is infinite.

Theorem 11.1 (Front-Top Theorem). *Let a front-top position*

$$U = B + F@(1,0,0) + T@(0,0,\omega)$$

be given. If $H(\square F, 1 \times 2)$ is finite then U is an N-position. If $H(\square F, 1 \times 2)$ is infinite then

$$U \text{ is a P-position} \quad \Longleftrightarrow \quad G_{\mathcal{F}}(F) = G(T).$$

(Here, as in the Side-Top Theorem earlier, $G_{\mathcal{F}}$ denotes restricted Grundy value.)

The proof is virtually identical to the Side-Top case. A more complicated Base-Top Theorem holds, where finite base material may be added in all directions around the brick, but this is all we need.

Defining an ordinal subtraction operation helps with the decompositions to be used in this section. For ordinals $b \leq z$, let $b \backslash z$ be the unique ordinal t such that $b \uplus t = z$, where "$\uplus$" denotes ordinal addition. For instance, $\omega \backslash (\omega \cdot (c+1) + d) = \omega \cdot c + d$ while $\omega \backslash z = z$ if $z \geq \omega^2$. In Chomp terms, $b \backslash z$ is the amount of a z-high tower that extends above base height b.

Returning to subpositions of $2 \times 2 \times \Omega$, introduce the bird's-eye view notation

$$\begin{bmatrix} v z \\ u y \end{bmatrix} = (1 \times 1 \times u) + (1 \times 2 \times v) + (2 \times 1 \times y) + (2 \times 2 \times z)$$

where $u \geq v \geq y \geq z$ without loss of generality. Thus we are looking at the position down the third axis — the origin is at the lower left corner, the first axis goes right, the second up the page. Since this section involves many decompositions, extend the bird's-eye notation also to

$$\begin{bmatrix} v \\ u \end{bmatrix} = 1 \times 1 \times u + 1 \times 2 \times v$$

so that

$$\begin{bmatrix} v z \\ u y \end{bmatrix} = \begin{bmatrix} v \\ u \end{bmatrix} + \begin{bmatrix} z \\ y \end{bmatrix} @(1, 0, 0).$$

The finite P-positions in $2 \times 2 \times \Omega$ have a nice closed form, unlike the three-wide case. They are found by repeated application of the Fundamental Theorem, giving

$$\begin{bmatrix} v & 0 \\ v+10 & \end{bmatrix} \text{ for } 0 \leq v < \omega, \qquad \begin{bmatrix} 10 \\ 11 \end{bmatrix},$$

$$\begin{bmatrix} v & 1 \\ v+21 & \end{bmatrix} \text{ for } 1 \leq v < \omega, \qquad \begin{bmatrix} 21 \\ 22 \end{bmatrix},$$

$$\begin{bmatrix} v & 2 \\ v+32 & \end{bmatrix} \text{ for } 2 \leq v < \omega, \qquad \begin{bmatrix} 32 \\ 33 \end{bmatrix},$$

etc.

We now classify the subpositions of $2 \times 2 \times \Omega$ into six types and characterize the P-positions for each type. For the classification, write uniquely $u = \hat{u} + \omega \cdot u_1 + u_0$ with $\hat{u}$ a multiple of ω^2 and u_1, u_0 finite; similarly for v, y, z.

TYPE A: $z < \omega$, $z = y$.

TYPE B: $z < \omega$, $z < y$.

TYPE C: $\omega \leq z < \omega^3$, $\hat{z} = \hat{v}$, $y < \omega^\omega$ (the third condition actually follows from the first two).

TYPE D: $\omega \leq z < \omega^3$, $\hat{z} < \hat{v}$, $y < \omega^\omega$.

TYPE E: $\omega \leq z < \omega^3$, $y \geq \omega^\omega$.

TYPE F: $z \geq \omega^3$.

Further, a type C, D, or E position is *short* if $z < \omega^2$ and *tall* if $z \geq \omega^2$. A type D position is *thick* if $z \backslash u < \omega^2 \cdot 2$ and *thin* if $z \backslash u \geq \omega^2 \cdot 2$.

The following properties characterize the P-subpositions of $2 \times 2 \times \Omega$.

PA1: A type A position with $v < \omega$ is a P-position if and only if $u = v + z + 1$.

PA2: A type A position with $v \geq \omega$ is a P-position if and only if

$$G(\omega \backslash u, \omega \backslash v) = z.$$

(This is the notation introduced earlier for Grundy values of two-wide positions.)

PB1: The P-positions of type B are $\begin{bmatrix} z+1 & z \\ z+1 & z+1 \end{bmatrix}$ for $z < \omega$.

PB2: If $z \geq \omega$ and h is finite then both $\begin{bmatrix} v & z \\ u & y \end{bmatrix} - \overline{(0,0,h+1)}$ and $\begin{bmatrix} v & z \\ u & y \end{bmatrix} - \overline{(1,1,h)}$ are N-positions.

Properties PA1 and PB1 follow from finite iteration of the Fundamental Theorem, as already observed. PB2 follows immediately from PB1 by noting that answering bite $\overline{(0,0,h+1)}$ with bite $\overline{(1,1,h)}$ and vice versa leaves a P-position.

PA2 follows from the Front-Top Theorem. The restricted Grundy value $G_{\mathcal{F}}(z,z) = z$ for the front portion $(1 \times 2 \times z)@(1,0,0)$ of a PA2 position is a consequence of the finite P-positions enumerated by PA1 and PB1.

The following property PA3 combines with PB2 to analyze all bites of finite height in superpositions of $2 \times 2 \times \omega$ (types C through F). With the one exception noted in PA3, all such bites leave N-positions.

PA3: If $z \geq \omega$ and h is finite then $\begin{bmatrix} v & z \\ u & y \end{bmatrix} - \overline{(0,1,h)}$ and $\begin{bmatrix} v & z \\ u & y \end{bmatrix} - \overline{(1,0,h)}$ are N-positions unless $\begin{bmatrix} v & z \\ u & y \end{bmatrix}$ is a short type C position with $u_1 = v_1$.

To see this, note that PA3 is a statement about bites that leave two columns of finite height. (In the case of bite $\overline{(0,1,h)}$, follow it by a mirror-reflection to restore our symmetry-breaking assumption $v \geq y$ before continuing the analysis.) The P-positions of this form are described by PA2: they satisfy $G(\omega \backslash u, \omega \backslash v) = h < \omega$. By two-wide Grundy value results, $G_2(\omega \backslash u, \omega \backslash v) < \omega$ if and only if $u < \omega^2$ and $u_1 = v_1$. This is precisely the exception in PA3; under any other conditions the bite leaves an N-position.

We next characterize P-positions of types C and D. The argument here is long and intricate, so the reader may just want to read through the results PC1 and PD1 and then skip onward to the discussion of type E.

The characterizations of type C and type D P-positions are proved by transfinite induction on Chomp positions, noting that all properties of a position are proved from properties of strictly smaller positions reached by biting into it. The induction hypothesis is the conjunction of all propositions PCn and PDn, including pending auxiliary ones. The basis of the induction is provided by propositions PAn and PBn.

Every type C position with $v \geq y$ can be written uniquely as

$$\begin{bmatrix} v & z \\ u & y \end{bmatrix} = B + T@(0,0,h),$$

where

$$h = \begin{cases} \omega & \text{if } \hat{z} = 0, \\ \omega^2 \cdot k & \text{if } \hat{z} = \omega^2 \cdot k, \, 0 < k < \omega, \end{cases}$$

$$B = \begin{bmatrix} hh \\ hh \end{bmatrix}, \quad T \in \left\{ \begin{bmatrix} vz \\ uy \end{bmatrix} : \hat{v} = 0 \right\}.$$

The B and T in this decomposition are called the base and top pieces.

Proposition PC1 now characterizes type C P-positions up to a restricted Grundy value calculation on the comparatively small top piece. After establishing PC1 and PD1, we will briefly consider some specifics of the calculation.

PC1: A type C Position as just decomposed is a P-position if and only if $T \in \mathcal{T}$ and $G_{\mathcal{T}}(T) = 0$, where

$$\mathcal{T} = \left\{ \begin{bmatrix} vz \\ uy \end{bmatrix} : u_1 > \max(v_1, y_1) \text{ and } \hat{u} = 0 \right\}.$$

Characterizing a type D P-position is accomplished by decomposing it into a base B, and two-wide front and top F and T. Specifically,

$$B = \begin{bmatrix} \omega^2 \omega \\ \omega^2 \omega \end{bmatrix}, \quad F = \begin{bmatrix} \omega \backslash z \\ \omega \backslash y \end{bmatrix}, \quad T = \begin{bmatrix} \omega^2 \backslash v \\ \omega^2 \backslash u \end{bmatrix}.$$

Thus $\begin{bmatrix} v & z \\ u & y \end{bmatrix} = B + F@(1, 0, \omega) + T@(0, 0, \omega^2)$.

PD1: A type D position as just decomposed is a P-position if and only if $G(F) = G(T)$.

Proving PC1 and PD1 requires several auxiliary propositions.

PC2: A type C position with $\hat{u} = \hat{v}$ and $u_1 = v_1$ is an N-position.

PC3: In a type C position with either $\hat{u} > \hat{v}$ or both $\hat{u} = \hat{v}$ and $u_1 > v_1$, every bite that intersects the base B leaves an N-position.

The proofs of PC2 and PC3 are briefly deferred.

PC4: If a short type C position is a P-position, then $u < \omega^2$.

For PC4, the Size Lemma gives $u \leq 1 + v + y + z$. In a short type C position, $\hat{v} = \hat{y} = \hat{z} = 0$ and the result follows.

PD2: If a type D position has $\hat{u} = \hat{v} = \hat{y}$, then it is an N-position.

To see PD2 note that the givens imply $G(T) < \omega^2$ in the type D decomposition, while $G(F) \geq \omega^2$. Thus PD2 follows from the inductive hypothesis on PD1.

PD3: If a type D position $\begin{bmatrix} v & z \\ u & y \end{bmatrix}$ satisfies $\hat{z} < \hat{y}$ and $\hat{v} < \hat{u}$ then

$$G(F) = G(T) \iff G(\tilde{F}) = G(\tilde{T}),$$

where (using row vectors to indicate decomposition in the other direction)

$$\tilde{F} = \begin{bmatrix} \omega \backslash v \omega \backslash z \end{bmatrix}, \quad \tilde{T} = \begin{bmatrix} \omega^2 \backslash u \omega^2 \backslash y \end{bmatrix}.$$

That is, for a type D position where neither the shortest nor the tallest column has the same ω^2-coefficient as either intermediate column, the decomposition

to determine whether it is a P-position can be made in either direction. PD3
follows from 2-wide Grundy value results.

We now prove PC1 from these auxiliary propositions and then resume the
deferred proofs of PC2 and PC3. The PC1 proof uses the decomposition of a
type C position into top and base pieces T and B.

Consider the set of type C positions with base B of a particular height h,
$B = 2 \times 2 \times h$. Let $\mathcal{T}(h)$ denote the set of top pieces of such positions for which
every bite intersecting B leaves an N-position. Then it follows easily from the
definition of restricted Grundy value that the P-positions of type C are precisely
those of the form $B + T@(0,0,h)$, where $T \in \mathcal{T}(h)$ and $G_{\mathcal{T}(h)}(T) = 0$.

Now it follows from PC2 and PC3 that

$$\mathcal{T}(h) = \left\{ \begin{bmatrix} v z \\ u y \end{bmatrix} : \hat{v} = 0 \text{ and either } \hat{u} > 0 \text{ or } u_1 > \max(v_1, y_1) \right\}$$

for all values of h that occur. Thus $\mathcal{T}(h)$ is independent of h, and whether a
type C position is a P-position depends solely on its top piece. Notice that $\mathcal{T}(h)$
(for any h) differs from the set $\mathcal{T}$ in PC1 only in allowing $\hat{u} > \hat{v}$ in the top piece.
This is because the restricted Grundy value calculation does not in itself rule out
the possibility $\hat{u} > \hat{v}$ in $\mathcal{T}(h)$. But PC4 excludes $\hat{u} > \hat{v}$ directly for short type
C P-positions, and $\mathcal{T}(h)$ being independent of h excludes $\hat{u} > \hat{v}$ for tall type C
P-positions. Thus for each h, $\mathcal{T}(h) \cap \{[\begin{smallmatrix} v & z \\ u & y \end{smallmatrix}] : \hat{u} = 0\} = \mathcal{T}$, completing the proof
of PC1.

To prove PC2, let $k = G(h \backslash u, h \backslash v)$, where h is the height of base B in the
type C decomposition; k is finite by two-wide results since $u_1 = v_1$ and $h \backslash u < \omega^2$.
We show that PC2 describes an N-position by finding a bite that leaves a P-
position. For a short type C position, the bite $\overline{(1, 0, k)}$ leaves a P-position of
type A by PA2. For a tall type C position, let $h = \omega^2 \cdot m$, $h' = \omega^2 \cdot (m - 1)$,
and $h'' = w \uplus h'$, where $\uplus$ denotes ordinal addition. Then bite $\overline{(1, 0, h'' + c)}$
leaves a P-position of type D by PD1, where c is the unique value satisfying
$G(c, c) = k$. In the decomposition of this type D P-position, $F = \begin{bmatrix} h'+c \\ h'+c \end{bmatrix}$,
$T = \begin{bmatrix} h'+(h\backslash v) \\ h'+(h\backslash u) \end{bmatrix} = \begin{bmatrix} \omega^2 \backslash v \\ \omega^2 \backslash u \end{bmatrix}$, and $G(F) = G(T) = \omega \cdot (m - 1) + k$.

To prove PC3, we sketch the cases showing that every bite $\overline{(i, j, k)}$ intersecting
the base in a type PC3 position leaves an N-position. PB2 and PA3 show this
for all finite k, covering all bites in short type C positions, so consider only
$k \geq \omega$ and tall type C positions; also, $k < h$ since the bite intersects the base.
If $(i, j) = (0, 0)$, we get a type C N-position by PC2. If $(i, j) = (1, 1)$, we get a
type D N-position by PD2.

If $(i, j) = (1, 0)$, we get a type D position with a decomposition such that
$G(T) > G(F)$, an N-position by PD1. To see why $G(T) > G(F)$ here, note that
for some $m < h$ and $f < \omega^2$,

$$T = \begin{bmatrix} \omega^2 \cdot (h - 1) + (h\backslash v) \\ \omega^2 \cdot (h - 1) + (h\backslash u) \end{bmatrix} = \begin{bmatrix} \omega^2 \backslash v \\ \omega^2 \backslash u \end{bmatrix}, \quad F = \begin{bmatrix} \omega^2 \cdot m + f \\ \omega^2 \cdot m + f \end{bmatrix}.$$

We have $G(h\backslash u, h\backslash v) \geq \omega$ since $u_1 > v_1$ or $\hat{u} > \hat{v}$, and $G(f, f) < \omega$ since $f < \omega^2$, hence $G(T) = \omega \cdot (h-1) + G(h\backslash u, h\backslash v) \geq \omega \cdot h$, and $G(F) = \omega \cdot m + G(f, f) < \omega \cdot h$. Finally, the result for $(i, j) = (0, 1)$ follows symmetrically from $(i, j) = (1, 0)$ by interchanging v and y.

We now prove PD1, that a type D position is a P-position if and only if $G(F) = G(T)$ in its decomposition. We first show that every type D position with $G(F) \neq G(T)$ can be bitten to a position with $G(F') = G(T')$, where the primes mean "after the bite." If $G(F) > G(T)$, find a bite (known to exist by 2-wide Grundy value results) that bites F without touching T, leaving a type D position with $G(F') = G(T)$ and $T' = T$. If $G(T) > G(F)$, find a bite that bites T without touching F, leaving a type D position with $G(T') = G(F)$ and $F' = F$. Case analysis (details omitted) confirms this can be done.

To finish PD1 we sketch the cases showing that every bite $\overline{(i, j, h)}$ from a type D position with $G(F) = G(T)$ leaves an N-position. PB2 and PA3 show this for all finite h, so take $h \geq \omega$. If $(i, j) = (0, 0)$, any h leaving a type C position (reducing two, three, or four columns) leaves an N-position by PC2. Otherwise a type D position is left (by reducing one, two, or three columns). Reducing three columns to type D leaves an N-position by PD2. Reducing one or two columns to type D leaves an N-position by PD1, since $G(T') \neq G(T)$ and $G(F') = G(F)$. If $(i, j) = (1, 1)$ or $(i, j) = (1, 0)$, we get a type D N-position by PD1 since $G(F') \neq G(F)$ but $G(T') = G(T)$. This leaves the final case $(i, j) = (0, 1)$. If $\hat{h} = \hat{z} = \hat{y}$, we get a type C N-position by PC4. If $\hat{z} < \hat{h}$ and $h \geq y$, we get $G(T') \neq G(T)$ but $G(F') = G(F)$, an N-position by PD1. If $\hat{z} < \hat{h} = \hat{v}$ and $h < y$, we get $G(T') = G(\tilde{T})$ but $G(F') \neq G(\tilde{F})$, an N-position by PD3 and PD1 (recall that the "$\sim$" notation arises from cleaving in the other direction). If $\hat{z} < \hat{h} < \hat{v}$ and $h < y$, we get $G(F') < G(F) \uplus \omega^2$ (since $h < y$ and $\hat{h} \leq \hat{y}$) and $G(T') \geq G(T) \uplus \omega^2$ (since $\hat{y} < \hat{v}$, by PD3), where $\uplus$ denotes ordinal addition and $F = [\begin{smallmatrix} z \\ y \end{smallmatrix}]$, $T = [\begin{smallmatrix} v \\ u \end{smallmatrix}]$, $F' = [hz]$, $T' = [uy]$. Since $G(F) \uplus \omega^2 = G(T) \uplus \omega^2$ follows from $G(F) = G(T)$, this gives $G(F') < G(T')$, an N-position by PD1. Finally, if $\hat{h} < \hat{z}$ or if $\hat{h} = \hat{z} < \hat{y}$, we get $G(T') >= \omega^2$ (since $\hat{y} < \hat{u}$) and $G(F') < \omega^2$, an N-position by PD1.

We now briefly consider specific examples of type C P-positions, which so far have only been characterized up to a restricted Grundy value calculation on positions with 2×2 cross section and height less than ω^2.

Type C P-positions $[\begin{smallmatrix} v & z \\ u & y \end{smallmatrix}]$ with $z < \omega \cdot 2$ (i.e., with $z_1 = 1$) are well behaved. They are just those positions with $u_1 + z_1 = v_1 + y_1 + 1$ whose finite top pieces have Nim sum 0. Top pieces at the same level (coefficient of ω) must have their num sum component computed together, so they can be 2-wide Grundy values, or even a 3-column piece with L-shaped cross section (which is computed by the suitable restricted Grundy value function).

Type C P-positions with $z_1 = 2$ start to get interesting. For example, P-positions $[\begin{smallmatrix} \omega \cdot 2 + b & \omega \cdot 2 + d \\ \omega \cdot 3 + a & \omega \cdot 2 + c \end{smallmatrix}]$ are given by precisely those $Q = [\begin{smallmatrix} b & d \\ a & c \end{smallmatrix}] \in \mathcal{Q}$ that satisfy

$G_Q(Q) = 0$, where

$$Q = \left\{ \begin{bmatrix} bd \\ ac \end{bmatrix} : a \oplus b \equiv a \oplus c \equiv 1 \,(\text{mod } 3), d \leq \min(b, c) \right\}.$$

This fact combines with 2-wide Grundy results and number theory to show that given b, c, and $d \leq \min(b, c)$, then there is a P-position

$$\begin{bmatrix} \omega \cdot 2 + b\omega \cdot 2 + d \\ \omega \cdot 3 + a\omega \cdot 2 + c \end{bmatrix} \quad \text{for some } a$$

if and only if $b \oplus c$ is not a power of 2. When $b \oplus c$ is a power of 2, the Fundamental Theorem finds a similar P-position but with taller highest column,

$$\begin{bmatrix} \omega \cdot 2 + b\omega \cdot 2 + d \\ \omega \cdot 5 + a\omega \cdot 2 + c \end{bmatrix} \quad \text{for some } a.$$

Type C P-positions with $z_1 = 2$ satisfy the conditions

$$u_1 + z_1 = v_1 + y_1 + 1 \quad \text{if } v_1 \text{ or } y_1 \text{ is odd or } v_1 = y_1 = 2,$$
$$u_1 + z_1 = v_1 + y_1 + 3 \quad \text{if } v_1 \text{ and } y_1 \text{ are even.}$$

Type C P-positions with $z_1 > 2$ become increasingly more complex. They can be found by finite calculation, but we don't prove this here. The principles justifying this claim are the subject of ongoing research.

For a type E positions $[\begin{smallmatrix} v & z \\ u & y \end{smallmatrix}]$, define the base and top to be

$$B = \begin{bmatrix} \omega^\omega & \omega \\ \omega^\omega & \omega^\omega \end{bmatrix}, \quad T = \begin{bmatrix} \omega^\omega \backslash v & 0 \\ \omega^\omega \backslash u\omega^\omega \backslash y \end{bmatrix},$$

so that $[\begin{smallmatrix} v & z \\ u & y \end{smallmatrix}] = B + T@(0, 0, \omega^\omega) + (1 \times 1 \times (\omega \backslash z))@(1, 1, \omega)$. Then

PE1: A type E position is a P-position if and only if $G(T) = \omega \backslash z$.

This is a variant of the Front-Top Theorem, with a higher, differently shaped top and an elevated "front." The winning strategy is: anwer any bite into the front or the top with a pairing strategy; the rest of the analysis in this section shows that any other bite leaves an N-position, so answer the bite appropriately.

Property PE1 isn't entirely satisfactory. It refers to the Grundy value of the top, which has an L-shaped cross-section; we haven't discussed such positions in general yet, though the section on P-ordered positions gives the special case result $G([\begin{smallmatrix} a & 0 \\ a+b & a \end{smallmatrix}]) = a + b$.

Finally, type F is the simplest to characterize.

PF1: A type F position is a P-position if and only if it is $2 \times 2 \times \omega^3$.

This is shown by answering any bite $\overline{(i, j, h)}$ into $2 \times 2 \times \omega^3$ with another bite that gives a P-position. By properites PB2 and PA3, we may assume $h \geq \omega$.

If the bite is $\overline{(0, 0, h)}$ with $\omega \leq h < \omega^2$, then the answering bite is

$$\overline{(1, 0, G(2 \times (\omega \backslash h)))},$$

giving a front-top position of type A. If the bite is $\overline{(0,0,h)}$ with $\omega^2 \le h < \omega^2 \cdot 2$, then the answering bite is $\overline{(1,0,\omega + (\omega\backslash h))}$, giving a short thick position of type D with $F = T$ in the decomposition. If the bite is $\overline{(0,0,h)}$ with $\omega^2 \cdot 2 \le h$, then the answering bite is $\overline{(1,0,\omega\backslash h)}$, giving a tall thick position of type D with $F = T$. If the bite is $\overline{(1,1,h)}$, then the answering bite is $\overline{(1,0,\omega^2 + h)}$, giving a thin position of type D with $T = \left[\begin{smallmatrix}\omega^3 \\ \omega^3\end{smallmatrix}\right]$, $F = \left[\begin{smallmatrix}a \\ \omega^2+a\end{smallmatrix}\right]$ for some $a < \omega^3$, and $G(T) = G(F) = \omega^2$. If the bite is $\overline{(1,0,h)}$, then the answering bite is $\overline{(0,0,\omega^2 + h)}$, giving a thick position of type D. The final case of bite $\overline{(0,1,h)}$ is symmetric.

We conclude the section with some examples of P-positions of the different types.

Finite type A: $\left[\begin{smallmatrix}1 & 1 \\ 3 & 1\end{smallmatrix}\right]$, $\left[\begin{smallmatrix}2 & 2 \\ 5 & 2\end{smallmatrix}\right]$, $\left[\begin{smallmatrix}3 & 2 \\ 6 & 2\end{smallmatrix}\right]$.

Type B: $\left[\begin{smallmatrix}2 & 1 \\ 2 & 2\end{smallmatrix}\right]$, $\left[\begin{smallmatrix}3 & 2 \\ 3 & 3\end{smallmatrix}\right]$.

Front-top type A: $\left[\begin{smallmatrix}\omega & 0 \\ \omega & 0\end{smallmatrix}\right]$, $\left[\begin{smallmatrix}\omega & 1 \\ \omega+1 & 1\end{smallmatrix}\right]$. The latter has top $T = \left[\begin{smallmatrix}0 \\ 1\end{smallmatrix}\right]$ and front $F = \left[\begin{smallmatrix}1 \\ 1\end{smallmatrix}\right]$, with $G(T) = G_{\mathcal{F}}(F) = 1$.

Short type C: $\left[\begin{smallmatrix}\omega & \omega \\ \omega\cdot 2 & \omega\end{smallmatrix}\right]$, $\left[\begin{smallmatrix}\omega+1 & \omega \\ \omega\cdot 2+1 & \omega\end{smallmatrix}\right]$.

Tall type C: $\left[\begin{smallmatrix}\omega^2 & \omega^2 \\ \omega^2+\omega & \omega^2\end{smallmatrix}\right]$, $\left[\begin{smallmatrix}\omega^2+1 & \omega^2 \\ \omega^2+\omega+1 & \omega^2\end{smallmatrix}\right]$.

Thick type D: $\left[\begin{smallmatrix}\omega^2 & \omega \\ \omega^2 & \omega\end{smallmatrix}\right]$, $\left[\begin{smallmatrix}\omega^2 & \omega \\ \omega^2+1 & \omega+1\end{smallmatrix}\right]$, $\left[\begin{smallmatrix}\omega^2\cdot 2 & \omega^2 \\ \omega^2\cdot 2 & \omega^2\end{smallmatrix}\right]$ with $T = F = \left[\begin{smallmatrix}\omega^2 \\ \omega^2\end{smallmatrix}\right]$.

Thin type D: $\left[\begin{smallmatrix}\omega^2 & \omega \\ \omega^2\cdot 2 & \omega^2\end{smallmatrix}\right]$, $\left[\begin{smallmatrix}\omega^2\cdot 2 & \omega^2 \\ \omega^2\cdot 3 & \omega^2\cdot 2\end{smallmatrix}\right]$, $\left[\begin{smallmatrix}\omega^2\cdot 2 & \omega \\ \omega^2\cdot 2 & \omega\cdot 2\end{smallmatrix}\right]$ with $T = \left[\begin{smallmatrix}\omega^2 \\ \omega^2\end{smallmatrix}\right]$ and $F = \left[\begin{smallmatrix}0 \\ \omega\end{smallmatrix}\right]$, $\left[\begin{smallmatrix}\omega^3 & \omega \\ \omega^3 & \omega^2\end{smallmatrix}\right]$ with $T = \left[\begin{smallmatrix}\omega^3 \\ \omega^3\end{smallmatrix}\right]$ and $F = \left[\begin{smallmatrix}0 \\ \omega^2\end{smallmatrix}\right]$, $\left[\begin{smallmatrix}\omega^3 & \omega^2 \\ \omega^3 & \omega^2\cdot 2\end{smallmatrix}\right]$ with $T = \left[\begin{smallmatrix}\omega^3 \\ \omega^3\end{smallmatrix}\right]$ and $F = \left[\begin{smallmatrix}\omega^2 \\ \omega^2\cdot 2\end{smallmatrix}\right]$.

Type E: $\left[\begin{smallmatrix}\omega^\omega & \omega \\ \omega^\omega & \omega^\omega\end{smallmatrix}\right]$, $\left[\begin{smallmatrix}\omega^\omega+\omega & \omega+1 \\ \omega^\omega+\omega & \omega^\omega\end{smallmatrix}\right]$, $\left[\begin{smallmatrix}\omega^\omega & \omega^2 \\ \omega^\omega+\omega^2 & \omega^\omega\end{smallmatrix}\right]$, $\left[\begin{smallmatrix}\omega^\omega+\omega^3 & \omega^2 \\ \omega^\omega+\omega^3 & \omega^\omega\end{smallmatrix}\right]$.

Type F: $\left[\begin{smallmatrix}\omega^3 & \omega^3 \\ \omega^3 & \omega^3\end{smallmatrix}\right]$.

12. Open Questions

In all cases we have examined to date, any two Chomp positions whose sizes agree past some power of ω (i.e., both sizes are $\sum_{i \ge i_0} \omega^i \cdot a_i$ plus possibly different lower order terms) also have Grundy values agreeing past the same power of ω (i.e., both values are $\sum_{i \ge i_0} \omega^i \cdot b_i$ plus possibly different lower order terms). We conjecture that this property always holds. Specifically, for any Chomp position X and ordinal i, define the stratification $\mathrm{strat}(X, \omega^i)$ to be the Chomp position obtained from X by deleting all rectangles of size less than ω^i in the construction of $\mathrm{size}(X)$. For ordinal j, define $\mathrm{strat}(j, \omega^i) = \mathrm{strat}(1 \times j, \omega^i)$.

Conjecture 12.1 (Stratification Conjecture). *For all Chomp positions X and Y and ordinals i, if $strat(X, w^i) = strat(Y, w^i)$ then*

$$strat(G(X), w^i) = strat(G(Y), w^i).$$

We would like to know which sets of Chomp positions have computable subsets of P-positions.

A winning strategy for a set of Chomp positions can be viewed as a pair of functions, one which identifies P-positions and N-positions, and another which identifies winning moves by mapping each N-position to one or more P-positions reachable from it in one bite. A complete analysis of a set of P-positions would give a winning strategy for each subposition of any member of the set. For instance, the discussion of two-wide Chomp gave a complete analysis of $(1 \times 2 \times \Omega) + (\Omega \times 1 \times 1)$. The discussion of P-ordered positions such as $\omega \times \omega \times 2 \times 2$ gave a winning strategy, but not a complete analysis.

It is not difficult to show that the set of P-positions contained in any Chomp position with a finite part and either two one-wide transfinite stalks or one two-wide transfinite stalk is recursive.

We are confident that the set of P-positions in $3 \times \Omega$ is recursive (though this has not been fully proved) and consider it very likely that the set of P-positions in $(1 \times 3 \times \Omega) + (\Omega \times 1 \times 1)$ is recursive.

However we don't know the Grundy value $G(4 \times \omega)$, or even whether it is computable. Put another way, we don't know if $(1 \times 4 \times \omega) + (\omega^2 \times 1 \times 1)$ has a recursive set of P-positions. The sets

$$\{(a, b, g) : G((4 \times a) + (3 \times b) + (2 \times \omega)) = g\},$$
$$\{(a, b, g) : G((4 \times a) + (3 \times b)) = g\}$$

are recursive, but the set

$$L = \{g : G((4 \times a) + (2 \times \omega)) = g \text{ or } G(4 \times a) = g \text{ for some } a < \omega\}$$

is recursively enumerable and not known to be recursive. This is of interest because $G(4 \times \omega) = \text{mex}(L)$. We know L contains all $g < 46$, we don't know if $46 \in L$, but if $46 \in L$ then $46 = G((4 \times a) + (2 \times \omega))$ for some $a > 480$. If $\text{mex}(L)$ is infinite then we believe $G(4 \times \omega) = \omega \cdot 2$.

13. Conclusion

Extending Chomp from the naturals to the ordinals gives it a pleasing structure.

The main tools used here are size, the Fundamental Theorem, pairing strategies, "change of venue" arguments, and stratification. The Fundamental Theorem extends any position in one direction by any nonempty base to produce a P-position, leading to constructions that find Grundy values and create certain extensions with any given Grundy value. Pairing orthogonal summands with the same Grundy value creates a P-position, as does taking the cartesian product of an arbitrary P-position and any product of P-ordered P-positions. Stratification estimates a Grundy value by looking at the dominant piece of a position, while change of venue arguments switch strategies in response to bites that alter a position's large structure.

Results include the Grundy values of all two-wide positions, a list of all three-wide P-positions, and a list of some three-dimensional P-positions with a 2-by-2 base. In particular, the boxes $2 \times \omega$, $3 \times \omega^\omega$, $2 \times 2 \times \omega^3$, and $\omega \times \omega \times \omega \times 2 \times 2 \times 2$ are all P-positions.

We briefly touched on the computability of sets of P-positions, giving one example at the boundary of our current knowledge, a candidate for a small uncomputable set.

References

[Ga74] David Gale, *A Curious Nim-type Game*, Amer. Math. Monthly **81**:8 (1974), 876–879.

[Ga93] David Gale, Mathematical Entertainments, Math. Intelligencer **15**:3 (Summer 1993), 59–60.

[Ga96] David Gale, Mathematical Entertainments, Math. Intelligencer **18**:3 (Summer 1996), 26.

[Ga98] David Gale, "Tracking the Automatic Ant and Other Mathematical Explorations," Springer-Verlag, 1998, Chapter 11.

[Sch52] Fred. Schuh, "The Game of Divisors," *Nieuw Tijdschrift voor Wiskunde* **39** (1952), 299–304.

[BCG82a] Elwyn R. Berlekamp, John H. Conway, Richard K. Guy, *Winning Ways for your Mathematical Plays, Volume 1: Games in General*, Academic Press, 1982.

[BCG82b] Elwyn R. Berlekamp, John H. Conway, Richard K. Guy, *Winning Ways for your Mathematical Plays, Volume 2: Games in Particular*, Academic Press, 1982.

SCOTT HUDDLESTON
INTEL CORPORATION, JF4-451
5200 NE ELAM YOUNG PKWY
HILLSBORO, OR 97124
UNITED STATES
scotth@ichips.intel.com

JERRY SHURMAN
MATHEMATICS DEPARTMENT
REED COLLEGE
PORTLAND, OR 97202
UNITED STATES
jerry@reed.edu

More Games of No Chance
MSRI Publications
Volume **42**, 2002

A Memory Efficient Retrograde Algorithm and Its Application To Chinese Chess Endgames

REN WU AND DONALD F. BEAL

ABSTRACT. We present an improved, memory efficient retrograde algorithm we developed during our research on solving Chinese chess endgames. This domain-independent retrograde algorithm, along with a carefully designed domain-specific indexing function, has enabled us to solve many interesting Chinese chess endgame on standard consumer class hardware.

We also report some of the most interesting results here. Some of these are real surprises for human Chinese chess experts. For example, the aegp-aaee[1] ending is a theoretical win, not as previously believed, a draw. Human analysis for this endgame over many years by top players has been proved to be wrong.

1. Introduction

Endgame databases have several benefits. First, the knowledge they provide about the game is perfect knowledge. Second, the databases, because of their complete knowledge about certain domains, are a useful background for Artificial Intelligence research, especially in machine learning. Third, the databases often provide knowledge beyond that achieved by humans, (and increasingly, beyond that achievable by humans). Many endgame databases in many games have been constructed since Ströhlein's pioneering work (Ströhlein, 1970). In Chess, Thompson (1986) generated almost all 5-men chess endgame databases and made them widely available in CD format. Databases construction also enable Gasser (1996) to solve the game of Nine Men's Morris. Perhaps the most impressive endgame database construction so far is Schaeffer's work (Schaeffer et al., 1994) on Checkers. His program created all 8 men endgame databases and comprised more than 440 billion (4.4×10^{11}) positions. The databases played a very important role in Chinook's success.

---

[1]One side with King, one Assistant, one Elephant, one Gunner, and one Pawn against another side with King, two Assistants, and two Elephants.

We started our work on constructing Chinese chess endgame databases back in 1992. We have constructed many Chinese chess endgame databases since. We wanted our program to solve as many endgames as possible using only moderate hardware we had access to. So a major effort was made to improve the retrograde algorithm as well as examine ways to reduce the size of the database.

Armed with careful analysis to reduce the databases' size and our fast, memory efficient retrograde algorithm, we were able to solve one class of very interesting Chinese chess endgames. In this class of endgames, one side has no attacking pieces left but have various defending pieces, while the other side has various attacking pieces. In this paper, we describe our new retrograde algorithm, the database indexing we employed to reduce the size of the database, as well as the results we found, include the result for the aegp-aaee endgame. The database reveals a surprise for human players.

2. Fast, Memory Efficient Retrograde Algorithm

The retrograde algorithm used to construct the databases demand large computing resources. Thompson's 1986 algorithm (Thompson, 1986) needs two bits of RAM for every board configuration, or random access to disc files, and it only gives one side's result. Stiller's algorithms (Stiller, 1995) need a highly-parallel supercomputer. Others also need similarly large amounts of memory or need a very long time to build a moderate database. In chess for example, Edwards reported that a 5-man endgame database took 89 days on a 486 PC (Edwards, 1996). Six-men chess databases have to deal with approximately 64 billion positions, although symmetries reduce this, and careful indexing to eliminate illegal positions can reduce this to 6 billion for some positions (Nalimov, 2000). Database construction is hardware-limited. This will remain so indefinitely. Although hardware advances continually bring computations that were previously out of reach into the realm of practicality, each additional piece multiplies the size of the task (of creating a complete set of databases) by around three orders of magnitude. Hence efficient algorithms, as well as the indexing methods that minimize hardware resources will always be desired. We present our fast and memory efficient algorithm here first, and discuss the indexing methods in the next section.

2.1. Previous Work. The first widely-known description of the basic retrograde algorithm was by Ströhlein (1970). It was independently re-invented by many people, including Clarke (1977) and Thompson (1986, 1996). Herik and Herschberg (1985) give a tutorial introduction to the retrograde concept. Thompson's first paper (1986) gives a brief but clear outline of his algorithm. His second paper (Thompson, 1996) gives details of an improved algorithm he used to solve 6-piece chess endgames.

These algorithms all assume one side is stronger than another and only calculate the distance to conversion (DTC), which is the number of moves the stronger side needs to either mate the weaker side or transfer into a known win position in a subgame. There are two problems here:

- The algorithm assumes that if the stronger side does not win a position, it must be a draw. This is only true for some endgames. Other endgames need a separate database to be constructed for the other side. And even with two separate databases, one has to be very careful when using this database system. Some endgame types now need one database to be probed, others need two, and there are additional technical inconveniences and hazards, in using the databases after construction.
- Sometimes the distance to mate (DTM) may be preferred to the distance to conversion.

To address these problems, variations on the basic algorithm have been used. Wu and Beal (1993) designed an algorithm that retains full information for both sides when generating databases. Edwards (1996) also designed an algorithm which can generate chess endgame database containing both side's distance to mate/loss. However, these algorithms usually need access the main database during the construction, and so require much more memory than Thompson's algorithm. To use these algorithms, one either has to run it on very high-end machine, or wait while it takes very long time to run. For example it took Edwards' program more than 89 days to build a 5 piece chess endgame database on a 486 PC (Edwards, 1996). In the most recent work on chess endgame KQQKQQ Nalimov had to use a 500 MHz server machine with 2GB of memory to build this 1.2 billion entries database (Nalimov 1999, 2000).

2.2. Fast, Memory Efficient Retrograde Algorithm We devised an algorithm which only needs one bit per position like Thompson's algorithm, but can still generate full information for both sides, like other algorithms. Furthermore, it can compute either distance to mate (DTM) or distance to conversion (DTC), controlled by a boolean variable.

We present our new retrograde algorithm on the next page in a pseudo C format:

The algorithm generates a pair of databases, one for each side. And the result gives the exact distances for win/lose/draw. In other words, it generates full information about the endgame.

The algorithm itself can be understood in terms of relatively simple operations on bitmaps, in the style first used by Thompson (1986). The summary below gives the memory and access requirement for this algorithm.

- DoInitialize() uses sequential access to database W, B and bitmap S. It also uses random access for bitmap R.
- Load() use sequential access to one database and one bitmap.

```
DATABASE W, B;   // full databases (i.e. depth to win/loss for each position),
                 // W for White-to-move positions, B for Black-to-move,
                 // sequential access only
SBITS    S;      // sequential access only bitmap
RBITS    R;      // random access bitmap

void TopLevel
{
  DoInitialize();
  n = 0;                              // depth to mate or conversion
  while (!DoneWhite() && !DoneBlack())
  {
    if (!DoneWhite())                 // last pass added new positions
    {
      S = Load(W, WIN_IN_N(n));       // S = WTM win_in_n
      R = Predecessor(S);             // R = BTM predecessors of S
      S = Load(B, UNKNOWN);           // S = BTM unknown
      S = S & R;                      // S = BTM maylose_in_n
      R = Load(W, WIN_<=_N(n));       // R = WTM win_in_n or less
      S = ProveSuccessor(S, R);       // S = BTM lose_in_n
      B = Add(S, LOSE_IN_N(n));       // B += S
      if (dtm)                        // distance_to_mate?
        S = Load(B, LOSE_IN_N(n));    // S = BTM lose_in_n
      R = Predecessor(S);             // R = WTM maybe win_in_n+1
      S = Load(W, UNKNOWN);           // S = WTM unknown
      S = S & R;                      // S = WTM win_in_n+1
      W = Add(S, WIN_IN_N(n+1));      // W += S
    }
    if (!DoneBlack())                 // done for BTM?
    {
      S = Load(B, WIN_IN_N(n));       // S = BTM win_in_n
      R = Predecessor(S);             // R = WTM predecessors of S
      S = Load(W, UNKNOWN);           // S = WTM unknown
      S = S & R;                      // S = WTM maylose_in_n
      R = Load(B, WIN_<=_N(n));       // R = BTM win_in_n or less
      S = ProveSuccessor(S, R);       // S = WTM lose_in_n
      W = Add(S, LOSE_IN_N(n));       // W += S
      if (dtm)                        // distance_to_mate?
        S = Load(W, LOSE_IN_N(n));    // S = WTM lose_in_n
      R = Predecessor(S);             // R = BTM maybe win_in_n+1
      S = Load(B, UNKNOWN);           // S = BTM unknown
      S = S & R;                      // S = BTM win_in_n+1
      B = Add(S, WIN_IN_N(n+1));      // B += S
    }
    n = n + 1;
  }
}
```

- Predecessor() use sequential access for bitmap S and random access for bitmap R.
- ProveSuccessor() use sequential access for bitmap S and random access for bitmap R.
- Add() uses sequential access for one database and bitmap S.

Thus the peak memory requirement is only random access for one bitmap and sequential access for one bitmap and two databases. Because the sequential file access is many times faster than random access, we keep the random access bitmaps to a minimum, and require just one bitmap in memory. Apart from the chess/Chinese chess specific routines, the rest is just simple load and boolean operations over the files.

The algorithm is a generic one, and can be used to construct databases for other games. For example, to construct a 5-men pawn-less chess endgame database, 15MB RAM is sufficient to avoid random disc access. The algorithm will enable such databases to be built on a modest desktop PC in the matter of hours.

3. Reducing the Size of the Database

The Chinese chess board has 90 squares, so the simplest formula for the size of a sufficiently-large database is simply 90^n, where n is the number of the pieces in that endgame. For almost all interesting endgames, this is too large for today's technology to handle. However, the size of database can be greatly reduced by careful analysis of geometric and combinational symmetries, and game-specific details. In the following sections, we detail the methods we used to reduce the size of our database. Our methods may be compared with those of Nalimov and Heinz (2000) for western chess.

3.1. Limiting the Pieces' Placement to Legal Squares. In Chinese chess, certain type of pieces can only move inside certain parts of the board. For example, the king can only be inside the palace, the assistants can only move in the five squares inside the palace, the elephants only have seven squares, and the pawn can only move forward before it crosses the river. Table 1 gives each kind of piece and its possible squares.

So a direct way to calculate the size of a more compact database is to enumerate all possible squares for every pieces in that endgame. This brings the size down considerably, but again for most interesting endgames, it is still too big for the hardware we have access to. Fortunately, there are further ways to reduce the size, as described in the next few sections.

3.2. Vertical Symmetry. The Chinese chess board has vertical symmetry and this gives us a reduction factor of almost 2. It is almost rather than exact, because it is possible for all pieces to be in the centre file, and such positions are their own mirror position, leading to no reduction. Moreover, there is a

Name	Notation	Squares
King	k	9
Assistant	a	5
Elephant	e	7
Horse	h	90
Chariot	c	90
Gunner	g	90
Pawn	p	55

Table 1.

significant processing cost to obtain the full reduction, because the program must process all pieces in turn to reflect all piece locations, and this operation has to be re-done every time a move is made. The retrograde analysis program is both space intensive and compute intensive, so this is unwelcome. If we simplify the processing to only consider the first piece in the symmetry reduction, we achieve about half the possible reduction, as shown in Table 2.

3.3. Multiple Piece Symmetry. If in an endgame, there is more than one piece of the same type, we can exchange these pieces' places without altering the position. In other words, the pieces of the same type together have a single contribution to the database. The size of the contribution can be determined by the combinatorial arithmetic as follows.

If an endgame has n of the same type of piece and this type of piece can only move in m Squares, and then the size of this contribution is:

So for the endgames with one side having two of the same type of pieces, this reduces its size by more than half, and the saving is even greater if one side has more than two of the same type of piece.

Name	Notation	Total	VSR	Saving
King	k	9	6	33.33%
Assistant	a	5	3	40.00%
Elephant	e	7	4	42.86%
Horse	h	90	50	44.44%
Chariot	c	90	50	44.44%
Gunner	g	90	50	44.44%
Pawn	p	55	31	43.64%

Table 2.

Name	Size	VSR	Saving
kaa	70	38	45.71%
ka	40	21	47.50%
kae	275	138	49.82%
ke	62	32	48.39%
kee	183	96	47.54%
ee	21	12	42.86%
hh	4005	2045	48.94%
cc	4005	2045	48.94%
gg	4005	2045	48.94%
rp2	1485	765	48.48%
bp2	1485	765	48.48%
rp3	26235	13135	49.93%
bp3	26235	13135	49.93%

Table 3.

3.4. Piece Grouping. In Chinese chess, both king and assistants can only move inside the palace, and this allows us another possible way to reduce the database's size. We can consider a few different type of pieces together, or piece grouping. These pieces will be considered together and form a single contribution. This offers increased reduction over applying space-enumeration and symmetry to the pieces separately, because the group enumeration can eliminate impossible positions in which pieces occupy the same square. This saving is significant for Chinese chess.

Taking a king and two assistants (kaa) as an example, we can enumerate all possible patterns for these three pieces in the palace, and assign a unique id to each pattern. It turns out that kaa only has 70 different patterns, which compares to $9 \times 10 = 90$ if we do it separately.

Moreover, we can combine the piece group enumerations with vertical symmetry. In the case of kaa, 32 out of possible 70 patterns are just mirrors of others. In other words, there are only 38 different patterns if we take the vertical symmetry into account. The saving here is $32/70 = 0.457$, which is better than the savings we get from the best single-piece vertical symmetry reduction.

There are more situations in which a few pieces can be considered together and give a single contribution. Table 3 lists more of the possible piece groupings that can be utilised.

Note the kae contribution in Table 3. kae has 275 possible patterns , and 138 if we consider the vertical symmetry reduction. The saving here is 0.498, which is slightly greater than the kaa contribution, and very close to the full saving of 0.5 that represents the theoretical symmetry limit. So in practice, we should always

endgame	database size	endgame	database size
h-aaee	646,380	egp-aaee	232,848,000
c-aaee	646,380	aagp-aaee	276,507,000
g-aaee	646,380	eegp-aaee	698,544,000
hp-aaee	35,550,900	aegp-aaee	1,004,157,000
cp-aaee	35,550,900	agg-aaee	123,634,350
gp-aaee	35,550,900	egg-aaee	188,395,200
hh-aaee	27,055,350	aagg-aaee	223,719,300
cc-aaee	27,055,350	eegg-aaee	565,185,600
gg-aaee	27,055,350	aegg-aaee	812,454,300
hc-aaee	58,174,200	pp-aaee	10,120,950
hg-aaee	58,174,200	hpp-aaee	910,885,500
cg-aaee	58,174,200	gpp-aaee	910,885,500
agp-aaee	152,806,500	cpp-aaee	910,885,500

Table 4.

use the group with maximum savings to incorporate the symmetry reduction, rather than apply separate symmetry reductions.

4. Results from the Database

We concentrate on the most interesting class of endgames. In these endgames, one side has no attacking pieces left but only defensive pieces, while the other side has various attacking pieces. Players always are keen to know what kind of piece combinations are enough to win, how hard or how easy it is to win, and how long it will take to win.

The perfect knowledge contained in the databases is invaluable for anyone who is serious about Chinese chess. Moreover, our databases show that current human understanding about this kind of endgames is far from perfect. Table 4 lists the endgames that we have solved.

To build any endgame database, the program has to build all its subgame databases first. For example, c-aaee endgame have 9 subgames, k-k, c-k, c-a, c-e, c-aa, c-ee, c-ae, c-aae, and c-aee. For the endgames we list above, if we count all subgames as well, there will be 378 in total, with about 35 billion entries in the resulting databases. Our construction program solves all these subgames automatically.

Our results are summarized on the next page, with some comments made in the next few paragrpahs. In the table, "db size" (database size) is the size of the database for this endgame. If we take out the illegal positions, and disregard the various symmetry reductions, we have a more human-like classification. We call this the "real size". This is more useful to humans when we talk about a particular endgame. All the percentage data are based on "real size" in this

| database | db size | real size | DTC | RTM(%) | | BTM(%) | |
				W	D	D	L
h-aaee	646,380	3,363,048	1	5.58	94.42	99.88	0.12
c-aaee	646,380	3,363,048	18	84.60	15.40	48.56	51.44
g-aaee	646,380	3,363,048	1	7.68	92.32	99.63	0.37
hp-aaee	35,550,900	169,290,216	31	31.06	68.94	94.47	5.53
(highpawn)		59,694,192	31	35.47	64.53	92.33	7.67
cp-aaee	35,550,900	169,290,216	14	99.99	0.01	8.12	91.88
(highpawn)		59,694,192	9	100.00	0.00	7.05	92.95
gp-aaee	35,550,900	169,290,216	12	14.15	85.85	99.31	0.69
(highpawn)		59,694,192	8	12.52	87.48	99.12	0.88
hh-aaee	27,055,350	284,348,544	17	99.34	0.66	12.70	87.30
cc-aaee	27,055,350	284,348,544	3	100.00	0.00	2.26	97.74
gg-aaee	27,055,350	284,348,544	29	40.49	59.51	92.23	7.77
hc-aaee	58,174,200	284,348,544	9	99.98	0.02	7.43	92.57
hg-aaee	58,174,200	284,348,544	22	99.40	0.60	12.73	87.27
cg-aaee	58,174,200	284,348,544	7	100.00	0.00	7.30	92.70
agp-aaee	152,806,500	765,989,832	32	44.95	55.05	86.32	13.68
(highpawn)		269,533,904	32	60.22	39.78	75.66	24.34
egp-aaee	232,848,000	1154,955,840	26	26.99	73.01	95.60	4.40
(highpawn)		409,481,824	26	39.12	60.88	89.93	10.07
aagp-aaee	276,507,000	2730,979,776	39	45.43	54.57	86.02	13.98
(highpawn)		958,979,200	39	60.74	39.26	75.22	24.78
eegp-aaee	698,544,000	6752,096,208	34	27.01	72.99	95.62	4.38
(highpawn)		2407,076,064	34	39.06	60.94	90.07	9.93
aegp-aaee	1004,157,000	5211,916,080	95	70.52	29.48	50.51	49.49
(highpawn)		1844,080,608	73	99.60	0.40	10.91	89.09
agg-aaee	123,634,350	1270,822,608	20	99.65	0.35	12.37	87.63
egg-aaee	188,395,200	1928,810,256	42	53.36	46.64	82.02	17.98
aagg-aaee	223,719,300	4474,606,752	21	99.64	0.36	12.54	87.46
eegg-aaee	565,185,600	11210,239,584	29	99.45	0.55	12.93	87.07
aegg-aaee	812,454,300	8595,899,712	20	99.64	0.36	12.53	87.47
pp-aaee	10,120,950	99,964,368	11	12.44	87.56	99.19	0.81
hpp-aaee	910,885,500	8338,089,024	34	96.72	3.28	23.93	76.07
gpp-aaee	910,885,500	8338,089,024	40	89.54	10.46	34.34	65.66
cpp-aaee	910,885,500	8338,089,024	12	100.00	0.00	5.77	94.23
aegp-aae	334,719,000	907,504,776	36	94.11	5.89	22.16	77.84
gpp-ee	113,395,950	549,454,356	39	94.03	5.97	23.92	76.08
hpp-aee	520,506,000	2,414,172,384	46	98.67	1.33	10.79	89.21
aagp-aee	152,806,500	786,450,264	49	85.37	14.63	31.46	68.54
egg-aee	107,654,400	541,173,192	51	30.59	69.41	95.66	4.34

paper. DTC is the maximum distance for the stronger side to capture a piece and transfer to a known winning subgame. W, D, and L stand for percentage of wins, draws, and losses. The L column under RTM and the W column under BTM are omitted since the values are zero in every case.

4.1. One Major Piece. From the first section of the table on page 221, we can see that when the attacking side has only one major piece, the game is usually a draw. However if the major piece is a chariot, the stronger side does have some chance to win, especially if it is the stronger side move first. In other word, the right to move, or initiative is very important here. The winning rate is 84.60% if the stronger side moves first, and drops to 51.44% if the opponent moves first.

4.2. One Major Piece Plus a Pawn. These results are given in the second section of the table on page 221. The second row for each endgame is the statistics if we assume the pawn is a high pawn. Most Chinese chess textbooks use this term to help classify endgames.

Chariot plus a pawn is too much to defend, and this comes with no surprise, because one can always use the pawn to exchange a minor piece, and a chariot against the rest of defense is a sure win.

Gunner plus a pawn is proved to be not enough to win. The gunner needs helps from minor pieces to really take effect, and in this case there are no helping pieces.

Horse plus a pawn is a very powerful combination, and it requires a very high level of skill to play well. The results here show that horse and pawn is not enough to break the best defense, even though it does have some chances.

4.3. Two Major Pieces. See third section of the table. If the stronger side have two major pieces, the advantage is usually overwhelming, except if both major pieces are gunners. Gunners need help from minor pieces, and lack of those minor pieces prevents the win.

4.4. One Gunner, One Pawn Plus Some Minor Pieces. See fourth section of the table. If we have a gunner, a pawn, what else we need to win? Having a single minor piece, either an Assistant, or an Elephant is not enough to secure a win. Two minor pieces of same type are also not enough. One has to have one minor piece of each kind to win, and if the pawn is a high pawn, the winning is guaranteed. This is a real surprise discovery, and will be discussed in detail at next section.

4.5. Two Gunners Plus Some Minor Pieces. See fifth section of the table. Two gunners plus a Assistant is already enough for win. And so two Assistants, or one Assistant and one Elephant, is also a win. But two gunners plus an Elephant is a totally different story. It most likely will end up as a draw, especially if the opponent moves first.

4.6. Two Pawns Plus One Major Piece. See penultimate section of the table. Two pawns without any major piece is not enough to win. However, one major piece and two pawns are usually too much to defend. Chariot and one pawn is already a sure win, two pawns can only make the winning sooner. Horse and two pawns are also a win, no matter how bad the pawn's positions are. Gunner and two pawns can also be regarded as a win, but the maximum DTC of 40 indicate it can be hard to play this endgame sometime.

4.7. Some of the Hard Subgames. We list some of the hard subgame results in the last section of the table on page 221. Results for all subgames can be found online at http://www.msri.org/publications/books/Book40/files/wu-beal.txt.

These endgames are mostly winning for stronger side except for egg-aee. However, the distance-to-conversion is rather long and so it can be hard to play them accurately.

5. The aegp-aaee Endgame

Among the many Chinese chess endgames we solved so far, the most interesting discovery is the aegp-aaee endgame. It is interesting not only because humans had made a thorough investigation on this endgame and it is very well known amongst experts, but also because our research reveals that this endgame is a theoretical win for the stronger side (aegp), contrary to current human belief. Furthermore, the maximum distance to conversion (the maximum number of moves the stronger side need to capture the first piece) is 95, which is 35 more than the maximum move allowed in official Chinese chess rules.

5.1. Human Analyses. Gunner and Pawn is a well-known class of endgames, where the strong side has only one Gunner and one Pawn as the attack force, plus a few defending pieces, while the opponent usually have the best defence possible, that is have both elephants, and assistants. One of the very early Chinese chess endgame books, the *Shi Qin Ya Qu*, first published in 1570, has already a few positions in this class. The most famous work on this class of endgame in human history, however, has to be the excellent work by Chen Lianrong at 1930s. His book, *Pao Bin Endgames*, is dedicated to this class of endgame, and has been regarded as a milestone in modern Chinese chess endgame theory.

In his book, he showed that the aeegp-aaee endgame is theory win for the stronger side. That was a real surprise for the Chinese chess community at that time, because the common belief at that time was that even aaeegp-aaee is a draw game! This is one of the biggest contribution of his book.

However, our research has produced a stronger result by showing that even with one elephant, the stronger side has already enough to win, even though it can take as many as 95 moves to capture the first piece!

5.2. Computer Analysis. To produce the computer analysis, the program ran on a Pentium Pro 200 machine with 128 MB memory under Windows NT 4.0. It took about 92 hours to generate the database for this endgame containing both sides full information using the distance-to-conversion (DTC) metric.

Here is a further breakdown of the surprise result, that aegp-aaee is actually a mostly-winning endgame. It is a win provided that the pawn has passed the river or can pass the river safely, and is not an old-pawn2.

```
Database:      110011-220000      Database Size: 1004157000

                   RTM                       BTM
Illegal     244406400 ( 24.34%)      265487568 ( 26.44%)
Lose                0 (  0.00%)      358203335 ( 35.67%)
Draw        290685444 ( 28.95%)      380466097 ( 37.89%)
Win         469065156 ( 46.71%)              0 (  0.00%)
Total      1004157000 (100.00%)     1004157000 (100.00%)

Maximum Distance-to-conversion:   95
```

If we take out the illegal positions, and disregard the various symmetry reductions, we have a better, or more human like, classification, as follows:

```
Database:      110011-220000      Real size:     5211916080

                   RTM                       BTM
Lose                0 (  0.00%)     2579230888 ( 49.49%)
Draw       1536511488 ( 29.48%)     2632685192 ( 50.51%)
Win        3675404592 ( 70.52%)              0 (  0.00%)
Total      5211916080 (100.00%)     5211916080 (100.00%)

Maximum Distance-to-conversion:   95
```

If we assume the pawn is a high-pawn, we have:

```
Database:      110011-220000      HighPawn:      1844080608

                   RTM                       BTM
Lose                0 (  0.00%)     1642833536 ( 89.09%)
Draw          7324704 (  0.40%)      201247072 ( 10.91%)
Win        1836755904 ( 99.60%)              0 (  0.00%)
Total      1844080608 (100.00%)     1844080608 (100.00%)

Maximum Distance-to-conversion:   73
```

We give a position with the maximum distance to conversion on the next page. We use Chinese notation, where each side counts the column from his right hand side, '=' means move side ways, '+' means move forward and '−' means move backward. The first letter is the type of moving piece, followed by a number indicating which column the moving piece resides in. In case there is

3a1Ge2/9/e2ak4/9/9/9/2P6/4E4/3K5/5A3/r ---- 407963625

RTM   407963625

1. G4-8 k5=6 2. A4+5 k6-1 3. A5+4 k6=5 4. G4=5 k5=6 5. G5=2 a4-5 6. G2=4 a5+6
7. E5+3 e7+9 8. G4=3 k6-1(k6=5,a4+5,a6-5,e1-3,e1+3) 9. K6=5(K6-1,G3+1,G3+2,
G3-1,P7+1) k6+1(a4+5,a6-5,e1-3,e1+3) 10. K5-1(G3+1,G3+2,G3-1,P7+1) k6-1(a6-5,
e1-3,e1+3) 11. G3+1(G3+2,G3-1,P7+1) k6+1(a6-5,e1-3,e1+3) 12. G3+1(G3-2,P7+1)
k6-1(a6-5,e1-3,e1+3) 13. G3-2(P7+1) k6+1(a6-5) 14. K5=4(P7+1) a6-5 15. P7+1
a5+4 16. G3=4 k6=5 17. G4=6 k5-1 18. G6=3 k5+1 19. A4-5 k5=4 20. K4=5(A5+6,
G3-1) a4-5 21. A5+6 a5+4 22. K5+1(G3-1) a4-5 23. G3-1 a5+4 24. K5=6 a4-5 25. E3-5
(G3=5) a5-6(e9-7,e9+7) 26. G3=5 a4+5(e9-7,e9+7) 27. K6=5 k4-1(e9-7, e9+7) 28. E5-7
k4=5(e9-7,e9+7,e1-3) 29. E7+9 k5=4(e9-7,e9+7,e1-3) 30. K5=6 (G5=9) k4+1(a5+6,
e9-7,e9+7,e1-3) 31. G5=9 e1-3 32. K6-1(G9+1,P7+1) k4-1(e9-7,e9+7) 33. P7+1 k4=5
(a5+6,e9-7,e9+7,e3+5) 34. A6-5(E9+7,G9=7,G9+1,P7+1,P7=6) a5-4(e9-7,e9+7) 35.
G9=7(P7+1) e3+1(e3+5) 36. P7+1(P7=6) a6+5 37. K6=5(E9+7,P7=6) k5=6 38. E9+7
(P7=6) e1-3 39. P7=6 e3+5 40. P6=5 k6+1(a5+4,e9-7,e9+7,e5-7) 41. G7+1(G7+2,G7+3,
P5=4) k6-1(e9-7) 42. P5=4 k6=5(k6+1,a5+4,e9-7,e9+7,e5-7) 43. P4=3 k5=6(e9-7,e9+7)
44. A5+4(P3=2) k6=5(a5+4,e9-7,e9+7,e5-7) 45. P3=2 k5=6(a5-6,a5+4,a5+6,e9-7,e5-7,
e5+7) 46. P2=1 e9-7(e9+7) 47. K5=4 e5+7 48. G7=4 k6=5 49. G4=3(G4=8,G4=9) e7-5
50. G3=2 k5=6 51. G2+8 k6+1 52. E7-5 e5+3(e5+7) 53. G2-8 k6-1 54. G2=4 k6=5 55.
G4=8 a5-6 56. K4=5 a4+5(e7+5,e3-1) 57. K5=6 a5+4(e7+5) 58. G8+8(P1=2) e7+5(e3-5,
e3-1) 59. P1=2 a6+5(e5-7,e5+7,e3-1) 60. P2=3 k5=6(e5+7) 61. P3=4 k6+1 62. P4=5
k6-1(e5+7) 63. K6=5(E5+3,P5=6) e5+7 64. K5=4(E5+3,P5=6) e3-1 65. E5+3(P5=6)
e7-9 66. P5=6 e9+7 67. G8-8 e7-5 68. P6=5 e1-3 69. G8+8 k6+1 70. P5=4 e5+3(e3+1)
71. A4-5(E3-5) e3+5 72. A5-6(E3-5) k6-1(a5-6,e5-3) 73. K4+1 k6+1(e3-1) 74. K4+1
k6-1(e3-1) 75. K4=5(A6+5) k6=5(e3-1) 76. A6+5 e3-1 77. P4=5 e5+7 78. K5=4 a5+6
79. G8-9 e1+3 80. K4=5(G8=5) k5=6(e3-5,e3-1,e7-5) 81. A5+4(G8=5,G8=4) k6=5
(a4-5,e7-5) 82. G8=6(G8=4) e7-9 83. K5=6 a4-5 84. A4-5(P5=4) k5=6(a5-6,e3-1,e9-7,
e9+7) 85. A5-4(P5=4) k6=5(k6+1,e3-5,e3-1,e9-7,e9+7) 86. P5=4 k5=6(e3-5,e9-7) 87.
P4=3 k6=5(k6+1,e3-5,e3-1,e9-7,e9+7) 88. G6=5(P3+1) k5=6(e3-5) 89. P3+1 e3-5
(e9-7,e9+7) 90. K6=5 e9-7(e9+7) 91. K5=4 e5-3(e5+3,e5+7) 92. K4-1(E3-1,G5+1)
e3+1(e7+9) 93. A4+5(E3-5) k6+1 94. P3+1 k6-1 95. G5+8

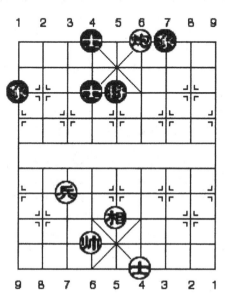

more than one same type of pieces in same column, 'F' and 'B' can be used to specify "front" or "back". The last number can be the relative ranks this move take, if the move is within the same column, or it can be the target column, if the target square is in a different column.

6. Conclusion

In this paper, we have described a improved, memory efficient retrograde algorithm. We have also outlined some methods we employed to reduce the size of the Chinese Chess endgame databases. Then we give the results for some of the Chinese chess endgames. One example in detail is the aegp-aaee endgame. The discovery that the aegp-aaee is a winning endgame is very interesting. Our research has shown that human understanding about this endgame is still far from perfect.

And the maximum distance of 95 will certainly be a challenge to human capacity, if not beyond it. This result is likely to shock the Chinese chess community, and makes a significant addition to knowledge of the world's most popular game.

References

[1] Clarke, M. R. B. (1977). A Quantitative Study of King and Pawn against King. Advances in Computer Chess 1, 108-118. Edinburgh University Press, Edinburgh.

[2] Edwards, S. J. (1996). An Examination of the endgame KBBKN. ICCA Journal, Vol. 19, No. 1. 24-32.

[3] Gasser, R. (1996). Solving Nine Men's Morris. Games of No Chance (ed. R. J. Nowakowski), pp. 101-113. MSRI Publications, v29, CUP, Cambridge, England. ISBN 0-521-64652-9.

[4] Herik, H. J. and Herschberg, I. S. (1985). The Construction of an Omniscient endgame data base. ICCA Journal, Vol. 8, No.2, 66-87.

[5] Lake, R. , Schaeffer J., and Lu, P. Solving Large Retrograde Analysis Problems Using a Network of Workstations. Advances in Computer Chess 7, Maastricht, Netherlands, 1994, 135-162.

[6] Nalimov, E. V. and Heinz, E. A. (2000). Space-Efficient Indexing of Endgame Databases for Chess. Advances in Computer Chess 9. (eds. H. J. van den Herik and B. Monien)

[7] Nalimov, E. V., Wirth C. and Haworth G. Mc. C., (2000) KQQKQQ and the Kasparov-World Game, ICCA Journal, Vol 22, No. 4, pp.195-212

[8] Stiller, L. B. (1995). Exploiting Symmetry On Parallel Architectures. Ph.D. thesis. The John Hopkins University, Baltimore, Maryland.

[9] Ströhlein, T. (1970). Untersuchungen über kombinatorische Spiele. Dissertation, Fakultät für Allgemeine Wissenschaften der Technischen Hochschule München.

[10] Thompson, K. (1986). Retrograde Analysis of Certain Endgames. ICCA Journal, Vol. 9, No. 3. 131-139.

[11] Thompson, K. (1996). 6-Piece Endgames.

[12] Wu, R and Beal, D. F. (1993). Retrograde Analysis of some Chinese Chess Endgames. Technical Report. QMW 1993.

[13] Xu, Zhi (1570) Shi Qin Ya Qu.

[14] Chen, Lianrong (1930). Pao Bin Endgames.

REN WU
  ren_wu@hp.com

DONALD F. BEAL
DEPARTMENT OF COMPUTER SCIENCE
QUEEN MARY & WESTFIELD COLLEGE
LONDON E1 4NS
UNITED KINGDOM
  don@dcs.qmw.ac.uk

The New Classics

More Games of No Chance
MSRI Publications
Volume **42**, 2002

The 4G4G4G4 Problems and Solutions

ELWYN BERLEKAMP

ABSTRACT. This paper discusses a chess problem, a checkers problem, a Go problem, a Domineering problem, and the sum of all four of these problems. These challenging problems were originally entitled *Four Games for Gardner* and presented at *Gathering for Gardner, IV*. The solutions of these problems illustrate the power of extended thermography and the notion of rich environments, the relevance and utility of a broad theory of games which may include kos and other loopy positions, and the robustness of this theory to a variety of interpretations of the rules. It also demonstrates the relevance of this branch of mathematics to the classical board games.

Introduction

An enthusiastic group of puzzlers, magicians, and mathematical game buffs held weekend gatherings in Atlanta in January or February of 1993, 1996, 1998, and 2000. These meetings, which honor Martin Gardner, the well-known author and former Scientific American games columnist, are now called "Gatherings for Gardner". A collection of papers presented at the earlier gatherings was published by [?]. The problems shown in Figure 1 were presented at the fourth such gathering in February 2000. The problem statement appears in [?]. The present paper presents solutions to the problems that appeared in "Four Games for Gardner" for Gathering for Gardner IV. The solutions require some background in combinatorial game theory and thermography, topics with which the readers of this volume are assumed to be familiar.

Superficially, in Figure 1 there appears to be one problem in each of four different well-known games: Go, Domineering, checkers and chess. In each of the four games, the reader is to play white (horizontal in domineering). But lurking below the surface is a more interesting and much deeper problem which occurs if we play the sum of all four games added together.

There may be disputes about how to interpret the rules. Do they matter? In Go, purists might quibble over whether to use the North American, Chinese, or

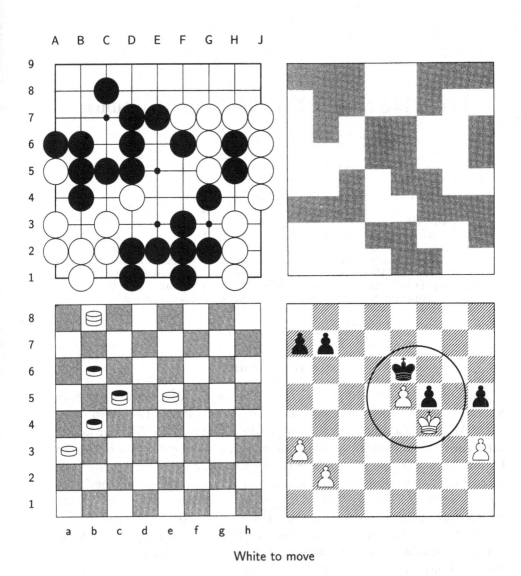

White to move

Figure 1. Initial position (in the Domineering board, the shaded areas are out of play).

Japanese version of the superko rule. (As in nearly all Go positions, it makes no difference here.) In chess, one might elect either the conventional rules or the simpler rules of "mortal chess", in which the circled kings and pawns are removed from the board and (as in checkers) the game ends when one player is unable to move, at which point his opponent is declared to be the winner. Again, in Figure 1 it makes no difference whether one uses the conventional or the oversimplified rule to determine who wins at chess.

The sum of all four games might also require clarification as to whether certain game-specific rules are to be interpreted locally or globally. In particular:

- The ko rule in Go forbids the recapture of a ko until after another move has been made elsewhere. Must this "elsewhere" be "elsewhere on the Go board", or can one play a chess move as a kothreat?
- The "compulsory capture" rule in checkers forbids noncapturing moves when a capture is available. Does it forbid only noncapturing checkers moves, or moves on all four boards?

Yet another question relates to the overall objective. Do we still seek to checkmate the opposing chess king, or to get the last move as in checkers and Domineering, or to surround the more territory on the Go board? Although Go players are accustomed to passing and then counting score, the American Go Association rules require that a player must pay a one-point fee for each pass. This fee is conventionally paid by returning one of your opponents' stones that you have captured to the pot. After both players have made such passes, they might very well elect to count score by alternately filling in their own territories to see who runs out of moves first. Thus, playing Go with the last-move-wins rule often has no effect on the players' strategies.

In Figure 1, it turns out that *none* of these legalistic questions really matters, because there is a uniform solution. White has a strategy to win the sum no matter which interpretation of the rules one's opponent might select. To simplify the discussion, we assume that the overall goal is to get the last legal move. However, this winning strategy is rather difficult to find without using a substantial body of knowledge about combinatorial game theory.

The notion of temperature turns out to be the key concept underlying all known analyses of problems of this type. Heuristically, the temperature of a move is a measure of the average amount one would gain by playing that move in an asymptotically large game. Rudimentary notions of temperature can be detected in [?; ?]. The modern version for loopfree games first appeared in [?] and *Winning Ways* [?]. Extensions needed to handle kos appeared in [?]. Refinements and reformulations continue.

The present problem is intended to encourage comparisons of temperatures among different games.

As expounded in [?], temperatures provide the basis of a strategy called *sentestrat*, which is optimum when the number of regions is asymptotically large.

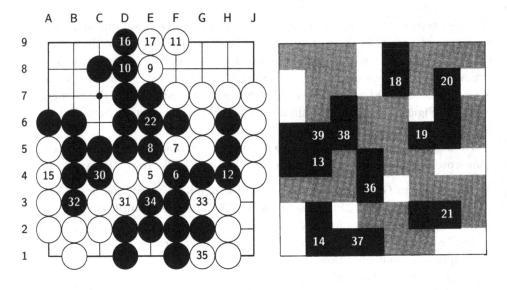

	CHECKERS			CHESS			
White	1:e5 d6	3:a3 a7		23:a4	25:a5	27:a6	29:b3
Black	2:c5 e7		4:e7 d6	24:h4	26:b5	28:b4	

Board	Chk	Go	Go	Go	Do	Do	Go	Do	Do	Do	Go	Chs	Go	Do	Do	Do
Moves	1–4	5–8	9–11	12	13	14	15–17	18	19	20–21	22	23–29	30–35	36–37	38	39
t	∞	>19	3	2	$\frac{9}{8}$	1	1	1	$\frac{3}{4}$	$\frac{1}{2}$	$\frac{1}{3}$	0	0	0	$-\frac{1}{4}$	$-\frac{1}{2}$

Figure 2. One orthodox line of play.

Any line of play consistent with sentestrat is called *orthodox*. *Orthodox account-ing* provides a method for analyzing other lines of play. It turns out that perfect play differs from orthodoxy only very rarely, and then usually only by making very small "sacrifices" that turn out to be investments that yield quick returns.

One orthodox line of play to the problem of Figure 1 is shown in Figure 2.

Although the line of play shown in Figure 2 is not perfect, it provides a baseline from which we are able to analyze the combined 4-board problem. But, let us first consider the four isolated warmup problems individually. We claim that in isolation, White should draw Checkers, and win at each of Go, Chess, and Domineering. When all four boards are treated as a single game played together, we claim that White can win, no matter how one chooses to interpret the rules.

Isolated Checkers

When the checkers game is played alone, White's best first move is from e5 to f6. This leads to a draw. If instead he plays as indicated in the caption below Figure 2, he loses the checkers game in only four moves. However, this position has value 0, which is unusual in checkers. If either player moves again, his opponent will play to a position in which he has an unlimited number of extra moves. In the spirit of Chapter 11 of *Winning Ways*, such a position is called on or off, depending on whether Black or White enjoys the unlimited advantage. (Actually, the on and off used here are bifurcations of the **on** and **off** in *Winning Ways*. Our on allows Left to move back and forth between two states, while Right has no moves. This models a lone checkers king who chooses to remain in a double corner. It also evades any conflict with the ko-ban rule.)

If White opens by playing the checker at e5 to f6 instead of d6, the position eventually becomes a Deathless Universal Draw (dud). The generic dud is the checkers position in which each side has a king located in a double corner. Since the dud always provides each an opportunity to move, neither player can win a game that contains a dud, no matter how strong his position might be elsewhere.

So, if White thinks he can win the sum of the Go, Domineering, and chess positions, then he should play the checkers position to 0 via the first four moves of the line of play depicted in Figure 2. If instead he thinks himself unable to win the sum of those three game positions, then he should open by playing the checkers position to a dud. This decision is more important than any finite number of extra moves that might be acquired by any imaginable capture on the Go board. The temperature of the checkers position is truly infinite.

Isolated Go

The Go moves shown in Figure 2, numbered 5, 6, 7, 8, 9, 10, 11, 12, 15, 16, 17, and 22, happen to alternate between White moves and Black moves. This also happens to be the correct sequence for the optimum line of play when the Go board is played in isolation.

Before move 5, the temperature is about 20 — to be precise, $19\frac{5}{12}$. Except for the Black stone at B4, none of the other stones located in rows 1–4 and columns A–G is yet safely connected to life. If either player can make a connection from his lower group through the central region around EF4–5 to his upper group, then he will kill his opponent's lower group and save his own. For example, if Black could play F5 in Figure 1, he would achieve this objective immediately, and all of White stones in the southwestern portion of the board would die.

This big issue is resolved in moves 5–8. Neither side is able to connect, but each side does succeed in preventing his opponent from connecting. The two opposing one-eyed groups are now destined to live, in seki. After move 8, the temperature drops dramatically, from about 20 to exactly 3.

(Alternative attempts to resolve this big issue prove less desirable. Thus, White's 5:F5?? is followed by 6:E4, 7:E5, 8:D3, 9:C4, 10:F4, so Black lives and kills White. If White's 5:E4 is met with Black's 6:F5?, we have 7:F4, 8:G3, 9:H4 ($t > 19$ again), 10:A4 (now $t = 3$), eventually leading to White at E5 and Black at E6; thus Black's small mistake at F5 has cost him $\frac{2}{3}$ point.)

Moves 9–11, at temperature 3, outline the boundary that divides the Black area in the northwest from the White area in the northeast. Move 12, at temperature 2, determines the fate of the two stones at H5–6. After move 12, the temperature of the Go board drops to 1, and after some plays elsewhere (moves 13–14), Go moves 15–17 are played at temperature 1. The node at E6 then has temperature $\frac{1}{3}$. After some more plays elsewhere reduce the ambient temperature to this level, Black 22:E6 reduces the temperature of the Go board to 0. Only dame remain. These are filled at moves 30–35. In this case, each of the Black plays at moves 30, 32, and 34 poses a grave threat that White wisely decides to answer. For example, if White elected to play either 33 or 35 elsewhere, then Black could continue with 36:C1, and then capture eleven Black stones with 38:A1. But this is prevented by White's 33 and 35, after which Black 36:C1? would be followed by White E1, capturing fourteen Black stones.

If this game were actually played in isolation, then after Black plays the stone at 22, White would begin playing at 31, 33, 35 while Black would play 30, 32, 34. Although the order in which these *dame* points are filled might differ, the result would be the same as shown in Figure 2.

After move 35 as shown in Figure 2, the empty points at ACE1 are *seki:* Neither player can play there without being subjected to a large loss. The other empty squares are all territory. Black has 9 points at ABC9, AB8, ABC7, and C6, but White has 10 points at GHJ9, FGHJ8, and J321. So White wins by one point. When the Go board is played in isolation, Black cannot do anything to prevent White from winning in this way.

We would obtain the same result if we insisted on continuing to play until the loser is unable to move. After each player has placed 7 stones in appropriate places in his own territory, only the seki and a few isolated empty nodes, called eyes, will remain. Playing into the opponent's territory only prolongs the game, because the opponent will eventually capture such stones. Following AGA rules, he will subsequently use up one turn returning each of them to the pot. We will eventually reach a position in which White has three eyes, and Black two. White then fills one of his eyes, and then Black's only remaining legal moves are disastrous. If he still refuses to resign, he must either fill in one of his own last two eyes, or play in the seki at nodes C1 or E1. No matter what he does, White will capture many Black stones on the next move. When play continues, White will fill in his new territory except for a few more eyes. Eventually the board will contain nothing but White stones and White eyes, and then Black will finally have no legal moves at all, because playing a suicidal move into an opponent's small eye is illegal in all dialects of Go rules.

Isolated Chess

All of the chess moves that occur in the line of play depicted in Figure 2 occur at an ambient temperature of 0. This means that the temperature of the entire four-board position is zero when the first chess move is made, and it is again zero when the last chess move is completed. The chess position (including the inactive kings and pawns shown in the circle in Figure 1) is among those analyzed mathematically by [?]. The line of play shown is canonical. Black cannot do anything to prevent White from winning in this way.

Isolated Domineering

The Domineering position shown in Figure 1 is the sum of several disjoint pieces, having canonical values $3 \mid \frac{3}{4}$, $1* \mid -1*$, $1 \mid -1$, $0 \parallel -1 \mid -2$, $0 \mid -1$, and $*$, with respective temperatures $\frac{9}{8}$, 1, 1, $\frac{3}{4}$, $\frac{1}{2}$, and 0. Black (Left) is vertical; White (Right) is horizontal. An orthodox strategy begins playing the regions in order of decreasing temperature, creating smaller regions with values $\frac{3}{4}$, $1*$, -1, 0, -1, and 0, respectively. Play will then continue. At $t = 0$, Right will play $1*$ to 1, and at $t = -\frac{1}{4}$, Left will play $\frac{3}{4}$ to $\frac{1}{2}$. At $t = -\frac{1}{2}$, Right will play $\frac{1}{2}$ to 1. All regions will then have values that are integers, and the sum of these values is 0. However, it is Left's turn to move, so Right wins.

To prove that Left cannot stop Right's win, we first observe that the values of the initial regions are all switches except for $1* \mid -1*$ (which is equal to the sum of two switches), and $0 \parallel -1 \mid -2$, whose incentives are confused only with the incentives of $1 \mid -1$ and $0 \mid -1$. So the only region which a canonical Left might possibly want to play out of its temperature-defined order is $0 \parallel -1 \mid -2$, and a simple calculation shows that this does her no good.

The Global Problem

We first notice that the line of play shown in Figure 2 led to a win for White. After move 38, only integers remain. White is one point ahead on the Go board ($9 - 10 = -1$), but one point behind on the Domineering board ($3 - 2 = +1$). However, Black's turn is next, so she will run out of moves just before White does.

However, Black made a fatal mistake at move 22. Although that play has temperature $\frac{1}{3}$, which is indeed hotter than any other move then available, Black should not fill this ko as long as she has enough kothreats remaining to ensure that she can win it. Black has several kothreats on the Go board. The sequence of moves later played at 30–35 might serve as Black kothreats, and after that Black has another threat at J3. Black also has two threats at the top of the board: F8 and then G9. And we shall find that Black also has one kothreat on the chessboard! So Black can act as komonster, attempting to leave the ko unfilled until all moves at $t \geq 0$ have been played. In this kofight, Black's goal should be to force White to play a move with negative temperature (such as at 38 in Figure 2), before Black fills the ko.

White, on the other hand, has no need to win this kofight. His goal is merely to force Black to fill the ko before the ambient temperature of the game goes negative. For this purpose, *any* White move that has nonnegative temperature can serve as a kothreat. That includes almost all White moves on the chessboard, two moves in Domineering (at plays numbered 36–37 in Figure 2), and all of the dame on the Go board at which White played during the sequence of moves numbered 30–35 in Figure 2.

Black achieves her goal relatively easily if the ko-ban rule is interpreted locally, because the Chess and Domineering boards contain many more kothreats for White than for Black. But a global interpretation of the ko-ban rule makes Black's problem much more challenging.

In general, some localities provide kothreats for one player or the other, and some offer "two-sided" kothreats that might be played by either player. Good Go strategy requires one to play out all two-sided kothreats *before* starting the ko, in order to prevent one's opponent from using them as kothreats. In the present problem, instead of filling the ko at move 22, Black could play the forcing sequence from 30–35. Detailed infinitesimal values of the various positions aren't very important now, because the bigger consideration is simply the number of moves during which the ambient temperature of the game remains at 0. White would like to drag this out. Black hopes to get this stage of the play (including all moves on chess board) over with while she still has one or more kothreats available on the Go board.

In the corrected line of play, moves 22–29 are skipped, and we will then continue the old numbering from the continuation at 30–35. Then, a good opening chess move for Black is 36:a5! The queenside of the chessboard is a two-sided kothreat. This Black move to a5 heats it up to temperature 1, a value of 2 | 0 to which White must respond. His best response is 37:b3, and Black can then continue with 38:b5. The combinatorial game theorist with no Go experience will surely be surprised at this line of queenside play. From a canonical perspective, White 37:b3 is dominated by White 37:a4, yielding an immediate value of $1 - 1 = 0$. But, because of the kofight, suspense and remoteness are now more important than canonical values, although temperatures remain very important. After Black 38:b5, the canonical value of the queenside is 0, but, until the kofight is concluded, it instead behaves like $*$. *Either* player can play here, and his opponent will then find it desirable to exchange pawn captures rather than to push.

White is then at last free to start the kofight at E6, and the game continues on the chess and Go, and Do (Domineering) boards like this:

Move	39	40	41[1]	42	43	44	45	46	47	48	49	50	51	52	53	54	55	56
White	E6		J2		h4		a3b4		ko		G8		Do		ko		H9	
Black		J3		ko		b4		a5b4		F8		ko		Do		G9		ko

The two zero-temperature moves on the Domineering board were taken at moves 51 and 52, and so at move 57, the hottest move available for White is the Domineering move at temperature $-\frac{1}{4}$. White plays there at move 57, and Black then plays 58:E6, filling the ko and leaving only moves of temperature -1. The score is tied, but White must now play next and lose.

So the line of play shown in Figure 2 is incorrect; Black could have won by adopting a different strategy at move 22.

However, any gurus reading this paper will notice that several other errors were committed before move 22. Moves 16 and 17 were both fatal mistakes, which, against a perfect opponent would have made the difference between winning and losing. Because of the kothreat potential, both players should refrain from playing in the vicinity of DE9 until after all other moves at temperature 1 have been taken, including the domino number 18. But the first fatal mistake was committed even earlier when White played the hanging connection at 11:F9. If he instead plays 11:F8, he denies Black the two kothreats at F8 and G9, and thereby wins the game.

The diligent reader is invited to confirm this assertion, which is not as obvious as it might seem to experienced Go players, most of whom intuitively prefer the solid connection to the hanging connection precisely because the ensuing position gives the opponent fewer kothreats. But other factors can assume more relevance. In particular, the hanging connection at F9 enables White to get the last move at temperature 1; the solid connection allows Black to get the last such move. As explained in detail by [?], this can be the decisive consideration, especially when there are no other moves of temperatures between 1 and 0. For example, in the isolated Go problem of Figure 1, the hanging connection wins and the solid connection loses, even though the ko configuration at $t = \frac{1}{3}$ is still present. In this case, the ko turns out to be less important than getting the last move at $t = 1$. However, when the same Go position of Figure 1 is played in the richer environment containing the Domineering positions, which have temperatures $\frac{3}{4}$, $\frac{1}{2}$, 0, and $-\frac{1}{4}$, then getting the last move at temperature 1 provides no advantage. What matters in this case is whether or not the master of the ko at $t = \frac{1}{3}$ can prolong the kofight until the ambient temperature becomes subzero.

The Isolated Go problem and the combined problem both have more than one solution. After Black 10, White has three plausible choices. 11:F8 forms a

---

[1]The novice Go player who ignores Black 40:J3 falls victim to 41:E6??, 42:J2, 43:J1, 44:J3, 45:J2, and Black 46:J3 captures 7 stones!

"solid connection"; 11:F9 forms a "hanging connection", or 11:E9 "descends" to the edge of the board. Any of these three moves would conclude play at $t = 3$. After the descent 11:E9, Black is able to resume playing in this region at a higher temperature than Black, via the following sequence:

Black	F8	D9
White	G8	F9

Thus, White's three choices at move 11 yield positions with the following properties at temperature 1:

Choice	chilled value	number of Black kothreats at $t < 1$
Solid	∗	0
Descent	**miny**	1
Hanging	**miny** | 0	2 (or 0)

So the merits of the descent are seen to lie somewhere between those of the solid connection and the hanging connection. The solid connection wins in the isolated Go problem, but not in the combined problem. The hanging connection wins in the combined problem, but not in the isolated Go problem. Surprisingly enough, the descent happens to win in both problems.

Conclusion

The 4G4G problems illustrate several general points:

(i) When the Go position is played in isolation, the hanging connection dominates because, when chilled, it has the best atomic weight. Playing there ensures getting the last move at temperature 1, which is decisive.

(ii) In the sum of all four games, the environment is sufficiently rich that getting the last move at temperature 1 does not matter. Getting more kothreats is now more important.

(iii) As Elkies [?] has shown, well-known infinitesimals like ↑ and ∗ can occur in chess.

(iv) **On**, **off**, and **dud** can occur in some very simple checkers positions.

(v) Temperature-based orthodox analysis is a very powerful tool for analyzing many kinds of combinatorial games.

(vi) Several different analytic methods can all be usefully applied to the same problem. A unified theory encompassing many of these methods appears in [Berlekamp 2002].

Acknowledgements

I am indebted to Bill Spight and Gin Hor Chan for help in composing and debugging these problems, and to Silvio Levy for assistance in the preparation of this paper.

References

[Berlekamp 1996] E. Berlekamp, "The economist's view of combinatorial games", pp. 365–405 in *Games of no chance* (Berkeley, 1994), edited by R. Nowakowski, Math. Sci. Res. Publ. **29**, Cambridge Univ. Press, Cambridge, 1996.

[Berlekamp 2002] E. Berlekamp, "Idempotents among partisan games", in *More games of no chance*, edited by R. Nowakowski, Math. Sci. Res. Publ. **42**, Cambridge Univ. Press, Cambridge, 2002.

[Berlekamp and Rodgers 1999] E. Berlekamp and T. Rodgers (editors), *The mathemagician and pied puzzler: A collection in tribute to Martin Gardner*, edited by E. Berlekamp and T. Rodgers, A K Peters Ltd., Natick (MA), 1999.

[Berlekamp and Wolfe 1994] E. Berlekamp and D. Wolfe, *Mathematical Go: Chilling gets the last point*, A K Peters, Wellesley (MA), 1994.

[Berlekamp et al. 1982] E. R. Berlekamp, J. H. Conway, and R. K. Guy, *Winning ways for your mathematical plays*, Academic Press, London, 1982. Second edition, A K Peters, Natick (MA), 2001.

[Conway 1976] J. H. Conway, *On numbers and games*, London Math. Soc. Monographs **6**, Academic Press, London, 1976. Reprinted by A K Peters, Natick (MA), 2001.

[Elkies 1996] N. Elkies, "On numbers and endgames: combinatorial game theory in chess endgames", pp. 135–150 in *Games of no chance* (Berkeley, 1994), edited by R. Nowakowski, Math. Sci. Res. Publ. **29**, Cambridge Univ. Press, Cambridge, 1996.

[Hanner 1959] O. Hanner, "Mean play of sums of positional games", *Pacific J. Math* **9** (1959), 81–99.

[Milnor 1953] J. Milnor, "Sums of positional games", pp. 291–301 in *Contributions to the theory of games*, edited by H. W. Kuhn and A. W. Tucker, Ann. Math. Studies **28**, Princeton Univ. Press, 1953.

[Rodgers and Wolfe 2001] T. Rodgers and D. S. Wolfe (editors), *A puzzler's tribute*, edited by T. Rodgers and D. S. Wolfe, A K Peters Ltd., Natick (MA), 2001.

ELWYN BERLEKAMP
MATHEMATICS DEPARTMENT
UNIVERSITY OF CALIFORNIA
BERKELEY, CA 94720-3840
berlek@math.berkeley.edu

More Games of No Chance
MSRI Publications
Volume **42**, 2002

Experiments in Computer Amazons

MARTIN MÜLLER AND THEODORE TEGOS

ABSTRACT. Amazons is a relatively new game with some similarities to the
ancient games of chess and Go. The game has become popular recently
with combinatorial games researchers as well as in the computer games
community. Amazons combines global full-board with local combinatorial
game features. In the opening and early middle game, the playing pieces
roam freely across the whole board, but later in the game they become
confined to one of several small independent areas.

A *line segment graph* is an abstract representation of a local Amazons
position. Many equivalent board positions can be mapped to the same
graph. We use line segment graphs to efficiently store a table of *defective
territories*, which are important for evaluating endgame positions precisely.
We describe the state of the art in the young field of computer Amazons,
using our own competitive program *Arrow* as an example. We also discuss
some unusual types of endgame and *zugzwang* positions that were discovered in the course of writing and testing the program.

1. Introduction

The game of Amazons was invented by Walter Zamkauskas. Two players with
four playing pieces each compete on a 10×10 board. Figure 1 shows the initial
position of the game. The pieces, called *queens* or *amazons*, move like chess
queens. After each move an amazon shoots an arrow, which travels in the same
way as a chess queen moves. The point where an arrow lands is *burned off* the
playing board, reducing the effective playing area. Neither queens nor arrows
can travel across a burned off square or another queen. The first player who
cannot move with any queen loses.

Amazons endgames share many characteristics with Go endgames, but avoid
the extra complexity of Go such as ko fights or the problem of determining
the safety of stones and territories. Just like Go, Amazons endgames are being
studied by combinatorial games researchers. Berlekamp and Snatzke have investigated play on sums of long narrow $n \times 2$ strips containing one amazon of each

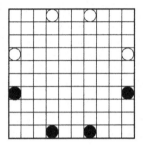

Figure 1. The playing board of Amazons.

player [1; 15]. Even though $n \times 2$ areas have a simple structure, sum game play is surprisingly subtle, and full combinatorial game values become very complex.

Amazons has also caught the attention of programmers who work on games such as shogi, Othello or Go. In 1998 and 1999, Hiroyuki Iida organized the first t wo computer Amazons championships, which were held at the Computer Games Research Institute of Shizuoka University [10; 13]. The winning program in both tournaments was *Yamazon*, written by the leading shogi programmer Hiroshi Yamashita [16]. Shortly after the second tournament, Michael Buro's *amsbot* won a *dream match* played on the GGS server [6] against the top four contestants. In summer 2000, another Amazons tournament was held at the Computer Games Olympiad in London. *8QP* by top chess programmer Johan de Koning convincingly won all its games in a field of six mostly new programs. *Yamazon* took second place. Yet another strong new program, *Amazong*, was developed by Jens Lieberum, who was the winner of the human Amazons tournament held at the combinatorial games workshop at MSRI in Berkeley in 2000.

This paper contributes to our understanding of the game of Amazons as follows: Section 2 discusses partitioning an Amazons board. In Section 3 we introduce the new concept of a *line segment graph*, which provides an abstract representation of a local Amazons position. Section 4 deals with *territories*, which are areas controlled by one player, and with the problem of *defective* territories. As an empirical result, we computed tables of small defective territories. Section 5 discusses *zugzwang* positions, which frequently occur in the game of Amazons. Section 6 contains a brief discussion of an evaluation function for computer Amazons, and Section 7 wraps up with a summary, pointers to Amazons resources and future work.

2. Board Partitioning in Amazons

The aim of board partitioning is to decompose a full-board Amazons position into a sum of independent subgames which can be analyzed locally. In the opening, creating such a partition is impossible, since the whole board is connected in many ways. In the late middle game and endgame, the board becomes more and more fragmented.

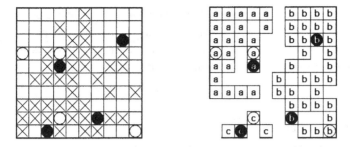

Figure 2. Board position and basic decomposition.

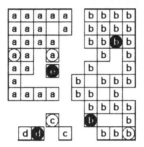

Figure 3. Partition using blocking amazons.

Starting from a given Amazons position, a basic board partition is given by the 8-connected components of all points that have not yet been burned off by arrows. Areas are independent since amazons can never move or shoot across a burned-off square into a different connected component. Figure 2 shows a board position and its basic decomposition. Burned-off squares are either indicated by a cross in the figures or omitted altogether.

Often, amazons which help to wall off some area as *territory* can be utilized to improve the board partitioning. Figure 3 shows such an improved decomposition for the same position as in Figure 2. The basic area c of Figure 2 has been divided into t wo areas c and d controlled by White and Black respectively. Area a has been further subdivided into a *dead* black stone e and a remaining area a controlled by t wo white amazons.

A *blocker* is an amazon that divides the board into t wo or more regions. Decomposition using blockers is not as strict as the basic decomposition into connected components, since blockers can move away. However, blockers often help to surround a territory. When they leave their position, they always have the option of shooting an arrow back at their origin square, and thereby keeping the partition intact.

3. Line Segment Graphs

For the analysis of Amazons positions, the fact that it is played on a rectangular grid does not matter. Even after taking into account the obvious symmetries of mirroring, rotation and translation, many positions that look different on a grid are in fact equivalent in terms of possible moves and game outcomes. Consider a representation of an Amazons board where all points on the same contiguous horizontal, vertical or diagonal line are joined by a line segment. All the structural information about an Amazons position is contained in two geometric primitives:

1. The points that are contained in each line segment, and
2. the relative order of the points on a line segment.

For brevity, we will often use the term line instead of line segment. The actual embedding of lines on a grid does not matter. For game play, of course, it must be known which points are occupied by amazons.

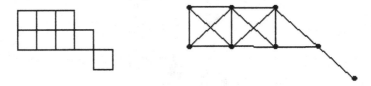

Figure 4. An Amazons region and its line segment graph.

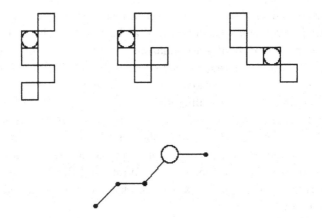

Figure 5. Amazons regions sharing the same line segment graph.

Example 1. Figure 4 shows an Amazons region and its underlying structure of line segments. Figure 5 illustrates that differently shaped regions on an Amazons board can be mapped to the same constellation of line segments.

The following definitions formally specify a *line segment graph (LSG)* and describe its basic properties. Let $V = \{v_0, \ldots v_k\}$ be a finite set of points.

Definition 1. A line segment over V is an ordered set of two or more points $[p_0, \ldots, p_n]$, $n > 0$, with the properties

1. Distinctness: $p_i \in V$ and $p_i = p_j \iff i = j$.
2. Ordering on a line: $[p_0, ..., p_n] = [q_0, ...q_m] \iff n = m$ and either $p_i = q_i$ or $p_i = q_{n-i}$ for all $i = 0 \ldots n$.

In words, only the relative order of points on the line is significant.

Definition 2. The *line distance* between points p_i and p_j on a line $[p_0, ..., p_n]$ is defined by $ld(p_i, p_j) = |i - j|$. $ld(p, q) = \infty$ if p, q are not on a line.

Definition 3. A *line segment graph* G is a pair $(V, L(V))$, where V is a finite set of points and $L(V)$ a set of line segments over V.

We'll call a LSG *well-formed* if it fulfills the following simple geometric conditions:

1. each point is contained in at least one line.
2. the unique intersection property: two distinct lines have either zero or one points in common. We write $l \cap k = p$ if p is the unique point contained in both l and k, and $l \cap k = \varnothing$ otherwise.

Definition 4. A *path* $P(p, q)$ between two points p, q in a line segment graph $(V, L(V))$ is a set of line segments $\{l_1 \ldots l_n\}$ such that $p = p_0, q = p_n, p_{i-1} \in l_i, p_i \in l_i, l_i \cap l_{i+1} = p_i$.
The distance between p and q along path $P(p, q)$ is defined as

$$d(p, q, P) = \sum_{i=1}^{n} ld(p_{i-1}, p_i).$$

The shortest distance between points in a LSG is defined as the minimum distance along all possible paths connecting p and q, and as ∞ in the case that no path exists.

Definition 5. A line segment graph $(V, L(V))$ is *connected* if there is a path between each pair of points.

LSG created from Amazons positions on a standard board obey further restrictions:

1. The shortest path property: if two points p, q are on the same line, $ld(p, q) < \infty$, then there is no shorter path between them: $d(p, q) = ld(p, q)$;
2. the grid restriction: each point is contained in at most four lines; and
3. the line length restriction: no line is longer than the board size.

3.1. Generating Small Line Segment Graphs. We have written programs that generate all Amazons positions up to size 10 that can fit within some rectangular box of size up to 56. We have also converted all these positions into LSG form. Table 1 shows the numbers and Figure 6 the ratio between the number of LSG and Amazons positions on a board. For both positions on the board and LSG the numbers have been minimized by eliminating isomorphic cases. Figure 7 shows all LSG up to size 4 that can be realized on a standard board.

Size	Board	LSG
1	1	1
2	2	1
3	5	3
4	22	11
5	94	42
6	524	199
7	3031	960
8	18769	4945
9	118069	25786
10	755047	137988

Table 1. Small regions and LSG.

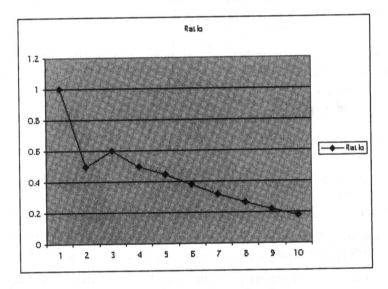

Figure 6. Ratio LSG/board positions.

3.2. Isomorphism Testing for Line Segment Graphs. In order to build and use a database of LSG, isomorphism of line segment graphs must be tested efficiently. The following mapping, proposed by Brendan McKay in a private communication, transforms the LSG isomorphism test problem into isomorphism testing for colored undirected "ordinary" graphs. The result is in a format suitable for input to *nauty*, McKay's efficient program for computing graph automorphisms and isomorphisms [11; 12]. Given an LSG $G = (V, L(V))$, create a colored graph $G' = (V', E, col)$ as follows:

1. Both points and lines of G become vertices of G': $V' = V \cup L(V)$. It is also possible and more efficient to omit all the lines of length 2 here.

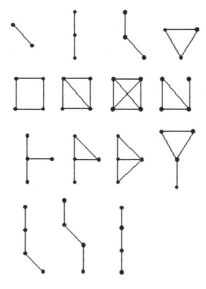

Figure 7. All LSG up to size 4.

2. In G', color all vertices from V with one color (green), and all vertices from $L(V)$ with a different color (red).
3. Construct the set of edges E: Connect green points if and only if they are adjacent points on some line. Furthermore, connect a green point representing a point v with a red point representing a line l if and only if $v \in l$.
4. Extend the mapping to graphs containing amazons by changing the coloring of the points which contain amazons from green to the color of the amazon.

It is easy to see that a given G' completely determines the underlying LSG G, even with the optimization that does not create red points for lines of length 2.

4. Filling Territory and Defective Areas

An area that contains only amazons of one color is a *territory* and represents a number of free moves for one player. The question is exactly how many moves? In most cases, a territory can be completely filled, and n empty squares yield n moves.

Definition 6. A *k-defective territory* provides k less moves than the number of empty squares in the territory. A k-defective territory is said to have k *defects*.

Example 2. Figure 8 shows the three smallest defective territories, namely t wo 1-defective areas with 2 empty squares and one 2-defective territory with 3 empty squares. In each case, the amazon can only move once because she has to shoot back to her current position, disconnecting herself from the remaining empty square(s).

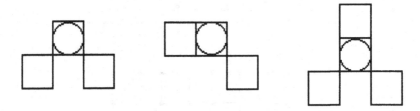

Figure 8. Small defective territories.

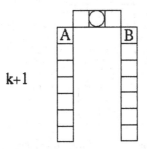

Figure 9. k-defective territory.

Example 3. Figure 9 illustrates that there are k-defective areas for any k. The amazon in the middle cannot avoid losing one of the long strips on the left or on the right. The furthest points that the queen can reach with its first shot are A and B, and all k or more points beyond the first shot are lost.

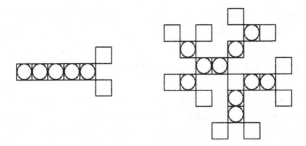

Figure 10. Defective territories containing many amazons.

Usually, two amazons in the same territory are much more powerful than a single one. However, there are small defective areas containing two or more amazons. Figure 10 shows some examples.

In theory, it is difficult to decide whether a given territory provides a certain number of moves. Michael Buro has recently proven that this is an NP-complete problem [5]. In practice, a table of all small defective areas combined with a simple filling heuristic for larger areas seems adequate.

4.1. Dead Ends.

Definition 7. A *dead end* in a LSG is a subgraph with the following property: if an amazon moves into the dead end (or already is in the dead end), she either

1. cannot fill the dead end region completely, or
2. has to disconnect the region from the rest of the board.

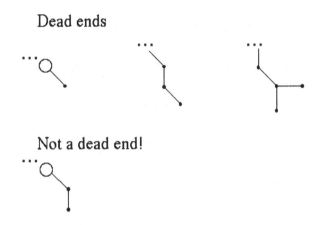

Figure 11. Dead ends.

Figure 11 shows some examples of dead ends, and one example which is not a dead end.

Definition 8. The *defect count* of a dead end is the minimum number of defects caused by a dead end if an amazon has to move out of the dead end. In the first t wo examples in Figure 11 the defect count is one while in the rightmost example the defect count is t wo.

LSG are certain to be defective if they contain more disjoint dead ends than amazons. To avoid a local defect in a dead end, some amazon's path has to end there. Therefore, each amazon can eliminate at most one dead end. Given a LSG containing n amazons and a number of disjoint dead ends, the sum of all but the n largest defect counts of dead ends is a lower bound on the number of defects of the LSG.

4.2. Small Defective Regions. We have built a database of all defective regions up to size 11 that can fit into a box of maximum size 56 on a 10×10 board, and again converted them to LSG format. Table 2 shows the number of defective board positions and LSG for areas up to size 11. Empirically, the number of LSG of a given size and defectiveness seems to grow exponentially slower than the number of such positions on the Amazons board. Figure 12 shows the ratios on a logarithmic scale. As a concrete example, Figure 13 shows all 12 3-defective Amazons areas of size 8 in our database and the single LSG that is equivalent to each of the 12 areas. Figure 14 shows all small highly defective

Size	1-defective	2-defective	3-defective	4-defective
3	2/1	-	-	-
4	2/1	1/1	-	-
5	37/6	-	-	-
6	236/12	23/3	-	-
7	2238/125	101/8	5/1	-
8	16442/666	1111/71	12/1	-
9	125797/4610	12974/370	274/14	-
10	929277/28500	137976/2665	5464/70	42/3
11	6747413/186564	1208467/17974	193410/1310	1188/17

Table 2. Small defective regions and LSG.

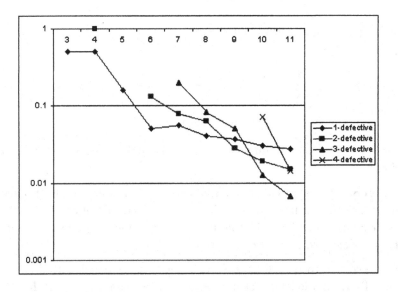

Figure 12. Ratio of number of LSG and number of regions on board.

LSG from the database: 1-defective LSG of size 3, 4 and 5, 2-defective LSG of size 4 and 6, 3-defective LSG of size 7 and 8 and 4-defective LSG of size 10.

5. Zugzwang in Amazons

Zugzwang positions in Amazons are interesting because in contrast to most other games, they have to be played out, and it is nontrivial to play them well.

5.1. Definition and Examples of Zugzwang Positions.

Example 4. The left side of Figure 15 shows an example of *zugzwang* in Amazons. A white blocker sits on an articulation point of its territory. If White is to move, one side or the other of the territory below is lost, since the white

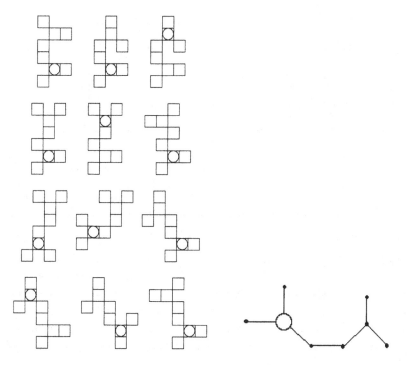

Figure 13. 3-defective Amazons areas and LSG of size 8.

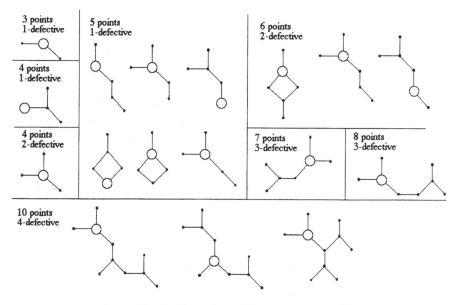

Figure 14. Table of all small highly defective LSG.

queen must shoot back to its origin square to prevent black from moving in there. White would much prefer if Black moved first since Black can only retreat to the territory above and block the entrance, thereby giving White control over

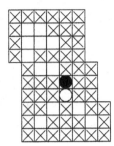 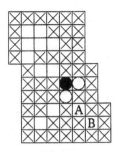

Figure 15. Zugzwang (left), and a blocker that need not block (right).

the complete area on both sides below. The combinatorial game value of this zugzwang position is $3|6 = 4$. Black is guaranteed four more moves than White locally, but can get more if White is forced to move first.

Example 5. The position shown on the right side in Figure 15 looks similar to the zugzwang on the left side but it is different. White can obtain all points by refusing to defend the blocking point. In this position, White can move to A and shoot at B, then *plod* along and fill the whole area in right-to-left order. The black amazon can never move down, since it is tied to the defense of its own, bigger, territory by the other white queen. The combinatorial game value of this position is 4 as well, but White can never get into zugzwang.

Definition 9. We define a *simple zugzwang* position in Amazons as a game $a|b$ with a, b integers such that $a < b - 1$.

By the *Simplicity Rule* [2] of combinatorial game theory, the value of a zugzwang position $a|b$ is the simplest number x such that $a < x < b$. For example, $-2|1 = 0$, $-10| - 2 = -3$ and $3|6 = 4$. In Amazons, the first player to get into zugzwang can still win the overall game. Zugzwang positions do *not* affect the main question of which player can win a sum game. In a full board situation involving one or more zugzwangs, it is sufficient to treat the zugzwangs just like an ordinary number.

Zugzwangs do matter if we want to determine the exact score with optimal play. Optimizing the score is important since it is used as the tiebreaking method in tournaments.

Example 6. Figure 16 illustrates a different t ypeof zugzwang. A white amazon defends an area W of size w, a black amazon defends an area B of size b, and both have the option of either retreating to their area or trading it in for a common region C of size c. An amazon who retreats gives the area C to the opponent, while an amazon who moves into C loses control over her own area. In the example in Figure 16, $w = 8, b = 9$ and $c = 3$. Black's options are $B_1 = b - 1 - w - c$ (retreat) and $B_2 = c - 1 - w$ (trade), while White's options are $W_1 = b + c - (w - 1)$ for retreating and $W_2 = b - (c - 1)$ for trading. In the example, $G = \{B_1, B_2 | W_1, W_2\} = \{-3, -6|5, 7\} = 0$. In this case both players

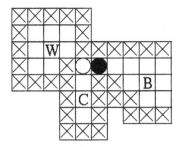

Figure 16. Another example of zugzwang.

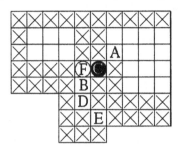

Figure 17. No zugzwang, hot position.

prefer to retreat if they get into zugzwang. If one player's territory is smaller, or if the bottom area is bigger, then it would become better to trade. An example is $w = 8, b = 5$ and $c = 3$, where Black's trading option -6 dominates the retreating option -7.

Example 7. Figure 17 shows a situation that looks very similar to Figure 16, but is a hot game with value 8|7. Black has a good move to A, shooting to B. This neutralizes the points D and E and leaves C as Black's privilege. On the other hand, White has a skilful move to B, shooting at D. Because of the path $B - C - A$ leading into the black area, Black cannot afford to move forward to F. White can later move back to F and gains.

Figure 18. A zugzwang situation that leads to hot intermediate positions.

5.2. Finding Small Zugzwang Positions. We have searched our database of small Amazons positions for zugzwang situations. Candidates were identified by performing t wo minimax searches, one with each player going first. We analyzed the cases where moving first leads to a worse score than moving second. Some of the zugzwangs we found involve intermediate hot positions. For example, the

Size 6
zugzwangs

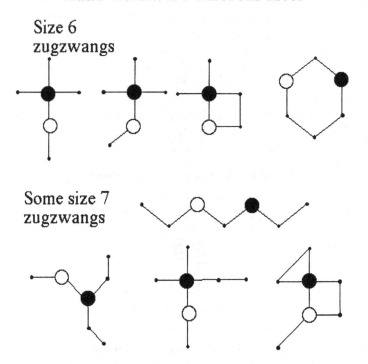

Some size 7
zugzwangs

Figure 19. Some small zugzwang positions.

game value of the hexagon in Figure 18 is $0| - 2||2|0 = 0$, with intermediate hot positions $0| - 2$ and $2|0$. Figure 19 shows the four smallest zugzwang positions, which have size 6, and a few representative examples of the many zugzwang positions of size 7.

5.3. Playing Sums of Zugzwang Positions. By the *Number Avoidance Theorem* of combinatorial game theory [2], zugzwangs will never be played while there are any noninteger games left. Moreover, the game loser will eventually be forced to use up all the zugzwangs that provide free moves (positive integers for Black, negative integers for White). In the very end the loser will also be forced to play all the zugzwangs that evaluate to zero and collect all the penalties there.

The interesting case are the zugzwangs that are nonzero integers for the winner. From now on, let's assume without loss of generality that Black is the winner. If Black can win a sum game containing a zugzwang such as $3|6 = 4$ in Figure 15 without making the move from 4 to 3 in this subgame, then White will be forced to move to 6, giving Black extra profit. Black wants to use up as few zugzwangs as possible, and maximize the gain when White is forced to play them in the end.

Define the gain for Black in a simple zugzwang position $z = a|b = a + 1, 0 \leq a < b - 1$, by $gain(z) = z^R - z - 1 = b - a - 2$ (and the gain for White in a position $z = a|b = b - 1, 0 \geq b > a + 1$, by $gain(z) = z - z^L - 1 = b - a - 2$). Let G be a sum game won by Black, and let k be the largest integer such that

$G - k$ is still a win for Black. Let $\{z_i\}$ be the set of zugzwang positions in G that evaluate to positive integers not greater than k. Black wants to win the game while leaving the largest overall gain, so the problem is to find the subset $I \subseteq \{1, \ldots, n\}$ such that $\sum_{i \in I} z_i \leq k$ and $\sum_{i \in I} gain(z_i)$ is maximized. This is exactly the well-known *knapsack problem*. This problem is NP-complete if the input is given in the form of numbers represented by their usual encoding which has logarithmic size. However, the problem can be solved in pseudopolynomial time [7] by dynamic programming if the input size is proportional to the size of the numbers. This second representation seems more appropriate here since each number n is derived from an Amazons area of size about n.

6. Programs that Play Amazons

6.1. Search. The large branching factor of Amazons in the opening and middle game strongly limits the search depth of full-width search programs. There are over 2000 possible starting moves on the 10×10 board, and although this number decreases quickly in the opening, there are several hundred legal moves during most of the game. Experiments have shown that as in most other games, search depth is strongly correlated to playing strength. Amazons seems to be a very suitable game for experimenting with selective search methods. Can deep search of a few promising move candidates be superior to shallow full-width search? Currently, the verdict is not clear. While some of the top programs are selective, others use a brute force approach, or vary their selectiveness depending on the game stage. Our programs *Arrow* and *Antiope* currently offer both options, selective and full-width search. In *Arrow*, in each position all moves are generated, executed and evaluated statically by the function explained in Section 6.2. In selective search mode only the top 10 to 15 moves in the static evaluation are further expanded in the search. A more systematic approach to selective search, the *MultiProbCut* algorithm [4], is used in Buro's *amsbot* program [3].

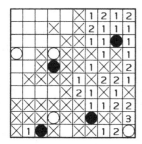 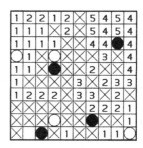

Figure 20. Min-distance function for Black and White.

6.2. Evaluation. Like most Amazons programs, our program *Arrow* uses a simple *min-distance* heuristic evaluation. Unlike most programs, *Arrow* does not use any other features in its static evaluation function. In *min-distance* evaluation, the player who can reach a square more quickly with a queen, ignoring arrow shots, owns that square. Figure 20 shows the min-distance function for both colors in a late endgame position.

The evaluation computes the following function:

1. for each point p, compare $db = \text{dist}(p, \text{Black})$ and $dw = \text{dist}(p, \text{White})$;
2. if $db < dw$: Black point;
3. else if $db > dw$: White point;
4. else point is neutral.

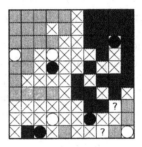

Figure 21. Evaluation using min-distance.

Figure 21 shows the evaluation of the example position. Squares marked with '?' are neutral. This algorithm is simple-minded but fast. It can be implemented efficiently by using bitmap operations. It correctly evaluates enclosed territories and regions that are somewhat open but well-controlled by one player. It also gives reasonable results in well-balanced situations such as fights with 1 vs 1 and 2 vs 2 amazons. The method fails in open and in unbalanced situations such as 1 vs 2 and 1 vs 3 amazons, since it optimistically propagates pieces in all directions at once. A single amazon that tries to defend a large territory can be badly outplayed by two or more attackers approaching from different sides.

We have tried a number of more sophisticated evaluation functions including features such as mobility and blockability. However, in test matches with equal time limits the simple but fast evaluation above always beat its more refined but slower competitors. A similar pattern became apparent in the Second Computer-Amazons championship in which *Arrow* participated and took third place. Most of the other programs used a better evaluation function including mobility terms, but searched less deeply. *Arrow* would typically get a bad position in the opening and early middle game, but use its greater search depth and specialized endgame knowledge about blockers to turn the games around later. However, against aggressive human players and programs, such as Jens Lieberum's *Amazong*, *Arrow* gets into trouble quickly. Because of the missing mobility term, it lets its amazons be blocked too easily in the opening.

7. Summary and Outlook

This paper made a number of conceptual and empirical contributions to the study of the game of Amazons. We have defined the concept of *line segment graphs*, an abstract representation that maps many equivalent Amazons positions on a real board into one graph. This representation is shown to achieve excellent compression for storing tables of exceptional *defective* territories. We have also analyzed zugzwang positions in Amazons, which tend to be a source of confusion not only for beginners. Optimal play in a sum of zugzwang positions is shown to be equivalent to solving the well-known *knapsack problem*.

7.1. Amazons Resources. Amazons is still a young game, and we are not aware of any books dealing with the game, or of any player's organizations. Some recent articles about the game are listed in the references [1; 5; 8; 10; 13]. A few opportunities exist for playing on Internet game servers or by email [6; 14]. For exchanging game records, the SGF file format has been extended to include Amazons [9]. We maintain an Amazons web site with links to all known other sites at http://www.cs.ualberta.ca/~tegos/amazons/index.html.

7.2. Future Work.

- We considered only simple zugzwangs in the form of games $a|b$ with integers a and b. However, playing into a zugzwang can lead to a hot game as in Figure 18. We found our small zugzwang positions in Figure 19 by a simple pair of minimax searches. It is conceivable that such minimax values are distorted if the follow-up positions contain further zugzwangs. It would be interesting to find such cases or prove that they cannot exist, and develop a complete theory of zugzwang in Amazons.

- Determine the asymptotic growth ratio of LSG and defective LSG.

- Determine the complexity of isomorphism testing for LSG. This could be easier than general graph isomorphism testing because of the extra structure and constraints.

- 4×4 Amazons is a second player win, and can be solved easily by brute-force search. Using a specialized endgame solver, we have recently shown that 5×5 Amazons is a first player win. The 6×6 case is much harder, and it is unclear whether the first or second player wins. Solve Amazons on 6×6 boards.

References

[1] E. Berlekamp. Sums of $N \times 2$ Amazons. In *Institute of Mathematics Statistics Lecture Notes*, number 35 in Monograph Series, pages 1–34, 2000.

[2] E. Berlekamp, J. Conway, and R. Guy. *Winning Ways*. Academic Press, London, 1982.

[3] M. Buro. Personal communication, 2000.

[4] M. Buro. Experiments with Multi-ProbCut and a new high-quality evaluation function for Othello. In J. van den Herik and H. Iida, editors, *Games in AI Research*, pages 77–96, Maastricht, 2000. Universiteit Maastricht.

[5] M. Buro. Simple Amazons endgames and their connection to Hamilton circuits in cubic subgrid graphs. NECI Technical Report #71. To appear in proceedings of Second International Conference on Computers and Games, 2000.

[6] I. Durdanovic. Generic Game Server (GGS). http://external.nj.nec.com/homepages/igord/gsa-ggs.htm, 1999.

[7] M. Garey and D. Johnson. *Computers and Intractability: A Guide to the theory of NP-Completeness*. Freeman, San Francisco, 1979.

[8] T. Hashimoto, Y. Kajihara, H. Iida, and J. Yoshimura. An evaluation function for Amazon. In *Advances in Computer Games 9*. 2000. To appear.

[9] A. Hollosi. Smart Game Format for Amazons, 2000. Available at http://www.red-bean.com/sgf/amazons.html.

[10] H. Iida and M. Müller. Report on the second open computer-Amazons championship. *ICGA Journal*, 23(1):51–54, 2000.

[11] B. McKay. Practical graph isomorphism. *Congressus Numerantium*, 30:45–87, 1981.

[12] B. McKay. The nauty page, 2000. Available at http://cs.anu.edu.au:80/people/bdm/nauty/.

[13] N.Sasaki and H.Iida. Report on the first open computer-amazon championship. *ICCA Journal*, 22(1):41–44, 1999.

[14] R. Rognlie. Play by e-mail server for Amazons, 1999. Available at http://www.gamerz.net/pbmserv/amazons.html.

[15] R. G. Snatzke. Exhaustive search in the game Amazons. In this volume, 2001.

[16] H. Yamashita. Hiroshi's computer shogi and Go, 1999. http://plaza15.mbn.or.jp/~yss/index.html. Yamashita's homepage contains some Amazons information and a link to download his *Yamazon* program.

MARTIN MÜLLER
DEPARTMENT OF COMPUTING SCIENCE
UNIVERSITY OF ALBERTA
EDMONTON, ALBERTA T6G 2E8
CANADA
    mmueller@cs.ualberta.ca

THEODORE TEGOS
DEPARTMENT OF COMPUTING SCIENCE
UNIVERSITY OF ALBERTA
EDMONTON, ALBERTA T6G 2E8
CANADA
    tegos@cs.ualberta.ca

More Games of No Chance
MSRI Publications
Volume **42**, 2002

Exhaustive Search in the Game Amazons

RAYMOND GEORG SNATZKE

ABSTRACT. Amazons is a young "real world" game that fulfills all defining constraints of combinatorial game theory with its original rule set. We present a program to evaluate small Amazons positions with a given number of amazons on game boards of restricted sizes with canonical combinatorial game theory values. The program does not use an analytical approach, instead it relies on exhaustive search in a bottom-up strategy. Here it is applied on all positions on game boards which fit into an underlying game board of size 11 by 2.

The results show that even under these restrictions Amazons offers a wide spread of game theoretic values, including some very interesting ones. Also the canonical forms of the values can be very complicated.

1. Introduction

"The Game of the Amazons" — or simply "Amazons" — is a relatively new star on the sky of abstract strategic two-person games, invented in 1988 by Argentinian Walter Zamkauskas. Compared with other non-classical games it seems to be well done — easy rules, many interesting choices for each move, a big range of different tactics and strategies to employ, challenging even after many games played. It has yet to prove if it can stand the comparison to classics like chess and go.

Amazons has already built a solid base of players and followers. It has especially produced interest in programmers, leading to several Amazons programming competitions including a yearly world-championship and a tournament at the Mind Sports Olympiad 2000, see for example [Iida and Müller 2000] and [Hensgens and Uiterwijk 2000]. Already there are about a dozen different Amazons programs competing at these tournaments. Some of these programs play strong enough to beat average human players easily.

In contrast to most games of this type combinatorial game theory can be applied directly to Amazons without any changes to the rules. Amazons also often decomposes into at least two independent subgames when the endgame is played out, making the application of combinatorial game theory worthwhile.

The first analysis of Amazons with the means of combinatorial game theory was done by E. Berlekamp, who looked at positions with one amazon per player on boards of size 2 by n [Berlekamp 2000]. Berlekamp calculated only the thermographs for these positions, but even these proved to be quite difficult to analyze.

The idea presented in this article is to produce a database of canonical values in Amazons, in order to find as many different values as possible, to look, how complex they can become and to search for patterns to generalize. A lot of this work still has to be done, but up to now the results show that Amazons is not disappointing our expectations: The canonical forms are complex, offer a wide range of values, and patterns seem to be hard to find.

The approach used to get the canonical data is one of brute computational force, calculating the values of all possible positions from small to big boards. The program searched in 2 by n boards, to walk alongside Berlekamp's thermographic results, checking its own correctness while filling in the large and empty canonical wastelands within the thermographical results.

2. Amazons

2.1. Rules. Amazons is played by two players, White and Black, who interchange moves. White moves first, but according to the standards in combinatorial game theory we assign positive numbers to Black and negative numbers to White.

Amazons is played on a game board of ten times ten squares, with four amazons per player.

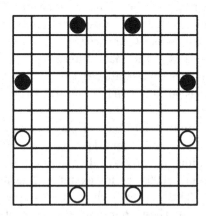

Figure 1. Amazons starting position.

A move consists of two parts, moving an amazon and shooting an arrow with that same amazon. Amazon and arrow both move like a chess queen, diagonally, horizontally or vertically as far as the player wants and no obstacle blocks the way. Neither amazon nor arrow may move onto or over a square which holds another amazon or arrow of either color.

Arrows never move again, working as blocks for the rest of the game, thereby eliminating their square. This limits the game to a maximum of 92 moves. The player who moves last wins.

2.2. Course of a Game. A game of Amazons unfolds in a t ypical manner. While more and more squares are blocked by arrows, the mobility of the amazons is restricted and the board is divided by firm walls of arrows (marked by crosses in the figures) into several independent rooms. The game practically comes to an end when all these independent, cut-off rooms contain only amazons of one player who thereby normally claims as many more moves as the rooms controlled by his amazons have free squares.

Figure 2 shows an end-position of Amazons. White has no chance to break into the large middle area of the game board if Black plays carefully. Therefore White resigned in this position, because Black will be able to claim larger territories with more remaining moves.

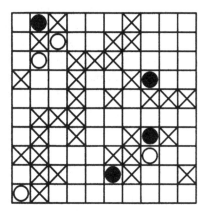

Figure 2. White to move resigns.

In rare cases a room is structured in such a way that the inhabiting amazons cannot move to all of the remaining squares and therefore some moves are lost. The smallest and most t ypical example of such a "defective" shape is given in Figure 3. Wherever the amazon moves, she has to shoot back on and block

Figure 3. One move left for Black.

the middle square, thereby losing the possibility to move to the third remaining square.

More thorough and extensive information about defective shapes in Amazons can be found in [Müller and Tegos 2002].

2.3. Amazons and Combinatorial Game Theory. It is easy to see that Amazons fulfills all conditions that define a combinatorial game as given in *Winning Ways* [Berlekamp, Conway and Guy 1982]:

(i) There are exactly two players.

(ii) The game has a finite number of distinctive positions.

(iii) From every position each player has a finite set of moves which lead to different other positions.

(iv) Both players take turns moving.

(v) The first player unable to move loses.

(vi) The game ends after a finite number of moves.

(vii) Amazons is a game of complete information.

(viii) There is no element of chance.

Additionally Amazons does the favor to decompose into a sum of several smaller games, making the application of combinatorial game theory not only possible but also very useful. This decomposition happens in practically all Amazons games, although in some cases in a trivial way, where only one of the smaller games contains amazons of both players, therefore yielding a sum of one hot game plus several integers. In other cases the barrier between two parts of the game board is not complete but very difficult to cross, making the application of combinatorial game theory practically possible despite the imperfect boundary.

3. Analyzing Amazons

3.1. General Idea. The approach used here to analyze Amazons is to build a complete database of canonical combinatorial game theory values for Amazons positions within a certain set of small game boards. In order to be able to crosscheck the electronically achieved data with the results from E. Berlekamp it was decided to calculate the game theoretic values of all game boards up to at least size 2 by 11 with one amazon per player.

Because the goal was to get the values for the complete set of game boards, a bottom-up strategy was used, first evaluating all possible Amazons positions for game board size m before proceeding to the positions of game board size $m + 1$. This way the results from smaller game boards were used to determine the values of the larger game boards.

3.2. The Algorithm. The strategy implemented is straightforward. Construct every game board, every position, make every move, look-up the values for the resulting smaller positions, calculate the result for the actual position by eliminating dominated and reversing reversible options. The algorithm in Figure 4 shows this in more detail but may also cloud the essence of the programming strategy.

Obviously this algorithm needs many subroutines, for example to construct the gameboards, set up the positions, make the moves and so on. Additionally it also needs an extensive software package to resolve the calculations of combinatorial game theory and thermography.

4. Technical Aspects

4.1. Programming Language. To choose Java as programming language may be surprising. This was done for personal reasons, but also because the problem is exponential anyway, so even a large speed advantage of another language over Java (such as C) probably would not have allowed to calculate much further. As it turned out the chosen size we wanted to reach — 22 — was just in the range where the choice of the language mattered and the computation with JDK 1.1 took nearly unacceptable 4 months the first time. On a rerun with JDK 1.3 that was necessary because of errors in the code the time dropped to approximately one month.

On the other hand Java lended structure and stability to the program. This greatly eased the programming task itself. Added to this Java is platform independent, which turned out to be very useful several times. Altogether the decision for Java was not regretted.

4.2. Machinery. More of a problem than speed was space. A lot of the programming work went into saving space, as all of the data for smaller game boards and positions had to be available in RAM where it was constantly requested for calculating the game theoretic values of positions on bigger game boards. 500 megabyte of RAM were just enough for the task. Most of the object structure typical for Java of the data had to be destroyed for space saving reasons and was only temporarily rebuilt when the data was needed for calculation.

The first, error prone run of the program was conducted on a Pentium III with 500 megahertz processing speed and 512 megabyte RAM. The rerun was again done on a Pentium III with 500 megahertz, but this time with 1.5 gigabyte of RAM.

4.3. Statistics. Table 1 shows the number of different game boards for the corresponding number of squares on the game board. Mirrored, turned or shifted boards are already excluded. The numbers behave as expected, rising rapidly up to a maximum of 14549 different boards for 14 squares and dropping back to

for $m = 1$ to 22 (size of game boards, number of squares)

construct the set B_m of all diagonally connected game boards with m squares that fit into an underlying board of size 2 by 11

eliminate all boards in B_m which are identical except for being mirrored or turned or shifted

for every $b \in B_m$

construct the set P_b of all possible positions on b with one black amazon or one black and one white amazon, excluding positions which are identical except for the colors of the amazons[1]

for every $p \in P_b$

construct the set L_p and the set R_p of all possible moves for Black and White in p

set the cgt-value $v(p)$ of p to 0

for every $mo \in L_p$

determine the set P_{mo} of all the resulting Amazons positions after move mo[2]

set the cgt-value $v(mo)$ to 0[3]

for every $r \in P_{mo}$

look-up the cgt-value $v(r)$ of r[4]

calculate $v(mo) = v(mo) + v(r)$

insert $v(mo)$ as new left option of $v(p)$

for every $mo \in R_p$

determine the set P_{mo} of all the resulting Amazons positions after move mo

set the cgt-value $v(mo)$ to 0

for every $r \in P_{mo}$

look-up the cgt-value $v(r)$ of r

calculate $v(mo) = v(mo) + v(r)$

insert $v(mo)$ as new right option of $v(p)$

compute the canonical form[5] and the thermographic data of $v(p)$ and save them to disc together with the corresponding moves in L_p and R_p

Figure 4. The basic algorithm. Notes are on the next page.

[1] This step is really processed twice, not once: First for positions with only one black amazon, then for positions with one black and one white amazon.

[2] This may be more than one position because the board may split apart after burning out one square.

[3] v(mo) is the sum of the cgt-values of the resulting positions after move mo. So it is not just the value of one single position.

[4] This value exists because it was already computed before — remember that the size of the board has shrunk. If the size of the board is zero, its cgt-value is 0, too.

[5] Calculating the canonical form especially means eliminating dominated and reversing reversible options.

Figure 4 (notes).

one board with 22 squares as there obviously is just one board with 22 squares that fits into an underlying 11 by 2 board.

size of game boards	number of game boards	number of positions
2	2	2
3	4	12
4	11	66
5	20	200
6	51	765
7	108	2268
8	267	7476
9	609	21924
10	1485	66825
11	3500	192500
12	7429	490314
13	12056	940368
14	14549	1323959
15	12792	1343160
16	8333	999960
17	3961	538696
18	1391	212823
19	340	58140
20	61	11590
21	6	1260
22	1	231
sum	66976	6212539

Table 1. Number of game boards and Amazons positions.

The table also shows the number of different Amazons positions on all of these game boards for every board size for one black and one white amazon. Cases where only the positions of the t wo amazons are exchanged are excluded, therefore the number of different positions for a board of size m is $\frac{1}{2}m(m-1)$. Not excluded are therefore positions which are symmetrical because of symmetries within the board itself, for instance the t wo positions of Figure 5.

Figure 5. Two "different" positions on a symmetric game board.

The computing time to calculate the game theoretic value for a position also behaved as expected, growing exponentially with the size of the game board, as did the coding space necessary to save the results.

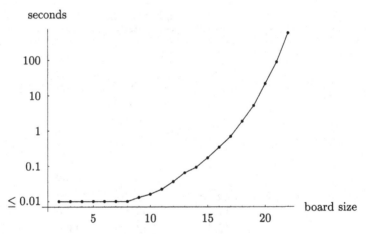

Figure 6. Average computing time per Amazons position, in seconds, as a function of the game-board size.

There are different reasons for the sharp growth in computing time. One is obviously the size of the game board, as larger game boards allow more moves to be made and therefore to produce more complex game theoretic values. These values provide the second reason, as these more complex values for size m, say, now appear in the calculation of the values for size $m+1$, slowing down the process and making the resulting values for size $m+1$ even more complex.

A third reason is the underlying game board of size 11 by 2. The boards of size $m+1$ not only have one more square than the boards of size m, these boards are also more compact. For boards up to size 16 or so it is usual to be stretched thinly, like in Figure 16. Boards with such a shape don't allow many moves by the amazons no matter how big they are.

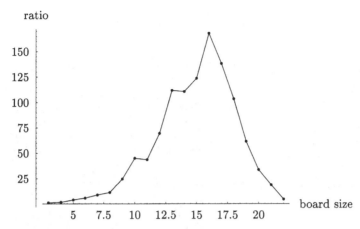

Figure 7. Rising complexity of compact game boards: ratio worst-case/average-case of number of subgames.

The sole game board with 22 squares is an extreme opposite: It is a simple rectangle and every amazon has plenty of different moves available. This behaviour can be clearly seen in Figure 7 which depicts the ratio between the maximum number of subgames for any Amazons position for a given board size with the average number of subgames for all positions on all boards of this size.[1]

What happens in Figure 7 is the disappearance of the average case, which is the thinly stretched game board for smaller sizes. From size 17 onwards all the boards get more compact and therefore the previously worst case now slowly becomes the average case.

5. Game-Theoretic Results

5.1. Amazons is Difficult. E. Berlekamp already proved that the depth of the canonical subgame-tree of an Amazons position can be approximately three quarters of the size of the game board [Berlekamp 2000]. Many of the results this program calculated show that Amazons positions can also have a very broad canonical subgame-tree. One example is the canonical form of one of the positions of Berlekamp's $1000 puzzle (cited in [Berlekamp 2000]) in Figure 8.

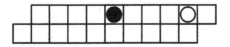

Figure 8. Second game of Berlekamp's G4G3 Problem.

---

[1]The number of subgames of a position is the number of canonical options in the game theoretic value of a position until a numeric value is reached.

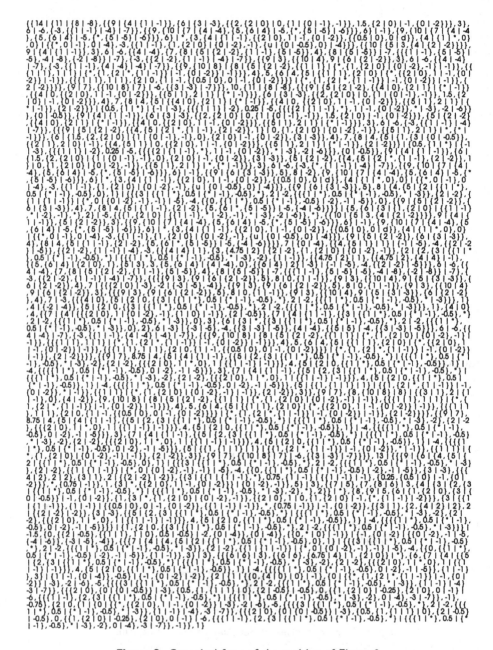

Figure 9. Canonical form of the position of Figure 8.

This position has just about 4000 subgames. But it already has a relatively deep subgame tree, which with its longest path needs 15 steps to reach a number. Black in this position has 2 different canonical options, White has 5.

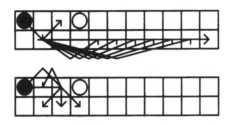

Figure 10. A position with 100833 subgames.

Another position with 100833 subgames is shown in Figure 10. The arrows show all the 13 different canonical moves for Black in this position. The canonical subgame-tree has depth 16, which is only very slightly higher than the lower bound for the maximal depth found by Berlekamp.

In Figure 11 you can see a symmetrical position which gives each amazon 12 different canonical options, together 24.

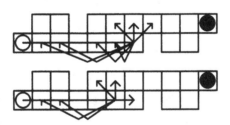

Figure 11. Twelve canonical options for each player.

5.2. Thermographic Results. While the canonical form of Amazons positions on some boards may be nearly impossible to handle, the thermographic data of all computed positions remains simple. Just consider the thermographs for each of the three positions given in Figures 8–11, shown in Figures 12–14.

In fact not one of these more than 6 million Amazons positions has a thermograph where either of its two walls has more than 10 sectors, that is, parts with changing slope, and this includes subzero thermography (see [Berlekamp 1996]), which has no special information value. The average number of sectors is around 2.5 per wall. And while the complexity of the canonical data grows exponentially with the size of the game board, the complexity of the thermographs remains constant from boardsize 15 upward, the worst case growing not at all.

All the thermographs calculated by the program, up to the time of writing, were compared and found to be the same as those found by E. Berlekamp

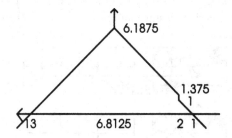

Figure 12. Thermograph of game in Figure 8.

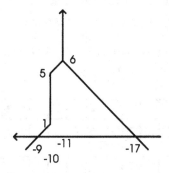

Figure 13. Thermograph of game in Figure 10.

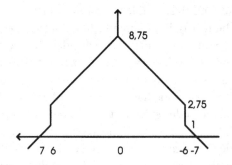

Figure 14. Thermograph of game in Figure 11.

[Berlekamp 2000]. Many canonical forms obviously are out of reach for manual confirmation or testing.

5.3. Special Cases of Interest. This computer aided approach to analyze Amazons also revealed some interesting game theoretic values that were previously unknown in Amazons.

5.3.1. $\frac{7}{8}$ The smallest fraction known before running the program was a quarter, shown in Figure 15.

Figure 15. Position with value $\frac{1}{4}$.

The highest denominator in a numeric value found by this program is sixteen. Figure 16 shows the smallest board within this setting that produces a sixteenth.

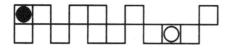

Figure 16. Position with value $-\frac{1}{16}$.

The smallest position which yields an eighth is shown in Figure 17. Interestingly this position does not have value $\frac{1}{8}$ but $\frac{7}{8}$. And the canonical moves of this position are really intriguing.

A game of Amazons normally follows a certain parity pattern: When the first player is to move, there is an even number of squares left. The first player moves and burns out a square, leaving an odd number of squares for the second player, who after his move and shot leaves again the first player with an even number of squares and so on. In this position Black with every move and shot closes off an additional square, thereby changing parity.

This position is also a good example to illuminate the complexity of Amazons. Changing a small detail can alter the value of a position quite a bit. In Figure 17 the dangling leftmost square of the board seems to be irrelevant: It is never moved upon, never shot into, it is just being cutoff somewhere in the course of canonical moves.

Moving that square one space upward basically adds $\frac{1}{8}$ to the game value, making it infinitesimally close to 1. The canonical moves following from this changed position are shown in Figure 18.

In Figure 18, if Black is to move twice in a row, a new option opens up for him. Instead of moving back to where he started, Black can now move even further to the left and shoot back to his starting square in Figure 18, yielding a

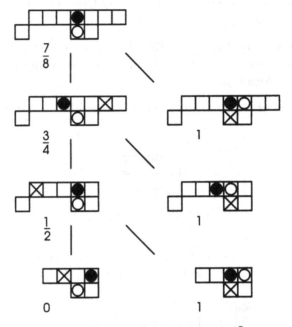

Figure 17. Canonical line of play for $\frac{7}{8}$.

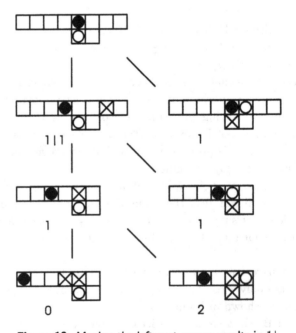

Figure 18. Moving the leftmost square results in $1\downarrow$.

solid 1 instead of $\frac{1}{2}$. If Black tries this trick in Figure 17 he ends up in Figure 19, getting only 0, not $\frac{1}{2}$. The reason is that in this position he has no winning move, because Black is left with a defective board like in Figure 3.

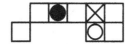

Figure 19. A defective shape.

5.3.2. ∗2 The program also found a position with value ∗2, which is unexpected for games that are not impartial. While ∗ offers just one option to each player and therefore is rather common, any other nimber can only be obtained by allowing more than one infinitesimal option to both players, which must also be exactly the same for both players. $\ast 2 = \{0, \ast \| 0, \ast\}$ is the smallest of these cases, but even it is hard to imagine in a partisan game like Amazons. Figure 20 shows it.

That this position is a ∗2 becomes clear only after several simplifications. Black and White cannot move to a zero directly. Instead, their first move that will eventually become a zero is reversible, so their opponent will move it instantly to another, hot position. From this position now the first player can move to zero. Figure 21 shows the trick.

From the original position both players can reach a ∗ after just one move, but the following position is a ∗ only because of a reversible move hidden one step further down the tree, see Figure 22.

Because of all these problems it would probably be quite difficult to determine the value of this position without electronic help. To construct a position of value ∗2 by hand seems to be outright impossible.

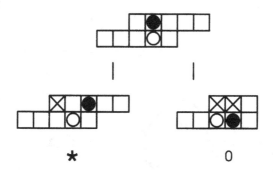

Figure 20. A nimber in Amazons.

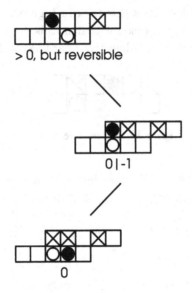

> 0, but reversible

0 | -1

0

Figure 21. Explaining the 0 in ∗2.

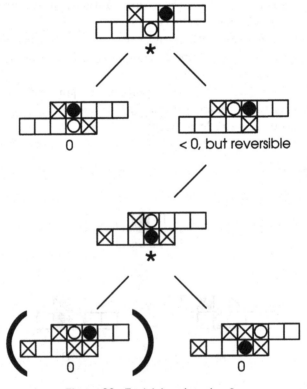

∗

0

< 0, but reversible

∗

0   0

Figure 22. Explaining the ∗ in ∗2.

6. Further Work

Positions like those presented in the preceeding section are some of the bonuses of the brute force approach. Neither the $\frac{7}{8}$ nor the $*2$ have nearly rectangular board shape. That the $*2$ actually is a $*2$ could also not be discerned by looking at the thermograph of the position.

6.1. Other Special Cases. The logical next step is to search the data calculated by the program for other infinitesimals like tinies, minies and other nimbers. $*3$ can now be easily constructed in Amazons, as $*3 = *2 + *$. Nevertheless it would be very interesting to know if a $*3$ can exist on its own on the Amazons board, that is, within one board where all squares are connected. The next theoretical challenge is the $*4$. But every nimber higher than $*2$ would come as a big surprise, because these positions need even more completely symmetrical and infinitesimal options for both players.

6.2. Patterns. More interesting would be to find patterns for constructing positions with certain values. After reaching $\frac{1}{16}$ it seems likely that there is some way to construct all fractions with dyadic denominators. But the structure of the sixteenths found up to now makes it also likely that this construction is very hard to find. Other patterns of interest would be sequences of certain infinitesimals like ups and tinies.

Finally — at least for this article — another way to reach new values or patterns could be to change the shape of the boards or to add more amazons to the fray. The last option obviously is devastating in respect to computing power, but the possibilities abound, too.

References

[Berlekamp 1996] E. R. Berlekamp, "The Economist's View of Combinatorial Games" in *Games of No Chance*, (edited by R. Nowakowski), Cambridge University Press, Cambridge, 1996.

[Berlekamp 2000] E. R. Berlekamp, Sums of $N \times 2$ Amazons, *Institute of Mathematics Statistics Lecture Notes–Monograph Series* **35** (2000), 1–34.

[Berlekamp, Conway and Guy 1982] E. R. Berlekamp, J. H. Conway and R. K. Guy, *Winning Ways for Your Mathematical Plays*, Academic Press, London, 1982.

[Hensgens and Uiterwijk 2000] P. Hensgens and J. Uiterwijk, "8QP wins Amazons Tournament", *ICGA Journal*, Vol. 23 (2000), No. 3, 179–181.

[Iida and Müller 2000] H. Iida and M. Müller, "Report on the Second Open Computer-Amazons Championship", *ICGA Journal*, Vol. 23 (2000), No. 1, 51–54.

[Müller and Tegos 2002] M. Müller and T. Tegos, "Experiments in Computer Amazons", in this volume.

RAYMOND GEORG SNATZKE
ALEX GUGLER STRASSE 13
D-83666 SCHAFTLACH
GERMANY
  RGSnatzke@aol.com

More Games of No Chance
MSRI Publications
Volume **42**, 2002

Two-player Games on Cellular Automata

AVIEZRI S. FRAENKEL

ABSTRACT. Cellular automata games have traditionally been 0-player or solitaire games. We define a two-player cellular automata game played on a finite cyclic digraph $G = (V, E)$. Each vertex assumes a weight $w \in \{0, 1\}$. A move consists of selecting a vertex u with $w(u) = 1$ and *firing* it, i.e., complementing its weight and that of a *selected* neighborhood of u. The player first making all weights 0 wins, and the opponent loses. If there is no last move, the outcome is a draw. The main part of the paper consists of constructing a strategy. The 3-fold motivation for exploring these games stems from complexity considerations in combinatorial game theory, extending the hitherto $\leq$ 1-player cellular automata games to two-player games, and the theory of linear error correcting codes.

1. Introduction

Cellular Automata Games have traditionally been 0-player games such as Conway's *Life*, or solitaire games played on a grid or digraph $G = (V, E)$. (This includes undirected graphs, since every undirected edge $\{u, v\}$ can be interpreted as the pair of directed edges (u, v) and (v, u).) Each cell or vertex of the graph can assume a finite number of possible *states*. The set of all states is the *alphabet*. We restrict attention to the binary alphabet $\{0, 1\}$. A *position* is an assignment of states to all the vertices. There is a local transition rule from one position to another: pick a vertex u and *fire* it, i.e., complement it together with its neighborhood $F(u) = \{v \in V : (u, v) \in E\}$. The aim is to move from a given position (such as all 1s) to a target position (such as all 0s). In many of these games *any* order of the moves produces the same result, so the outcome depends on the set of moves, not on the sequence of moves. Two commercial manifestations are *Lights Out* manufactured by Tiger Electronics, and *Merlin Magic Square* by Parker Brothers (but Arthur–Merlin games are something else again). Quite a bit is known about such solitaire games. Background and theory can be found

---

Key words and phrases. two-player cellular automata games, generalized Sprague–Grundy function, games-strategy.

Invited one-hour talk at second MSRI Workshop on Combinatorial Games, July, 2000.

e.g., in [7], [20], [25], [26], [27], [29], [31], [32], [33], [34], [35], [36]. Incidentally, related but different solitaires are *chip firing games*, see e.g., [3], [24], [2].

What seems to be new is to extend such solitaire games to two-player games, where the player first achieving 0s on all the vertices wins and the opponent loses. If there is no last move, the outcome is a draw. In this context it seems best to restrict the players to firing only a vertex in state 1.

Specifically, we play a two-player cellular automata game on a finite cyclic digraph G with an initial distribution of *weights* $w \in \{0,1\}$ on the vertices. For the purposes of the present paper it is convenient to agree that a digraph is cyclic if it *may* contain cycles, but no loops. We put $w(u) = 1$ if u is in state 1, otherwise $w(u) = 0$. The two players alternate in selecting a vertex u with $w(u) = 1$ and *firing* it, i.e., "complementing" it together with a selected neighborhood $N(u)$ of vertices. By complementing we mean that $w(u)$ switches to 0, and $w(v)$ reverses its parity for every vertex $v \in N(u)$. The player making the last move wins (after which all vertices have weight 0), and the opponent loses. If there is no last move, i.e., there is always a vertex with weight 1, the outcome is a draw. A precise definition of the games is given in § 3.

Our aim is to provide a strategy for two-player cellular automata games. The game graph of cellular automata games is exponentially large. For the special case where $N(u)$ is restricted to a single vertex, we can provide a polynomial strategy. For small digraphs, even some cases of large digraphs, as we shall see, the "γ-function" (defined below) can be found by inspection for any fixed size of $N(u)$, leading to an optimal strategy.

2. Preliminaries

For achieving our aim, we need to compute the generalized Sprague–Grundy function γ polynomially on the game graph $\mathbf{G}$ induced by G. The γ function has been defined in [30]. See also [4], Ch. 11. The following simplified definition appears in [18] Definition 1, see also [11] Sect. 3.

Given a cyclic digraph $G = (V, E)$. The set $F(u)$ of *followers* of $u \in V$ is defined by $F(u) = \{v \in V : (u, v) \in E\}$. If $F(u) = \varnothing$, then u is a *leaf*. The *Generalized Sprague–Grundy function* γ is a mapping $\gamma: V \to \mathbb{Z}^0 \cup \{\infty\}$, where the symbol ∞ indicates a value larger than any natural number. If $\gamma(u) = \infty$, we also say that $\gamma(u)$ is infinite. We wish to define γ also on certain subsets of vertices. Specifically: $\gamma(F(u)) = \{\gamma(v) < \infty : v \in F(u)\}$. If $\gamma(u) = \infty$ and if we denote the set $\gamma(F(u))$ by K for brevity, then we also write $\gamma(u) = \infty(K)$. Next we define equality of $\gamma(u)$ and $\gamma(v)$: if $\gamma(u) = k$ and $\gamma(v) = \ell$ then $\gamma(u) = \gamma(v)$ if one of the following holds: (a) $k = \ell < \infty$; (b) $k = \infty(K)$, $\ell = \infty(L)$ and $K = L$. We also use the notations

$$V_i = \{u \in V : \gamma(u) = i\} \ (i \in \mathbb{Z}^0), \quad V^f = \{u \in V : \gamma(u) < \infty\}, \quad V^\infty = V \setminus V^f,$$

$$\gamma'(u) = \operatorname{mex} \gamma\big(F(u)\big) \ ,$$

where for any subset $S \subset \mathbb{Z}_{\geq 0}$, $S \neq \mathbb{Z}_{\geq 0}$, we define $\operatorname{mex} S := \min(\mathbb{Z}_{\geq 0} \backslash S) = \text{least}$ nonnegative integer not in S.

We need some device to tell the winner where to move to. This device is a counter function, as used in the following definition. For realizing an optimal strategy, we will normally select a follower of least counter function value with specified γ-value. If only local information is available, or the subgraph is embedded in a larger one, we may not know to which seemingly optimal follower to move. The counter function is the guide in these cases. We remark that it also enables one to prove assertions by induction.

Definition 1. Given a cyclic digraph $G = (V, E)$. A function $\gamma: V \to \mathbb{Z}^0 \cup \{\infty\}$ is a γ-*function* with *counter function* $c: V^f \to J$, where J is any infinite well-ordered set, if the following three conditions hold:

A. If $\gamma(u) < \infty$, then $\gamma(u) = \gamma'(u)$.
B. If there exists $v \in F(u)$ with $\gamma(v) > \gamma(u)$, then there exists $w \in F(v)$ satisfying $\gamma(w) = \gamma(u)$ and $c(w) < c(u)$.
C. If $\gamma(u) = \infty$, then there is $v \in F(u)$ with $\gamma(v) = \infty(K)$ such that $\gamma'(u) \notin K$.

Remarks.

- In **B** we have necessarily $u \in V^f$; and we may have $\gamma(v) = \infty$ as in **C**.
- To make condition **C** more accessible, we state it also in the following equivalent form:

  C'. If for every $v \in F(u)$ with $\gamma(v) = \infty$ there is $w \in F(v)$ with $\gamma(w) = \gamma'(u)$, then $\gamma(u) < \infty$.

- If condition **C'** is satisfied, then $\gamma(u) < \infty$, and so by **A**, $\gamma(w) = \gamma'(u) = \gamma(u)$.
- To keep the notation simple, we write $\infty(0)$, $\infty(1)$, $\infty(0, 1)$ etc., for $\infty(\{0\})$, $\infty(\{1\})$, $\infty(\{0, 1\})$, etc.
- γ exists uniquely on every finite cyclic digraph.

We next formulate an algorithm for computing γ. Initially a special symbol ν is attached to the label $\ell(u)$ of every vertex u, where $\ell(u) = \nu$ means that u has no label. We also introduce the notation $V_\nu = \{u \in V : \ell(u) = \nu\}$.

Algorithm GSG for computing the Generalized Sprague–Grundy function for a given finite cyclic digraph $G = (V, E)$.

1. (Initialize labels and counter.) Put $i \leftarrow 0$, $m \leftarrow 0$, $\ell(u) \leftarrow \nu$ for all $u \in V$.
2. (Label and counter.) As long as there exists $u \in V_\nu$ such that no follower of u is labeled i and every follower of u which is either unlabeled or labeled ∞ has a follower labeled i, put $\ell(u) \leftarrow i$, $c(u) \leftarrow m$, $m \leftarrow m + 1$.
3. (∞-label.) For every $u \in V_\nu$ which has no follower labeled i, put $\ell(u) \leftarrow \infty$.
4. (Increase label.) If $V_\nu \neq \varnothing$, put $i \leftarrow i + 1$ and return to 2; otherwise end.

We then have $\gamma(u) = \ell(u)$. A realization of Algorithm GSG performs a depth-first (endorder) traversal of the digraph for each finite ℓ-value. Letting $\gamma_{\max} = \max\{\gamma(u) : u \in V^f\}$, we evidently have $\gamma_{\max} < |V|$. Hence the number of steps of the algorithm is bounded by $O\big((|V| + |E|)|V|\big)$. For a connected digraph the complexity of the entire algorithm is thus $O(|V||E|)$.

Informally, the *sum* of a finite collection of games is a game in which a move consists of selecting one of the component games and making a move in it. Formally, let $\{\Gamma_1, \ldots, \Gamma_m\}$ be a finite disjoint collection of games with game-graphs $\{G_1 = (V_1, E_1), \ldots, G_m = (V_m, E_m)\}$, which may have cycles or may be infinite. Then the *sum-game* $\Gamma = \Gamma_1 + \ldots + \Gamma_m$ is the 2-player game in which a position has the form $(u_1, \ldots, u_m)$ with $u_i \in V_i$ for all i, and a move consists of selecting some Γ_i and making a legal move $u_i \to v_i$ in it $((u_i, v_i) \in E_i)$.

The *sum-graph* $G = G_1 + \ldots + G_m$ is the digraph $\mathbf{G} = (\mathbf{V}, \mathbf{E})$ defined as follows:

$$\mathbf{V} = \big\{(u_1, \ldots, u_m) : u_i \in V_i \quad i \in \{1, \ldots, m\}\big\}.$$

If $\mathbf{u} = (u_1, \ldots, u_m)$, $\mathbf{v} = (v_1, \ldots, v_m) \in \mathbf{V}$, then $(\mathbf{u}, \mathbf{v}) \in \mathbf{E}$ if there is some $j \in \{1, \ldots, m\}$ such that $v_j \in F(u_j)$, that is, $(u_j, v_j) \in E_j$, and $u_i = v_i$ for all $i \neq j$.

The *generalized Nim sum* $\oplus$ of a finite number of nonnegative integers is their sum over $\mathbf{GF}(2)$, also called *exclusive or* (XOR). Further, for any nonnegative integer a and subsets $K, L \subseteq \mathbb{Z}^0$ we have $a \oplus \infty(K) = \infty(K) \oplus a = \infty(K \oplus a)$, and $\infty(K) \oplus \infty(L) = \infty(\varnothing)$.

Notation. The generalized Nim sum is denoted by $\sum'$. Thus, $\sum'^h_{i=1} u_i$ is the generalized Nim sum of $u_1, \ldots, u_h$.

The important observation is that if $\mathbf{u} = (u_1, \ldots, u_m) \in \mathbf{V}$ is any position in a game graph, then $\gamma(\mathbf{u}) = \sum'^m_{i=1} \gamma(u_i)$ (see [18], Theorem 5). It enables one to compute polynomially the strategy on the exponentially large sum graph for a special case of our games.

Informally, a *P-position* is any position u from which the *P*revious player can force a win, that is, the opponent of the player moving from u. An *N-position* is any position v from which the *N*ext player can force a win, that is, the player who moves from v. The next player can win by moving to a *P*-position. A *D-position* is any position from which neither player can win, but both have a nonlosing next move, namely, moving to some *D*-position. Denote the set of all *P*-positions by $\mathcal{P}$, all *N*-positions by $\mathcal{N}$ and all *D*-positions by $\mathcal{D}$. The connection between γ on the sum of one or a finite number of disjoint games and P, N, D is given by:

$$\mathcal{P} = \{\mathbf{u} \in \mathbf{V} : \gamma(\mathbf{u}) = 0\}, \quad \mathcal{D} = \{\mathbf{u} \in \mathbf{V} : \gamma(\mathbf{u}) = \infty(L), \ 0 \notin L\}, \quad (1)$$

$$\mathcal{N} = \{\mathbf{u} \in \mathbf{V} : 0 < \gamma(\mathbf{u}) < \infty\} \cup \{\mathbf{u} \in \mathbf{V} : \gamma(\mathbf{u}) = \infty(L), \ 0 \in L\}, \quad (2)$$

and

for every $\mathbf{u} \in \mathcal{P}$ and every $\mathbf{v} \in F(\mathbf{u})$ there is $\mathbf{w} \in F(\mathbf{v}) \cap \mathcal{P}$ with $c(\mathbf{w}) < c(\mathbf{u})$.

3. Idiosyncrasies of the Exponentially Large Game-Graph

Given a finite digraph $G = (V, E)$, also called *groundgraph*, order V in some way, say

$$V = \{z_0, \dots, z_{n-1}\}.$$

This ordering, with $|V| = n$, is assumed throughout.

In its general form, the family of two-player cellular automata games played on G depends on integer parameters $(q(0), \dots, q(n-1))$ such that $1 \le q(k) \le |F(z_k)|$ for $0 \le k \le n - 1$. A move from z_k consists of firing some neighborhood of z_k of size $q(k)$. This family will now be modeled by means of a *game graph*.

Let $\mathbf{G} = (\mathbf{V}, \mathbf{E})$ denote the following game graph of the two-player cellular automata game played on $G = (V, E)$. The digraph $\mathbf{G}$ is also called the *cellular automata graph* or *game graph* of G. Any vertex in $\mathbf{G}$ can be described in the form $\mathbf{u} = (u^0, \dots, u^{n-1})$ over the field GF(2), where $u^k = 1$ if $w(z_k) = 1$, $u^k = 0$ if $w(z_k) = 0$. In particular, $\Phi = (0, \dots, 0)$ is a leaf of $\mathbf{V}$, and $|\mathbf{V}| = 2^n$.

Note that $\mathbf{V}$ is an abelian group under the addition $\oplus$ of GF(2), which is Nim-addition, with identity Φ. Every nonzero element has order 2. Moreover, $\mathbf{V}$ is a vector space over GF(2) satisfying $1\mathbf{u} = \mathbf{u}$ for all $\mathbf{u} \in \mathbf{V}$. For $i \in \{0, \dots, n-1\}$, define unit vectors $\mathbf{z}_i = (z_i^0, \dots, z_i^{n-1})$ with $z_i^j = 1$ if $i = j$; $z_i^j = 0$ otherwise. They span the vector space. In particular, for any $\mathbf{u} = (u^0, \dots, u^{n-1}) \in \mathbf{V}$ we can write, $\mathbf{u} = \sum_{i=0}^{n-1} u^i \mathbf{z}_i = \sum_{i=0}^{'n-1} u^i \mathbf{z}_i$.

For defining $\mathbf{E}$, let $\mathbf{u} \in \mathbf{V}$ and let $0 \le k \le n - 1$. For $0 \le q = q(k) \le |F(z_k)|$, let $F^q(z_k) \subseteq F(z_k)$ be any subset of $F(z_k)$ satisfying

$$|F^q(z_k)| = q. \tag{3}$$

Define

$$(\mathbf{u}, \mathbf{v}) \in \mathbf{E} \ \text{ if } u^k = 1, \ q > 0, \ \text{and } \mathbf{v} = \mathbf{u} \oplus \mathbf{z}_k \oplus {\sum_{z_\ell \in F^q(z_k)}}' \mathbf{z}_\ell, \tag{4}$$

for every $F^q(z_k)$ satisfying (3).

Informally, an edge $(\mathbf{u}, \mathbf{v})$ reflects the firing of u^k in $\mathbf{u}$ (with $u^k = 1$), i.e., the complementing of the weights of z_k and $F^q(z_k)$. Such an edge exists for every $F^q(z_k)$ satisfying (3). Note that if $z_k \in G$ is a leaf, then there is no move from z_k, since then $q = 0$.

If (4) holds, we also write $\mathbf{v} = F_k^q(\mathbf{u})$. The set of all followers of $\mathbf{u}$ is

$$F(\mathbf{u}) = \bigcup_{u^k=1} \ \bigcup_{F^q(z_k) \subseteq F(z_k)} F_k^q(\mathbf{u}).$$

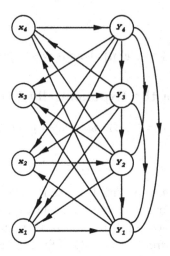

Figure 1. Playing on a parametrized digraph. © 2000

Example 1. Play on the digraph $G(p)$ which depends on a parameter $p \in \mathbb{Z}^+$. It has vertex set $\{x_1, \ldots, x_p, y_1, \ldots, y_p\}$, and edges:

$$F(x_i) = y_i \quad \text{for} \quad i = 1, \ldots, p,$$

$$F(y_k) = \{y_i \colon 1 \le i < k\} \cup \{x_j \colon 1 \le j \le p \text{ and } j \ne k\} \text{ for } k = 1, \ldots, p.$$

Figure 1 depicts $G(4)$. Suppose we fire the selected vertex u and complement precisely any *two* of its options. Let's play on $G(7)$ with $w(x_7) = w(y_1) = \ldots = w(y_7) = 1$, where all the other vertices have weight 0. What's the nature of this position?

The reader can verify that though the groundgraph $G(p)$ has no leaf, the gamegraph $\mathbf{G}(p)$ has no γ-value ∞. Using step 2 of Algorithm GSG, inspection of $G(p)$ implies that any collection of x_i is in $\mathbf{V_0}$. Moreover, $\gamma(y_i) =$ the ith *odious* number, where the odious numbers are those positive integers whose binary representations have an odd number of 1-bits. Incidentally, odious numbers arise in the analysis of other games, such as Grundy's game, Kayles, Mock Turtles, Turnips. See [1]. They arose earlier in a certain two-way splitting of the nonnegative integers [23] (but without this odious terminology!). Thus $\gamma(x_7 y_1 \ldots y_7) = 0 \oplus 1 \oplus 2 \oplus 4 \oplus 8 \oplus 15 \oplus 16 \oplus 32 = 48$. So either firing y_7 and complementing y_1, y_2, or firing y_6 and complementing y_1, y_3 reduces γ to 0 and so is a winning move.

Definition 2. For $s \in \mathbb{Z}^+$, an *s-game* on a digraph $G = (V, E)$ is a two-player cellular automata game on G satisfying $q \le s$ for all $k \in \{0, \ldots, n-1\}$ (q as in (3)). An *s-regular game* is an s-game such that for all k we have $q = s$ if $|F(z_k)| \ge s$, otherwise $q = |F(z_k)|$. (Thus the game played in Example 1 is a 2-regular game on $G(7)$. Of course if $s \ge \max_{u \in V} |F(u)|$, then firing u in an s-regular game entails complementing all of $F(u)$, for all $u \in V$, in addition to complementing u, unless u is a leaf.)

Remark. The strategy of a cellular automata game on a finite *acyclic* digraph is the same as that of a sum-game on $F^q(z_k)$, i.e., a game without complementation of the 1s of $F^q(z_k)$. This follows from the fact that $a \oplus a = 0$ for any nonnegative integer a. Only the length of play may be longer for the classical case. Thus 1-regular play is equivalent to play without interaction, as far as the strategy is concerned. With increasing s, s-regular games seem to get more difficult. For example, 2-regular play on a Nim-pile gives $g(0) = g(2) = 0$, $g(1) = 1$, $g(i+1) = i$-th odious number ($i \geq 2$).

Our aim is to compute γ on **G**. This function provides a strategy for sums of games some or all of whose components are two-player cellular automata games.

Lemma 1. *Let* $\mathbf{u}_1, \ldots, \mathbf{u}_h \in V$, $\mathbf{u} = \sum_{i=1}^{\prime h} \mathbf{u}_i$. *Then,*

(i) $F(\mathbf{u}) \subseteq \bigcup_{j=1}^{h} \left(F(\mathbf{u}_j) \oplus \sum_{i \neq j}^{\prime} \mathbf{u}_i \right) \subseteq F(\mathbf{u}) \cup F^{-1}(\mathbf{u})$.

(ii) *Let* $\mathbf{v}_j = F_k^q(\mathbf{u}_j)$, $\mathbf{v} = \mathbf{v}_j \oplus \sum_{i \neq j}^{\prime} \mathbf{u}_i$. *Then* $\mathbf{u} \in F(\mathbf{v})$ *if and only if either*

(a) $u^k = 0$, *or*

(b) *for some* $s \neq k$, $u^s = 0$ *and*

$$\mathbf{z}_k \oplus \sum_{z_\ell \in F^q(z_k)}^{\prime} \mathbf{z}_\ell = \mathbf{z}_s \oplus \sum_{z_t \in F^q(z_s)}^{\prime} \mathbf{z}_t. \tag{5}$$

Before proving the lemma, we single out the special case $h = 2$ of (i).

Corollary 1. *Let* $\mathbf{u}_1, \mathbf{u}_2 \in V$. *Then*

$$F(\mathbf{u}_1 \oplus \mathbf{u}_2) \subseteq \left(\mathbf{u}_1 \oplus F(\mathbf{u}_2) \right) \cup \left(F(\mathbf{u}_1) \oplus \mathbf{u}_2 \right) \tag{6}$$

$$\subseteq F(\mathbf{u}_1 \oplus \mathbf{u}_2) \cup F^{-1}(\mathbf{u}_1 \oplus \mathbf{u}_2). \tag{7}$$

Notes.

(i) Intuitively, (6) is explained by the similarity between sum-graphs and cellular automata games mentioned in the above remark. The intuition for the appearance of F^{-1} in (7) stems from the observation that $\mathbf{v} = \mathbf{u} \oplus \mathbf{z}_k \oplus \sum_{z_\ell \in F^q(z_k)}^{\prime} \mathbf{z}_\ell$ of (4) is consistent with both $\mathbf{v} \in F(\mathbf{u})$ and $\mathbf{u} \in F(\mathbf{v})$. In fact, if $\mathbf{v}$ is in the set on the right hand side of (6), then say, $\mathbf{v} \in F(\mathbf{u}_1) \oplus \mathbf{u}_2$, so for some $k \in \{0, \ldots, n-1\}$ with $u_1^k = 1$ we have,

$$\mathbf{v} = \mathbf{u} \oplus \mathbf{z}_k \oplus \sum_{z_\ell \in F^q(z_k)}^{\prime} \mathbf{z}_\ell,$$

and there are two cases: (I) $u^k = 1$, then $\mathbf{v} \in F(\mathbf{u})$, or (II) $u^k = 0$. But then $v^k = 1$, so $\mathbf{u} \in F(\mathbf{v})$.

(ii) Equality (5) is consistent with both $z_k \in F^q(z_s)$ and $z_s \in F^q(z_k)$.

Example 2. Consider a 2-play on $G(2)$ (defined in Example 1). Let $\mathbf{u}_1 = x_1 y_2$ (meaning that $w(x_1) = w(y_2) = 1$ and all other weights are 0), $\mathbf{u}_2 = y_1 y_2$. Then $\mathbf{u}_1 \oplus \mathbf{u}_2 = x_1 y_1$, $F(x_1 y_1) = \{\Phi, x_1 x_2\}$, $F^{-1}(x_1 y_1) = \{y_2\}$, $F(\mathbf{u}_1) = \{y_1 y_2, y_1\}$, $F(\mathbf{u}_2) = \{x_2 y_2, x_1\}$, $\mathbf{u}_1 \oplus F(\mathbf{u}_2) = \{x_1 x_2, y_2\}$, $F(\mathbf{u}_1) \oplus \mathbf{u}_2 = \{\Phi, y_2\}$. We see that Corollary 1 is satisfied.

Proof of Lemma 1. Let $\mathbf{v} \in F(\mathbf{u})$. Then $\mathbf{v} = F_k^q(\mathbf{u})$, $u^k = 1$ for some $0 \le k < n$, $F^q(z_k) \subseteq F(z_k)$. Hence $u_j^k = 1$ for some $1 \le j \le h$, and so

$$
\mathbf{v} = \mathbf{u} \oplus \mathbf{z}_k \oplus {\sum}'_{z_\ell \in F^q(z_k)} \mathbf{z}_\ell = \mathbf{u}_j \oplus \mathbf{z}_k \oplus {\sum}'_{z_\ell \in F^q(z_k)} \mathbf{z}_\ell \oplus {\sum}'_{i \ne j} \mathbf{u}_i
$$
$$
= \left(F_k^q(\mathbf{u}_j) \oplus {\sum}'_{i \ne j} \mathbf{u}_i \right) \in \left(F(\mathbf{u}_j) \oplus {\sum}'_{i \ne j} \mathbf{u}_i \right),
$$

proving the left hand side of (i).

Now let $\mathbf{v} \in \bigcup_{j=1}^{h} \left(F(\mathbf{u}_j) \oplus \sum'_{i \ne j} \mathbf{u}_i \right)$. Then $\mathbf{v} = F_k^q(\mathbf{u}_j) \oplus \sum'_{i \ne j} \mathbf{u}_i$ for some $1 \le j \le h$, $0 \le k < n$, $F^q(z_k) \subseteq F(z_k)$. Substituting $F_k^q(\mathbf{u}_j) = \mathbf{u}_j \oplus \mathbf{z}_k \oplus \sum'_{z_\ell \in F^q(z_k)} \mathbf{z}_\ell$, we get,

$$
\mathbf{v} = \mathbf{u} \oplus \mathbf{z}_k \oplus {\sum}'_{z_\ell \in F^q(z_k)} \mathbf{z}_\ell \qquad (u_j^k = 1). \tag{8}
$$

If $u^k = 1$, then $\mathbf{v} \in F(\mathbf{u})$. If $u^k = 0$, then (8) implies $v^k = 1$ and

$$
\mathbf{u} = \mathbf{v} \oplus \mathbf{z}_k \oplus {\sum}'_{z_\ell \in F^q(z_k)} \mathbf{z}_\ell, \tag{9}
$$

so $\mathbf{u} \in F(\mathbf{v})$, proving the right hand side of (i).

For (ii) we have $\mathbf{v}_j = F_k^q(\mathbf{u}_j)$ for some j, $\mathbf{v} = \mathbf{v}_j \oplus \sum'_{i \ne j} \mathbf{u}_i$. Then both (8) and (9) are valid. Suppose first that (a) holds. Then $v^k = 1$ by (8), and so $\mathbf{u} = F_k^q(\mathbf{v})$ by (9). If (b) holds, then (5) and (9) imply

$$
\mathbf{u} = \mathbf{v} \oplus \mathbf{z}_k \oplus {\sum}'_{z_\ell \in F^q(z_k)} \mathbf{z}_\ell = \mathbf{v} \oplus \mathbf{z}_s \oplus {\sum}'_{z_t \in F^q(z_s)} \mathbf{z}_t. \tag{10}
$$

Therefore $u^s = 0$ implies $v^s = 1$. Hence $\mathbf{u} = F_s^q(\mathbf{v})$.

Conversely, suppose that $\mathbf{u} \in F(\mathbf{v})$, say $\mathbf{u} = F_s^q(\mathbf{v})$. If $s = k$, then $v^k = 1$ so (a) holds by (9). If $s \ne k$, then $\mathbf{u} = F_s^q(\mathbf{v})$ and (9) imply (10), so (5) holds and also $u^s = 0$, i.e., (b) holds. $\qquad \Box$

4. The Additivity of γ

The first key observation for getting a handle at two-player cellular automata games is that γ on $\mathbf{G}$ is, essentially, additive.

Theorem 1. *Let $\mathbf{G} = (\mathbf{V}, \mathbf{E})$ be the cellular automata graph of the finite cyclic digraph $G = (V, E)$. Then $\gamma(\mathbf{u}_1 \oplus \mathbf{u}_2) = \gamma(\mathbf{u}_1) \oplus \gamma(\mathbf{u}_2)$ if $\mathbf{u}_1 \in \mathbf{V}^f$ or $\mathbf{u}_2 \in \mathbf{V}^f$.*

Proof. We use the notation $(\mathbf{v}_1, \mathbf{v}_2) \in \mathcal{F}(\mathbf{u}_1, \mathbf{u}_2)$ if either $\mathbf{v}_1 = \mathbf{u}_1$, $\mathbf{v}_2 \in F(\mathbf{u}_2)$, or $\mathbf{v}_1 \in F(\mathbf{u}_1)$, $\mathbf{v}_2 = \mathbf{u}_2$, i.e.,

$$
(\mathbf{v}_1, \mathbf{v}_2) \in \mathcal{F}(\mathbf{u}_1, \mathbf{u}_2) \quad \text{if} \quad (\mathbf{v}_1, \mathbf{v}_2) \in \big(\mathbf{u}_1, F(\mathbf{u}_2) \big) \cup \big(F(\mathbf{u}_1), \mathbf{u}_2 \big).
$$

Note that $\mathcal{F}$ is not a follower in $\mathbf{G}$, but rather in the sum-graph $\mathbf{G} + \mathbf{G}$. It is natural to consider $\mathcal{F}$, because $\sigma = \gamma(\mathbf{u}_1) \oplus \gamma(\mathbf{u}_2)$ is the γ-function of the sum $\mathbf{G} + \mathbf{G}$ (see [11], Sect. 3).

(a). We assume $(\mathbf{u}_1, \mathbf{u}_2) \in \mathbf{V}^f \times \mathbf{V}^f$. Let

$$T = \{(\mathbf{u}_1, \mathbf{u}_2) \in \mathbf{V}^f \times \mathbf{V}^f \colon \gamma(\mathbf{u}_1 \oplus \mathbf{u}_2) \neq \gamma(\mathbf{u}_1) \oplus \gamma(\mathbf{u}_2)\},$$

$$t = \min_{(\mathbf{u}_1, \mathbf{u}_2) \in T} \big(\gamma(\mathbf{u}_1 \oplus \mathbf{u}_2), \ \gamma(\mathbf{u}_1) \oplus \gamma(\mathbf{u}_2)\big).$$

Further, we define

$$U = \{(\mathbf{u}_1, \mathbf{u}_2) \in T \colon \gamma(\mathbf{u}_1) \oplus \gamma(\mathbf{u}_2) = t\}.$$

We first show that $T \neq \varnothing$ implies $U \neq \varnothing$.

If there is $(\mathbf{u}_1, \mathbf{u}_2) \in T$ such that $\gamma(\mathbf{u}_1 \oplus \mathbf{u}_2) = t$, then $t < \gamma(\mathbf{u}_1) \oplus \gamma(\mathbf{u}_2) < \infty$. Since σ is a γ-function, **A** of Definition 1, implies that there exists $(\mathbf{v}_1, \mathbf{v}_2) \in \mathcal{F}(\mathbf{u}_1, \mathbf{u}_2)$ such that $\gamma(\mathbf{v}_1) \oplus \gamma(\mathbf{v}_2) = t$. Now

$$\mathbf{v}_1 \oplus \mathbf{v}_2 \in \big(\mathbf{u}_1 \oplus F(\mathbf{u}_2)\big) \cup \big(F(\mathbf{u}_1) \oplus \mathbf{u}_2\big) \subseteq F(\mathbf{u}_1 \oplus \mathbf{u}_2) \cup F^{-1}(\mathbf{u}_1 \oplus \mathbf{u}_2)$$

by (7). Since $\gamma(\mathbf{u}_1 \oplus \mathbf{u}_2) = t$, it thus follows that $\gamma(\mathbf{v}_1 \oplus \mathbf{v}_2) > t$, so $(\mathbf{v}_1, \mathbf{v}_2) \in T$.

We have just shown that $T \neq \varnothing$ implies $U \neq \varnothing$. Next we show that $T = \varnothing$.

Pick $(\mathbf{u}_1, \mathbf{u}_2) \in U$ with $c(\mathbf{u}_1) + c(\mathbf{u}_2)$ minimum, where c is a monotonic counter function on $\mathbf{V}^f$. Then $c(\mathbf{u}_1) + c(\mathbf{u}_2)$ is a counter function for the γ-function σ on $\mathbf{G} + \mathbf{G}$ (see [18] Theorem 5). Note that

$$\gamma(\mathbf{u}_1) \oplus \gamma(\mathbf{u}_2) = t, \qquad \gamma(\mathbf{u}_1 \oplus \mathbf{u}_2) > t. \tag{11}$$

(i) Suppose that there exists $\mathbf{v} \in F(\mathbf{u}_1 \oplus \mathbf{u}_2)$ such that $\gamma(\mathbf{v}) = t$. By (6) there exists $(\mathbf{v}_1, \mathbf{v}_2) \in \mathcal{F}(\mathbf{u}_1, \mathbf{u}_2)$ such that $\mathbf{v} = \mathbf{v}_1 \oplus \mathbf{v}_2$, so $\gamma(\mathbf{v}_1 \oplus \mathbf{v}_2) = t$. Now the definition of $\mathcal{F}$, the minimality of t and the equality of (11) imply $\gamma(\mathbf{v}_1) \oplus \gamma(\mathbf{v}_2) > t$. Since σ is a γ-function, **B** of Definition 1 implies existence of $(\mathbf{w}_1, \mathbf{w}_2) \in \mathcal{F}(\mathbf{v}_1, \mathbf{v}_2)$ such that $\gamma(\mathbf{w}_1) \oplus \gamma(\mathbf{w}_2) = t$, $c(\mathbf{w}_1) + c(\mathbf{w}_2) < c(\mathbf{u}_1) + c(\mathbf{u}_2)$. By (7),

$$\mathbf{w}_1 \oplus \mathbf{w}_2 \in \big(\mathbf{v}_1 \oplus F(\mathbf{v}_2)\big) \cup \big(F(\mathbf{v}_1) \oplus \mathbf{v}_2\big) \subseteq F(\mathbf{v}_1 \oplus \mathbf{v}_2) \cup F^{-1}(\mathbf{v}_1 \oplus \mathbf{v}_2).$$

Since $\gamma(\mathbf{v}_1 \oplus \mathbf{v}_2) = t$, we thus have $\gamma(\mathbf{w}_1 \oplus \mathbf{w}_2) > t$, so $(\mathbf{w}_1, \mathbf{w}_2) \in U$, contradicting the minimality of $c(\mathbf{u}_1) + c(\mathbf{u}_2)$.

(ii) Suppose that $\mathbf{v} \in F(\mathbf{u}_1 \oplus \mathbf{u}_2)$ implies $\gamma(\mathbf{v}) \neq t$. Then **A** of Definition 1 implies $\gamma(\mathbf{u}_1 \oplus \mathbf{u}_2) = \infty$. By **A** applied to σ and the equality in (11), for every $j \in \{0, \ldots, t-1\}$ there exists $(\mathbf{v}_1, \mathbf{v}_2) \in \mathcal{F}(\mathbf{u}_1, \mathbf{u}_2)$ such that $\gamma(\mathbf{v}_1) \oplus \gamma(\mathbf{v}_2) = j$. By the minimality of t, also $\gamma(\mathbf{v}_1 \oplus \mathbf{v}_2) = j$. As above, by (7), $\mathbf{v}_1 \oplus \mathbf{v}_2 \in F(\mathbf{u}_1 \oplus \mathbf{u}_2) \cup F^{-1}(\mathbf{u}_1 \oplus \mathbf{u}_2)$. If $\mathbf{u}_1 \oplus \mathbf{u}_2 \in F(\mathbf{v}_1 \oplus \mathbf{v}_2)$, then there exists $\mathbf{w} \in F(\mathbf{u}_1 \oplus \mathbf{u}_2)$ with $\gamma(\mathbf{w}) = j$. Hence in any case, $\gamma'(\mathbf{u}_1 \oplus \mathbf{u}_2) = t$. (This holds also if $t = 0$.) By **C** there exists $\mathbf{v} \in F(\mathbf{u}_1 \oplus \mathbf{u}_2)$ such that $\gamma(\mathbf{v}) = \infty(L)$, $t \notin L$. By (6), there exists $(\mathbf{v}_1, \mathbf{v}_2) \in \mathcal{F}(\mathbf{u}_1, \mathbf{u}_2)$ such that $\mathbf{v} = \mathbf{v}_1 \oplus \mathbf{v}_2$. Thus,

$$\gamma(\mathbf{v}_1 \oplus \mathbf{v}_2) = \infty(L), \qquad t \notin L, \tag{12}$$

and $\gamma(\mathbf{v}_1) \oplus \gamma(\mathbf{v}_2) > t$ by the equality in (11). As in (i) we deduce existence of $(\mathbf{w}_1, \mathbf{w}_2) \in \mathcal{F}(\mathbf{v}_1, \mathbf{v}_2)$ such that $\gamma(\mathbf{w}_1) \oplus \gamma(\mathbf{w}_2) = t$, $c(\mathbf{w}_1) + c(\mathbf{w}_2) < c(\mathbf{u}_1) + c(\mathbf{u}_2)$. By (7) either $\mathbf{w}_1 \oplus \mathbf{w}_2 \in F(\mathbf{v}_1 \oplus \mathbf{v}_2)$ or $\mathbf{v}_1 \oplus \mathbf{v}_2 \in F(\mathbf{w}_1 \oplus \mathbf{w}_2)$. In the former case, $\gamma(\mathbf{w}_1 \oplus \mathbf{w}_2) > t$ by (12). In the latter case, if $\gamma(\mathbf{w}_1 \oplus \mathbf{w}_2) = t$, then $\mathbf{B}$ implies existence of $\mathbf{y} \in F(\mathbf{v}_1 \oplus \mathbf{v}_2)$ such that $\gamma(\mathbf{y}) = t$, contradicting (12). Thus in either case $(\mathbf{w}_1, \mathbf{w}_2) \in U$, contradicting the minimality of $c(\mathbf{u}_1) + c(\mathbf{u}_2)$. Thus $U = T = \varnothing$.

(b). We now assume, without loss of generality, $\gamma(\mathbf{u}_1) < \infty$, $\gamma(\mathbf{u}_2) = \infty$. If $\gamma(\mathbf{u}_1 \oplus \mathbf{u}_2) < \infty$, then by (a),

$$\gamma(\mathbf{u}_2) = \gamma(\mathbf{u}_1 \oplus (\mathbf{u}_1 \oplus \mathbf{u}_2)) = \gamma(\mathbf{u}_1) \oplus \gamma(\mathbf{u}_1 \oplus \mathbf{u}_2) < \infty,$$

a contradiction. Hence $\gamma(\mathbf{u}_1 \oplus \mathbf{u}_2) = \infty(M)$ for some set M. If the theorem's assertion is false, then there exist $\mathbf{u}_1, \mathbf{u}_2$ with $\gamma(\mathbf{u}_1) < \infty$, $\gamma(\mathbf{u}_2) = \infty(L)$ and $c(\mathbf{u}_1)$ minimum such that $\gamma(\mathbf{u}_1 \oplus \mathbf{u}_2) \neq \gamma(\mathbf{u}_1) \oplus \gamma(\mathbf{u}_2)$, i.e., $M \neq \gamma(\mathbf{u}_1) \oplus L$.

Let $d \in \gamma(\mathbf{u}_1) \oplus L$. Then $d = \gamma(u_1) \oplus d_1$, where $d_1 \in L$. Let $\mathbf{v}_2 \in F(\mathbf{u}_2)$ satisfy $\gamma(\mathbf{v}_2) = d_1$. Then $\gamma(\mathbf{u}_1 \oplus \mathbf{v}_2) = \gamma(\mathbf{u}_1) \oplus d_1 = d$ by (a). By (7), $\mathbf{u}_1 \oplus \mathbf{v}_2 \in \mathbf{u}_1 \oplus F(u_2) \subseteq F(\mathbf{u}_1 \oplus \mathbf{u}_2) \cup F^{-1}(\mathbf{u}_1 \oplus \mathbf{u}_2)$. Hence $\mathbf{B}$ implies $d \in M$.

Now let $d \in M$ and $\mathbf{v} \in F(\mathbf{u}_1 \oplus \mathbf{u}_2)$ satisfy $\gamma(\mathbf{v}) = d$. By (6), $\mathbf{v} \in (\mathbf{u}_1 \oplus F(\mathbf{u}_2)) \cup (F(\mathbf{u}_1) \oplus \mathbf{u}_2)$. We consider two cases.

(i) $\mathbf{v} = \mathbf{u}_1 \oplus \mathbf{v}_2$ with $\mathbf{v}_2 \in F(\mathbf{u}_2)$. Then by (a), $\gamma(\mathbf{v}_2) = \gamma(\mathbf{u}_1) \oplus \gamma(\mathbf{v}) = \gamma(\mathbf{u}_1) \oplus d \in L$. Hence $d \in \gamma(\mathbf{u}_1) \oplus L$.

(ii) $\mathbf{v} = \mathbf{v}_1 \oplus \mathbf{u}_2$ with $\mathbf{v}_1 \in F(\mathbf{u}_1)$. As at the beginning of (b) we conclude $\gamma(\mathbf{v}_1) = \infty$. By $\mathbf{B}$ there exists $\mathbf{w}_1 \in F(\mathbf{v}_1)$ such that $\gamma(\mathbf{w}_1) = \gamma(\mathbf{u}_1)$, $c(\mathbf{w}_1) < c(\mathbf{u}_1)$. Let $\mathbf{w} = \mathbf{w}_1 \oplus \mathbf{u}_2$. The minimality of $c(\mathbf{u}_1)$ implies

$$\gamma(\mathbf{w}) = \gamma(\mathbf{w}_1 \oplus \mathbf{u}_2) = \gamma(\mathbf{w}_1) \oplus \gamma(\mathbf{u}_2) = \gamma(\mathbf{u}_1) \oplus \gamma(\mathbf{u}_2) = \infty(\gamma(\mathbf{u}_1) \oplus L).$$

By (7), $\mathbf{w} \in F(\mathbf{v}_1) \oplus \mathbf{u}_2 \subseteq F(\mathbf{v}) \cup F^{-1}(\mathbf{v})$. If $\mathbf{v} \in F(\mathbf{w})$, then $\gamma(\mathbf{v}) = d \in \gamma(\mathbf{u}_1) \oplus L$. If $\mathbf{w} \in F(\mathbf{v})$, the same holds by $\mathbf{B}$. Thus in all cases $M = \gamma(\mathbf{u}_1) \oplus L$, a contradiction. $\square$

5. The Structure of γ

Denote by $\mathrm{GF}(2)^t := (\mathrm{GF}(2))^t$ the vector space of all t-dimensional binary vectors over $\mathrm{GF}(2)$ under $\oplus$. It is often convenient to identify $\mathrm{GF}(2)^t$ with the set of integers in the interval $[0, 2^t - 1]$. We now give very precise information about the structure of $\mathbf{V}^f$.

Theorem 2. Let $\mathbf{G} = (\mathbf{V}, \mathbf{E})$ be the cellular automata graph of the finite cyclic digraph $G = (V, E)$. Then $\mathbf{V}^f$ and $\mathbf{V}_0$ are linear subspaces of $\mathbf{V}$. Moreover, γ is a homomorphism from $\mathbf{V}^f$ onto $\mathrm{GF}(2)^t$ for some $t \in \mathbb{Z}^0$ with kernel $\mathbf{V}_0$ and quotient space $\mathbf{V}^f / \mathbf{V}_0 = \{\mathbf{V}_i : 0 \leq i < 2^t\}$, $\dim(\mathbf{V}^f) = m + t$, where $m = \dim(\mathbf{V}_0)$.

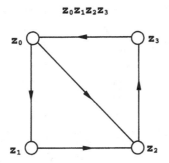

Figure 2. Illustrating Theorem 2. © 2000

Example 3. Consider a 2-regular game played on the digraph depicted in Fig. 2. We adopt a decimal encoding of the vertices of $\mathbf{G}$: $z_i = 2^i$ for all $i \in \mathbb{Z}^0$. Thus 5 means that there are tokens on vertices z_0 and z_2 only. Using inspection and step 2 of Algorithm GSG, we see that $\gamma(1) = 0$ (the position in $\mathbf{G}$ with $w(z_0) = 1$, $w(z_i) = 0$ for $i > 0$). Also $\gamma(4) = \gamma(10) = 0$. Hence by linearity, $\mathbf{V}_0 = \{\Phi, 1, 4, 10, 5, 11, 14, 15\}$. We further note that $\gamma(2) = 1$. Hence we get the coset $\mathbf{V}_1 = 2 \oplus \mathbf{V}_0 = \{2, 3, 6, 7, 8, 9, 12, 13\}$. Also $m = 3$, $t = 1$, and $\mathbf{V}^f$ is spanned by the basis vectors $\beta^q = \{1, 4, 10, 2\}$, $\dim(\mathbf{V}^f) = n = 4$. This illustrates the nice structure of the general case: the number of vertices of $\mathbf{V}$ assuming γ-value i is the same for all i from 0 to some maximum value which is necessarily a power of 2 less 1.

Proof. Let $\mathbf{u}, \mathbf{v} \in \mathbf{V}^f$. Then $\mathbf{u} \oplus \mathbf{v} \in \mathbf{V}^f$ by Theorem 1, and also $\Phi \in \mathbf{V}^f$. Thus $\mathbf{V}^f$ is a subspace of $\mathbf{V}$.

Let t be the smallest nonnegative integer such that $\gamma(\mathbf{u}) \leq 2^t - 1$ for all $\mathbf{u} \in \mathbf{V}^f$. Hence, if $t \geq 1$, there is some $\mathbf{v} \in \mathbf{V}^f$ such that $\gamma(\mathbf{v}) \geq 2^{t-1}$. Then the "1's complement" $2^t - 1 - \gamma(\mathbf{v}) < \gamma(\mathbf{v})$. By **A** of Definition 1, there exists $\mathbf{w} \in F(\mathbf{v})$ such that $\gamma(\mathbf{w}) = 2^t - 1 - \gamma(\mathbf{v})$. By Theorem 1, $\gamma(\mathbf{v} \oplus \mathbf{w}) = \gamma(\mathbf{v}) \oplus \gamma(\mathbf{w}) = 2^t - 1$. Thus by **A**, every integer in $[0, 2^t - 1]$ is assumed as a γ-value by some $\mathbf{u} \in \mathbf{V}$. This last property holds trivially also for $t = 0$. Hence γ is onto. It is a homomorphism $\mathbf{V}^f \to GF(2)^t$ by Theorem 1 and since $\gamma(1\mathbf{u}) = \gamma(\mathbf{u}) = 1\gamma(\mathbf{u})$, $\gamma(0\mathbf{u}) = \gamma(\Phi) = 0 = 0\gamma(\mathbf{u})$.

By linear algebra we have the isomorphism $GF(2)^t \cong \mathbf{V}^f / \mathbf{V}_0$, where $\mathbf{V}_0$ is the kernel, so $t = \dim(\mathbf{V}^f / \mathbf{V}_0)$. Hence $\mathbf{V}_0$ is a subspace of $\mathbf{V}^f$, and so also of $\mathbf{V}$. Let $m = \dim(\mathbf{V}_0)$. Then $\dim(\mathbf{V}^f) = m + t$. The elements of $\mathbf{V}^f / \mathbf{V}_0$ are the cosets $\mathbf{V}_i = \mathbf{w} \oplus \mathbf{V}_0$ for any $\mathbf{w} \in \mathbf{V}_i$ and every integer $i \in [0, 2^t - 1]$. $\qquad\square$

Clearly $\mathbf{V}^\infty$ is not a linear subspace: $\mathbf{u} \oplus \mathbf{u} = \Phi$ for every $\mathbf{u} \in \mathbf{V}^\infty$, but $\Phi \in \mathbf{V}^f$. For revealing also the structure of $\mathbf{V}^\infty$ we thus have to embark on a different course. We extend the homomorphism $\gamma \colon \mathbf{V}^f \to GF(2)^t$ to a homomorphism ρ on the entire space $\mathbf{V}$. Since any $\mathbf{u} \in \mathbf{V}$ can be written as a linear combination of the unit vectors, i.e., $\mathbf{u} = \Sigma_{i=0}^{'n-1} \varepsilon_i \mathbf{z}_i$ ($\varepsilon_i \in \{0, 1\}$, $0 \leq i < n$), but some or all of the γ-values of the $\mathbf{z}_i$ may be ∞, the extended homomorphism ρ will then permit to compute $\rho(\mathbf{u}) = \Sigma_{i=0}^{\prime n-1} \varepsilon_i \rho(\mathbf{z}_i)$, such that $\mathbf{u} \in \mathbf{V}^f$ if $\rho(\mathbf{u}) = \gamma(\mathbf{u})$; and we will arrange things so that $\rho(\mathbf{u}) > \gamma(\mathbf{v})$ for all $\mathbf{v} \in \mathbf{V}^f$ if $\mathbf{u} \in \mathbf{V}^\infty$. Since the homomorphism γ maps an $(m+t)$-dimensional space onto a t-dimensional space and we wish to preserve the kernel $\mathbf{V}_0$ in ρ, we will make ρ map the n-dimensional space $\mathbf{V}$ onto an $(n-m)$-dimensional space $\mathbf{W}$, preserving the reduction by m dimensions. (The extended part will then actually be an isomorphism.)

Lemma 2. *Let $\mathbf{V}$ be the n-dimensional vector space over $\mathrm{GF}(2)$ of the cellular automata game on $G = (V, E)$ with $\dim(\mathbf{V}_0) = m$. There exists a homomorphism ρ mapping $\mathbf{V}$ onto $\mathrm{GF}(2)^{n-m}$ with kernel $\mathbf{V}_0$ such that $\mathbf{u} \in \mathbf{V}^f$ if $\rho(\mathbf{u}) = \gamma(\mathbf{u})$, and $\mathbf{u} \in \mathbf{V}^\infty$ if $\rho(\mathbf{u}) > \gamma(\mathbf{v})$ for all $\mathbf{v} \in \mathbf{V}^f$.*

Proof. From linear algebra, there exists an $(n-m-t)$-dimensional subspace $\mathbf{W}$ of $\mathbf{V}$, such that $\mathbf{V}$ is the direct sum of $\mathbf{V}^f$ and $\mathbf{W}$. Thus, every $\mathbf{u} \in \mathbf{V}$ can be written uniquely in the form

$$\mathbf{u} = \mathbf{w} \oplus \mathbf{v}, \qquad \mathbf{w} \in \mathbf{W}, \qquad \mathbf{v} \in \mathbf{V}^f. \tag{13}$$

Let $I \colon \mathbf{W} \to \mathrm{GF}(2)^{n-m-t}$ be any isomorphism, and define

$$\rho(\mathbf{u}) = \big(I(\mathbf{w}), \gamma(\mathbf{v})\big), \quad \big(I(\mathbf{w}) \in \mathrm{GF}(2)^{n-m-t}, \quad \gamma(\mathbf{v}) \in \mathrm{GF}(2)^t\big),$$

which is well-defined in view of the uniqueness of the representation (13). Then $\rho \colon \mathbf{V} \to \mathrm{GF}(2)^{n-m}$ is a homomorphism, since if $\mathbf{u}' = \mathbf{w}' \oplus \mathbf{v}'$, $\mathbf{w}' \in \mathbf{W}$, $\mathbf{v}' \in \mathbf{V}^f$, then

$$\mathbf{u} \oplus \mathbf{u}' = (\mathbf{w} \oplus \mathbf{w}') \oplus (\mathbf{v} \oplus \mathbf{v}'), \qquad \mathbf{w} \oplus \mathbf{w}' \in \mathbf{W}, \qquad \mathbf{v} \oplus \mathbf{v}' \in \mathbf{V}^f,$$

so

$$\begin{aligned} \rho(\mathbf{u} \oplus \mathbf{u}') &= \big(I(\mathbf{w} \oplus \mathbf{w}'), \gamma(\mathbf{v} \oplus \mathbf{v}')\big) = \big(I(\mathbf{w}) \oplus I(\mathbf{w}'), \gamma(\mathbf{v}) \oplus \gamma(\mathbf{v}')\big) \\ &= \big(I(\mathbf{w}), \gamma(\mathbf{v})\big) \oplus \big(I(\mathbf{w}'), \gamma(\mathbf{v}')\big) = \rho(\mathbf{u}) \oplus \rho(\mathbf{u}'), \end{aligned}$$

and

$$\rho(1\mathbf{u}) = \rho(\mathbf{u}) = 1\rho(\mathbf{u}), \qquad \rho(0\mathbf{u}) = \rho(\Phi) = \big(I(\Phi), \gamma(\Phi)\big) = 0 = 0\rho(\mathbf{u}).$$

Finally, $\mathbf{u} \in \mathbf{V}^f$ if $\mathbf{u} = \Phi \oplus \mathbf{u}$ with $\Phi \in \mathbf{W}$, $\mathbf{u} \in \mathbf{V}^f$, and then $\rho(\mathbf{u}) = (\Phi, \gamma(\mathbf{u}))$ with numerical value $\gamma(\mathbf{u})$. For $\mathbf{u} \in \mathbf{V}^\infty$, $I(\mathbf{w}) \neq \Phi$, so the numerical value of the binary vector $(I(\mathbf{w}), \gamma(\mathbf{v}))$ is larger than that of $\gamma(\mathbf{v})$ for all $\mathbf{v} \in \mathbf{V}^f$. $\square$

Example 4. We play a 3-regular game on $G(7)$ (recall that $G(p)$ was introduced in Example 1, and $G(4)$ is displayed in Fig. 1). Inspection shows that all collections of an even number of x_i are in $\mathbf{V}_0$; and $\mathbf{V}^f$ consists precisely of all collections of an even number of vertices with weight 1. Furthermore, $\gamma(x_j y_j) =$ smallest nonnegative integer not the Nim sum of at most three $\gamma(x_i y_i)$ for $i < j$. Also $\gamma(x_j y_j) = \gamma(x_i y_j)$ for all i. Thus $\{\gamma(x_1 y_i)\}_{i=1}^7 = \{1, 2, 4, 8, 15, 16, 32\}$. Further, $\mathbf{W} = \mathcal{L}(x_1) = \{\Phi, x_1\}$ is the complement of $\mathbf{V}^f$, and $n - m - t =$

$1 = \dim(\mathbf{W})$ with basis $\beta^\infty = \{x_1\}$, where $\mathcal{L}$ denotes the linear span. Also $m = \dim \mathbf{V}_0 = 7$ since $\mathbf{V}_0$ contains the 2^6 subsets of an even number of x_i as well as $y_1 y_2 y_3 y_4 y_5$. Thus $t = 6$, which is consistent with $\gamma(x_7 y_7) = 32$, since the value 64 is not attained. Moreover, we can put $\rho(x_i) = 64$ for all i, and $\{\rho(y_i)\}_{i=1}^7 = \{65, 66, 68, 72, 79, 80, 96\}$. Since $t = 6$ and the ρ-values on the singletons all have a nonzero bit in a position $> 2^t - 1$, the γ-value of any collection with an odd (even) number of vertices from G has value ∞ ($< \infty$). For example, $\rho(x_7 y_1 \ldots y_7) = \gamma(x_7 y_1 \ldots y_7) = 1 \oplus 2 \oplus 4 \oplus 8 \oplus 15 \oplus 16 \oplus 32 = 48$. So firing y_7 and complementing y_6 and any two of the x_i $(i < 7)$ is a winning move. Incidentally, the sequence $\{1, 2, 4, 8, 15, 16, 32, 51, \ldots\}$ has been used in [1] for a special case of the game "Turning Turtles".

This a posteriori verification is not very satisfactory, but at this stage the example nevertheless illustrates nicely Theorem 2 and Lemma 2. In §6, where we construct ρ by embedding its computation in an algorithm for computing $\mathbf{V}_0$, $\mathbf{V}^f$ and $\mathbf{W}$, we will see how to compute a priori results such as these. The problem right now is that despite the precise information about the structure of $\mathbf{V}^f$, $\mathbf{V}_0$ and $\mathbf{V}^\infty$, the computation of say, $\mathbf{V}_0$, is exponential, as we may have to scan the 2^n vectors of $\mathbf{V}$ for membership in $\mathbf{V}_0$. Actually only a polynomial fragment of the 2^n vectors has to be examined, as we will see in the next section.

6. Sparse Vectors Suffice

A vector $\mathbf{u} \in GF(2)^n$ has *weight* i, if it has precisely i 1-bits, i.e., $\sum_{k=1}^n u^k = i$. We write $w(\mathbf{u}) = i$ if $\mathbf{u}$ has weight i. This terminology is standard in coding theory. There should be no confusion between the weights $w(u^i)$, $w(\mathbf{u})$ and the vector $\mathbf{w} \in \mathbf{V}$.

The second key observation conducive to producing a strategy for cellular automata games is that for s-regular games, to which we now confine ourselves, $\mathbf{V}_0$, $\mathbf{V}^f$ and ρ can be computed by restricting attention to the linear span of vectors of weight at most $2(s + 1)$.

We begin with some notation.

$$Z_i = \bigcup_{h=1}^i \{\mathbf{u} \in \mathbf{V} : w(\mathbf{u}) = h\} \qquad \text{(vectors of weight } \leq i\text{)}$$

$$Z_i^f = Z_i \cap \mathbf{V}^f$$

$$S = Z_1 \cap \{\mathbf{u} \in \mathbf{V} : F(\mathbf{u}) = \varnothing\} \cup \{\Phi\} \qquad \text{(leaves of weight } \leq 1\text{)}$$

$$L = \{\mathbf{u} \in \mathbf{V} : F(\mathbf{u}) = \varnothing\} \qquad \text{(set of all leaves)}$$

$$Q = (Z_{2(s+1)} \cap \mathbf{V}_0) \cup S$$

$$g(n, s) = (s + 1)\binom{n-1}{s} - s$$

$$\phi(s) = \max(3s + 2, 2s + 4) = \begin{cases} 6 & \text{if } s = 1, \\ 3s + 2 & \text{if } s > 1. \end{cases}$$

Theorem 3. *Let* $\mathbf{G} = (\mathbf{V}, \mathbf{E})$ *be the cellular automata graph of the finite digraph* $G = (V, E)$. *Then*

(i) $L = \mathcal{L}(S)$ *(linear span of* S *over* GF (2))
(ii) $\mathbf{V}_0 = \mathcal{L}(Q)$
(iii) $\mathbf{V}^f = \mathcal{L}(Q \cup Z_{s+1}^f)$
(iv) $\{\gamma(\mathbf{u}): \mathbf{u} \in \mathbf{V}^f\} = \{\gamma(\mathbf{u}): \mathbf{u} \in (Z_{s+1}^f \cup \{\Phi\})\}$
(v) $t \le \lfloor \log_2(1 + g(n, s)) \rfloor$ $(n \ge s + 1)$, *where* $2^t - 1$ *is the maximum value of* γ *on* $\mathbf{V}^f$.

Comment. Since every vector in S, Q or $Q \cup Z_{s+1}^f$ has weight at most $2(s+1)$, it follows that $L, \mathbf{V}_0, \mathbf{V}^f$ can all be computed from a set of vectors of weight at most $2(s+1)$.

Proof. (i) Follows from the definition of a leaf.

(ii) Clearly $\mathcal{L}(Q) \subseteq \mathbf{V}_0$. If the result is false, let $\mathbf{u}$ with $c(\mathbf{u})$ minimal satisfy $\mathbf{u} \in \mathbf{V}_0$, $\mathbf{u} \notin \mathcal{L}(Q)$. In particular, $\mathbf{u} \notin \mathcal{L}(S)$, and so by (i), $\mathbf{u}$ is not a leaf. Hence there is $\mathbf{v} \in F(\mathbf{u})$. By $\mathbf{A}$ of Definition 1, $\gamma(\mathbf{v}) > 0$, and by $\mathbf{B}$, there exists $\mathbf{w} \in F(\mathbf{v}) \cap \mathbf{V}_0$ with $c(\mathbf{w}) < c(\mathbf{u})$. By the minimality of $c(\mathbf{u})$, we have $\mathbf{w} \in \mathcal{L}(Q)$. Now $\mathbf{v} = \mathbf{u} \oplus \mathbf{z}_k \oplus \sum'_{z_\ell \in F^q(z_k)} \mathbf{z}_\ell$ and $\mathbf{w} = \mathbf{v} \oplus \mathbf{z}_r \oplus \sum'_{z_h \in F^q(z_r)} \mathbf{z}_h$ say, so letting $\mathbf{y} = \mathbf{w} \oplus \mathbf{u}$ we have $\mathbf{y} = \mathbf{z}_k \oplus \sum'_{z_\ell \in F^q(z_k)} \mathbf{z}_\ell \oplus \mathbf{z}_r \oplus \sum'_{z_h \in F^q(z_r)} \mathbf{z}_h$, thus $w(\mathbf{y}) \in \{0, \ldots, 2(s+1)\}$. Hence $\mathbf{y} \in Q \subseteq \mathcal{L}(Q)$. Then $\mathbf{u} = \mathbf{w} \oplus \mathbf{y} \in \mathcal{L}(Q)$, since $\mathcal{L}(Q)$ is a subspace, which is a contradiction.

(iii) Clearly $\mathcal{L}(Q \cup Z_{s+1}^f) \subseteq \mathbf{V}^f$. Let $\mathbf{u} \in \mathbf{V}^f$. If $\mathbf{u} \in \mathbf{V}_0$, then by (ii), $\mathbf{u} \in \mathcal{L}(Q) \subseteq \mathcal{L}(Q \cup Z_{s+1}^f)$. Otherwise, let $\mathbf{v} \in F(\mathbf{u}) \cap \mathbf{V}_0$, $\mathbf{w} = \mathbf{u} \oplus \mathbf{v} = \mathbf{z}_k \oplus \sum'_{z_\ell \in F^s(z_k)} \mathbf{z}_\ell$. Since $\mathbf{w} \ne \Phi$, we have $\mathbf{w} \in Z_{s+1}^f$. By (ii), $\mathbf{v} \in \mathcal{L}(Q)$. Hence $\mathbf{u} = \mathbf{v} \oplus \mathbf{w} \in \mathcal{L}(Q \cup Z_{s+1}^f)$.

(iv) Denote the left hand set by S_ℓ and the right hand set by S_r. Clearly $S_r \subseteq S_\ell$. Let $j \in S_\ell$. If $j = 0$, then $j = \gamma(\Phi) \in S_r$. Otherwise, pick $\mathbf{u} \in \mathbf{V}_j$. There exists $\mathbf{v} \in F(\mathbf{u}) \cap \mathbf{V}_0$. Let $\mathbf{w} = \mathbf{u} \oplus \mathbf{v}$. Then $\gamma(\mathbf{w}) = j$, and $\mathbf{w} \in Z_{s+1}^f \in S_r$.

(v) Let $\mathbf{u} \in \mathbf{V}^f$ have maximum γ-value. By (iv) we may assume that $\mathbf{u} = \mathbf{z}_k \oplus \sum'_{z_\ell \in F^s(z_k)} \mathbf{z}_\ell \in Z_{s+1}^f$. Any of the $s+1$ tokens can fire into at most s followers in $n - 1$ locations. At least s of the followers are identical. Thus the outdegree of $\mathbf{u}$ is at most $g(n, s)$. Then $2^t - 1 \le g(n, s)$, which implies the result. $\quad\square$

For the special case of 1-regular games ($s = q = 1$), we can do a little better. This was done in [17]. For the sake of completeness, we reproduce it here, in a more transparent form. Define

$$Y_i = \{\mathbf{u} \in \mathbf{V}: w(\mathbf{u}) = i\} \quad \text{(vectors of weight } i\text{)}$$
$$Y_i^f = Y_i \cap \mathbf{V}^f$$

$$S = Y_1 \cap \{\mathbf{u} \in \mathbf{V} : F(\mathbf{u}) = \varnothing\} \cup \{\Phi\} \qquad \text{(leaves of weight} \leq 1)$$
$$L = \{\mathbf{u} \in \mathbf{V} : F(\mathbf{u}) = \varnothing\} \qquad \text{(set of all leaves)}$$
$$Q_i = Y_i \cap \mathbf{V}_0$$
$$Q = Q_2 \cup Q_4 \cup S.$$

Theorem 4. *Let* $\mathbf{G} = (\mathbf{V}, \mathbf{E})$ *be the cellular automata graph of the finite digraph* $G = (V, E)$. *Then*

(i) $L = \mathcal{L}(S)$ *(linear span of* S *over* GF (2))
(ii) $\mathbf{V}_0 = \mathcal{L}(Q)$
(iii) $\mathbf{V}^f = \mathcal{L}(Q \cup Y_2^f)$
(iv) $\{\gamma(\mathbf{u}) : \mathbf{u} \in \mathbf{V}^f\} = \{\gamma(\mathbf{u}) : \mathbf{u} \in (Y_2^f \cup \{\Phi\})\}$
(v) $t \leq \lceil \log_2 n \rceil$ $(n \geq 2)$, *where* $2^t - 1$ *is the maximum value of* γ *on* $\mathbf{V}^f$.

Proof. The proof is very similar to that of Theorem 3 and is therefore omitted.
$\square$

7. An $O(n^6)$ Algorithm for γ for the Case $q = 1$

We shall now consolidate the results of the previous sections into a polynomial algorithm for computing γ for the cellular automata graph $\mathbf{G} = (\mathbf{V}, \mathbf{E})$ of a digraph $G = (V, E)$ in polynomial time for the special case $q = 1$. We begin with some further notation.

$\beta_0 = \{B_0, \ldots, B_{m-1}\}$ is a basis for $\mathbf{V}_0$, where $B_i \in Q$ $(0 \leq i < m)$.
$\beta^f = \{B_m, \ldots, B_{m+t-1}\} \cup \beta_0$ is a basis for $\mathbf{V}^f$, $B_{m+i} \in Y_{q+1}^f$ $(0 \leq i < t)$.
$\beta^\infty = \{B_{m+t}, \ldots, B_{n-1}\}$ is a basis for the complement $\mathbf{W}$ of $\mathbf{V}^f$ in $\mathbf{V}$, so $\beta = \beta^f \cup \beta^\infty$ is a basis for $\mathbf{V}$.
$\mathbf{G}^{[2]} = (\mathbf{V}^{[2]}, \mathbf{E}^{[2]})$ is the subgraph of $\mathbf{G}$ induced by $Y_2 \cup S$.
$\mathbf{G}^{[4]} = (\mathbf{V}^{[4]}, \mathbf{E}^{[4]})$ is the subgraph of $\mathbf{G}$ induced by $Y_4 \cup Y_2 \cup S$.

Algorithm CEL (cellular) for computing γ on $\mathbf{G} = (\mathbf{V}, \mathbf{E})$ for $q = 1$ (without constructing the exponentially large $\mathbf{G}$!).
Input: Digraph $G = (V, E)$ with $|V| = n$.
Output: An $n \times n$ matrix $B = (\beta^f, \beta^\infty)$, and m, t, B^{-1}. The n columns of the bottom $n - m$ rows of B^{-1} constitute the homomorphism (see Lemma 2, §5) $\rho(\mathbf{z}_i) = (\varepsilon_i^m, \ldots, \varepsilon_i^{n-1}) \in GF(2)^{n-m}$ $(0 \leq i < n)$.

Notes.

1. For any vector $\mathbf{v} = (\delta^0, \ldots, \delta^{n-1}) \in \mathbf{V}$, we have $\mathbf{v} = \sum_{i=0}^{n-1} \delta^i \mathbf{z}_i$, so $\rho(\mathbf{v}) = \sum_{i=0}^{n-1} \delta^i \rho(\mathbf{z}_i)$ can thus be computed polynomially. Using Lemma 2 we then have either $\gamma(\mathbf{v}) = \rho(\mathbf{v})$, or $\gamma(\mathbf{v}) = \infty$. This polynomial computation is the main thrust of Algorithm CEL. It permits computing γ on all of $\mathbf{V}$.

2. The equation $Bx = v$ has the solution $x = B^{-1}v = (\varepsilon^0, \ldots, \varepsilon^{n-1}) \in GF(2)^n$, so v can be represented as a linear combination of the basis consisting of the columns of B:

$$v = \varepsilon^0 B_0 \oplus \varepsilon^1 B_1 \oplus \cdots \oplus \varepsilon^{n-1} B_{n-1}.$$

This fact is used in the next section for forcing a win for $q = 1$ in polynomial time.

3. Notice that both $z_0, \ldots, z_{n-1}$ and $B_0, \ldots, B_{n-1}$ are bases of $\mathbf{V}$. The former is more convenient for the ρ-computation, and the latter for forcing a win.

Procedure: (i) Construct $\mathbf{G}^{[2]}$ and $\mathbf{G}^{[4]}$ ($\mathbf{G}^{[2]}$ has $O(n^2)$ vertices and $O(n^3)$ edges; $\mathbf{G}^{[4]}$ has $O(n^4)$ vertices and $O(n^5)$ edges).

(ii) Apply the first iteration ($i = 0$) of Algorithm GSG (§ 2) to $\mathbf{G}^{[4]}$. Store the resulting set $Q = \{q_1, \ldots, q_p\}$ of vectors in $\mathbf{V}_0$ together with their counter values c ($O(n^5)$ steps). (We may omit from Q the vector Φ and any other vectors which obviously depend linearly on the rest.)

(iii) Apply Algorithm GSG to $\mathbf{G}^{[2]}$. The largest γ-value will be $2^t - 1$ for some $t \in \mathbb{Z}^0$. Store vectors $v_1, \ldots, v_t$ with $\gamma(v_i) = 2^{i-1}$, $w(v_i) = 2$ ($1 \leq i \leq t$) and t, together with their monotonic counter values c, such that

$$\min\{c(v) : v \in \mathbf{V}^{[2]}\} > \max\{c(u) : u \in \mathbf{V}^{[4]}\}$$

($O(n^5)$steps).

(iv) Construct the matrix $A = (q_1, \ldots, q_p, v_1, \ldots, v_t, z_0, \ldots, z_{n-1})$, where the z_i are the unit vectors ($O(n^5)$ steps, since $p = O(n^4)$, so A has order $n \times O(n^4)$).

(v) Transform A into a row-echelon matrix E, using elementary row operations ($O(n^6)$ steps; the $z_0, \ldots, z_{n-1}$ of A then become B^{-1} — see e.g., [22], Ch. 7, § 47).

(vi) Let $1 \leq i_1 < \cdots < i_n \leq p + t + n$ be the indices of the unit vectors of E. Then m is the largest subscript s such that $i_s \leq p$. Let $B = (B_0, \ldots, B_{n-1})$ be the matrix consisting of the columns $A_{i_1}, \ldots, A_{i_n}$ of A. Store

$$\beta_0 = \{B_0, \ldots, B_{m-1}\},$$
$$\beta^f = \{B_m, \ldots, B_{m+t-1}\} \cup \beta_0,$$
$$\beta = \{B_{m+t}, \ldots, B_{n-1}\} \cup \beta^f,$$

and the matrix $B^{-1} = (E_{p+t}, \ldots, E_{p+t+n-1})$ (the last n columns of E) ($O(n^4)$). Compute $B^{-1}z_i = (\varepsilon_i^0, \ldots, \varepsilon_i^{n-1})$; store $\rho(z_i) = (\varepsilon_i^m, \ldots, \varepsilon_i^{n-1})$ ($0 \leq i < n$); the numerical values of the columns consisting of the last $n - m$ rows of B^{-1}; ($O(n^3)$ steps.) End.

Example 5. Play a 1-regular game on the digraph depicted in Fig. 3. We apply Algorithm CEL to it. The output of step (ii) is $Q = (0, 5, 10, 15)$, and step (iii) yields $v_1 = \{3\}$ ($t = 1$). The following constitutes steps (iv) and (v), where $\sim$ denotes equivalence under elementary row operations; we omitted the 0-vector

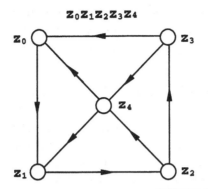

Figure 3. Play a 1-regular game. © 2000

from A.

$$A = \begin{pmatrix} 1 & 0 & 1 & 1 & 1 & 0 & 0 & 0 & 0 \\ 0 & 1 & 1 & 1 & 0 & 1 & 0 & 0 & 0 \\ 1 & 0 & 1 & 0 & 0 & 0 & 1 & 0 & 0 \\ 0 & 1 & 1 & 0 & 0 & 0 & 0 & 1 & 0 \\ 0 & 0 & 0 & 0 & 0 & 0 & 0 & 0 & 1 \end{pmatrix} \sim \begin{pmatrix} 1 & 0 & 1 & 1 & 1 & 0 & 0 & 0 & 0 \\ 0 & 1 & 1 & 1 & 0 & 1 & 0 & 0 & 0 \\ 0 & 0 & 0 & 1 & 1 & 0 & 1 & 0 & 0 \\ 0 & 0 & 0 & 1 & 0 & 1 & 0 & 1 & 0 \\ 0 & 0 & 0 & 0 & 0 & 0 & 0 & 0 & 1 \end{pmatrix}$$

$$\sim \begin{pmatrix} 1 & 0 & 1 & 0 & 0 & 0 & 1 & 0 & 0 \\ 0 & 1 & 1 & 0 & 1 & 1 & 1 & 0 & 0 \\ 0 & 0 & 0 & 1 & 1 & 0 & 1 & 0 & 0 \\ 0 & 0 & 0 & 0 & 1 & 1 & 1 & 1 & 0 \\ 0 & 0 & 0 & 0 & 0 & 0 & 0 & 0 & 1 \end{pmatrix} \sim \begin{pmatrix} 1 & 0 & 1 & 0 & 0 & 0 & 1 & 0 & 0 \\ 0 & 1 & 1 & 0 & 0 & 0 & 0 & 1 & 0 \\ 0 & 0 & 0 & 1 & 0 & 1 & 0 & 1 & 0 \\ 0 & 0 & 0 & 0 & 1 & 1 & 1 & 1 & 0 \\ 0 & 0 & 0 & 0 & 0 & 0 & 0 & 0 & 1 \end{pmatrix}$$

$$= E.$$

The last $n = 5$ columns of E constitute B^{-1} e.g., by [22], Ch. 7, §47. The unit vectors are in columns 1, 2, 4, 5, 9. Since $p = 3$, we then have $m = 2$. Columns 1, 2, 4, 5, 9 of A constitute B. Thus,

$$B = \begin{pmatrix} 1 & 0 & 1 & 1 & 0 \\ 0 & 1 & 1 & 0 & 0 \\ 1 & 0 & 0 & 0 & 0 \\ 0 & 1 & 0 & 0 & 0 \\ 0 & 0 & 0 & 0 & 1 \end{pmatrix}, \qquad B^{-1} = \begin{pmatrix} 0 & 0 & 1 & 0 & 0 \\ 0 & 0 & 0 & 1 & 0 \\ 0 & 1 & 0 & 1 & 0 \\ 1 & 1 & 1 & 1 & 0 \\ 0 & 0 & 0 & 0 & 1 \end{pmatrix},$$

and $\big(\rho(z_0), \rho(z_1), \rho(z_2), \rho(z_3), \rho(z_4)\big) = (2, 3, 2, 3, 4)$, which are the bottom $n - m = 3$ rows of B^{-1}. Already in Example 4 we saw that these values determine the ρ-values of all positions of the game. Since the 5th and 9th columns of A contain 1 and 16 respectively, we have $\mathbf{W} = \{0, 1, 16, 17\}$.

Furthermore, for say $\mathbf{v} = z_3 = 8$, the equation $B\mathbf{x} = \mathbf{v}$ has the solution $\mathbf{x} = B^{-1}\mathbf{v} = 14$, hence $\mathbf{v}$ is the following linear combination of the B-column vectors: $\mathbf{v} = 10 \oplus 3 \oplus 1 = 8$, and $\rho(\mathbf{v}) = \rho(10) \oplus \rho(3) \oplus \rho(1) = 0 \oplus 1 \oplus 2 = 3$. Since $\rho(\mathbf{v}) > 2^t - 1 = 1$, $\gamma(\mathbf{v}) = \infty$. Similarly, for $\mathbf{v} = z_2 \oplus z_3 = 12$, the equation $B\mathbf{x} = \mathbf{v}$ has the solution $x = 7$, hence $\mathbf{v} = 5 \oplus 10 \oplus 3 = 12$, and

$\rho(\mathbf{v}) = \rho(5) \oplus \rho(10) \oplus \rho(3) = 0 \oplus 0 \oplus 1 = 1$, so $\gamma(\mathbf{v}) = 1$. But a simpler computation of $\rho(\mathbf{v})$ in these two examples is to express $\mathbf{v}$ in terms of the unit vectors $\mathbf{z}_i$ and then apply the homomorphism ρ on them, as was done in Example 5.2: $\rho(8) = 3$, so $\gamma(\mathbf{v}) = \infty$; $\rho(12) = \rho(4 \oplus 8) = \rho(4) \oplus \rho(8) = 2 \oplus 3 = 1$, so $\gamma(12) = 1$.

Example 6. Suppose we put $w(z_4) = w(z_1) = 1$ for the digraph of Fig. 3, $w(z_1) = 1$ for the digraph of Fig. 2, and all other weights 0 on both digraphs. Play the sum of the 1-regular game on the digraph of Fig. 3 and the 2-regular game on the digraph of Fig. 2. We see that the γ-value of the former is $\infty(0,1)$ and of the latter 1, and $1 \oplus \infty(0,1) = \infty(1 \oplus (0,1)) = \infty(1,0)$. By (2) this is an N-position. The unique winning move is to to fire z_4 and complement $w(z_0)$.

For proving validity of Algorithm CEL, we need an auxiliary result. We first define a special type of subgraph which is closed under the operation of taking followers, i.e., it contains the followers of all of its vertices.

Definition 3. Let $G = (V, E)$ be any digraph. A subset $U \subseteq V$ is a *restriction* of V, if $F(U) \subseteq U$, where $F(U) = \{v \in V : (u, v) \in E,\ u \in U\}$.

Lemma 3. (Restriction Principle). *Let $G = (V, E)$ be a digraph, U a restriction of V, G_1 the subgraph of G induced by U. Then the γ-function computed on G_1 alone* (without considering G) *is identical with the γ-function on G, restricted to U.*

Proof. Let γ_1 be the γ-function of G restricted to G_1. Since $u \in U$ implies $F_G(u) \subseteq U$, Definition 1 implies that γ_1 is a Generalized Sprague–Grundy function on G_1. Since also γ computed on G_1 is, we have $\gamma_1 = \gamma$ by the uniqueness of γ. $\qquad\square$

Validity Proof of Algorithm CEL. The vertex set $\mathbf{V}^{[2]}$ of $\mathbf{G}^{[2]}$ is clearly a restriction of $\mathbf{V}$ $(q = 1)$; and $\mathbf{V}^{[4]}$ of $\mathbf{G}^{[4]}$ is also a restriction of $\mathbf{V}$. By Lemma 3, all the γ-values computed in steps (ii) and (iii) are γ-values of $\mathbf{G}$. By Theorem 4, these computed values generate γ on all of $\mathbf{G}$.

Note that the n unit vectors of A guarantee that A has rank n. The matrix E constructed in step (v) has the claimed properties by linear algebra properties over $GF(2)$. In particular, the product of the elementary row operation matrices operated on the unit matrix $I = (\mathbf{z}_0, \ldots, \mathbf{z}_{n-1})$ — which is the tail end of A — is B^{-1} as claimed. Since $\mathbf{V}_0 = \mathcal{L}(Q)$, the value of m is as stated in step (vi). Since the vectors $\mathbf{v}_1, \ldots, \mathbf{v}_t$ are linearly independent, $\dim \mathbf{V}^f = m + t$.

Finally, we show that for any $\mathbf{u} \in \mathbf{V}$, if $B^{-1}\mathbf{u} = \varepsilon = (\varepsilon^0, \ldots, \varepsilon^{n-1}) \in GF(2)^n$, then $\rho(\mathbf{u}) = (\varepsilon^m, \ldots, \varepsilon^{n-1}) \in GF(2)^{n-m}$ is a homomorphism from $\mathbf{V}$ onto $GF(2)^{n-m}$ with kernel $\mathbf{V}_0$ such that $\mathbf{u} \in \mathbf{V}^f$ if and only if $\rho(\mathbf{u}) \le 2^t - 1$, whence $\rho(\mathbf{u}) = \gamma(\mathbf{u})$. Let $\mathbf{v} \in \mathbf{V}$ and $B^{-1}\mathbf{v} = \delta = (\delta^0, \ldots, \delta^{n-1}) \in GF(2)^n$.

$$\rho(\mathbf{u}) \oplus \rho(\mathbf{v}) = (\varepsilon^m, \ldots, \varepsilon^{n-1}) \oplus (\delta^m, \ldots, \delta^{n-1}) = (\varepsilon^m \oplus \delta^m, \ldots, \varepsilon^{n-1} \oplus \delta^{n-1}).$$

Now

$$\varepsilon \oplus \delta = (\varepsilon^0, \ldots, \varepsilon^{n-1}) \oplus (\delta^0, \ldots, \delta^{n-1}) = B^{-1}\mathbf{u} \oplus B^{-1}\mathbf{v} = B^{-1}(\mathbf{u} \oplus \mathbf{v})$$
$$= (\varepsilon^0 \oplus \delta^0, \ldots, \varepsilon^{n-1} \oplus \delta^{n-1}).$$

Hence $\rho(\mathbf{u} \oplus \mathbf{v}) = (\varepsilon^m \oplus \delta^m, \ldots, \varepsilon^{n-1} \oplus \delta^{n-1}) = \rho(\mathbf{u}) \oplus \rho(\mathbf{v})$, so ρ is a homomorphism $\mathbf{V} \to GF(2)^{n-m}$. It is onto, since for any $\varepsilon = (\varepsilon^0, \ldots, \varepsilon^{n-1}) \in \mathbf{V}$, the equation $B\varepsilon = \mathbf{u} \in \mathbf{V}$ has the solution $\varepsilon = B^{-1}\mathbf{u}$, and $\rho(\mathbf{u}) = (\varepsilon^m, \ldots, \varepsilon^{n-1})$. By linear algebra, $B^{-1}\mathbf{u} = (\varepsilon^0, \ldots, \varepsilon^{n-1})$ implies $\mathbf{u} = \varepsilon^0 B_0 \oplus \cdots \oplus \varepsilon^{n-1} B_{n-1}$. Since $\beta_0 = (B_0, \ldots, B_{m-1})$ is a basis of $\mathbf{V}_0$, we see that $\varepsilon^m = \cdots = \varepsilon^{n-1} = 0$ if and only if $\mathbf{u} \in \mathbf{V}_0$. Hence the kernel is $\mathbf{V}_0$. The same argument shows that $\mathbf{u} \in \mathbf{V}^f$ if and only if $\varepsilon^{m+t} = \cdots = \varepsilon^{n-1} = 0$ if and only if $\rho(\mathbf{u}) = \gamma(\mathbf{u})$.

8. Forcing a Win in Cellular Automata Games for 1-Regular Games

By using ρ, computed in the previous section in $O(n^6)$ steps, we can compute the γ-function for every $\mathbf{u} \in \mathbf{V}$ for any cellular automata game-graph $\mathbf{G} = (\mathbf{V}, \mathbf{E})$ played on a digraph $G = (V, E)$ with $|V| = n$, which determines the P-, N- and D-membership for every $\mathbf{u} \in \mathbf{V}$ by (1) and (2). However, our method only constructed an $O(n^4)$ fragment of the exponentially-large game-graph. In particular, we don't have a counter function for all $\mathbf{u} \in \mathbf{V}^f$, so the question arises, given any N-position in $\mathbf{G}$, how can we insure a win in a finite number of moves.

If $\mathbf{u} \in \mathcal{D}$, then $\mathbf{v} \in F(\mathbf{u}) \cap \mathcal{D}$ can be found polynomially by scanning $F(\mathbf{u})$ $(|F(\mathbf{u})| < n^2)$: compute $\rho(\mathbf{v})$ for $\mathbf{v} \in F(\mathbf{u})$. If $\gamma(\mathbf{v}) = \infty$, then compute $\rho(\mathbf{w})$ for $\mathbf{w} \in F(\mathbf{v})$, to determine K such that $\gamma(\mathbf{v}) = \infty(K)$. Similarly, if $\mathbf{u} \in \mathcal{N}$, then $\mathbf{v} \in F(\mathbf{u}) \cap \mathcal{P}$ can be found polynomially. This, however, does not guarantee a win because of possible cycling and never reaching a leaf. The strategy of moving from $\mathbf{u} \in \mathcal{N}$ to any $F(\mathbf{u}) \cap \mathcal{P}$ guarantees a nonlosing outcome, nevertheless.

In this section we show how to force a win from a position $\mathbf{u} \in \mathcal{N}$ in polynomial time, including the case when the cellular automata game is a component in a sum of finitely many games. Let

$$\mathbf{R}_j = Q \bigcup_{i=1}^{j} (Y_2 \cap \mathbf{V}_{2^i}) \; (0 \le j < t), \text{ so } \mathbf{R}_0 = Q.$$

Informally, this third key idea is this. Given any $\mathbf{u} \in \mathbf{V}^f$, we can write $\mathbf{u} = \sum_{i=1}^{'h} \mathbf{u}_i$ with $\mathbf{u}_i \in \mathbf{R}_{t-1}$ $(1 \le i \le h)$, where the $\mathbf{u}_i$ can be computed polynomially using the matrix B produced by Algorithm CEL. We then say that $\widetilde{u} = \{\mathbf{u}_1, \ldots, \mathbf{u}_h\}$ *represents* $\mathbf{u}$. Moreover, we can define a counter function $\widetilde{c}$ on representations $\widetilde{u}$ and arrange that the winner moves to a sequence of positions $\mathbf{u}^0, \mathbf{u}^1, \ldots$ with $\widetilde{c}(\widetilde{u}^0) > \widetilde{c}(\widetilde{u}^1) > \cdots (\widetilde{u}^i = (\mathbf{u}_1^i, \ldots, \mathbf{u}_{h_i}^i))$, leading to a win. In doing this we will be confronted, analogously to Lemma 1 (§3) and the proof of Theorem 1, by the possibility of encountering a predecessor of a representation instead of the descendant we are seeking. That is, $\tilde{c}(\tilde{u}^i) > \tilde{c}(\tilde{u}^{i+1})$ with $\tilde{u}^{i+1} \in \tilde{F}(\tilde{u}^i)$ implies $\mathbf{u}^{i+1} \in F(\mathbf{u}^i)) \cup F^{-1}(\mathbf{u}^i))$, where $\tilde{F}$ is a follower function for representations, defined below, and $\mathbf{u}^i = \sum_{j=1}^{\prime h_i} \mathbf{u}_j^i$.

Definition 4. (i) Let $\mathbf{R} \subseteq \mathbf{V}$. A *representation* $\tilde{u}$ of $\mathbf{u} \in \mathbf{V}$ over $\mathbf{R}$ is a subset $\tilde{u} = \{\mathbf{u}_1, \ldots, \mathbf{u}_h\} \subseteq \mathbf{R}$ of distinct elements $\mathbf{u}_i$. If $\mathbf{R}$ is either indicated by the context or irrelevant, we may say simply that $\tilde{u}$ is a *representation* of $\mathbf{u} = \sum_{j=1}^{\prime h} \mathbf{u}_j$, omitting over $\mathbf{R}$. The empty representation is denoted by $\tilde{\varnothing}$.

(ii) For $\tilde{u} = \{\mathbf{u}_1, \ldots, \mathbf{u}_h\} \subseteq \mathbf{R}$, a follower function for representations is given by

$$\tilde{F}(\tilde{u}; \mathbf{u}_j, \mathbf{v}_j) = (\tilde{u} \ominus \{\mathbf{u}_j, \mathbf{v}_j\}),$$

where $\mathbf{u}_j \in \tilde{u}$, $\mathbf{v}_j \in F_k^q(\mathbf{u}_j)$ for any $1 \leq j \leq h$, and where $\ominus$ denotes the symmetric difference: $\tilde{x}_1 \ominus \tilde{x}_2 = (\tilde{x}_1 \cup \tilde{x}_2) - (\tilde{x}_1 \cap \tilde{x}_2)$ $(\tilde{x}_1, \tilde{x}_2 \in \mathbf{R})$. Thus,

$$\tilde{F}(\tilde{u}; \mathbf{u}_j, \mathbf{v}_j) = \begin{cases} \{\mathbf{u}_1, \ldots, \mathbf{u}_{j-1}, \mathbf{u}_{j+1}, \ldots, \mathbf{u}_h, \mathbf{v}_j\} & \text{if } \mathbf{v}_j \notin \tilde{u} \\ \{\mathbf{u}_1, \ldots, \mathbf{u}_{i-1}, \mathbf{u}_{i+1}, \ldots, \mathbf{u}_{j-1}, \mathbf{u}_{j+1}, \ldots, \mathbf{u}_h\} & \text{if } \mathbf{v}_j = \mathbf{u}_i \in \tilde{u}. \end{cases}$$

(iii) We also define the set of all representation followers of $\tilde{u}$ by

$$\tilde{F}(\tilde{u}) = \bigcup_{j=1}^{h} \bigcup_{\mathbf{v}_j \in F(\mathbf{u}_j)} \tilde{F}(\tilde{u}; \mathbf{u}_j, \mathbf{v}_j).$$

Notation. Let $\tilde{u} = \{\mathbf{u}_1, \ldots, \mathbf{u}_h\} \subseteq \mathbf{R}$, $\mathbf{v}_j \in F_k^k(\mathbf{u}_j)$. We put

$$\mu(\tilde{u}) = \sum_{i=1}^{h}{}' \mathbf{u}_i, \quad \mu(\tilde{F}(\tilde{u}; \mathbf{u}_j, \mathbf{v}_j)) = \mathbf{v}_j \oplus \sum_{i \neq j}{}' \mathbf{u}_i \in \mu(\tilde{F}(\tilde{u})),$$

where

$$\mu(\tilde{F}(\tilde{u})) = \bigcup_{j=1}^{h} (F(\mathbf{u}_j) \oplus \sum_{i \neq j}{}' \mathbf{u}_i), \quad \mu(\tilde{F}(\tilde{u})) = \bigcup_{j=1}^{h} (F(\mathbf{u}_j) \oplus \sum_{i \neq j}{}' \mathbf{u}_i).$$

Notes. (i) We see that $\tilde{F}(\tilde{u}; \mathbf{u}_j, \mathbf{v}_j)$ is a representation, namely the representation of $\mu(\tilde{u}) \oplus \mathbf{u}_j \oplus \mathbf{v}_j = \mathbf{v}_j \oplus \sum_{i \neq j}' \mathbf{u}_i$

(ii) Let $0 \leq k < 2^t$. Every $\mathbf{u} \in \mathbf{V}_k$ has a representation over $\mathbf{R}_j$ where $j = \lceil \log_2(k+1) \rceil$. Such a representation can be constructed polynomially by computing $B^{-1}\mathbf{u}$ (see the Note at the beginning of Algorithm CEL). The elements of this representation have γ-values which are either 0 or nonnegative powers of 2.

(iii) If $\tilde{u} = \tilde{\varnothing}$, then $\mu(\tilde{u}) = \Phi$.

Lemma 1 implies directly:

(a) $F(\mu(\tilde{u})) \subseteq \mu(\tilde{F}(\tilde{u}))$.

(b) $\mu\big(\widetilde{F}(\widetilde{u})\big) \subseteq F\big(\mu(\widetilde{u})\big) \cup F^{-1}\big(\mu(\widetilde{u})\big)$.

(c) Let $\widetilde{v} = \widetilde{F}(\widetilde{u}; \mathbf{u}_j, \mathbf{v}_j)$, where $\mathbf{v}_j = F_k^q(\mathbf{u}_j)$. Then $\mu(\widetilde{u}) \in F\big(\mu(\widetilde{v})\big)$ if and only if either: (a) $u^k = 0$, or (b) for some $s \neq k$, $u^s = 0$ and $\mathbf{z}_k \oplus \sum'_{z_\ell \in F^q(z_k)} \mathbf{z}_\ell = \mathbf{z}_s \oplus \sum'_{z_t \in F^q(z_s)} \mathbf{z}_t$. (Note that $\mu(\widetilde{u}) = (u^0, \ldots, u^{n-1})$, $\mu(\widetilde{v}) = \mathbf{v}_j \oplus \sum'_{i \neq j} \mathbf{u}_i$.)

Example 7. We refer back to Example 5 (Fig. 3), §7, and let $\mathbf{u} = 15 \in \mathbf{V}_0$. The equation $B\mathbf{x} = \mathbf{u}$ has then the solution $\mathbf{x} = B^{-1}\mathbf{u} = (11000)$, so $\mathbf{u}$ has a representation $\widetilde{u} = \{5, 10\}$ over Q (with $\mathbf{u} = 15$). Now

$$\widetilde{F}(\widetilde{u}) = \bigcup_{j=1}^{h} \bigcup_{\mathbf{v}_j \in F(\mathbf{u}_j)} \widetilde{F}(\widetilde{u}; \mathbf{u}_j, \mathbf{v}_j)$$

$$= \widetilde{F}(\widetilde{u}; 5, 6) \cup \widetilde{F}(\widetilde{u}; 5, 9) \cup \widetilde{F}(\widetilde{u}; 5, 17) \cup \widetilde{F}(\widetilde{u}; 10, 3) \cup \widetilde{F}(\widetilde{u}; 10, 12) \cup \widetilde{F}(\widetilde{u}; 10, 18)$$

$$= \{6, 10\} \cup \{9, 10\} \cup \{10, 17\} \cup \{3, 5\} \cup \{5, 12\} \cup \{5, 18\}.$$

Thus, $\mu\big(\widetilde{F}(\widetilde{u})\big) = \{12, 3, 27, 6, 9, 23\}$. As can be seen directly from Fig. 3, $F\big(\mu(\widetilde{u})\big) = F(15) = \{3, 6, 9, 12, 23, 27\}$ so $F\big(\mu(\widetilde{u})\big) = \mu\big(\widetilde{F}(\widetilde{u})\big)$ in this case.

Recall that in steps (ii) and (iii) of Algorithm CEL, a counter function c on $\mathbf{G}^{[4]}$ and on $\mathbf{G}^{[2]}$ was computed. Since the vertex sets $\mathbf{V}^{[4]}$ of $\mathbf{G}^{[4]}$ and $\mathbf{V}^{[2]}$ of $\mathbf{G}^{[2]}$ have sizes $O(n^4)$ and $O(n^2)$ respectively, c has values bounded by $O(n^4)$ and $O(n^2)$ for these two cases. For $\widetilde{u} = \{\mathbf{u}_1, \ldots, \mathbf{u}_h\} \subseteq \mathbf{R}_{t-1}$, define $\widetilde{c}(\widetilde{u}) = \sum_{i=1}^{h} c(\mathbf{u}_i)$. We have $h \leq n$, since the $\mathbf{u}_i$ are computed using the $n \times n$ matrix B (Note 2, §7). Thus, if $\widetilde{u} \subseteq \beta_0$, then the values of $\widetilde{c}$ are bounded by $O(n^5)$, and if $\widetilde{u} \subseteq \beta^f \setminus \beta_0$ then the values of $\widetilde{c}$ are bounded by $O(n^3)$.

Suppose now that we are given a sum of r games, one of which is a cellular automata game played on a finite cyclic digraph $G(V, E)$. It follows from (1) and (2) that for winning the sum by means of a move in G, it suffices if this winning move is of one of the following two types:

(i) Given $\mathbf{u} \in \mathbf{V}_p$ and $\mathbf{v} \in F(\mathbf{u})$ with $\gamma(\mathbf{v}) > p$, move to $\mathbf{w} \in F(\mathbf{v}) \cap \mathbf{V}_p$ with $\widetilde{c}(\widetilde{w}) < \widetilde{c}(\widetilde{u})$.

(ii) Given $\mathbf{u} \in \mathbf{V}_\ell$ and $p < \ell$; or $\gamma(\mathbf{u}) = \infty(K)$ with $p \in K$. Move to $\mathbf{v} \in F(\mathbf{u}) \cap \mathbf{V}_p$ such that $\widetilde{c}(\widetilde{v}) < \widetilde{c}(\widetilde{u})$, where we put $\widetilde{c}(\widetilde{u}) = \infty$ if $\gamma(\mathbf{u}) = \infty$.

These moves in G lead to a win when the *other* $r - 1$ component games in the sum have generalized Nim sum value p.

What's the complexity of computing these moves? The more complicated of these cases is (i). We deal with it in Theorem 5, and then summarize it, together with case (ii), in an overall strategy described in the proof of Theorem 6.

Theorem 5. *For any integer $0 \leq p \leq 2^t - 1$, letting $j = \lceil \log_2(p + 1) \rceil$, a function $\Psi \colon (\mathbf{R}_j, \mathbf{V}) \to \mathbf{R}_j$ can be computed in polynomial time, such that if $\widetilde{u} = \{\mathbf{u}_1, \ldots, \mathbf{u}_h\} \subseteq \mathbf{R}_j$ is a representation of $\mu(\widetilde{u}) \in \mathbf{V}_p$ and $\mu(\widetilde{v}) \in F\big(\mu(\widetilde{u})\big)$ with $\gamma(\mu(\widetilde{v})) > p$, then $\Psi\big(\widetilde{u}, \mu(\widetilde{v})\big) = \widetilde{w} \subseteq \mathbf{R}_j$, where $\mu(\widetilde{w}) \in F\big(\mu(\widetilde{v})\big) \cap \mathbf{V}_p$ and $\widetilde{c}(\widetilde{w}) < \widetilde{c}(\widetilde{u})$.*

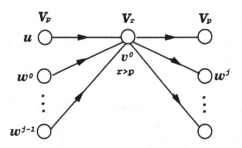

Figure 4. Illustrating Theorem 5.

Proof. Let $\mathbf{v}^0 \in F\big(\mu(\widetilde{u})\big)$ with $\gamma(\mathbf{v}^0) > \gamma(\mathbf{u})$. By (a), $\mathbf{v}^0 \in \mu\big(\widetilde{F}(\widetilde{u})\big)$, say $\mathbf{v}^0 = \mu\big(\widetilde{F}(\widetilde{u}; \mathbf{u}_j, \mathbf{v}_j)\big)$. Since $w(\mathbf{u}_j) \leq 4$, the computation of j, k such that $\mathbf{v}_j = F_k^q(\mathbf{u}_j) = \mathbf{v}^0 \oplus \sum_{i \neq j}' \mathbf{u}_i$ takes $O(n^2)$ steps. For simplicity of notation, assume that $j = 1$. Since $\gamma(\mathbf{v}^0) > \gamma(\mathbf{u})$ and $\{\mathbf{u}_1, \ldots, \mathbf{u}_h\}$ is a representation over $\mathbf{R}_j$, i.e., its elements have γ-values 0 or distinct powers of 2, it follows that $\gamma(\mathbf{v}_1) > \gamma(\mathbf{u}_1)$ and $\mathbf{v}_1 \neq \mathbf{u}_i$ ($1 \leq i \leq h$), so $\widetilde{v}^0 = \{\mathbf{v}_1, \mathbf{u}_2, \ldots, \mathbf{u}_h\}$ is a representation of $\mathbf{v}^0 = \mathbf{v}_1 \oplus \sum_{i=2}'^{h} \mathbf{u}_i$ over $\mathbf{V}$. Also $\mathbf{v}_1 \in \mathbf{V}^{[4]}$, hence $\mathbf{v}_1$ has only $O(n)$ followers, so we can compute $\mathbf{w}_1 \in F(\mathbf{v}_1) \cap \mathbf{V}^{[4]}$ with $\gamma(\mathbf{w}_1) = \gamma(\mathbf{u}_1)$ and $c(\mathbf{w}_1) < c(\mathbf{u}_1)$ in $O(n)$ steps. Hence $\widetilde{c}(\widetilde{w}^0) < \widetilde{c}(\widetilde{u})$, where $\widetilde{w}^0 = \{\mathbf{w}_1, \mathbf{u}_2, \ldots, \mathbf{u}_h\}$ if $\mathbf{w}_1 \neq \mathbf{u}_i$ ($2 \leq i \leq h$), or $\widetilde{w}^0 = \{\mathbf{u}_2, \ldots, \mathbf{u}_{i-1}, \mathbf{u}_{i+1}, \ldots, \mathbf{u}_h\}$ otherwise, and in any case $\widetilde{w}^0 \subseteq \widetilde{F}(\widetilde{v}^0) \cap \mathbf{R}_j$, so $\mu(\widetilde{w}^0) \in \mu\big(\widetilde{F}(\widetilde{v}^0)\big)$. Hence by (b), $\mu(\widetilde{w}^0) \in F\big(\mu(\widetilde{v}^0)\big) \cup F^{-1}\big(\mu(\widetilde{v}^0)\big)$.

If $\mu(\widetilde{w}^0) \in F\big(\mu(\widetilde{v}^0)\big)$, we let $\Psi(\widetilde{u}, \mu(\widetilde{v}^0)) = \widetilde{w}^0$, which satisfies the desired requirements. If $\mu(\widetilde{v}^0) \in F\big(\mu(\widetilde{w}^0)\big)$, we replace the ancestor $\mu(\widetilde{u})$ of $\mu(\widetilde{v}^0)$ by its ancestor $\mu(\widetilde{w}^0)$ with representation $\widetilde{w}^0$. A representation $\widetilde{v}_1$ of $\mu(\widetilde{v}^0) = \mu(\widetilde{v}^1)$ can be obtained from $\widetilde{w}^0$ based on (a), as at the beginning of this proof. As before we get $\widetilde{w}^1 \subseteq \widetilde{F}(\widetilde{v}^1) \cap \mathbf{R}_j$ with $\widetilde{c}(\widetilde{w}^1) < \widetilde{c}(\widetilde{w}^0)$ and $\mu(\widetilde{w}^1) \in F\big(\mu(\widetilde{v}^1)\big) \cup F^{-1}\big(\mu(\widetilde{v}^1)\big)$.

This process thus leads to the formation of two sequences $\widetilde{v}^0, \widetilde{v}^1, \ldots;$ $\widetilde{w}^0, \widetilde{w}^1, \ldots$, where $\mu(\widetilde{v}^0) = \mu(\widetilde{v}^i)$ ($i = 1, 2, \ldots$), $\widetilde{w}^i \subseteq \mathbf{R}_j$, $\mu(\widetilde{w}^i) \in F\big(\mu(\widetilde{v}^i)\big) \cup F^{-1}\big(\mu(\widetilde{v}^i)\big)$ ($i = 1, 2, \ldots$). Since $\widetilde{c}(\widetilde{w}^0) > \widetilde{c}(\widetilde{w}_1) > \cdots$, these sequences must be finite. In fact, each sequence has at most $O(n^5)$ terms. Since this process keeps producing a new sequence term if $\mu(\widetilde{w}^i) \in F^{-1}\big(\mu(\widetilde{v}^i)\big)$, there exists $j = O(n^5)$ such that $\mu(\widetilde{w}^j) \in F\big(\mu(\widetilde{v}^j)\big)$. We then define $\Psi(\widetilde{u}, \mu(\widetilde{v})) = \widetilde{w}^j$, which satisfies the desired requirements. The process is indicated schematically in Fig. 4.

Finally, it can be decided in $O(n)$ steps whether $\mu(\widetilde{w}^i) \in F\big(\mu(\widetilde{v}^i)\big)$ or $\mu(\widetilde{v}^i) \in F\big(\mu(\widetilde{w}^i)\big)$ by using (c). $\qquad\square$

Example 8. Continuing Example 6, suppose player I moves from $\mathbf{u} = 15 \in \mathbf{V}_0$ with $\widetilde{u} = \{5, 10\}$ ($\mathbf{u}_1 = 5$, $\mathbf{u}_2 = 10$) to $\mathbf{v}^0 = 6 \in \mu\big(\widetilde{F}(\widetilde{u})\big)$. Then $\widetilde{v}^0 = \widetilde{F}(\widetilde{u}; 10, 3) = \{5, 3\} = \{\mathbf{u}_1, \mathbf{v}_2\}$ ($\mathbf{v}_2 = 3$, $\gamma(\mathbf{v}_2) > \gamma(\mathbf{u}_2)$), and $\mu(\widetilde{v}^0) = \mathbf{v}^0$. Now $\mathbf{w}_2 \in F(\mathbf{v}_2)$ with $\gamma(\mathbf{w}_2) = \gamma(\mathbf{u}_2)$ and $c(\mathbf{w}_2) < c(\mathbf{u}_2)$ is clearly satisfied by $\mathbf{w}_2 = \Phi$. Thus $\widetilde{w}^0 = \{\mathbf{u}_1\}$, $\mu(\widetilde{w}^0) = 5$. Since $5 \in F^{-1}(6)$, we replace the ancestor $\mathbf{u}$ of $\mathbf{v}^0$ by $\mu(\widetilde{w}^0) = 5$ with representation $\widetilde{w}^0 = \{\mathbf{u}_1\}$, pretending that play began from $\mu(\widetilde{w}^0) = 5$, rather than from $\mathbf{u} = 15$. Then $\widetilde{v}^1 = \widetilde{F}(\widetilde{w}^0; 5, 6) = \{6\}$. Now $\widetilde{w}^1 = \widetilde{\varnothing} \in F\left(\mu(\widetilde{v}^1)\right)$ with $\mu(\widetilde{w}^1) = \varnothing$. Thus $\Psi\left(\widetilde{u}^0, \mu(\widetilde{v}^0)\right) = \widetilde{w}^1$, and so player II moves to $\mu(\widetilde{w}^1) \in \mathbf{V}_0$.

Given a sum of r games, one of which is a two-player cellular automata game played on a finite cyclic digraph $G = (V, E)$. Given an N-position of the sum, consider the subset M of moves in the cellular automata game which realize a win. How large is M? What's the complexity of computing it?

Theorem 6. *Given an N-position in a sum of r games containing a two-player cellular automata game played on a finite cyclic digraph $G = (V, E)$ with $|V| = n$. The subset M of moves on G leading to a win has size $O(n^5)$, and its computation needs $O(n^6)$ steps.*

Proof. Apply Algorithm CEL to G $\left(O(n^6)\right)$. Given an N-position of the sum, we may assume that a winning move is of type (ii). So we have to move from a vertex u in G, which corresponds to $\mathbf{u} \in \mathbf{V}_\ell$ or to $\gamma(\mathbf{u}) = \infty(K)$, $p \in K$ in the cellular automata game-graph, to $\mathbf{v} \in F(\mathbf{u}) \cap \mathbf{V}_p$.

Assume first $\mathbf{u} \in \mathbf{V}_\ell$. Compute $B^{-1}\mathbf{u}$ to get a representation $\widetilde{u} = \{\mathbf{u}_1, \ldots, \mathbf{u}_h\} \subseteq \mathbf{R}_s$ with $\mu(\widetilde{u}) = \sum_{i=1}^{'h} \mathbf{u}_i$, where $s = \lceil \log_2(\ell + 1) \rceil$ $\left(O(n^2)\right)$. For a move of type (ii), let $\mathbf{v} \in F\left(\mu(\widetilde{u})\right) \cap \mathbf{V}_p$. By (a), $\mathbf{v} \in \mu\left(\widetilde{F}(\widetilde{u})\right)$, say $\mathbf{v} = \mu\left(\widetilde{F}(\widetilde{u}; \mathbf{u}_1, \mathbf{v}_1)\right)$. As we saw at the beginning of the proof of Theorem 5, the computation of $\mathbf{v}_1 = F_k^1(\mathbf{u}_1)$ takes $O(n^2)$ steps. It can always be arranged so that $\gamma(\mathbf{v}_1) < \gamma(\mathbf{u}_1)$. Also $w(\mathbf{v}_1) \leq 4$. Thus $\widetilde{v}_1 = \{\mathbf{v}_1\}$ is a representation. Replacing $\mathbf{u}_1$ by $\mathbf{v}_1$ in $\widetilde{u}$ and deleting $\mathbf{v}_1$ if $\mathbf{v}_1 = \mathbf{u}_i$ for some i, we get a representation $\widetilde{v}$ of $\mathbf{v}$ over $\mathbf{R}_j$, where $j = \lceil \log_2(p + 1) \rceil$ $\left(O(n)\right)$. Since $p < \ell$ and c is monotonic we have $\widetilde{c}(\widetilde{v}) < \widetilde{c}(\widetilde{u})$.

Secondly assume $\gamma(\mathbf{u}) = \infty(K)$. Scan the $O(n^2)$ followers of $\mathbf{u}$, to locate one, say $\mathbf{v}$, which is in $\mathbf{V}_p$. Compute $B^{-1}\mathbf{v}$ to yield a representation $\widetilde{v} = \{\mathbf{v}_1, \ldots, \mathbf{v}_h\} \subseteq \mathbf{R}_j$ with $\mathbf{v} = \mu(\widetilde{v}) = \sum_{i=1}^{'h} \mathbf{v}_i$, where $j = \lceil \log_2(p + 1) \rceil$ $\left(O(n^4)\right)$. By definition, $\widetilde{c}(\widetilde{v}) < \widetilde{c}(\widetilde{u})$.

In any subsequent move of type (ii) we compute the new representation from the previous one as was done above for the case $\mathbf{u} \in \mathbf{V}_\ell$, where $\widetilde{v}$ was computed from $\widetilde{u}$ in $O(n)$ steps.

For a move of type (i), assume that player II moves from $\mathbf{u}^i = \mu(\widetilde{u}^i) \in \mathbf{V}_p$ with $\widetilde{u}^i \subseteq \mathbf{R}_j$ $\left(j = \lceil \log_2(p + 1) \rceil\right)$ to $\mathbf{v}^i \in F(\mathbf{u}^i)$ with $\gamma(\mathbf{v}^i) > \gamma(\mathbf{u}^i)$. Then player I computes $\widetilde{u}^{i+1} = \Psi(\widetilde{u}_i, \mathbf{v}^i)$ $\left(O(n)\right)$ and moves to $\mathbf{u}^{i+1} = \mu(\widetilde{u}^{i+1}) \in F(\mathbf{v}^i) \cap \mathbf{V}_p$ such that $\widetilde{c}(\widetilde{u}^{i+1}) < \widetilde{c}(\widetilde{u}^i)$. This can be done as we saw in Theorem 5.

Thus $\widetilde{c}$ decreases strictly for both a move of type (i) and of type (ii). Since $\widetilde{c}(\widetilde{u}) = O(n^5)$, player I can win in $O(n^5)$ moves made in the cellular automata game, for whatever sequence of moves of type (i) and (ii) is taken. This is in addition to any other moves in the other sum components. Since each computation of one move of type (ii) and of Ψ requires $O(n)$ steps, the entire computation time for player I in the cellular automata game is $O(n^6)$ steps. $\square$

9. Epilogue

A collection of two-player cellular automata games with only a minimum of the underlying theory can be found in [12], [13].

An obvious remaining question is whether two-player cellular automata games have a polynomial strategy also for every $q > 1$. We have, in fact, provided a polynomial infrastructure for the general case, in the sense that everything up to the end of §6 is consistent with a polynomial strategy for all $q \geq 1$. But $V^{[q+1]}$ and $V^{[2(q+1)]}$ are not restrictions of V when $q > 1$, so we cannot apply Lemma 3. Therefore we cannot prove polynomiality for $q > 1$ in the same way we used for 1-regular games.

The special case of 1-regular games are the so-called *annihilation games*, analyzed in [9], [15], [17]. The present paper is a generalization of [17], and §7 and §8 of the present paper are simplifications and clarifications of the corresponding parts there. For annihilation games it is natural to consider a vertex with weight 1 to be occupied by a token, and one with weight 0 to be unoccupied. Two tokens are then mutually annihilated on impact, hence the name. Misère play (in which the player making the last move loses, and the opponent wins) of annihilation games was investigated in [8]. Our motivation for examining annihilation games, suggested to us by John Conway, was to create games which exhibit some interaction between tokens, yet still have a polynomial strategy.

Annihilation games are "barely" polynomial, in several senses. Their complexity is $O(|V|^6)$, and just about any perturbation of them yields Pspace-hard games (see [16], [14], [19]). Moreover, the polynomial computation of a winning move may require a "strategy in the broad sense" (see [11], §4).

Kalmár [21] and Smith [30] defined a *strategy in the wide sense* to be a strategy which depends on the present position and on all its antecedents, from the beginning of play. Having defined this notion, both authors concluded that it seems logical that it suffices to consider a *strategy in the narrow sense*, which is a strategy that depends only on the present position (the terminology *Markoff strategy* suggests itself here). They then promptly restricted attention to strategies in the narrow sense.

Let us define a *strategy in the broad sense* to be a strategy that depends on the present position v and on *all* its predecessors $u \in F^{-1}(v)$, whether or not such u is a position in the play of the game. This notion, if anything, seems to be even less needed than a strategy in the wide sense.

Yet, in §8, we did employ a strategy in the broad sense, for computing a winning move in polynomial time. It was needed, since the counter function associated with γ was computed only for a small subgraph of size $O(n^4)$ of the game-graph of size $O(2^n)$, in order to preserve polynomiality. This suggests the possibility that a polynomial strategy in the narrow sense may not exist; but we have not proved anything like this. We only report that we didn't find a polynomial strategy in the narrow sense, and that perhaps the polynomial strategy in the broad sense used here suggested itself precisely because the game is "barely" polynomial, so to speak.

Annihilation games can lead to linear error correcting codes [10], but their Hamming distance is ≤ 4. The current work was motivated by the desire to create games which naturally induce codes of Hamming distance > 4. Cellular automata games may provide such codes, a topic to be taken up elsewhere. In practice, the computation of $\mathbf{V}_0$, which is all that's needed for the codes, can often be done by inspection. The best codes may be derived from a digraph which is a simplification of that of Figure 1, where $\mathbf{V}_0$ can also be computed easily. The lexicodes method [6], [5], [28] for deriving codes related to games is plainly exponential.

Another motivation was to further explore polynomial games with "token interactions". Last but not least was the desire to create two-player cellular automata games, which traditionally have been solitaire and 0-player only.

In conclusion, we have have extended cellular automata games to two-player games in a natural way and given a strategy for them. Four key ideas were used for doing this: (I). Showing that γ is essentially additive: $\gamma(\mathbf{u}_1 \oplus \mathbf{u}_2) = \gamma(\mathbf{u}_1) \oplus \gamma(\mathbf{u}_2)$ if $\mathbf{u}_1 \in \mathbf{V}^f$ or $\mathbf{u}_2 \in \mathbf{V}^f$. (II). Showing that $\mathbf{V}_0$, $\mathbf{V}^f$ and γ can be computed by restricting attention to the linear span of "sparse" vectors of polynomial size. (III). Providing an algorithm to compute the sparse vector space. (IV). Computing a winning move. Whereas (I) and (II) and (IV) are polynomial for all two-player cellular automata games, this has been shown for (III) only for the special case $q = 1$ (annihilation games). So the main open question is the complexity status of (III) for $q > 1$. Another question is what happens when loops are permitted in the groundgraph. This question and the polynomiality of (III) have been settled for the case of annihilation games in [17].

Finally, we point out that our strategy also solves a two-player cellular automata game with a modified move-rule: Instead of firing a neighborhood of z_k of size $q(k)$ (see the beginning of § 3), we fire a neighborhood of z_k of size at most $q(k)$. Indeed, we can reduce our original game to the modified one by adjoining to every z_k with $q(k) > 0$, $q(k) - 1$ edges to leaves and playing a regular two-player cellular automata game on the modified game. See also [10], Remark 4.2.

Acknowledgment

Thanks to Ofer Rahat who went over the paper with a fine-tooth comb, leading to a number of improvements.

Note added in proof The present work has indeed led to a much improved algorithm for producing *lexicodes*. Previously known algorithms had complexity exponential in n. The new algorithm has complexity $O(n^{d-1})$, where n is the size of the code, and d its distance. See A. S. Fraenkel and O. Rahat, "Complexity of error-correcting codes derived from combinatorial games"; preprint available from my homepage.

References

[1] E. R. Berlekamp, J. H. Conway and R. K. Guy [1982], *Winning Ways for Your Mathematical Plays* (two volumes), Academic Press, London.

[2] N. L. Biggs [1999], Chip-firing and the critical group of a graph, *J. Algebr. Comb.* **9**, 25–45.

[3] A. Björner and L. Lovász [1992], Chip-firing on directed graphs, *J. Algebr. Comb.* **1**, 305–328.

[4] J. H. Conway [1976], *On Numbers and Games*, Academic Press, London.

[5] J. H. Conway [1990], Integral lexicographic codes, *Discrete Math.* **83**, 219–235.

[6] J. H. Conway and N. J. A. Sloane [1986], Lexicographic codes: error-correcting codes from game theory, *IEEE Trans. Inform. Theory* **IT-32**, 337–348.

[7] Y. Dodis and P. Winkler [2001], Universal configurations in light-flipping games, *Proc. 12th Annual ACM-SIAM Sympos. on Discrete Algorithms* (Washington, DC, 2001), ACM, New York, pp. 926–927.

[8] T. S. Ferguson [1984], Misère annihilation games, *J. Combin. Theory* (Ser. A) **37**, 205–230.

[9] A. S. Fraenkel [1974], Combinatorial games with an annihilation rule, in: The Influence of Computing on Mathematical Research and Education, *Proc. Symp. Appl. Math.* (J. P. LaSalle, ed.), Vol. 20, Amer. Math. Soc., Providence, RI, pp. 87–91.

[10] A. S. Fraenkel [1996], Error-correcting codes derived from combinatorial games, in: *Games of No Chance* (R. J. Nowakowski, ed.), MSRI Publications **29**, Cambridge University Press, Cambridge, pp. 417–431.

[11] A. S. Fraenkel [1996], Scenic trails ascending from sea-level Nim to alpine chess, in: *Games of No Chance* (R. J. Nowakowski, ed.), MSRI Publications **29**, Cambridge University Press, Cambridge, pp. 13–42.

[12] A. S. Fraenkel [2001], Virus versus mankind, *Proc. 2nd Intern. Conference on Computers and Games CG'2000* (T. Marsland and I. Frank, eds.), Vol. 2063, Hamamatsu, Japan, Oct. 2000, Lecture Notes in Computer Science, Springer, pp. 204–213.

[13] A. S. Fraenkel [2002], Mathematical chats between two physicists, in: *Puzzler's Tribute: a Feast for the Mind* (D. Wolfe and T. Rodgers, eds.), honoring Martin Gardner, A K Peters, Natick, MA., pp. 383–386.

[14] A. S. Fraenkel and E. Goldschmidt [1987], Pspace-hardness of some combinatorial games, *J. Combin. Theory* (Ser. A) **46**, 21–38.

[15] A. S. Fraenkel and Y. Yesha [1976], Theory of annihilation games, *Bull. Amer. Math. Soc.* **82**, 775–777.

[16] A. S. Fraenkel and Y. Yesha [1979], Complexity of problems in games, graphs and algebraic equations, *Discrete Appl. Math.* **1**, 15–30.

[17] A. S. Fraenkel and Y. Yesha [1982], Theory of annihilation games — I, *J. Combin. Theory* (Ser. B) **33**, 60–86.

[18] A. S. Fraenkel and Y. Yesha [1986], The generalized Sprague–Grundy function and its invariance under certain mappings, *J. Combin. Theory* (Ser. A) **43**, 165–177.

[19] A. S. Goldstein and E. M. Reingold [1995], The complexity of pursuit on a graph, *Theoret. Comput. Sci.* (Math Games) **143**, 93–112.

[20] E. Goles [1991], Sand piles, combinatorial games and cellular automata, *Math. Appl.* **64**, 101–121.

[21] L. Kalmár [1928], Zur Theorie der abstrakten Spiele, *Acta Sci. Math. Univ. Szeged* **4**, 65–85.

[22] J. L. Kelley [1965], Algebra, A Modern Introduction, Van Nostrand, Princeton, NJ.

[23] J. Lambek and L. Moser [1959], On some two way classifications of the integers, *Can. Math. Bull.* **2**, 85–89.

[24] C. M. López [1997], Chip firing and the Tutte polynomial, *Ann. of Comb.* **1**, 253–259.

[25] Ó. Martín-Sánchez and C. Pareja-Flores [2001], Two reflected analyses of lights out, *Math. Mag.* **74**, 295–304.

[26] J. Missigman and R. Weida [2001], An easy solution to mini lights out, *Math. Mag.* **74**, 57-59.

[27] D. H. Pelletier [1987], Merlin's magic square, *Amer. Math. Monthly* **94**, 143–150.

[28] V. Pless [1991], Games and codes, in: *Combinatorial Games*, Proc. Symp. Appl. Math. (R. K. Guy, ed.), Vol. 43, Amer. Math. Soc., Providence, RI, pp. 101–110.

[29] A. Shen [2000], Lights out, *Math. Intelligencer* **22**, 20-21.

[30] C. A. B. Smith [1966], Graphs and composite games, *J. Combin. Theory* **1**, 51–81. Reprinted in slightly modified form in: *A Seminar on Graph Theory* (F. Harary, ed.), Holt, Rinehart and Winston, New York, NY, 1967.

[31] D. L. Stock [1989], Merlin's magic square revisited, *Amer. Math. Monthly* **96**, 608–610.

[32] K. Sutner [1988], On σ-automata, *Complex Systems* **2**, 1–28.

[33] K. Sutner [1989], Linear cellular automata and the Garden-of-Eden, *Math. Intelligencer* **11**, 49–53.

[34] K. Sutner [1990], The σ-game and cellular automata, *Amer. Math. Monthly* **97**, 24–34.

[35] K. Sutner [1995], On the computational complexity of finite cellular automata, *J. Comput. System Sci.* **50**, 87–97.

[36] G. P. Tollisen and T. Lengyel [2000], Color switching games, *Ars Combin.* **56**, 223-234.

AVIEZRI S. FRAENKEL
DEPARTMENT OF COMPUTER SCIENCE AND APPLIED MATHEMATICS
THE WEIZMANN INSTITUTE OF SCIENCE
REHOVOT 76100
ISRAEL
fraenkel@wisdom.weizmann.ac.il
http://www.wisdom.weizmann.ac.il/~fraenkel

More Games of No Chance
MSRI Publications
Volume **42**, 2002

Who Wins Domineering on Rectangular Boards?

MICHAEL LACHMANN, CRISTOPHER MOORE, AND
IVAN RAPAPORT

ABSTRACT. Using mostly elementary considerations, we find out who wins
the game of Domineering on all rectangular boards of width 2, 3, 5, and 7.
We obtain bounds on other boards as well, and prove the existence of
polynomial-time strategies for playing on all boards of width 2, 3, 4, 5, 7,
9, and 11. We also comment briefly on toroidal and cylindrical boards.

1. Introduction

Domineering or Crosscram is a game invented by Göran Andersson and introduced to the public in [1]. Two players, say Vera and Hepzibah, have vertical
and horizontal dominoes respectively. They start with a board consisting of
some subset of the square lattice and take turns placing dominoes until one of
them can no longer move. For instance, the 2×2 board is a win for the first
player, since whoever places a domino there makes another space for herself while
blocking the other player's moves.

A beautiful theory of combinatorial games of this kind, where both players
have perfect information, is expounded in [2; 3]. Much of its power comes from
dividing a game into smaller subgames, where a player has to choose which subgame to make a move in. Such a combination is called a *disjunctive sum*. In
Domineering this happens by dividing the remaining space into several components, so that each player must choose in which component to place a domino.

Each game is either a win for Vera, regardless of who goes first, or Hepzibah
regardless of who goes first, or the first player regardless of who it is, or the
second regardless of who it is. These correspond to a value G which is positive,
negative, fuzzy, or zero, i.e. $G > 0$, $G < 0$, $G \parallel 0$, or $G = 0$. (By convention Vera
and Hepzibah are the left and right players, and wins for them are positive and
negative respectively.) However, we will often abbreviate these values as $G = V$,
H, 1st, or 2nd. We hope this will not confuse the reader too much.

In this paper, we find who wins Domineering on all rectangles, cylinders, and tori of width 2, 3, 5, and 7. We also obtain bounds on boards of width 4, 7, 9, and 11, and partial results on many others. We also comment briefly on toroidal and cylindrical boards.

Note that this is a much coarser question than calculating the actual game-theoretic values of these boards, which determine how they act when disjunctively summed with other games. Berlekamp [4] has found exact values for $2 \times n$ rectangles with n odd, and approximate values to within an infinitesimal or 'ish' (which unfortunately can change who wins in unusual situations) for other positions of width 2 and 3. In terms of who wins, the 8×8 board and many other small boards were recently solved by Breuker, Uiterwijk and van den Herik using a computer search with a good system of transposition tables in [6; 7; 8]. We make use of these results below.

2. $2 \times n$ Boards

On boards of width 2, it is natural to consider dividing it into two smaller boards of width 2. At first glance, Vera (the vertical player) has greater power, since she can choose where to do this. However, she can only take full advantage of this if she goes first. Hepzibah (the horizontal player) has a greater power, since whether she goes first or second, she can divide a game into two simply by *not* placing a domino across their boundary. We will see that, for sufficiently large n, this gives Hepzibah the upper hand.

We will abbreviate the value of the $2 \times n$ game as $[n]$.

Let's look at what happens when Vera goes first, and divides a board of length $m + n + 1$ into one of length m and one of length n. If she can win on both these games, i.e. if $[m] = [n] = V$, clearly she wins. If $[m] = V$ and $[n] = $ 2nd, Hepzibah will eventually lose in $[m]$ and be forced to play in $[n]$, whereupon Vera replies there and wins. Finally, if $[m] = [n] = $ 2nd, Vera replies to Hepzibah in both and wins. Since Vera can win if she goes first, $[m + n + 1]$ must be a win either for the first player or for V. This gives us the following table for combining boards of lengths m and n into boards of length $m + n + 1$:

$[m + n + 1]$	2nd	V	
2nd	1st or V	1st or V	
V	1st or V	1st or V	(2–1)

This table can be summarized by the equation

$$\text{If } [m] \geq 0 \text{ and } [n] \geq 0, \text{ then } [m + n + 1] \parallel> 0. \qquad (2\text{–}2)$$

Hepzibah has a similar set of tools at her disposal. By declining to ever place a domino across their boundaries, she can effectively play $[m+n]$ as a sum of $[m]$ and $[n]$ for whichever m and n are the most convenient. If Hepzibah goes first, she can win whenever $[m] = $ 1st and either $[n] = $ 2nd or $[n] = H$, by playing first in $[m]$ and replying to Vera in $[n]$. If $[m] = [n] = H$ she wins whether she goes first or second, and if $[m] = $ 2nd and $[n] = H$, the same is true since she plays in $[n]$ and replies to Vera in $[m]$. Finally, if $[m] = [n] = $ 2nd, she can win if she goes second by replying to Vera in both games. This gives the table

$[m + n]$	1st	2nd	H	
1st	?	1st or H	1st or H	(2–3)
2nd	1st or H	2nd or H	H	
H	1st or H	H	H	

which can be summarized by the equation

$$[m + n] \le [m] + [n].$$ (2–4)

This simply states that refusing to play across a vertical boundary can only make things harder for Hepzibah.

These two tables alone, in conjunction with some search by hand and by computer, allow us to determine the following values. Values derived from smaller games using Tables 2–1 and 2–3 are shown in plain, while those found in other ways, such as David Wolfe's Gamesman's Toolkit [5], our own search program, or Berlekamp's solution for odd lengths [4] are shown in bold.

0	**2nd**	10	1st	20	H	30	H
1	**V**	11	1st	21	H	31	**H**
2	**1st**	12	H	22	**H**	32	H
3	**1st**	13	**2nd**	23	1st	33	H
4	**H**	14	1st	24	H	34	H
5	**V**	15	1st	25	H	35	H
6	1st	16	H	26	H	36	H
7	1st	17	H	27	1st	37	H
8	H	18	**1st**	28	H	38	H
9	**V**	19	1st	29	H	39	H

In fact, $[n]$ is a win for Hepzibah for all $n \ge 28$.

Some discussion is in order. Once we know that $[4] = H$, we have $[4k] = H$ for all $k \ge 1$ by Table 2–3. Combining Tables 2–1 and 2–3 gives $[6] = [7] = $ 1st, since these are both 1st or V and 1st or H. A similar argument gives $[10] = [11] = $ 1st and $[14] = [15] = $ 1st, once we learn through search that $[13] = $ 2nd (which is rather surprising, and breaks an apparent periodicity of order 4).

Combining $[13]$ with multiples of 4 and with itself gives $[13+4k] = [26+4k] = [39+4k] = H$ for $k \ge 1$. Since $26 = 24 + 2 = 13 + 13$, we have $[26] = H$ since it is both 1st or H and 2nd or H. Then $[39] = H$ since 39=26+13.

Since $19 = 9 + 9 + 1 = 17 + 2$, we have $[19] = $ 1st since it is both 1st or V and 1st or H. Similarly $23 = 9 + 13 + 1 = 21 + 2$ and $27 = 13 + 13 + 1 = 25 + 2$ so

[23] = [27] = 1st. A computer search gives [22] = H, and since $35 = 22 + 13$ we have [35] = H.

So far, we have gotten away without using the real power of game theory. However, for [31] we have found no elementary proof, and it is too large for our search program. Therefore, we turn to Berlekamp's beautiful solution for $2 \times n$ Domineering when n is odd (Ref. [4]), evaluate it with the Gamesman's Toolkit [5], and find the following (see Refs. [2; 3; 4] for notation):

$$[31] = \tfrac{1}{2} - 15 \cdot \left(\tfrac{1}{4} + \int^{3/4} * \right) + \int^{3/4} \int_{1/2}^{1/2*} 3\tfrac{7}{8} = \{2 \,|\, 0 \,\|\, -\tfrac{1}{2} \,|\, -2\} \,|\, -\tfrac{5}{2} < 0 \quad (2\text{-}5)$$

Thus [31] is negative and a win for Hepzibah. This closes the last loophole, telling us who wins the $2 \times n$ game for all n.

3. Boards of Width 3, 4, 5, 7, 9, 11 and Others

The situation for rectangles of width 3 is much simpler. While Equation 2-2 no longer holds since Vera cannot divide the board in two with her first move, Equation 2-4 holds for all widths, since Hepzibah can choose not to cross a vertical boundary between two games. A quick search shows that $[3 \times n] = H$ for $n = 4, 5, 6$ and 7, so we have for width 3

$$[1] = V, \quad [2] = [3] = 1\text{st}, \quad \text{and } [n] = H \text{ for all } n \geq 4.$$

For width 5, we obtain

$$[1] = [3] = V, \quad [2] = [4] = H, \quad [5] = 2\text{nd}, \quad \text{and } [n] = H \text{ for all } n \geq 6.$$

For width 7, Breuker, Uiterwijk and van den Herik found by computer search (references [6; 7; 8]) that $[4] = [6] = [9] = [11] = H$. Then $[8] = [10] = H$, and combining this with searches on small boards we have

$$[1] = [3] = [5] = V, \quad [2] = [7] = 1\text{st}, \quad [4] = [6] = H, \quad \text{and } [n] = H \text{ for all } n \geq 8.$$

In all these cases, we were lucky enough that $[n] = 2\text{nd}$ or H for enough small n to generate all larger n by addition. This becomes progressively rarer for larger widths. However, we have some partial results on other widths. For width 4, Uiterwijk and van den Herik (Refs. [7; 8]) found by computer search that $[8] = [10] = [12] = [14] = H$, so $[n] = H$ for all even $n \geq 8$. They also found that $[15] = [17] = H$, so

$$[4 \times n] = H \text{ for all } n \geq 22.$$

This leaves $[4 \times 19]$ and $[4 \times 21]$ as the only unsolved boards of width 4.

As a general method, whenever we can find a length for which Hepzibah wins by some positive number of moves (rather than by an infinitesimal), then she wins on any board long enough to contain a sufficient number of copies of this one to overcome whatever advantage Vera might have on smaller boards. Game-theoretically, if $[n] < -r$, then $[m] < [m \bmod n] - r\lfloor m/n \rfloor$, so then $[m] < 0$ whenever $m \geq n(1/r) \max_{l < n} [l]$.

For width 9, for instance, we have $[1] = 4$, $[2] = \frac{3}{2} \,|\, 0\, \|\, -\frac{1}{2} \,|\, -\frac{5}{2}$, $[3] = 5\,|\,3\,\|\,\frac{11}{4}\,|\,\frac{1}{4}$, and $[4] \leq [2] + [2] = 1\,|\,-\frac{1}{2}\,\|\,-1\,|\,-\frac{5}{2} < -\frac{1}{2}$. By summing these, we have $[23] \leq 11 \cdot [2] + [1] < 0$, and in general

$$[9 \times n] = H \text{ for all } n \geq 22.$$

Similarly, for width 11 we have $[1] = 5$ and $[2] = 1\,|\,\left\{\frac{1}{2}\,|\,-1\,\|\,-\frac{3}{2}\,|\,-\frac{7}{2}\right\}$. Then $[8] \leq [2] + [2] + [2] + [2] = 1\,|\,-\frac{1}{2}\,\|\,-1\,|\,-\frac{5}{2} < -\frac{1}{2}$ and $[16] \leq [8] + [8] \leq -\frac{3}{2}$, so

$$[11 \times n] = H \text{ for all } n \geq 56.$$

Unfortunately, for all other widths greater than 7, either $[2]$ or $[2] + [2]$ is positive and $[3]$ is as well, so without some way to calculate values for $[4]$ or more we can't establish this kind of bound. Nor do we know of any length for which Hepzibah wins on width 8, or a proof that she wins any board of width 6 by a positive amount.

To get results on boards of other widths, we can use a variety of tricks. First of all, just as Hepzibah can choose to cross a vertical boundary between games, Vera can choose not to cross a horizontal one. Thus Equation 2–4 is one of a dual pair,

$$[m \times (n_1 + n_2)] \leq [m \times n_1] + [m \times n_2] \tag{3-1}$$

$$[(m_1 + m_2) \times n] \geq [m_1 \times n] + [m_2 \times n] \tag{3-2}$$

Another useful rule is that $[n \times n] = $ 1st or 2nd, since neither player can have an advantage on a square board. In fact, in game-theoretic terms $[n \times n] + [n \times n] = 0$, so if Hepzibah goes second she can win by mimicking Vera's move, rotated 90°, in the other board. More generally we have

If $[n \times n] = $ 1st, then $[n \times kn] = \begin{cases} \text{2nd or } H & \text{for even } k > 1, \\ \text{1st or } H & \text{for odd } k > 1, \end{cases}$

If $[n \times n] = $ 2nd, then $[n \times kn] = $ 2nd or H for all $k > 1$.

For instance, this tells us that $[6 \times 12] = $ 2nd or H, and since $[6 \times 4] = $ 1st and $[6 \times 8] = H$ (Ref. [6]) we also have $[6 \times 12] = $ 1st or H. Therefore $[6 \times 12] = H$ and

$$[6 \times (4 + 4k)] = H \text{ for all } k \geq 1.$$

We can also use our addition rules backward; since no two games can sum to a square in a way that gives an advantage to either player,

If $m < n$ and $[m \times n] = $ 2nd or V, then $[(n - m) \times n] \neq V$ \hfill (3-3)

and similarly for the dual version.

0	1	2	3	4	5	6	7	8	9	10	11	12	13	14	15	16	17	18	19	20	21	22	23	24	25	26	27	28	29	30	31	32	33
1	2	H	H	H	H	H	H	H	H	H	H	H	H	H	H	H	H	H	H	H	H	H	H	H	H	H	H	H	H	H	H	H	H
2	V	1	1	H	V	1	1	H	V	1	1	H	2	1	1	H	H	1	1	H	H	H	1	H	H	H	1	H	H	H	H	H	H
3	V	1	1	H	H	H	H	H	H	H	H	H	H	H	H	H	H	H	H	H	H	H	H	H	H	H	H	H	H	H	H	H	H
4	V	V	V	1	V	1	V	H	V	H	V	H	V	H	H	H	H	H	1h	H	1h	H	H	H	H	H	H	H	H	H	H	H	H
5	V	H	V	H	2	H	H	H	H	H	H	H	H	H	H	H	H	H	H	H	H	H	H	H	H	H	H	H	H	H	H	H	H
6	V	1	V	1	V	1	V	H	V	1	1	H	V	1h		H		1h	1h	H		1h	1h	H		1h	1h	H		1h	1h	H	
7	V	1	V	H	V	H	1	H	H	H	H	H	H	H	H	H	H	H	H	H	H	H	H	H	H	H	H	H	H	H	H	H	H
8	V	V	V	V	V	V	V	1	V		V		V			2h								1h								2h	
9	V	H	V	H	V	H	V	H	1	H	1h	H	1h	H	1h	H	1h	H	1h	H	1h	H	H	H	H	H	H	H	H	H	H	H	H
10	V	1	V	V	V	1	V		V	12	1v		V							2h		1h				1h				–v			
11	V	1	V	H	V	1	V	H	1v	1h	12	H	–v	1h	1h	H	1h	1h	1h	H	1h	H	1h	H	1h	H	1h	H	1h	H	1h	H	1h
12	V	V	V	V	V	V	V		V		V	12	V											2h									
13	V	2	V	H	V	H	V	H	1v	H	–h	H	12	H	–v	H	1h	H	1h	H	1h	H	1h	H	1h	H	1h	H	1h	H	1h	H	1h
14	V	1	V	V	V	1v	V		V		1v		V	12														2h		1h			
15	V	1	V	V	V		V		1v		1v		–h		12															2h		1h	
16	V	V	V	V	V	V	V	2v	V		V		V			12																2h	
17	V	V	V	V	V		V		1v		1v		1v				12																
18	V	1	V	V	V	1v	V		V		1v		V					12															
19	V	1	V	1v	V	1v	V		1v		1v		1v						12														
20	V	V	V	V	V	V	V		V	2v	V		V							12													
21	V	V	V	1v	V		V		1v		1v		1v								12												
22	V	V	V	V	V	1v	V		V	1v	V		V									12											
23	V	1	V	V	V	1v	V		V		1v		1v										12										
24	V	V	V	V	V	V	V	1v	V		V	2v	V											12									
25	V	V	V	V	V		V		V		1v		1v												12								
26	V	V	V	V	V	1v	V		V	1v	V		V													12							
27	V	1	V	V	V	1v	V		V		1v		1v														12						
28	V	V	V	V	V	V	V		V		V		V	2v														12					
29	V	V	V	V	V		V		V		1v		1v																12				
30	V	V	V	V	V	1v	V		V	–h	V		V	1v	2v															12			
31	V	V	V	V	V	1v	V		V		1v		1v																		12		
32	V	V	V	V	V	V	V	2v	V		V		V		1v	2v																12	
33	V	V	V	V	V		V		V		1v		1v																				12

Figure 1. What we know so far about who wins Domineering on rectangular boards. 1, 2, H and V mean a win for the first player, second player, Hepzibah and Vera respectively. Things like "1h" mean either 1st or H (i.e. all we know is that Hepzibah wins if she goes first) and "-v" means that it is not a win for Vera all the time. Values outlined in black are those provided by search or other methods; all others are derived from these using our rules or by symmetry.

Using the results of Refs. [4] and [6], some computer searches of our own, and a program that propagates these rules as much as possible gives the table shown in Figure 1. It would be very nice to deduce who wins on some large squares; the 9×9 square is the largest known so far (Ref. [8]). We note that if $[9 \times 13] = 1\text{st}$ then $[13 \times 13] = 1\text{st}$ since $[4 \times 13] = V$.

4. Playing on Cylinders and Tori

On a torus, Hepzibah can choose not to play across a vertical boundary and Vera can choose not to play across a horizontal one. Thus cutting a torus, or pasting a rectangle along one pair of edges, to make a horizontal or vertical

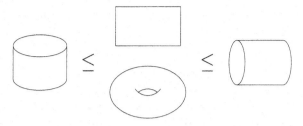

Figure 2. Inequalities between rectangular, cylinder, and toroidal boards of the same size. Cutting vertically can only hurt Hepzibah, while cutting horizontally can only hurt Vera.

cylinder gives the inequalities shown in Figure 2. Note that there is no obvious relation between the value of a rectangle and that of a torus of the same size.

While it is easy to find who wins on tori and cylinders of various small widths, we do the analysis here only for tori of width 2. Since Vera's move takes both squares in the same column, these boards are equivalent to horizontal cylinders like those shown on the left of Fig. 2 (or, for that matter, Möbius strips or Klein bottles). Therefore, if Hepzibah can win on the rectangle of length n, she can win here as well.

The second player has slightly more power here than she did in on the rectangle, since the first player has no control over the effect of her move. If Vera goes first, she simply converts a torus of length n into a rectangle of length $n - 1$, and if Hepzibah goes first, Vera can choose where to put the rectangle's vertical boundary, in essence choosing Hepzibah's first move for her. On the other hand, in the latter case Hepzibah gets to play again, and can treat the remainder of the game as the sum of t wo rectangles and a horizontal space.

These observations give the following table for tori of width 2:

$[n]_{\text{rect}}$	$[n - 1]_{\text{rect}}$	$[n]_{\text{torus}}$
H		H
1st	1st or H	H
1st	2nd or V	1st
2nd or V	1st or H	2nd or H

These and our table for $2 \times n$ rectangles determine $[n]_{\text{torus}}$ for all n except 5, 9, and 13. Vera loses all of these if she plays first, since she reduces the board to a rectangle which is a win for Hepzibah. For 5 and 9, Vera wins if Hepzibah plays first by playing in such a way that Hepzibah's domino is in the center of the resulting rectangle, creating a position which has zero value. Thus these boards are wins for the 2nd player. For 13, Hepzibah wins since (as the Toolkit tells us) all of Vera's replies to Hepzibah leave us in a negative position.

Our computer searches show that $n \times n$ tori are wins for the 2nd player when $n = 1$, 3, or 5, and for the 1st player when $n = 2$, 4, or 6. We conjecture that this alternation continues, and that square tori of odd and even size are wins for the 2nd and 1st players respectively. We note that a similar argument can be used to show that 9 is prime.

5. Polynomial-Time Strategies

While correctly playing the sum of many games is PSPACE-complete in general [9], the kinds of sums we have considered here are especially easy to play. For instance, if $[m]$ and $[n]$ are both wins for Hepzibah, she can win on $[m+n]$ by playing wherever she likes if she goes first, and replying to Vera in whichever game Vera chooses thereafter. Thus if we have strategies for both these games, we have a strategy for their sum. All our additive rules are of this kind.

Above, we showed for a number of widths that boards of any length can be reduced to sums of a finite number of lengths. Since each of these can be won with some finite strategy, and since sums of them can be played in a simple way, we have proved the following theorem:

Theorem 1. *For boards of width 2, 3, 4, 5, 7, 9, and 11, there exist polynomial-time strategies for playing on boards of any length.*

Note that we are not asking that the strategy produce optimum play, in which Hepzibah (or on small boards, 1st, 2nd or Vera) wins by as many moves as possible, but only that it tells her how to win.

In fact, we conjecture that this theorem is true for boards of any width. This would follow if for any m there exists an n such that Hepzibah wins by some positive number of moves, which in turn implies that there is some n' such that she wins on all boards longer than n'. A similar conjecture is made in [7]. Note, however, that this is not the same as saying that there is a single polynomial-time strategy for playing on boards of any size. The size or running time of the strategy could grow exponentially in m, even if it grows polynomially when m is held constant.

Acknowledgements

We thank Elwyn Berlekamp, Aviezri Fraenkel, and David Wolfe for helpful communications, and Jos Uiterwijk for sharing his group's recent results. I.R. also thanks the Santa Fe Institute for hosting his visit, and FONDECYT 1990616 for their support. Finally, C.M. thanks Molly Rose and Spootie the Cat.

References

[1] M. Gardner, "Mathematical Games." *Scientific American* **230** (1974) 106–108.

[2] J. H. Conway, *On Numbers and Games.* A K Peters, 2000.

[3] E. R. Berlekamp, J. H. Conway, and R. K. Guy, *Winning Ways.* A K Peters, 2001.

[4] E. R. Berlekamp, "Blockbusting and Domineering." *Journal of Combinatorial Theory Ser. A* **49** (1988) 67–116.

[5] D. Wolfe, "The Gamesman's Toolkit." In R. J. Nowakowski, Ed., *Games of No Chance.* Cambridge University Press, 1998.

[6] D. M. Breuker, J. W. H. M. Uiterwijk and H. J. van den Herik, "Solving 8 × 8 Domineering." *Theoretical Computer Science* bf 230 (2000) 195–206.

[7] J. W. H. M. Uiterwijk and H. J. van den Herik, "The Advantage of the Initiative." *Information Sciences* **122**:1 (2000) 43–58.

[8] See http://www.cs.unimaas.nl/~uiterwyk/Domineering_results.html.

[9] F. L. Morris, "Playing Disjunctive Sums is Polynomial Space Complete." *Int. Journal of Game Theory* **10** (1981) 195–205.

MICHAEL LACHMANN
SANTA FE INSTITUTE
1399 HYDE PARK ROAD
SANTA FE NM 87501
UNITED STATES
dirk@santafe.edu

CRISTOPHER MOORE
SANTA FE INSTITUTE and
COMPUTER SCIENCE DEPARTMENT AND DEPARTMENT OF PHYSICS AND ASTRONOMY
UNIVERSITY OF NEW MEXICO
ALBUQUERQUE NM 87131
UNITED STATES
moore@cs.unm.edu

IVAN RAPAPORT
DEPARTAMENTO DE INGENIERIA MATEMATICA
FACULTAD DE CIENCIAS FISICAS Y MATEMATICAS
UNIVERSIDAD DE CHILE
CASILLA 170/3, CORREO 3
SANTIAGO
CHILE
irapapor@dim.uchile.cl

More Games of No Chance
MSRI Publications
Volume **42**, 2002

Forcing Your Opponent to Stay in Control of a Loony Dots-and-Boxes Endgame

ELWYN BERLEKAMP AND KATHERINE SCOTT

ABSTRACT. The traditional children's pencil-and-paper game called Dots-and-Boxes is a contest to outscore the opponent by completing more boxes. It has long been known that winning strategies for certain types of positions in this game can be copied from the winning strategies for another game called Nimstring, which is played according to similar rules except that the Nimstring loser is whichever player completes the last box. Under certain common but restrictive conditions, one player (Right) achieves his optimal Dots-and-Boxes score, v, by playing so as to win the Nimstring game. An easily computed lower bound on v is known as the controlled value, cv. Previous results asserted that $v = cv$ if $cv \geq c/2$, where c is the total number of boxes in the game.

In this paper, we weaken this condition from $cv \geq c$ to $cv \geq 10$, and show this bound to be best possible.

Introduction to Loony Endgames and Controlled Value

The reader is assumed to be familiar with the game of Dots and Boxes. An excellent introduction can be found in [WW], [Nowakowski] or [D&B].

Some results about this game are more easily described in terms of the dual game, called Strings-and-Coins [D&B, Chapter 2]. This game is played on a graph G, whose nodes are called coins and whose branches are called strings. In this graph, the ends of each string are attached to two different coins or to a coin and the *ground*, which is a special uncapturable node. The *valence* of any coin is the number of strings attached to it. It has long been known that typical endgames reach a *loony* stage in which no coin has valence 1, and every coin of valence 2 is part of a chain of at least 3 such coins or of a loop of at least 4 such coins. When such a position occurs, the player who has just completed his turn is said to be *in control*. Let's call him Right. All moves now available to his opponent, called Left, are of the type called *loony*. Wherever Left plays, Right will then (possibly after capturing a few coins) have a choice between two options: (1) he can complete his turn by declining the last two or four coins of the chain or loop (respectively) which Left just began, thereby forcing her to play the first move elsewhere, or (2) he can capture all currently available coins and play the next move elsewhere himself. If Right chooses the former option, he is said to *retain control.*

When a loony endgame is played well, Right will score at least as many points as Left. Right strives to maximize the difference between their scores, while Left strives to minimize it. The *value* of the loony endgame G, denoted by $v(G)$, is Right's net margin of victory if both players play optimally. If Left has already acquired a lead of t points before the position G is reached, then Left can win the game if and only if $v(G) < t$.

If Right chooses to stay in control at least until the last or second-last turn of the game, and if Left chooses her best strategy knowing that this is how Right intends to play, then Right's resulting net score is called the *controlled value* of G, denoted by $cv(G)$. It is known [D&B, Chapter 10] that $v(G) \geq cv(G)$. The controlled value can often be computed much more quickly than the actual value. It also turns out that under certain conditions the two are known to be equal. This paper weakens those conditions substantially.

The Formula for Controlled Value

A *joint* is a coin with valence three or four. A coin which is immediately capturable has valence one. Other coins that are not joints have valence two. Let c be the number of coins in G, and let j be the number of joints. No matter what the valence of the ground is, we treat it as a special joint, even though it is counted in neither c nor j. Let v be the total valence of all joints (including the ground). The *excess valence* is $v - 2j$.

Recall from [D&B, Chapter 10] that the controlled value is given by the following formula:

$$\text{If } p > \tfrac{1}{4}, \text{ then } cv(G) = 8 + c + 4j - 2v - 8p,$$

where p is the maximum weighted number of node-disjoint loops in G, obtained by counting each loop as

1 if it excludes the ground,

$\frac{1}{2}$ if it includes the ground and at least 4 other nodes,

$\frac{1}{4}$ if it includes the ground and 3 other nodes.

Since the ground is a single uncapturable node, and independent chains are viewed as loops through the ground, at most one chain can be included in p. So $p = \frac{1}{4}$ only in the degenerate case in which G consists entirely of independent chains of length 3.

Overview of Conditions for $v(G) = cv(G)$

Let K be the subgraph of G consisting of the p node-disjoint loops. In [WW, pp. 543–544] it was shown that that Left minimizes Right's score by playing in such a way that all of the joints in each of these loops is eliminated before the loop is played. One easy way for Left to accomplish this is to play all other chains before playing K. This works if $cv(K)$ is sufficiently large. In particular, it is easy to see that

$$\text{if } cv(K) \geq c/2, \text{ then } v(G) = cv(G).$$

Here K is the subgraph of G which remains as the final stage of the loony endgame. The main result of the present paper concerns another latestage endgame subgraph, K, as well as another subgraph, H, which represents the position after a relatively short opening phase. We show that $v(G) = cv(G)$ under appropriate restrictions on H and K. The major improvement is that the former condition

$$cv(K) \geq c/2$$

is now replaced by the much weaker constraint

$$cv(K) \geq 10.$$

"Very Long" Defined

A *very long loop* is a loop containing at least 8 nodes. A *very long chain* is a chain containing at least 5 nodes. Loops of length at least 4 but less than 8, and chains of length 3 or 4 are called *moderate*. This is a minor but deliberate change from [D&B, p. 84], where chains of length 4 were considered *very long*.

We will see that Left has a rather powerful strategy which postpones playing all very long loops and very long chains until very near the end of the game, after she has extracted all of the profits she can get from playing 3-chains and moderate loops. But moderate loops can be interconnected with the remainder of the graph in ways which require Left to play with considerable care. A few of the simpler such configurations, including those shown in Figure 1, appeared in [D&B].

We now undertake a more thorough investigation of those subgraphs of a loony endgame which contain the moderate loops.

Theorem. *In a loony endgame position, a moderate loop can include at most one joint.*

Proof. If it had two or more joints A and B, these joints would partition the loop into two sets, one running clockwise from A to B, and another running clockwise from B to A. Since the entire moderate loop has at most 7 nodes, including A and B, then there are only 5 other nodes, and so at least one of these proper

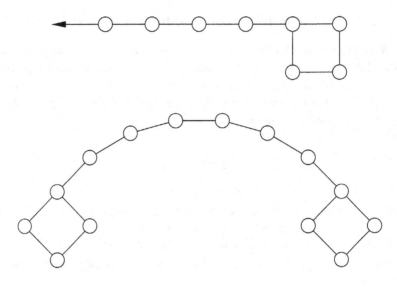

Figure 1. A dipper and earmuffs.

chains has at most 2 nodes. That means it is short and so the position is not loony.   □

Arbitrary strings-and-coins graphs might have joints with valence 5 or more, but such graphs lie beyond the scope of our present study. Since graphs arising from dots-and-boxes positions cannot have valence more than 4,

we henceforth assume that every joint has valence 3 or 4.

Definition. A pair of loops which each has only one joint, which they share, is called a *twin*.

A twin is always best-played as if it were a dipper. The loop of the dipper is the longer loop of the twin, and the handle of the dipper is the shorter loop of the twin.

Definition. A moderate loop with a joint, that is not part of a twin, is called a *trinket*. A trinket and the chain or chains connected to it form part of a *bracelet*. A bracelet contains only a single path of chains, each internal joint of which connects to a trinket. The bracelet may also include a chain at either of its ends which adjoins only one trinket; such a chain is called a *tag*.

In a loony endgame graph whose joints have valence at most 4, there are only four possible kinds of environments of moderate loops:

(1) isolated loops
(2) twins
(3) bracelets

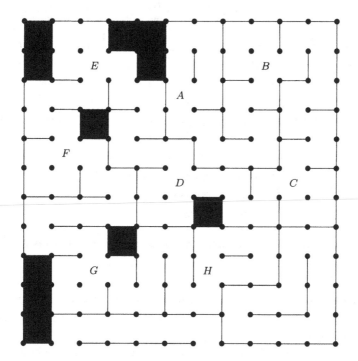

Figure 2. A loony endgame position. All 11 shaded boxes have been taken by
L.

It is possible for a bracelet to be cyclic, in the sense that the path between its
chains forms a cycle. A cyclic bracelet may be isolated, or it can have one joint
connecting it to other parts of the graph. In the latter case, the chains of the
bracelet which adjoin that joint are regarded as the bracelet's tags. If a cyclic
bracelet had two or more joints without trinkets, it would be regarded as two
separate bracelets, each of whose tags happened to share a joint with a tag of
another bracelet.

The loony endgame position of Figure 2 includes a bracelet, a 12-loop, and a
6-chain. The bracelet includes 8 joints, labeled A, B, C, D, E, F, G, and H.
This bracelet has no tags. The lengths of its chains and trinkets are labeled in
the diagram of Figure 3. The trinkets at A and H are 6-loops; all of the other
trinkets are 4-loops. The negative controlled value of each chain and trinket is
shown in this table:

$AB : -5$ $BC : -6$ $CD : -1$ $DE : -4$ $EF : -1$ $FG : -6$ $GH : -5$

$A : +2$ $B : +4$ $C : +4$ $D : +4$ $E : +4$ $F : +4$ $G : +4$ $H : +2$

In total, Left should enjoy a net gain of 28 points on the trinkets, but at a net
cost of 28 points on the bracelet's chains. So, altogether, the controlled value
of the entire bracelet is 0. The controlled value of the 12-loop in the southeast
corner of Figure 2 is $8 - 4 = 4$, and when the 6-chain along the bottom of
the figure is played on the very last turn, Right will take it all. This yields a

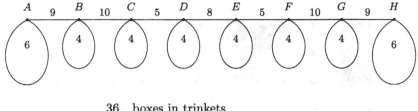

	36	boxes in trinkets
	+ 56	boxes in chains within the bracelet
	92	boxes in bracelet
	11	already taken
	12	loop in southeast corner
	+ 6	chain at bottom
	121	total boxes on the board

Figure 3. The bracelet corresponding to the position of Figure 2.

controlled value of 10 points for the entire position shown in Figure 2. Since Left is 11 points ahead, she can win unless at some point Right can afford to relinquish control.

A naive strategy for Left is to offer all of the chains within the bracelet before offering any of the trinkets. It has long been known that this naive Left strategy is optimal when the rest of the game includes such a large number of nodes on very long loops and very long chains that the controlled value is very big. However, in the present problem, there are 92 points within the bracelet, and the controlled value of the independent chains and loops elsewhere is only 10, which is far too small. The naive strategy won't work for the problem of Figure 2, because at some point along the way, Right will gain a lead that will enable him to win by relinquishing control.

Let N denote Left's net score, that is, the number of boxes Left has taken minus the number of boxes Right has taken. Assume (for now) that Right must remain in control. Suppose Left offers a 3-chain. We know that Right takes one box and declines two, resulting in a net profit of one box for Left. Let's take a closer look, though, at the changes in N move-by-move. Left offers the 3-chain and there is no change in N. Right takes one box and N decreases by one. Right declines the last two boxes with his next move and there is no change in N. Finally, Left takes these two boxes at once and N increases by two.

We know that a 3-chain results in a net increase in N of one, but it is not an immediate gain for Left. As can be seen in Figure 4, when Left offers a 3-chain, N first dips by one and then climbs by two.

More examples of how N changes move-by-move are given in Figure 5. When Left offers a 4-loop, N increases by four and never dips. When she offers a 5-chain, N eventually decreases by one but only after partially rebounding from an earlier dip of three.

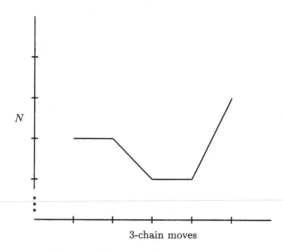

Figure 4. Move-by-move graph of 3-chain.

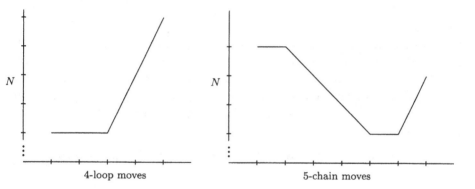

Figure 5. Move-by-move graphs of 4-loop and 5-chain.

The following flight analogy is useful when discussing Left's net score: let N be the altitude of a plane and let the move-by-move graph represent the plane's path over time. The plane gains altitude with little or no dipping when Left offers moderate loops and 3-chains. It loses altitude when she offers long loops and chains. More complex pieces can result in small gains in altitude with big dips. Other pieces, such as a 4-chain, might cause dips but not change the cruising altitude.

Recall that the controlled value, $cv(G)$, of a loony endgame G is the value of the game to Right if he remains in control (until the last or perhaps penultimate move). Since it is Right's choice whether or not to stay in control, the actual value of the game, $v(G)$, is always at least $cv(G)$. Right would have to concede control at some point in order to do better than $cv(G)$. Therefore,

if Left can force Right to stay in control, then $cv(G) = v(G)$.

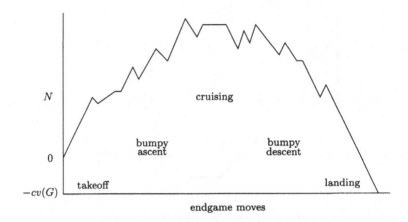

Figure 6. Successful flight for Left, with N measured relative to takeoff, $cv(G)$ above ground level.

Returning to our flight analogy, let the ground be at $N = -cv(G)$. If Right concedes control when $N = n > -cv(G)$, then if Left stays in control for the rest of the game, the plane ends at $n + (n + cv(G))$. Since $2n + cv(G) > -cv(G)$, Right cannot afford to lose control before $N = -cv(G)$. If ever $N < -cv(G)$, then Right will do better by giving up control. In order for Left to force Right to stay in control, the plane must not crash. By that we mean the flight path must maintain an altitude at least as high as the ground at $-cv(G)$.

To summarize:

> If Left's plane can take off, get to a sufficiently high cruising altitude and land, all without crashing, Left can force Right to stay in control.

Figure 6 illustrates a successful flight for Left with labeled flight segments.

We now return to the problem of Figure 2. Another Left strategy is to play through the bracelet in the following order: $AB, A, BC, B, CD, C, DE, D, EF,$ E, FG, F, GH, G, H. (Here each pair of letters refers to the chain between the respective joints, and each single letter refers to the trinket at that joint, which becomes a loop before it is played.) This is a special case of the *direct* strategy that Left can use to play any bracelet: play the chains in order from one end to the other, but intersperse these chain plays with plays on the trinkets (= loops), playing each loop as soon as it becomes detached.

Figure 7. Bumpty Direct Flight from AB to H, for the problem of Figure 2, shown move by move (refer to table on page 321).

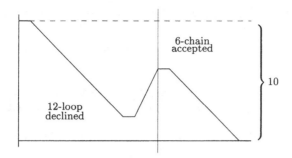

6-chain accepted

12-loop declined

10

Figure 8. A smooth landing.

Suppose Left plays the direct strategy on the bracelet of Figure 3, and also suppose that Right elects to stay in control by declining each piece of the bracelet. Then Left's net score, move by move as the bracelet is played, is shown in Figure 7. During the first move of any chain, the score remains constant as Left concludes her turn by offering that chain. For the next several moves, Left's net score then declines by one point per move as Right takes all but the last two boxes of the chain. Then the score stays constant as Right makes the double-dealing move which concludes his turn. On the last move of the chain, Left's net score increases by two points as he makes the double-cross. Left then concludes her turn by starting another piece of the bracelet with a move that leaves the net score unchanged. If Left has offered a loop, then after taking all but four of its boxes, Right concludes his turn with a move that offers four boxes at the end of the loop with a double double-deal. Left then finishes the play on the loop with two moves *each* of which increases Left's score by two points, after which Left begins play on another piece of the bracelet.

Portions of the flight path of Figure 7 which correspond to separate pieces of the bracelet are delineated by vertical lines. Notice that the relative score at each of these transitions between pieces is the same as one gets from the partial sums of the turn-by-turn analysis of the table on page 321:

$$-5 + 2 - 6 + 4 - 1 + 4 - 4 + 4 - 1 + 4 - 6 + 4 - 5 + 4 + 2 = 0.$$

Altogether, if Right stays in control while Left plays the bracelet in this way, then Left's relative net score at the end of the bracelet is the same as it was at the beginning. In Figure 2, this would mean that when the play of the bracelet was concluded, Left would again be ahead by 11 points, just as she was when the play of the bracelet began. Left would then go on to win the game by playing out the *landing* plan shown in Figure ??.

However, an astute Right will not play this way. Measured relative to the ground level, the altitude at the start and end of the bracelet's flight path is 10, but the altitude at the lowest point within BC is -1. **At this point the plane has crashed!** Right can thwart Left's "direct flight" strategy by accepting BC

rather than declining it. In the present example, that enables Right to win the game.

So Left's "direct flight" plan for playing the bracelet as indicated in Figure 7 is too bumpy. Let's be more specific:

Definition. A *k-point bump* in a flight is a descent of k followed by an ascent of k. A *k-point blip* is an ascent of k followed by a descent of k.

The flight plan of Figure 7 has an 11-point bump, beginning at the start of AB and ending one move prior to the end of E.

For some bracelets, the direct flight plan in the reverse direction can be more or less bumpy than the direct flight plan in the forward direction. However, as seen in Figure 3, this particular bracelet has palindrome symmetry, so the direct flight plan has the same bumpiness in either direction.

Extracting a Profitable Subbracelet

In the big bump of the flight plan of Figure 7, from the lowest point near the end of the segment BC, to the highest point at the beginning of FG, there is a net 12-point rise. This 12-point rise includes all of the segments B, CD, C, DE, D, EF, and E. Omitting the first two loops (B and C), we find the sequence CD, DE, D, EF, E with a net gain of 2. *These segments form a subbracelet with tags at both end.* Evidently, instead of playing in the *direct* order shown in Figure 7, Left could instead play this subbracelet first, in this order: CD, DE, D, EF, E. This subbracelet yields Left a gain of 2 and a maximum bump of 7.

So to win the game shown in Figure 7, Left should begin by playing the profitable subbracelet CD, DE, D, EF, and E. Since this flight plan stays above ground, Right does better to decline all five of those pieces than he can do by accepting any of them.

The play of the profitable subbracelet increases Left's lead from 11 points to 13. Left can then play each of the two residual bracelets, and then land safely as in Figure ??.

This example is readily generalized to the following theorem:

Theorem. *If the direct flight plan of a bracelet includes a bump of 11 or more, than the bracelet includes a playable profitable proper sub-bracelet.*

Definition. To be *playable*, a subbracelet must either have tags at both ends, or end with a trinket that is also an end of the original bracelet.

Note: The place to find this profitable subbracelet is evidently within the ascending part of the bump.

Proof. Except at the beginning (which might be either one or two consecutive chains) and the end (which might be either one or two consecutive loops), any bracelet's direct flight plan alternates between chains and loops. Each chain is a net descent, and each loop is a net ascent. Since we have assumed (without loss of generality) that every chain within the bracelet is very long, we know that every chain ends at a lower altitude than it began. So there is no loss of generality in assuming that the bump ends within a loop, which we keep as the last loop of the subbracelet. The bump's ascent must have begun from an elevation at least 11 points lower. So, starting from the beginning of this ascent, let us purge the next two loops. (If the ascent starts within a loop, we include that loop as one of the two which we purge.) The maximum ascent of the purged moves can be at most ten: 4 within each of the two purged loops, and possibly another 2 within the preceding chain within which the ascent might have begun. So the unpurged portion of the ascent still includes an ascent of at least one point. It remains to verify that these segments constitute a subbracelet with tags and both ends. In the direct flight plan, the initial tag of the subbracelet appeared between the first two loops, and the final tag appeared just before the final loop. The purged direct flight plan can be used as the direct flight plan for the subbracelet. □

Corollary 1. *If a bracelet includes no playable profitable subbracelets, its direct flight plan includes no bumps greater than 10.*

Corollary 2. *A minimal profitable sub-bracelet includes no bumps greater than 10.*

These corollaries will ensure the success of the following algorithm:

Algorithm (finding a smooth flight plan through any configuration of bracelets).

1. If any bracelet contains a profitable sub-bracelet, play a minimal profitable sub-bracelet.
2. If no bracelet contains any profitable sub-bracelet(s), proceed with a direct flight through any one of the bracelets.

Figure ?? shows a situation in which three identical bracelets have tags which share a common joint. Initially, each of these three bracelets has a profitable subbracelet. However, after such a profitable subbracelet is played, the profitable subbracelets of the other two disappear into an unprofitable merger.

Nevertheless, it is easily seen that

Theorem. *By using the preceding algorithm, Left can play any set of bracelets in such a way that the overall flight plan incurs no bump greater than 10.*

This 10 is also easily seen to be the best possible: One of the simplest non-degenerate bracelets is a pair of earmuffs with muffs that are 4-loops connected by a headband that is a chain of length 12 or more. The only feasible flight plan for this bracelet has a bump of 10.

So we are now ready to state our main result, which includes one more (final) term of technical jargon.

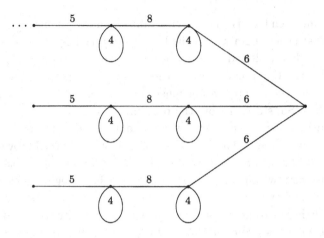

Figure 9. Interconnected bracelets.

Definition. A subgraph H of a graph G is said to be *compatible* if there is a maximal set of node-disjoint loops in H which is a subset of a maximal set of node-disjoint loops in G.

Theorem. *Let G be a loony endgame, which contains a compatible subgraph H, which contains a compatible subgraph K, such that*

$$cv(G) \geq 0,$$
$$cv(H) \geq 10,$$
$$cv(K) \geq 10,$$

and in which K contains no moderate loops or moderate chains, and where Left has a flight plan that ascends from G to H without crashing.
    Then $v(G) = cv(G)$.

DISCUSSION: In the example of Figure 2, $G = H =$ initial position, and K consists only of the 12-loop in the southeast and the 6-chain at the bottom.

In a typical general case, isolated loops, profitable twins, and some moderate chains can be played on the ascent en route from G to H. The route from H to K plays all of the bracelets. K contains all of the very long loops counted in the set which defined the controlled value, plus all very long chains not within bracelets.

Comments

Comment 1. For some graphs G Left can succeed by leaving some portions of bracelets within K. All that is really required of K is that $cv(K)$ be large enough to remain above the bumpiness from H to K, and that a safe landing from K be possible. If the graph contains unprofitable bracelets which aren't very bumpy (as surely happens, for example, if all of their trinkets are 6-loops

Before: After:

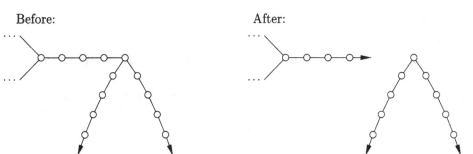

Figure 10. Detaching.

rather than 4-loops), then some such bracelets might be included in the early portion of the landing from K.

Comment 2. Ascending from G to H is usually a simple matter of playing isolated loops and moderate chains. However, if there aren't enough such profitable opportunities available, then some profitable but not-too-bumpy subbracelets might also be included in the latter portions of this ascent.

Comment 3. In general, deciding which 3-chains to play in order to ascend from G to H can be challenging. The difficulty is that playing such a chain can eliminate a joint, fusing other chains to possibly eliminate other 3-chains, or possibly extend the tag of some bracelet in a way which makes a subbracelet unprofitable.

Comment 4. In general, Left begins by finding a maximal set of node-disjoint loops in G. Although a slight (nonplanar) generalization of this problem is NP-hard, in practice it is easily solved by hand for any loony dots-and-boxes positions that can be constructed on boards of sizes up to at least 11×11. The grounded loop and the other very long loops in such a position will be played last, so all other chains connected to them can effectively be detached and viewed as terminating at the ground instead of at joints which are parts of these very long loops.

If there were any loop whose nodes were disjoint from these very long loops and the bracelets, then the supposedly maximal set of loops could have been increased, so it is apparent that after the very long loops and bracelets have been detached, the remaining graph forms a disjoint set of trees. If there are few or no isolated moderate loops or twins, Left may be eager to play as many moderate chains as possible. Within each tree, she can find a maximal set of moderate chains as follows:

Start: If there is a moderate chain which has one end grounded, play it.

If some joint has two or more chains of length at least 4 connecting it directly to the ground, detach all other chains from this joint, and return to start.

(The detachment is illustrated in Figure ??.)

End: When this point is reached, the tree contains no joints adjacent to the ground, so it must be empty.

However, this approach may be ill advised when bracelets are connected to these trees. The problem is that if a tag of a bracelet terminates at the same joint as a moderate chain of the tree, then playing the moderate chain might fuse the tag of the bracelet into a longer chain, thereby eliminating a profitable subbracelet that might have contributed more to the ascent then the moderate chain Left wanted to play.

Comment 5. When Dots-and-Boxes games are played on the 25-square 5 × 5 board, only rather short bracelets such as the dipper and earmuffs of Figure 1 appear, and even these are relatively rare. The most common loony endgames are merely sums of chains and loops of assorted lengths. But even a sum of chains and loops can be fairly complicated. An extensive study of such cases appears in Scott [2000].

References

[WW] BERLEKAMP, E. R.; CONWAY, J. H.; GUY, R. K. *Winning Ways for Your Mathematical Plays* (two volumes). Academic Press, London, 1982.

[Nowakowski] NOWAKOWSKI, R. "...Welter's game, sylver coinage, dots-and-boxes, ...", pp. 155–182 in *Combinatorial Games*, Proc. Symp. Appl. Math **43**, edited by Richard K. Guy. AMS, Providence, 1991

[D&B] BERLEKAMP, E. R. *The-dots-and boxes games: sophisticated child's play*, A K Peters, Natick (MA), 2000.

[Scott] SCOTT, K. *Loony endgames in dots & boxes*, Master's thesis, University of California, Berkeley, 2000.

ELWYN BERLEKAMP
MATHEMATICS DEPARTMENT
UNIVERSITY OF CALIFORNIA
BERKELEY, CA 94720-3840
    berlek@math.berkeley.edu

KATHERINE SCOTT

More Games of No Chance
MSRI Publications
Volume **42**, 2002

1×n Konane: A Summary of Results

ALICE CHAN AND ALICE TSAI

ABSTRACT. We look at $1 \times n$ Konane positions consisting of three solid patterns each separated by a single space and present forumulas for the values of certain positions.

Introduction

Combinatorial game theory [Berlekamp et al. 1982, Conway 1976] has discovered a fascinating array of mathematical structures that provide explicit winning strategies for many positions in a wide variety of games. The theory has been most successful on those games that tend to decompose into sums of smaller games. After assigning a game-theoretic mathematical value to each of the summands, these values are added to determine the value of the entire position. This approach has provided powerful new insights into a wide range of popular games, including Go [Berlekamp and Wolfe 1994], Dots-and-Boxes [Berlekamp 2000a], and even certain endgames in Chess [Elkies 1996].

Some positions on two-dimensional board games breakup into one-dimensional components, and the values of these components often have their own interesting structures. Examples include Blockbusting [Berlekamp 1988], Toads and Frogs [Erickson 1996] and Amazons [Berlekamp 2000b]. The same is true for Konane as suggested by previous analyses of one-dimensional positions [Ernst 1995 and Scott 1999].

Konane is an ancient Hawaiian game similar to checkers. It is played on an 18×18 board with black and white stones placed in an alternating fashion so that no two stones of the same color are in adjacent squares. Two adjacent pieces adjacent to the center of the board are removed to begin the game. A player moves by taking one of his stones and jumping, in the horizontal or vertical direction, over an adjacent opposing stone into an empty square. The jumped stone is removed. A player can make multiple jumps on his turn but cannot change direction mid turn. The first player who cannot make a move loses.

Our goal was to find values for all possible $1 \times n$ Konane positions that consist of three solid patterns each separated by a single space, which we call an "almost solid pattern with two spaces". We assumed that the positions were far away from the edges of the board so that there was no interference.

Let • represent a black stone, ○ a white stone and · represent an empty square. One example of a solid pattern is ○ • ○ • ○ and an example of an almost solid pattern with two spaces is ○ • · • ○ • · • ○ •. We represent positions with k stones in the first fragment, m stones in the second and n stones in the third with $\mathbf{S}(k, m, n)$. The convention is to start the left end of the position with a white stone.

Like almost solid patterns with one space, in which a segment's parity (odd or even in length) determines the value of a game, almost solid patterns with two spaces exhibit the same trend. Thus we categorize a game by its parity e.g. odd-odd-odd or even-odd-even, etc.

Odd-odd-odd and odd-even-odd games have been completely solved. For even-odd-even games and even-even-even games, we have found patterns and proved some of them. For the rest, we have discovered patterns and some partial proofs [Chan and Tsai 2000]. The value of $\mathbf{S}(k, m, n)$, for most values of the arguments, is a number or a number plus *, a common and well-known infinitesimal. However, for odd even even and even odd odd cases, we encounter more complicated infinitesimals.

Our proofs for the most part depend on the decomposition of games of the form $\mathbf{S}(k, m \cdot, n)$. These are games where there are two solid patterns separated by a single space followed by two spaces and another solid pattern, e.g. $\mathbf{S}(4, 2 \cdot, 3)$ = ○ • ○ • · • ○ · · • ○ •. From the data we have collected on this topic, we propose a new decomposition theorem. Specifically, we conjecture that such games decompose into $\mathbf{S}(k, m) + \mathbf{S}(n)$ except for the cases $\mathbf{S}(2j, 2j \cdot, 2k)$ where j is an odd integer and k is any non-negative integer. We have been able to prove only restricted cases of this conjecture. We have verified it empirically in many other cases, but it is still conceivable that there are some additional exceptions.

Here are some forumlas for $1 \times n$ Konane positions that play a part in our proofs.

Theorem. [Ernst 1995] *Let* $\mathbf{S}(n)$ *represent a solid pattern of stones of length n, beginning with a white stone. Then*

$$\mathbf{S}(2k+1) = k, \qquad \mathbf{S}(2k) = k \cdot *.$$

Theorem. [Scott 1999] *Let* $\mathbf{S}(n, m)$ *represent two solid patterns, of length n and m respectively, starting with a white stone and separated by a single space.*

$$\mathbf{S}(2j+1, 2k+1) = j+k,$$

$$\mathbf{S}(2j+1, 2k) = \begin{cases} k \cdot \uparrow + k \cdot * & \text{if } j = 0, \\ j-k & \text{if } j > k, \\ 2^{j-k-1} & \text{otherwise.} \end{cases}$$

For $j \le k$:
$$S(2j, 2k) = \begin{cases} j+k\cdot* & \text{if } j < k, \\ k & \text{if } j = k \text{ and } k \text{ is even,} \\ (k-1) & \text{if } j = k \text{ and } k \text{ is odd.} \end{cases}$$

The tables of data included in the paper are formatted as follows: for a given game $S(k, m, n)$, the numbers in the leftmost column are the k values, the numbers along the top are the n values and the number in the upper left had corner is the m value. We present a table for each of pattern of parities (mod2) of the arguments.

1. Odd-Odd-Odd $S(2i+1, 2j+1, 2k+1)$

3	1	3	5	7	9	11	13
1	1	2	3	4	5	6	7
3	2	3	4	5	6	7	8
5	3	4	5	6	7	8	9
7	4	5	6	7	8	9	10
9	5	6	7	8	9	10	11
11	6	7	8	9	10	11	12
13	7	8	9	10	11	12	13

5	1	3	5	7	9	11	13
1	2	3	4	5	6	7	8
3	3	4	5	6	7	8	9
5	4	5	6	7	8	9	10
7	5	6	7	8	9	10	11
9	6	7	8	9	10	11	12
11	7	8	9	10	11	12	13
13	8	9	10	11	12	13	14

Table 1. Sample values for odd-odd-odd games.

For all i, j, k, $S(2i+1, 2j+1, 2k+1) = i+j+k$.

White (right) has no move. Black (left) can either move to $S(2i-1, 2j+1, 2k+1)$ when $i > 0$ or to $S(2i+1, 2j+1, 2k-1)$ when $k > 0$. If both $i, k = 0$, black moves to $S(2j-1, 1)$ or $S(1, 2j-1)$ both of which have value $j-1$.

2. Odd-Even-Even $S(2i+1, 2j, 2k)$

4	2	4	6	8	10	12	14
1	1	2	2*	2	2*	2	2*
3	$\frac{5}{4}$	$\frac{9}{4}$	$\frac{5}{2}$	3	3*	3	3*
5	$\frac{3}{2}$	$\frac{5}{2}$	3	4	4*	4	4*
7	2	3	4	5	5*	5	5*
9	3	4	5	6	6*	6	6*
11	4	5	6	7	7*	7	7*
13	5	6	7	8	8*	8	8*

6	2	4	6	8	10	12	14
1	1↑3*	2↑3*	$\frac{5}{2}$	3	3*	3	3*
3	$\frac{9}{8}$	$\frac{17}{8}$	3	4	4*	4	4*
5	$\frac{5}{4}$	$\frac{9}{4}$	$\frac{17}{4}$	$\frac{9}{2}$	5	5	5*
7	$\frac{3}{2}$	$\frac{5}{2}$	$\frac{7}{2}$	$\frac{9}{2}$	5	6	6*
9	2	3	4	5	6	7	7*
11	3	4	5	6	7	8	8*
13	4	5	6	7	8	9	9*

Table 2. Sample values for odd-even-even games.

Conjecture:

$$S(2i+1, 2j, 2k) = \begin{cases} (j+i)+k \cdot * & \text{if } k \geq 2j, \\ k+i-j & \text{if } i > j \text{ and } k < 2j. \end{cases}$$

3. Odd-Odd-Even $S(2i+1, 2j+1, 2k)$

3	0	2	4	6	8	10	12	14
1	1	$\frac{1}{2}$	$\frac{1}{4}$	$\frac{1}{8}$	$\frac{1}{16}$	$\frac{1}{32}$	$\frac{1}{64}$	$\frac{1}{128}$
3	2	1	$\frac{1}{2}$	$\frac{3}{8}$	$\frac{1}{4}$	$\frac{1}{8}$	$\frac{1}{16}$	$\frac{1}{32}$
5	3	2	1	$\frac{3}{4}$	$\frac{1}{2}$	$\frac{1}{2}*$	$\frac{1}{2}$	$\frac{1}{2}*$
7	4	3	2	$\frac{3}{2}$	1	1*	1	1*
9	5	4	3	$\frac{5}{2}$	2	2*	2	2*
11	6	5	4	$\frac{7}{2}$	3	3*	3	3*
13	7	6	5	$\frac{9}{2}$	4	4*	4	4*

5	0	2	4	6	8	10	12	14
1	2	1	$\frac{1}{2}$	$\frac{1}{4}$	$\frac{1}{8}$	$\frac{1}{16}$	$\frac{1}{32}$	$\frac{1}{64}$
3	3	2	1	$\frac{1}{2}$	$\frac{1}{4}$	$\frac{3}{16}$	$\frac{1}{8}$	$\frac{1}{8}*$
5	4	3	2	1	$\frac{1}{2}$	$\frac{3}{8}$	$\frac{1}{4}$	$\frac{1}{4}*$
7	5	4	3	2	1	$\frac{3}{4}$	$\frac{1}{2}$	$\frac{1}{2}*$
9	6	5	4	3	2	$\frac{3}{2}$	1	1*
11	7	6	5	4	3	$\frac{5}{2}$	2	2*
13	8	7	6	5	4	$\frac{7}{2}$	3	3*

Table 3. Sample values for odd-odd-even games.

White has only one move to $S(2i+1, 2j+1, 2k-2)$. When $k > 2j$ black moves to $S(2i-1, 2j+1, 2k)$. When $k \leq 2j$, black moves to $S(2i+1, 2j+1 \bullet \cdot, 2k-2)$ which decomposes into the sum

$$S(2i+1, 2(j+1)) + S(2k-2) = \begin{cases} i-(j+1)+(k-1)\cdot * & \text{if } i > j+1, \\ 2^{i-(j+1)-1}+(k-1)\cdot * & \text{if } i \leq j+1. \end{cases}$$

4. Even-Odd-Even $S(2i, 2j+1, 2k)$

3	2	4	6	8	10	12	14
2	0	$\frac{1}{2}$	$\frac{3}{4}$	$\frac{7}{8}$	$\frac{15}{16}$	$\frac{31}{32}$	$\frac{63}{64}$
4	$\frac{1}{2}$	1	$\frac{3}{2}$	$\frac{7}{4}$	$\frac{15}{8}$	$\frac{31}{16}$	$\frac{63}{32}$
6	$\frac{3}{4}$	$\frac{3}{2}$	2	2*	2	2*	2
8	$\frac{7}{8}$	$\frac{7}{4}$	2*	2	2*	2	2*
10	$\frac{15}{16}$	$\frac{15}{8}$	2	2*	2	2*	2
12	$\frac{31}{32}$	$\frac{31}{16}$	2*	2	2*	2	2*
14	$\frac{63}{64}$	$\frac{63}{32}$	2	2*	2	2*	2

5	2	4	6	8	10	12	14
2	0	$\frac{1}{2}$	$\frac{3}{4}$	$\frac{7}{8}$	$\frac{15}{16}$	$\frac{31}{32}$	$\frac{63}{64}$
4	$\frac{1}{2}$	1	$\frac{3}{2}$	$\frac{7}{4}$	$\frac{15}{8}$	$\frac{31}{16}$	$\frac{63}{32}$
6	$\frac{3}{4}$	$\frac{3}{2}$	2	$\frac{5}{2}$	$\frac{5}{2}*$	$\frac{5}{2}$	$\frac{5}{2}*$
8	$\frac{7}{8}$	$\frac{7}{4}$	$\frac{5}{2}$	3	3*	3	3*
10	$\frac{15}{16}$	$\frac{15}{8}$	$\frac{5}{2}*$	3*	3	3*	3
12	$\frac{31}{32}$	$\frac{31}{16}$	$\frac{5}{2}$	3	3*	3	3*
14	$\frac{63}{64}$	$\frac{63}{32}$	$\frac{5}{2}*$	3*	3	3*	3

Table 4. Sample values for even-odd-even games.

7	2	4	6	8	10	12	14
2	-1	0	$\frac{1}{2}$	$\frac{3}{4}$	$\frac{7}{8}$	$\frac{15}{16}$	$\frac{31}{32}$
4	0	1	$\frac{3}{2}$	$\frac{7}{4}$	$\frac{15}{8}$	$\frac{31}{16}$	$\frac{63}{32}$
6	$\frac{1}{2}$	$\frac{3}{2}$	2	$\frac{5}{2}$	$\frac{11}{4}$	$\frac{23}{8}$	$\frac{47}{16}$
8	$\frac{3}{4}$	$\frac{7}{4}$	$\frac{5}{2}$	3	$\frac{7}{2}$	$\frac{15}{4}$	$\frac{31}{8}$
10	$\frac{7}{8}$	$\frac{15}{8}$	$\frac{11}{4}$	$\frac{7}{2}$	4	4*	4
12	$\frac{15}{16}$	$\frac{31}{16}$	$\frac{23}{8}$	$\frac{15}{4}$	4*	4	4*
14	$\frac{31}{32}$	$\frac{63}{32}$	$\frac{47}{16}$	$\frac{31}{8}$	4	4*	4

Table 4. Sample values for even-odd-even games (continued).

If $i, k \geq j$, $\mathbf{S}(2i, 2j+1, 2k) = (j+1) + (k+i) \cdot *$

Since $\mathbf{S}(2i, 2j, 2k) = -\mathbf{S}(2k, 2j, 2i)$, we'll assume that $k \geq i$. Black's best move is to $\mathbf{S}(2i, 2j+1, 2k-2)$. White moves to $\mathbf{S}(2i \circ \cdot, 2j-1, 2k)$ if $i \leq (j+1)$ and to $\mathbf{S}(2(i-1), \cdot \circ 2j, 2k)$ otherwise. If $k \neq (j-1)$ or j is not even, $\mathbf{S}(2i \circ \cdot, 2j-1, 2k)$ decomposes into

$$\mathbf{S}(2i+1) + \mathbf{S}(2j-1, 2k) = \begin{cases} i + k \cdot \downarrow + k \cdot * & \text{if } j = 2, \\ i - (j-1-k) & \text{if } (j-1) > k, \\ i - 2^{(j-1)-k-1} & \text{if } (j-1) \leq k. \end{cases}$$

If $k \neq (j+1)$ or k not odd, $\mathbf{S}(2(i-1), \cdot \circ 2j+1, 2k)$ decomposes into

$$\mathbf{S}(2(i-1)) + \mathbf{S}(2(j+1), 2k) = (i-1) \cdot *\!\!+ (j+1) + k \cdot *\!\!= (j+1) + (k+i-1) \cdot *$$

In the case where j is even (ex. $\mathbf{S}(2i, 5, 2k)$), the values are slightly different when $i = (j+1)$ since the right and left followers of the game do not decompose.

5. Odd-Even-Odd $\mathbf{S}(2i+1, 2j, 2k+1)$

2	1	3	5	7	9	11	13
1	*2	$-\frac{1}{2}$	-1	-2	-3	-4	-5
3	$\frac{1}{2}$	0	$-\frac{1}{2}$	-1	-2	-3	-4
5	1	$\frac{1}{2}$	0	$-\frac{1}{2}$	-1	-2	-3
7	2	1	$\frac{1}{2}$	0	$-\frac{1}{2}$	-1	-2
9	3	2	1	$\frac{1}{2}$	0	$-\frac{1}{2}$	-1
11	4	3	2	1	$\frac{1}{2}$	0	$-\frac{1}{2}$
13	5	4	3	2	1	$\frac{1}{2}$	0

Table 5. Sample values for odd-even-odd games.

4	1	3	5	7	9	11	13
1	0	$-\frac{1}{4}$	$-\frac{1}{2}$	-1	-2	-3	-4
3	$\frac{1}{4}$	0	$-\frac{1}{4}$	$-\frac{1}{2}$	-1	-2	-3
5	$\frac{1}{2}$	$\frac{1}{4}$	0	$-\frac{1}{4}$	$-\frac{1}{2}$	-1	-2
7	1	$\frac{1}{2}$	$\frac{1}{4}$	0	$-\frac{1}{4}$	$-\frac{1}{2}$	-1
9	2	1	$\frac{1}{2}$	$\frac{1}{4}$	0	$-\frac{1}{4}$	$-\frac{1}{2}$
11	3	2	1	$\frac{1}{2}$	$\frac{1}{4}$	0	$-\frac{1}{4}$
13	4	3	2	1	$\frac{1}{2}$	$\frac{1}{4}$	0

Table 6. Sample values for oddevenodd games (continued).

Assume $i \geq k$ since $S(2i+1, 2j, 2k+1) = -S(2k+1, 2j, 2i+1)$.

$$S(2i+1, 2j, 2k+1) = \begin{cases} 0 & \text{if } i = k \text{ (except in the case } S(1,2,1) = *2), \\ -2^{-k-1} & \text{for } 2k+1 = 2i+3 \text{ to } 2k+1 = 2i+1+j, \\ -k & \text{for } 2k+1 > 2i+1+j. \end{cases}$$

When $i = k = 0$: if j is even then $S(1, 2j, 1)$ has value $\{\ S(1, 2j-2\cdot, 2)\ |\ S(2, \cdot 2j-2, 1)\ \}$ otherwise the canonical value is $\{\ S(1, 2j-2) \mid S(2j-1, 1)\ \}$

When $i = 0$ and $k \neq 0$, $S(1, 2j, 2k+1)$ has value

$$\{\ S(2j-2, 2k+1) \mid S(1, 2j, 2k-1)\ \}.$$

If $j = 0$, white has no options.

When $i \neq 0$ and $k = 0$, $S(2i+1, 2j, 1)$ has value

$$\{\ S(2i-1, 2j, 1) \mid S(2i+1, 2j-2)\ \}.$$

If $j = 0$, black has no options.

When $i, k \neq 0$ $S(2i+1, 2j, 2k+1)$ has value

$$\{\ S(2i-1, 2j, 2k+1) \mid S(2i+1, 2j, 2k-1)\ \}.$$

6. Even-Even-Even $S(2i, 2j, 2k)$

For $j > 1$, $j \equiv 2 \bmod 4$ and $k \geq i$:

$$S(2i, 2j, 2k) = \begin{cases} (i-k)+j\cdot* & \text{if } i, k < j, \\ (j-k-1) & \text{if } k = j \text{ and } i < j, \\ (j-k)+k\cdot* & \text{if either } k \text{ or } i \text{ (but not both) is greater than } j. \end{cases}$$

Conjecture: For $i, k > j$, $S(2i, 2j, 2k) = S(2(i-j), 2, 2(k-j))$.

For $j > 2$, $j \equiv 0 \bmod 4$ and $4k \geq i$:

$$S(2i, 2j, 2k) = \begin{cases} i-k & \text{if } i, k < j, \\ (j-k)+k\cdot* & \text{if } i < j \text{ and } k > j. \end{cases}$$

2	2	4	6	8	10	12	14
2	*	0	$-\frac{1}{4}$	$-\frac{1}{2}$	$-\frac{1}{2}*$	$-\frac{1}{2}$	$-\frac{1}{2}*$
4	0	*	0	$-\frac{1}{4}$	$-\frac{1}{4}*$	$-\frac{1}{4}$	$-\frac{1}{4}*$
6	$\frac{1}{4}$	0	*	0	$-\frac{1}{16}$	$-\frac{1}{8}$	$-\frac{1}{8}*$
8	$\frac{1}{2}$	$\frac{1}{4}$	0	*	0	$-\frac{1}{16}$	$-\frac{1}{16}*$
10	$\frac{1}{2}*$	$\frac{1}{4}*$	$\frac{1}{16}$	0	*	0	$\frac{1}{64}$
12	$\frac{1}{2}$	$\frac{1}{4}$	$\frac{1}{8}$	$\frac{1}{16}$	0	*	0
14	$\frac{1}{2}*$	$\frac{1}{4}*$	$\frac{1}{8}*$	$\frac{1}{16}*$	$\frac{1}{16}$	0	*

6	2	4	6	8	10	12	14
2	*	$-1*$	-1	-2	$-2*$	-2	$-2*$
4	$1*$	*	0	-1	$-1*$	-1	$-1*$
6	1	0	*	0	$-\frac{1}{4}$	$-\frac{1}{2}$	$-\frac{1}{2}*$
8	2	1	0	*	0	$-\frac{1}{4}$	$-\frac{1}{4}*$
10	$2*$	$1*$	$\frac{1}{4}$	0	*	0	$-\frac{1}{16}$
12	2	1	$\frac{1}{2}$	$\frac{1}{4}$	0	*	0
14	$2*$	$1*$	$\frac{1}{2}*$	$\frac{1}{4}*$	$\frac{1}{16}$	0	*

Table 7. Sample values for even-even-even games. $S(2i, 2j, 2k)$ where $2j \equiv 2$ mod 4.

Conjecture: If $i, k > j$, $S(2i, 2j, 2k) = S(2(i-j), 4, 2(k-j))$.

Assume that $k \geq i$ since $S(2i, 2j, 2k) = -S(2k, 2j, 2i)$. In general, if $i < j$ and $k < j$, $S(2i, 2j, 2k)$ has canonical value $\{ S(2i, 2j-2, \cdot \bullet 2k) \mid S(2i \circ \cdot, 2(j-1), 2k)\}$.

$S(2i, 2(j-1), \cdot \bullet 2k)$ decomposes into the sum

$$S(2i, 2(j-1)) - S(2k+1) = i + (j-1) \cdot \!*\!- k$$

If j is not even or $k \neq (j-1)$, $S(2i \circ \cdot, 2(j\text{-}1), 2k)$ decomposes into the sum

$$S(2i+1) + S(2(j-1), 2k) = i - k + (j-1) \cdot *$$

So the value of the game is

$$\{ i-k+(j-1) \cdot *\mid (i-k)+(j-1) \cdot *\} = (i-k)+j \cdot *$$

If $i < j < k$, black moves to $S(2(i-1), 2j, 2k)$ and white moves to $S(2i, 2j \bullet \cdot, 2(k-1))$.

4	2	4	6	8	10	12	14
2	0	-1	$-1*$	-1	$-1*$	-1	$-1*$
4	1	0	$-\frac{1}{4}$	$-\frac{1}{2}$	$-\frac{1}{2}*$	$-\frac{1}{2}$	$-\frac{1}{2}*$
6	$1*$	$\frac{1}{4}$	0	$-\frac{1}{4}$	$-\frac{1}{4}*$	$-\frac{1}{4}$	$-\frac{1}{4}*$
8	1	$\frac{1}{2}$	$\frac{1}{4}$	0	$-\frac{1}{16}$	$-\frac{1}{8}$	$-\frac{1}{8}*$
10	$1*$	$\frac{1}{2}*$	$\frac{1}{4}*$	$\frac{1}{16}$	0	$-\frac{1}{16}$	$-\frac{1}{16}*$
12	1	$\frac{1}{2}$	$\frac{1}{4}$	$\frac{1}{8}$	$\frac{1}{16}$	0	$-\frac{1}{16}$
14	$1*$	$\frac{1}{2}*$	$\frac{1}{4}*$	$\frac{1}{8}*$	$\frac{1}{16}*$	$\frac{1}{16}$	0

8	2	4	6	8	10	12	14
2	0	-1	-2	-3	$-3*$	-3	$-3*$
4	1	0	-1	-2	$-2*$	-2	$-2*$
6	2	1	0	-1	$-1*$	-1	$-1*$
8	3	2	1	0	$-\frac{1}{4}$	$-\frac{1}{2}$	$-\frac{1}{2}*$
10	$3*$	$2*$	$1*$	$\frac{1}{4}$	0	$-\frac{1}{4}$	$-\frac{1}{4}*$
12	3	2	1	$\frac{1}{2}$	$\frac{1}{4}$	0	$-\frac{1}{16}$
14	$3*$	$2*$	$1*$	$\frac{1}{2}*$	$\frac{1}{4}*$	$\frac{1}{16}$	0

Table 8. Sample values for even-even-even games. $S(2i, 2j, 2k)$ where $2k \equiv 0$ mod 4.

$S(2i, 2j \bullet \cdot , 2(k-1))$ decomposes into the sum

$$S(2i, 2j+1)+S(2(k-1)) = \begin{cases} i-j+(k-1)\cdot * & \text{if } j > i, \\ -(2^{j-i-1})+(k-1)\cdot * & \text{otherwise.} \end{cases}$$

If $i, j > k$ the canonical value is $\{ S(2i-2, 2j, 2k) \mid S(2i, 2j, 2k-2) \}$.

Conclusion

An open question is to determine precisely which patterns of $S(k, m \cdot , n)$ decompose. That would eliminate the gaps in some of our proofs and help to complete the analysis for all $1 \times n$ games.

References

[Berlekamp et al. 1982] E. R. Berlekamp, J. H. Conway, and R. K. Guy, *Winning Ways For Your Mathematical Plays*, Academic Press, London, 1982.

[Berlekamp 1988] E. R. Berlekamp, "Blockbusting and Domineering", *J. Combinatorial Theory* A49 (1988), 67-116.

[Berlekamp 2000a] E. R. Berlekamp, *The Game of Dots and Boxes: sophisticated child's play*, A K Peters, Wellesley (MA), 2000.

[Berlekamp 2000b] E. R. Berlekamp, "Sums of $Nx2$ Amazons", Institute of Mathematics Statistics "Lecture Notes – Monograph Series" (2000), 35:1-34.

[Berlekamp and Wolfe 1994] E. R. Berlekamp and D. Wolfe, *Mathematical Go: Chilling Gets the Last Point*, A K Peters, Wellesley (MA), 1994.

[Chan and Tsai 2000] A. Chan and A. Tsai, "A Second Look at $1×n$ Konane", UC Berkeley, 2000.

[Conway 1976] J. H. Conway, *On Numbers and Games*, Academic Press, London, 1976.

[Elkies 1996] N. D. Elkies, "On Numbers and Endgames: Combinatorial Game Theory in Chess Endgames", pp. 135–150 in *Games of No Chance* (edited by R. J. Nowakowski), Cambridge University Press, Cambridge, 1996.

[Erickson 1996] J. Erickson, "New Toads and Frogs Results", pp. 299–310 in *Games of No Chance* (edited by R. J. Nowakowski), Cambridge University Press, Cambridge, 1996.

[Ernst 1995] M. D. Ernst, "Playing Konane Mathematically: A Combinatorial Game-Theoretic Analysis", *UMAP Journal* 16:2 (1995), 95–121.

[Scott 1999] K. Scott, "A Look at $1×n$ Konane", UC Berkeley, 1999.

ALICE CHAN
DEPARTMENT OF MATHEMATICS
77 MASSACHUSETTS AVENUE
CAMBRIDGE, MA 02139–4307
UNITED STATES
alicec@math.mit.edu

ALICE TSAI

More Games of No Chance
MSRI Publications
Volume **42**, 2002

One-Dimensional Peg Solitaire, and Duotaire

CRISTOPHER MOORE AND DAVID EPPSTEIN

ABSTRACT. We solve the problem of one-dimensional Peg Solitaire. In particular, we show that the set of configurations that can be reduced to a single peg forms a regular language, and that a linear-time algorithm exists for reducing any configuration to the minimum number of pegs.

We then look at the impartial two-player game, proposed by Ravikumar, where two players take turns making peg moves, and whichever player is left without a move loses. We calculate some simple nim-values and discuss when the game separates into a disjunctive sum of smaller games. In the version where a series of hops can be made in a single move, we show that neither the $\mathcal{P}$-positions nor the $\mathcal{N}$-positions (i.e. wins for the previous or next player) are described by a regular or context-free language.

1. Solitaire

Peg Solitaire is a game for one player. Each move consists of hopping a peg over another one, which is removed. The goal is to reduce the board to a single peg. The best-known forms of the game take place on cross-shaped or triangular boards, and it has been marketed as "Puzzle Pegs" and "Hi-Q." Discussions and various solutions can be found in [1; 2; 3; 4; 5].

In [6], Guy proposes one-dimensional Peg Solitaire as an open problem in the field of combinatorial games. Here we show that the set of solvable configurations forms a regular language, i.e. it can be recognized by a finite-state automaton. In fact, this was already shown in 1991 by Plambeck ([7], Introduction and Ch.5) and appeared as an exercise in a 1974 book of Manna [8]. More generally, B. Ravikumar showed that the set of solvable configurations on rectangular boards of any finite width is regular [9], although finding an explicit grammar seems to be difficult on boards of width greater than 2.

Thus there is little new about this result. However, it seems not to have appeared in print, so here it is.

Theorem 1. *The set of configurations that can be reduced to a single peg is the regular language 0^*L0^*, where*

$$L = 1 + 011 + 110$$
$$+ 11(01)^* \left[00 + 00(11)^+ + (11)^+00 + (11)^*1011 + 1101(11)^* \right] (10)^*11$$
$$+ 11(01)^*(11)^*01 + 10(11)^*(10)^*11. \tag{1-1}$$

Here 1 and 0 indicate a peg and a hole respectively, w^* means 'zero or more repetitions of w,' and $w^+ = ww^*$ means 'one or more repetitions of w.'

Proof. To prove the theorem, we follow Leibnitz [4] in starting with a single peg, which we denote

$$1$$

and playing the game in reverse. The first 'unhop' produces

$$011 \text{ or } 110$$

and the next

$$1101 \text{ or } 1011.$$

(As it turns out, 11 is the only configuration that cannot be reduced to a single peg without using a hole outside the initial set of pegs. Therefore, for all larger configurations we can ignore the 0's on each end.)

We take the second of these as our example. It has two ends, $10\ldots$ and $\ldots 11$. The latter can propagate itself indefinitely by unhopping to the right,

$$1010101011.$$

When the former unhops, two things happen; it becomes an end of the form $11\ldots$ and it leaves behind a space of two adjacent holes,

$$110010101011.$$

Furthermore, this is the only way to create a 00. We can move the 00 to the right by unhopping pegs into it,

$$111111110011.$$

However, since this leaves a solid block of 1's to its left, we cannot move the 00 back to the left. Any attempt to do so reduces it to a single hole,

$$111111101111.$$

Here we are using the fact that if a peg has another peg to its left, it can never unhop to its left. We prove this by induction: assume it is true for pairs of pegs farther left in the configuration. Since adding a peg never helps another peg unhop, we can assume that the two pegs have nothing but holes to their left. Unhopping the leftmost peg then produces 1101, and the original (rightmost) peg is still blocked, this time by a peg which itself cannot move for the same reason.

In fact, there can never be more than one 00, and there is no need to create one more than once, since after creating the first one the only way to create another end of the form 10... or ...01 is to move the 00 all the way through to the other side

$$111111111101$$

and another 00 created on the right end now might as well be the same one.

We can summarize, and say that any configuration with three or more pegs that can be reduced to a single peg can be obtained in reverse from a single peg by going through the following stages, or their mirror image:

1. We start with 1011. By unhopping the rightmost peg, we obtain $10(10)^*11$. If we like, we then
2. Unhop the leftmost peg one or more times, creating a pair of holes and obtaining $11(01)^*00(10)^*11$. We can then
3. Move the 00 to the right (say), obtaining $11(01)^*(11)^*00(10)^*11$. We can stop here, or
4. Move the 00 all the way to the right, obtaining $11(01)^*(11)^*01$, or
5. Fill the pair by unhopping from the left, obtaining $11(01)^*(11)^*1011(10)^*11$.

Equation (1-1) simply states that the set of configurations is the union of all of these plus 1, 011, and 110, with as many additional holes on either side as we like. Then 0^*L0^* is regular since it can be described by a regular expression [10], i.e. a finite expression using the operators $+$ and $*$.   □

Among other things, Theorem 1 allows us to calculate the number of distinct configurations with n pegs, which is

$$N(n) = \begin{cases} 1 & n = 1 \\ 1 & n = 2 \\ 2 & n = 3 \\ 15 - 7n + n^2 & n \geq 4,\ n \text{ even} \\ 16 - 7n + n^2 & n \geq 5,\ n \text{ odd} \end{cases}$$

Here we decline to count 011 and 110 as separate configurations, since many configurations have more than one way to reduce them.

We also have the corollary

Corollary 1. *There is a linear-time strategy for playing Peg Solitaire in one dimension.*

Proof. Our proof of Theorem 1 is constructive in that it tells us how to unhop from a single peg to any feasible configuration. We simply reverse this series of moves to play the game.   □

More generally, a configuration that can be reduced to k pegs must belong to the regular language $(0^*L0^*)^k$, since unhopping cannot interleave the pegs coming from different origins [7]. This leads to the following algorithm:

Theorem 2. *There is a linear-time strategy for reducing any one-dimensional Peg Solitaire configuration to the minimum possible number of pegs.*

Proof. Suppose we are given a string $c_0 c_1 c_2 \dots c_{n-1}$ where each $c_i \in \{0, 1\}$. Let $\mathcal{A}$ be a nondeterministic finite automaton (without ε-transitions) for $0^* L 0^*$, where A is the set of states in $\mathcal{A}$, s is the start state, and T is the set of accepting states. We then construct a directed acyclic graph G as follows: Let the vertices of G consist of all pairs (a, i) where $a \in A$ and $0 \le i \le n$. Draw an arc from (a, i) to $(b, i+1)$ in G whenever $\mathcal{A}$ makes a transition from state a to state b on symbol c_i. Also, draw an arc from (t, i) to (s, i) for any $t \in T$ and any $0 \le i \le n$. Since $|\mathcal{A}| = \mathcal{O}(1)$, $|G| = \mathcal{O}(n)$.

Then any path from $(s, 0)$ to (s, n) in G consists of n arcs of the form (a, i) to $(b, i+1)$, together with some number k of arcs of the form (t, i) to (s, i). Breaking the path into subpaths by removing all but the last arc of this second type corresponds to partitioning the input string into substrings of the form $0^* L 0^*$, so the length of the shortest path from $(s, 0)$ to (s, n) in G is $n + k$, where k is the minimum number of pegs to which the initial configuration can be reduced. Since G is a directed acyclic graph, we can find shortest paths from $(s, 0)$ by scanning the vertices (a, i) in order by i, resolving ties among vertices with equal i by scanning vertices (t, i) (with $t \in T$) earlier than vertex (s, i). When we scan a vertex, we compute its distance to $(s, 0)$ as one plus the minimum distance of any predecessor of the vertex. If the vertex is $(s, 0)$ itself, the distance is zero, and all other vertices $(a, 0)$ have no predecessors and infinite distance.

Thus we can find the optimal strategy for the initial configuration by forming G, computing its shortest path, using the location of the edges from (t, i) to (s, i) to partition the configuration into one-peg subconfigurations, and applying Corollary 1 to each subconfiguration. Since $|G| = \mathcal{O}(n)$, this algorithm runs in linear time. $\qquad \square$

In contrast to these results, Uehara and Iwata [11] showed that in two or more dimensions Peg Solitaire is NP-complete. However, the complexity of finding the minimum number of pegs to which a $k \times n$ configuration can be reduced, for bounded $k > 2$, remains open.

2. Duotaire

Ravikumar [9] has proposed an impartial two-player game, in which players take turns making Peg Solitaire moves, and whoever is left without a move loses. We call this game "Peg Duotaire." While he considered the version where each move consists of a single hop, in the spirit of the game we will start with the "multihop" version where a series of hops with a single peg can be made in a single move.

We recall the definition of the *Grundy number* or *nim-value* G of a position in an impartial game, namely the smallest non-negative integer not appearing among the nim-values of its options [4]. The $\mathcal{P}$-positions, in which the second (Previous) player can win, are those with nim-value zero: any move by the first (Next) player is to a position with a non-zero G, and the second player can then return it to a position with $G = 0$. This continues until we reach a position in which there are no moves, in which case $G = 0$ by definition; then Next is stuck, and Previous wins. Similarly, the $\mathcal{N}$-positions, in which the first player can win, are those for which $G \neq 0$.

The nim-value of a disjunctive sum of games, in which each move consists of a move in the game of the player's choice, is the *nim-sum*, or bitwise exclusive or (binary addition without carrying) of the nim-values of the individual games. We notate this $\oplus$, and for instance $4 \oplus 7 = 5$. Like many games, positions in Peg Duotaire often quickly reduce to a sum of simple positions:

Lemma 1. *In either version of Peg Duotaire, a position of the form* $x\,0(01)^*00\,y$ *is equal to the disjunctive sum of* $x0$ *and* $0y$.

Proof. Any attempt to cross this gap only creates a larger gap of the same form; for instance, a hop on the left end from $110(01)^n00$ yields $0(01)^{n+1}00$. Thus the two games cannot interact.   □

As in the Hawai'ian game of Konane [12], interaction across gaps of size 2 seems to be rare but by no means impossible. For instance, Previous can win from a position of the form $w00w$ by strategy stealing, i.e. copying each of Next's moves, unless Next can change the parity by hopping into the gap. In the multihop case, however, Previous can sometimes recover by hopping into the gap and over the peg Next has placed there:

Lemma 2. *In multihop Peg Duotaire, any palindrome of the form*

$$w\,010010\,w^R, \quad w\,01100110\,w^R, \quad or \quad w\,00(10)^*11100111(01)^*00\,w^R$$

is a $\mathcal{P}$-*position.*

Proof. Previous steals Next's strategy until Next hops into the gap. Previous then hops into the gap and over Next's peg, leaving a position of the form in Lemma 1. The games then separate and Previous can continue stealing Next's strategy, so the nim-value is $G(w0) \oplus G(0w^R) = 0$.

To show that this remains true even if Next tries to hop from $w = v11$, consider the following game:

$v11\,001110011100\,11v^R$	
$v00\,101110011100\,11v^R$	Next hops from the left
$v00\,101110011101\,00v^R$	Previous steals his strategy
$v00\,101001011101\,00v^R$	Next hops into the breach
$v00\,101010000101\,00v^R$	Previous hops twice

w	$G(0^* w 0^*)$
1^n	$\begin{cases} 0 & n \equiv 0 \text{ or } 1 \bmod 4 \\ 1 & n \equiv 2 \text{ or } 3 \bmod 4 \end{cases}$
$11(01)^n$	$n+1$
$111(01)^n$	$n+1$
$11(01)^n 1$	$n \oplus 1$
$\begin{aligned} & 11(01)^n 11 \\ & = 111(01)^n 1 \end{aligned}$	$\begin{cases} 3 & n = 1 \\ 4 & n = 2 \\ 2 & n = 3 \\ n+2 & n \geq 4 \end{cases}$
$111(01)^n 11, \; n > 0$	1
$11011(01)^n$	$(n+1) \oplus 1$
$1011(01)^n 1$	$n+2$
$(10)^m 11(01)^n$	$\max(m,n)+1$

Table 1. Some simple nim-values in multihop Peg Duotaire.

Now v and v^R are separated by two gaps of the form of Lemma 1. Since hopping from w and w^R into 010010 gives 0011001100, and since hopping into this gives 001110011100, and since hopping into 00(10)*11100111(01)*00 gives another word of the same form, we're done. □

Lemma 2 seems to be optimal, since 110111 00 111011 and 01111 00 11110 have nim-values 1 and 2 respectively. Nor does it hold in the single-hop version, since there 1011 00 1101 has nim-value 1.

Note that the more general statement that $G(x\,010010\,y) = G(x0) \oplus G(0y)$ is not true, since countering your opponent's jump into the gap is not always a winning move; for example, $G(1011\,010010\,1011) = 5$ even though $G(10110) \oplus G(01011) = 0$.

In fact, the player who desires an interaction across a 00 has more power here than in Konane, since she can hop into the gap from either or both sides. In Konane, on the other hand, each player can only move stones of their own color, which occur on sites of opposite parity, so that the player desiring an interaction must force the other player to enter the gap from the other side.

Using a combination of experimental math and inductive proof, the reader can confirm the nim-values of the multihop positions shown in Table 1. In these examples we assume there are holes to either side.

In the previous section, we showed that the set of winnable configurations in Peg Solitaire is recognizable by a finite-state automaton, i.e. is a regular language. In contrast to this, for the two-player version we can show the following, at least in the multihop case:

Theorem 3. *In multihop Peg Duotaire, neither the $\mathcal{P}$-positions nor the $\mathcal{N}$-positions are described by a regular or context-free language.*

Proof. Let P be the set of $\mathcal{P}$-positions. Since the nim-value of $011(01)^n0$ is $n+1$, the intersection of P with the regular language

$$L = 011(01)^*00011(01)^*00011(01)^*0$$

is

$$P \cap L = \left\{\, 011\,(01)^i\,00011\,(01)^j\,00011\,(01)^k\,0 \mid i \oplus j \oplus k = 0 \,\right\}$$

To simplify our argument, we run this through a finite-state transducer which the reader can easily construct, giving

$$P' = \left\{\, a^i b^j c^k \mid i \oplus j \oplus k = 0 \text{ and } i,j,k > 0 \,\right\}$$

It is easy to show that P' violates the Pumping Lemma for context-free languages [10] by considering the word $a^i b^j c^k$ where $i = 2^n$, $j = 2^n - 1$, and $k = 2^{n+1} - 1$ where n is sufficiently large. Since regular and context-free languages are closed under finite-state transduction and under intersection with a regular language, neither P' nor P is regular or context-free.

A more general argument applies to both P and the set of $\mathcal{N}$-positions $N = \bar{P}$. We define N' similarly to P'. Now the Parikh mapping, which counts the number of times each symbol appears in a word, sends any context-free language to a semilinear set [13]. This implies that the set

$$S = \{n \in \mathbb{N} \mid a^n b^{2n} c^{3n} \in P'\}$$

is eventually periodic. However, it is easy to see that this is

$$S = \{n \text{ does not have two consecutive 1's in its binary expansion}\}.$$

Suppose S is eventually periodic with period p, and let k be sufficiently large that 2^k is both in the periodic part of S and larger than $3p$. Then $2^k \in S$, but if $p \in S$ then $2^k + 3p \notin S$, while if $p \notin S$ then $2^k + p \notin S$. This gives a contradiction, and since S is not eventually periodic neither is its complement. Thus neither P nor N is regular or context-free. $\square$

We conjecture that Theorem 3 is true in the single-hop case as well. However, we have been unable to find a simple family of positions with arbitrarily large nim-values. The lexicographically first positions of various nim-values, which we found by computer search, are as shown in Table 2.

It is striking that the first positions with nim-values $2n$ and $2n+1$ coincide on fairly large initial substrings; this is most noticeable for $G = 10$ and 11, which coincide for the first 17 symbols. We do not know if this pattern continues; it would be especially interesting if some sub-family of these positions converged to an aperiodic sequence.

w	$G(0w0)$
1	0
11	1
1011	2
110111	3
11010111	4
11011010111	5
10110111001111	6
10110110010111011	7
1101101101101110111	8
1100110110111001101 0111	9
101101100110110110 1110111	10
1011011001101101110 011010111	11

Table 2. Lexicographically first positions of various nim-values.

In any case, as of now it is an open question whether there are positions in single-hop Peg Duotaire with arbitrarily large nim-values. We conjecture that there are, and offer the following conditional result:

Lemma 3. *If there are positions with arbitrarily large nim-values, then the set of $\mathcal{P}$-positions is not described by a regular language.*

Proof. Recall that a language is regular if and only if it has a finite number of equivalence classes, where we define u and v as equivalent if they can be followed by the same suffixes: $uw \in L$ if and only if $vw \in L$. Since $u000w \in P$ if and only if $u0$ and $0w$ have the same nim-value by Lemma 1, there is at least one equivalence class for every nim-value. $\square$

In fact, a computer search for inequivalent initial strings shows that there are at least 225980 equivalence classes for each nim-value. Since we can combine 1, 2, 4, and 8 to get any nim-value between 0 and 15, any deterministic finite automaton that recognizes the $\mathcal{P}$-positions must have at least 3615680 states.

We conjecture that single-hop Peg Duotaire is not described by a context-free language either. Of course, there could still be polynomial-time strategies for playing either or both versions of the one-dimensional game. One approach might be a divide-and-conquer algorithm, based on the fact that a boundary between two sites can be hopped over at most four times:

$$
\begin{array}{c|c}
1111 & 0111 \\
1100 & 1111 \\
1101 & 0011 \\
0000 & 1011 \\
0001 & 0000
\end{array}
$$

In two or more dimensions, it is tempting to think that either or both versions of Peg Duotaire are PSPACE-complete, since Solitaire is NP-complete [11].

Acknowledgements

We thank Elwyn Berlekamp, Aviezri Fraenkel, Michael Lachmann, Molly Rose, B. Sivakumar, and Spootie the Cat for helpful conversations, and the organizers of the 2000 MSRI Workshop on Combinatorial Games.

References

[1] M. Kraitchik, "Peg Solitaire," in *Mathematical Recreations*. W.W. Norton, New York, 1942.

[2] M. Gardner, "Peg Solitaire," in *The Unexpected Hanging and Other Mathematical Diversions*. Simon and Schuster, New York, 1969.

[3] R.W. Gosper, S. Brown, and M. Rayfield, Item 75, and M. Beeler, Item 76, in M. Beeler, R.W. Gosper, and R. Schroeppel, Eds., *HAKMEM*. MIT Artificial Intelligence Laboratory Memo AIM-239 (1972) 28–29.

[4] Chapter 23 in E.R. Berlekamp, J.H. Conway, and R.K. Guy, *Winning Ways*. Academic Press, 1982.

[5] J. Beasley, *The Ins and Outs of Peg Solitaire*. Oxford University Press, 1985.

[6] R.K. Guy, "Unsolved problems in combinatorial games." In R.J. Nowakowski, Ed., *Games of No Chance*. Cambridge University Press, 1998.

[7] E. Chang, S.J. Phillips and J.D. Ullman, "A programming and problem solving seminar." Stanford University Technical Report CS-TR-91-1350, February 1991. Available from http://elib.stanford.edu

[8] Z. Manna, *Mathematical Theory of Computation*. McGraw-Hill, 1974.

[9] B. Ravikumar, "Peg-solitaire, string rewriting systems and finite automata." *Proc. 8th Int. Symp. Algorithms and Computation,* Lecture Notes in Computer Science **1350**, Springer (1997) 233–242.

[10] J.E. Hopcroft and J.D. Ullman, *Introduction to Automata Theory, Languages, and Computation*. Addison-Wesley, 1979.

[11] R. Uehara and S. Iwata, "Generalized Hi-Q is NP-complete." *Trans IEICE 73*.

[12] W.S. Sizer, "Mathematical notions in preliterate societies." *Mathematical Intelligencer* **13** (1991) 53–59.

[13] R.J. Parikh, "On context-free languages." *Journal of the ACM* **13** (1966) 570–581.

CRISTOPHER MOORE
COMPUTER SCIENCE DEPARTMENT
UNIVERSITY OF NEW MEXICO
ALBUQUERQUE NM 87131
and SANTA FE INSTITUTE
1399 HYDE PARK ROAD
SANTA FE NM 87501
UNITED STATES
  moore@cs.unm.edu

DAVID EPPSTEIN
DEPARTMENT OF INFORMATION AND COMPUTER SCIENCE
UNIVERSITY OF CALIFORNIA
IRVINE, IRVINE CA 92697-3425
UNITED STATES
  eppstein@ics.uci.edu

More Games of No Chance
MSRI Publications
Volume **42**, 2002

Phutball Endgames are Hard

ERIK D. DEMAINE, MARTIN L. DEMAINE, AND DAVID EPPSTEIN

ABSTRACT. We show that, in John Conway's board game Phutball (or Philosopher's Football), it is NP-complete to determine whether the current player has a move that immediately wins the game. In contrast, the similar problems of determining whether there is an immediately winning move in checkers, or a move that kings a man, are both solvable in polynomial time.

1. Introduction

John Conway's game Phutball [1–3, 13], also known as Philosopher's Football, starts with a single black stone (the *ball*) placed at the center intersection of a rectangular grid such as a Go board. Two players sit on opposite sides of the board and take turns. On each turn, a player may either place a single white stone (a *man*) on any vacant intersection, or perform a sequence of *jumps*. To jump, the ball must be adjacent to one or more men. It is moved in a straight line (orthogonal or diagonal) to the first vacant intersection beyond the men, and the men so jumped are immediately removed (Figure 1). If a jump is performed, the same player may continue jumping as long as the ball continues to be adjacent to at least one man, or may end the turn at any point. Jumps are not obligatory: one can choose to place a man instead of jumping. The game is over when a jump sequence ends on or over the edge of the board closest to the opponent (the opponent's *goal line*) at which point the player who performed the jumps wins. It is legal for a jump sequence to step onto but not over one's own goal line. One of the interesting properties of Phutball is that any move could be played by either player, the only partiality in the game being the rule for determining the winner.

It is theoretically possible for a Phutball game to return to a previous position, so it may be necessary to add a loop-avoidance rule such as the one in Chess (three repetitions allow a player to claim a draw) or Go (certain repeated

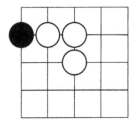 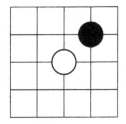

Figure 1. A jump in Phutball. Left: The situation prior to the jump. Right: The situation after jumping two men. The same player may then jump the remaining man.

positions are disallowed). However, repetitions do not seem to come up much in actual practice.

It is common in other board games[1] that the problems of determining the outcome of the game (with optimal play), or testing whether a given move preserves the correct outcome, are PSPACE-complete [5], or even EXPTIME-complete for loopy games such as Chess [8] and Go [11]. However, no such result is known for Phutball. Here we prove a different kind of complexity result: the problem of determining whether a player has a move that immediately wins the game (a mate in one, in chess terminology) is NP-complete. Such a result seems quite unusual, since in most games there are only a small number of legal moves, which could all be tested in polynomial time. The only similar result we are aware of is that, in Twixt, it is NP-complete to determine whether a player's points can be connected to form a winning chain [10]. However, the Twixt result seems to be less applicable to actual game play, since it depends on a player making a confusing tangle of his own points, which does not tend to occur in practice. The competition between both players in Phutball to form a favorable arrangement of men seems to lead much more naturally to complicated situations not unlike the ones occurring in our NP-completeness proof.

2. The NP-Completeness Proof

Testing for a winning jump sequence is clearly in NP, since a jump sequence can only be as long as the number of men on the board. As is standard for NP-completeness proofs, we prove that the problem is hard for NP by reducing a known NP-complete problem to it. For our known problem we choose 3-SAT: satisfiability of Boolean formulae in conjunctive normal form, with at most three variables per clause. We must show how to translate a 3-SAT instance into a Phutball position, in polynomial time, in such a way that the given formula is solvable precisely if there exists a winning path in the Phutball position.

---

[1]More precisely, since most games have a finite prescribed board size, these complexity results apply to generalizations in which arbitrarily large boards are allowed, and in which the complexity is measured in terms of the board size.

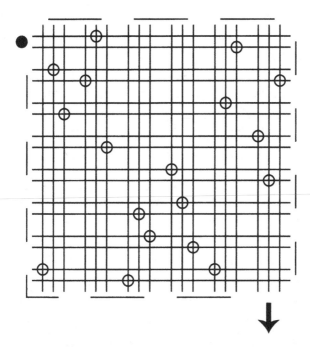

Figure 2. Overall plan of the NP-completeness reduction: a path zigzags through horizontal pairs of lines (representing variables) and vertical triples of lines (representing clauses). Certain variable-clause crossings are marked, representing an interaction between that variable and clause.

The overall structure of our translation is depicted in Figure 2, and a small complete example is shown in Figure 6. We form a Phutball position in which the possible jump sequences zigzag horizontally along pairs of lines, where each pair represents one of the variables in the given 3-SAT instance. The path then zigzags vertically up and down along triples of lines, where each triple represents one of the clauses in the 3-SAT instance. Thus, the potential winning paths are formed by choosing one of the two horizontal lines in each pair (corresponding to selecting a truth value for each variable) together with one of the three vertical lines in each triple (corresponding to selecting which of the three variables has a truth value that satisfies the clause). By convention, we associate paths through the upper of a pair of horizontal lines with assignments that set the corresponding variable to true, and paths through the lower of the pair with assignments that set the variable to false. The horizontal and vertical lines interact at certain marked crossings in a way that forces any path to correspond to a satisfying truth assignment.

We now detail each of the components of this structure.

FAN-IN AND FAN-OUT: At the ends of each pair or triple of lines, we need a
    configuration of men that allows paths along any member of the set of lines

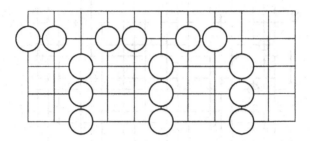

Figure 3. Configuration of men to allow a choice between three vertical lines. Similar configurations are used at the other end of each triple of lines, and at each end of pairs of horizontal lines.

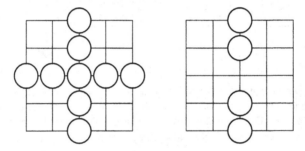

Figure 4. Left: configuration of men to allow horizontal and vertical lines to cross without interacting. Right: after the horizontal jump has been taken, the short gap in the vertical line still allows it to be traversed via a pair of jumps.

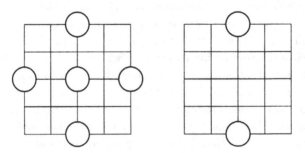

Figure 5. Left: configuration of men to allow horizontal and vertical lines to interact. Right: after the horizontal jump has been taken, the long gap in the vertical line prevents passage.

to converge, and then to diverge again at the next pair or triple. Figure 3 depicts such a configuration for the triples of vertical lines; the configuration for the horizontal lines is similar. Note that, if a jump sequence enters the configuration from the left, it can only exit through one of the three lines at the bottom. If a jump sequence enters via one of the three vertical lines, it can exit horizontally or on one of the other vertical lines. However, the possibility of using more than one line from a group does not cause a problem: a jump sequence that uses the second of two horizontal lines must get stuck at the other end of the line, and a jump sequence that uses two of three vertical lines must use all three lines and can be simplified to a sequence using only one of the three lines.

NON-INTERACTING LINE CROSSING: Figure 4 depicts a configuration of men that allows two lines to cross without interacting. A jump sequence entering along the horizontal or vertical line can and must exit along the same line, whether or not the other line has already been jumped.

INTERACTION: Figure 5 depicts a configuration of men forming an interaction between two lines. In the initial configuration, a jump sequence may follow either the horizontal or the vertical line. However, once the horizontal line has been jumped, it will no longer be possible to jump the vertical line.

Theorem 1. *Testing whether a Phutball position allows a winning jump sequence is NP-complete.*

Proof. As described above, we reduce 3-SAT to the given problem by forming a configuration of men with two horizontal lines of men for each variable, and three vertical lines for each clause. We connect these lines by the fan-in and fan-out gadget depicted in Figure 3. If variable i occurs as the jth term of clause k, we place an interaction gadget (Figure 5) at the point where the bottom horizontal line in the ith pair of horizontal lines crosses the jth vertical line in the kth triple of vertical lines. If instead the negation of variable i occurs in clause k, we place an interaction gadget similarly, but on the top horizontal line in the pair. At all other crossings of horizontal and vertical lines we place the crossing gadget depicted in Figure 4. Finally, we form a path of men from the final fan-in gadget (the arrow in Figure 2) to the goal line of the Phutball board.

The lines from any two adjacent interaction gadgets must be separated by four or more units, but other crossing types allow a three-unit separation. By choosing the order of the variables in each clause, we can make sure that the first variable differs from the last variable of the previous clause, avoiding any adjacencies between interaction gadgets. Thus, we can space all lines three units apart. If the 3-SAT instance has n variables and m clauses, the resulting Phutball board requires $6n + \mathcal{O}(1)$ rows and $9m + \mathcal{O}(1)$ columns, polynomial in the input size.

Finally, we must verify that the 3-SAT instance is solvable precisely if the Phutball instance has a winning jump sequence. Suppose first that the 3-SAT instance has a satisfying truth assignment; then we can form a jump sequence that uses the top horizontal line for every true variable, and the bottom line for every false variable. If a clause is satisfied by the jth of its three variables, we choose the jth of the three vertical lines for that clause. This forms a valid jump sequence: by the assumption that the given truth assignment satisfies the formula, the jump sequence uses at most one of every two lines in every interaction gadget. Conversely, suppose we have a winning jump sequence in the Phutball instance; then as discussed above it must use one of every two horizontal lines and one or three of every triple of vertical lines. We form a truth assignment by setting a variable to true if its upper line is used and false if its lower line is used. This must be a satisfying truth assignment: the vertical line used in each clause gadget must not have had its interaction gadget crossed horizontally, and so must correspond to a satisfying variable for the clause.   □

Figure 6 shows the complete reduction for a simple 3-SAT instance. We note that the Phutball instances created by this reduction only allow orthogonal jumps, so the rule in Phutball allowing diagonal jumps is not essential for our result.

3. Phutball and Checkers

Phutball is similar in many ways to Checkers. As in Phutball, Checkers players sit at opposite ends of a rectangular board, move pieces by sequences of jumps, remove jumped pieces, and attempt to move a piece onto the side of the board nearest the opponent. As in Phutball, the possibility of multiple jumps allows a Checkers player to have an exponential number of available moves. Checkers is PSPACE-complete [7] or EXPTIME-complete [12], depending on the termination rules, but these results rely on the difficulty of game tree search rather than the large number of moves available at any position. Does Checkers have the same sort of single-move NP-completeness as Phutball?

It is convenient to view Checkers as being played on a nonstandard diamond-shaped grid of square cells, with pieces that move horizontally and vertically, rather than the usual pattern of diagonal moves on a checkerboard (Figure 7). This view does not involve changing the rules of checkers nor even the geometric positions of the pieces, only the markings of the board on which they rest. Then, any jump preserves the parity of both the x- and y-coordinates of the jumping piece, so at most one fourth of the board's cells can be reached by jumps of a given piece (Figure 8, left).

For any given piece p, form a bipartite graph G_p by connecting the vacant positions that p can reach by jumping with the adjacent pieces of the opposite color that p can jump. If p is a king, this graph should be undirected, but otherwise it should be directed according to the requirement that the piece not move backwards. Note that each jumpable piece has degree two in this graph,

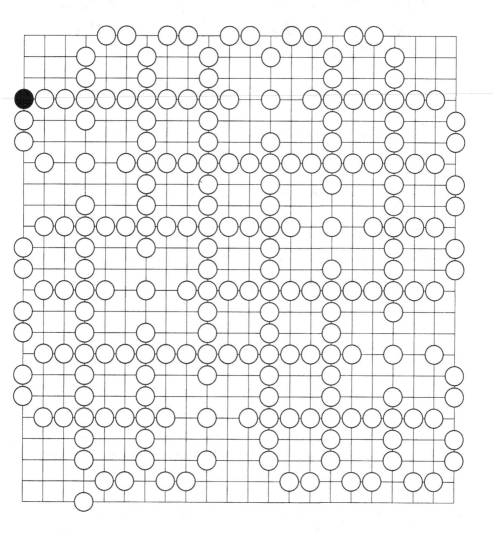

Figure 6. Complete translation of 3-SAT instance $(a \vee b \vee c) \wedge (\neg a \vee \neg b \vee \neg c)$.

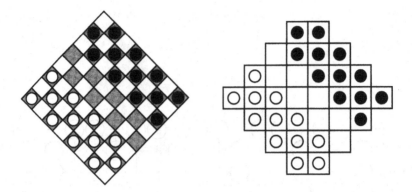

Figure 7. The checkerboard can equivalently be viewed as a diamond-shaped grid of orthogonally adjacent square cells.

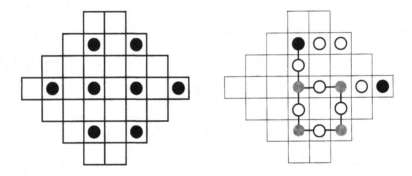

Figure 8. Left: only cells of the same parity can be reached by jumps. Right: graph G_p formed by connecting jumpable pieces with cells that can be reached by jumps from the upper black king.

so the possible sequences of jumps are simply the graph paths that begin at the given piece and end at a vacant square. Figure 8, right, depicts an example; note that opposing pieces that can not be jumped (because they are on cells of the wrong parity, or because an adjacent cell is occupied) are not included in G_p. Using this structure, it is not hard to show that Checkers moves are not complex:

Theorem 2. *For any Checkers position (on an arbitrary-size board), one can test in polynomial time whether a checker can become a king, or whether there is a move which wins the game by jumping all the opponent's pieces.*

Proof. Piece p can king precisely if there is a directed path in G_p from p to one of the squares along the opponent's side of the board. A winning move exists precisely if there exists a piece p for which G_p includes all opposing pieces and contains an Euler path starting at p; that is, precisely if G_p is connected and has at most one odd-degree vertex other than the initial location of p. □

The second claim in this theorem, testing for a one-move win, is also proved in [7]. That paper also shows that the analogous problem for a generalization of checkers to arbitrary graphs is NP-complete.

4. Discussion

We have shown that, in Phutball, the exponential number of jump sequences possible in a single move, together with the ways in which parts of a jump sequence can interfere with each other, leads to the high computational complexity of finding a winning move. In Checkers, there may be exponentially many jump sequences, but jumps can be performed independently of each other, so finding winning moves is easy. What about other games?

In particular, Fanorona [4,6] seems a natural candidate for study. In this game, capturing is performed in a different way, by moving a piece in one step towards or away from a contiguous line of the opponent's pieces. Board squares alternate between strong (allowing diagonal moves) and weak (allowing only orthogonal moves), and a piece making a sequence of captures must change direction at each step. Like Checkers (and unlike Phutball) the game is won by capturing all the opponent's pieces rather than by reaching some designated goal. Is finding a winning move in Fanorona hard? If so, a natural candidate for a reduction is the problem of finding Hamiltonian paths in grid graphs [9].

The complexity of determining the outcome of a general Phutball position remains open. We have not even proven that this problem is NP-hard, since even if no winning move exists in the positions we construct, the player to move may win in more than one move.

Acknowledgment

This work was inspired by various people, including John Conway and Richard Nowakowski, playing phutball at the MSRI Combinatorial Game Theory Research Workshop in July 2000.

References

[1] E. R. Berlekamp, J. H. Conway, and R. H. Guy. *Winning Ways for your Mathematical Plays*, vol. II, Games in Particular. Academic Press, 1982, pp. 688–691.

[2] M. Burke. Zillions of Games: Phutball. Web page, 1999, http://www.zillions-of-games.com/games/phutball.html.

[3] T. Cazenave. Résultats du Tournoi Berder99. Web page, September 1999, http://www.ai.univ-paris8.fr/~cazenave/Phutball/Phutball.html.

[4] D. Eppstein. Fanorona. Java program, 1997, http://www.ics.uci.edu/~eppstein/180a/projects/fanorona/.

[5] D. Eppstein. Computational Complexity of Games and Puzzles. Web page, 2000, http://www.ics.uci.edu/~eppstein/hard.html.

[6] L. Fox. *Fanorona: History, Rules and Strategy.* Northwest Corner, Inc., 1987.

[7] A. S. Fraenkel, M. R. Garey, D. S. Johnson, T. Schaefer, and Y. Yesha. The complexity of checkers on an $N \times N$ board — preliminary report. *Proc. 19th Symp. Foundations of Computer Science*, pp. 55–64. IEEE, 1978.

[8] A. S. Fraenkel and D. Lichtenstein. Computing a perfect strategy for $n \times n$ Chess requires time exponential in n. *J. Combinatorial Theory, Ser. A* 31:199–214, 1981.

[9] A. Itai, C. H. Papadimitriou, and J. L. Szwarcfiter. Hamilton paths in grid graphs. *SIAM J. Comput.* 11(4):676–686, November 1982.

[10] D. Mazzoni and K. Watkins. Uncrossed knight paths is NP-complete. Manuscript, http://www.mathematik.uni-bielefeld.de/~sillke/PROBLEMS/Twixt_Proof_Draft, October 1997.

[11] J. M. Robson. The complexity of Go. *Proc. IFIP*, pp. 413–417, 1983.

[12] J. M. Robson. N by N checkers is EXPTIME complete. *SIAM J. Comput.* 13(2):252–267, May 1984.

[13] J. Williams. Philosopher's Football. *Abstract Games* no. 3, pp. 10–14, Autumn 2000.

ERIK D. DEMAINE
MIT LABORATORY FOR COMPUTER SCIENCE
200 TECHNOLOGY SQUARE
CAMBRIDGE, MA 02139
UNITED STATES
edemaine@mit.edu

MARTIN L. DEMAINE
MIT LABORATORY FOR COMPUTER SCIENCE
200 TECHNOLOGY SQUARE
CAMBRIDGE, MA 02139
UNITED STATES
mdemaine@mit.edu

DAVID EPPSTEIN
UNIVERSITY OF CALIFORNIA, IRVINE
DEPARTMENT OF INFORMATION AND COMPUTER SCIENCE
IRVINE, CA 92697-3425
UNITED STATES
eppstein@ics.uci.edu

More Games of No Chance
MSRI Publications
Volume **42**, 2002

One-Dimensional Phutball

J. P. GROSSMAN AND RICHARD J. NOWAKOWSKI

ABSTRACT. We consider the game of one-dimensional phutball. We solve the case of a restricted version called Oddish Phutball by presenting an explicit strategy in terms of a potential function.

1. Introduction

J. H. Conway's *Philosophers Football*, otherwise known as *Phutball* [1], is played by two players, *Left* and *Right*, who move alternately. The game is usually played on a 19 by 19 board, starts with a *ball* on a square, one side is designated the Left goal line and the opposite side the Right goal line. A move consists of either placing a stone on an unoccupied square or jumping the ball over a (horizontal, vertical or diagonal) line of stones one end of which is adjacent to the square containing the ball. The stones are removed immediately after being jumped. Jumps can be chained and a player does not have to jump when one is available. A player wins by getting the ball on or over the opponent's goal line. Demaine, Demaine and Eppstein [2] have shown that deciding whether or not a player has a winning jump is NP complete.

In this paper we consider the 1-dimensional version of the game. For example:

$$\text{Left} \quad \cdot \; \cdot \; \diamond \; \diamond \; \bullet \; \diamond \; \diamond \; \cdot \; \diamond \; \cdot \; \cdot \; \cdot \quad \text{Right}$$

is a position on a finite linear strip of squares. Initially, there is a black stone, $\bullet$, which represents the ball. A player on his turn can either place a white stone, $\diamond$, on an empty square, $\cdot$, or jump the ball over a string of contiguous stones one end of which is adjacent to the ball; the ball ends on the next empty square. The stones are removed immediately upon jumping. A jump can be continued if there is another group adjacent to the ball's new position. Left wins by jumping the ball onto or over the rightmost square — Right's goalline; Right wins by jumping on over the leftmost square — Left's goalline.

Nowakowski was partially supported by a grant from the NSERC..

It would seem clear that

- Your position can only improve by having a stone placed between the ball and the opponent's goalline; and
- Your position can only improve if an empty square between the ball and the opponent's goalline is deleted.

However, both of these assertions are false as can be seen from the following examples. In all cases it is Left to move first and since Right can win by jumping if Left does not move the ball, Left's move must be a jump.

(1) Left wins: ◊ ● ◊ · ◊ · ◊ · · · ·

(2) Right wins: ◊ ● ◊ · ◊ ◊ ◊ · · · ·

(3) Right wins: ◊ ● ◊ · ◊ ◊ · · · ·

The reader may wish to check that in (1) Left loses if he jumps all three stones. The main continuation for (1) and (2) are given next, the reader may wish to try other variations. An arrow indicates the move was a jump and the numbers give the order of the placement of the stones.

(1): ◊ ● ◊ · ◊ · ◊ · · · ·   → ◊3 · 1 ● ◊ · 2 · ·

→ ◊ ◊ ◊ · ◊ · ◊ · 2 ● 1

Left jumps and wins

(2): ◊ ● ◊ · ◊ ◊ ◊ · · · ·   → ◊ · · · 3 · 1 ● 2 · 4

→ ◊ · 2 ● 1 · 3 · ◊ · ◊

Right jumps and wins

Also the obvious *Never jump backwards* heuristic does not always work.

In · · · · · · ◊ ◊ ◊ ◊ · ● ◊ ◊ · ◊ · · · · · · · · ·, Right's *only* winning move is to jump backwards to

· · · · · · ◊ ◊ ◊ ◊ · · · 4 · 2 ● 1 · 3 · · · ·

This splits into two options, again the final plays are left to the reader:

Option 1 → · · · · · · ◊ ◊ ◊ ◊ · · · ◊ · ◊ · 3 · 1 ● 2 · 4

→ · · · · · · ◊ ◊ ◊ ◊ · 2 ● 1 · 3 · · · · · ◊ · ◊

→ · · 4 · 2 ● 1 · 3 · · · · ◊ · ◊ · · · · · ◊ · ◊

→ · · ◊ · ◊ · 3 · 1 ● 2 · 4 ◊ · ◊ · · · · · ◊ · ◊

Option 2 → · · · · · · ◊ ◊ ◊ ◊ · · · 4 · 2 ● 1 · 3 · 5 · ·

→ · · · · · · ◊ ◊ ◊ ◊ · 2 ● 1 · 3 · ◊ · ◊ · ◊ · ·

→ · · 4 · 2 ● 1 · 3 · · · · ◊ · ◊ · ◊ · ◊ · ◊ · ·

→ · · ◊ · ◊ · 3 · 1 ● 2 · 4 ◊ · ◊ · ◊ · ◊ · ◊ · ·

The other main Right option is:

$$\cdots \cdots \diamond \diamond \diamond \diamond 1 \bullet \diamond \diamond \cdot \diamond \cdot 2 \cdots \cdots$$
$$\rightarrow \cdots 4 \cdot 2 \bullet 1 \cdot 3 \cdots \diamond \diamond \cdot \diamond \cdot 2 \cdots \cdots$$
$$\rightarrow \cdots \diamond \cdot \diamond \cdot 3 \cdot 1 \bullet 2 \cdot \diamond \diamond \cdot \diamond \cdot \diamond \cdots \cdots$$
$$\rightarrow \cdots \diamond \cdot \diamond \cdot \diamond \cdot \diamond \cdots \cdots \cdots \bullet \cdots \cdots$$

Note that on Right's second move it does her no good to jump backwards. Left responds by placing a stone to the right.

2. Oddish Phutball

We do not have a complete strategy for 1-dimensional Phutball and as the preceding examples show, it is a surprisingly difficult and often counter-intuitive game to analyze in full generality. However, if we place a small restriction on the moves allowed (one that naturally arises in actual play between two intelligent beings), then we do have a complete analysis.

Oddish Phutball is 1-dimensional Phutball in which the players will only place stones on the squares at an odd distance from the ball. We therefore eliminate all the squares at an even distance from the ball and we represent the ball by an arrow between two squares. This we call the *Oddish* board. For example, the 1-dimensional Phutball position

$$\cdots \diamond \cdots \diamond \cdot \diamond \bullet \diamond \cdot \diamond \cdots \cdots$$

on the Oddish board becomes

$$\cdot \diamond \cdot \diamond \diamond \diamond^{\downarrow} \diamond \diamond \cdots$$

On the Oddish board the arrow can only move past a set of consecutive stones, stopping anywhere between the stones. The stones passed over by the arrow are removed. The arrow cannot move past an empty square. The winning condition is to move the arrow past the last square on the Oddish board. In this simplified game, one might hypothesize that an optimal player will only jump when he can 'jump to a winning position', that is, jump far enough towards the opponent's goal line that the arrow will never again return to its position before this jump. It turns out that this hypothesis is correct. It follows that one can define a potential function for each player which is the number of stones they must place before they can jump to a winning position. Our analysis will consist of showing how to calculate this potential function and demonstrating that the player with the lower potential always has a winning strategy.

The Left *potential function*, p_L, takes into account only the squares of the Oddish board to the right of the ball. The following are the rules for the calculation.

(i) Summing: Compute the *left sum* by starting with a sum of 0 at the rightmost edge and, moving left, up to the square before the arrow:

   add 1 if the next square is empty; and

   subtract 1 if the next square is a stone, *except if the sum is 0, in which case the sum remains 0*;

(ii) The sum up to but not including the square immediately to the right of the arrow we denote by *psum* and the total sum we denote by *bsum*. The Left potential is then

$$p_L = \lfloor bsum/2 \rfloor + (1 - \delta_p^b)$$

where δ_p^b is the Kronecker delta function which is 1 if $bsum = psum$ (and this can only happen if they both equal 0) and it equals 0 otherwise.

The Right potential, p_R, is calculated similarly.

In calculating Right's potential for the Oddish position

$$\cdot \cdot \diamond \diamond \diamond^{\downarrow} \diamond \diamond \diamond \cdot \cdot \cdot \cdot \diamond \cdot$$

$psum = bsum = 0$ therefore $p_R = 0$. For Left, $psum = 2$ and $bsum = 1$ therefore $p_L = 1$.

Theorem 1. *In a game of Oddish Phutball, Left can win if $p_L < p_R$; Right can win if $p_L > p_R$; it is a first player win if $p_L = p_R$.*

To prove this result we need several observations.

Lemma 2. *In an Oddish Phutball position:*

(i) *If $p_L = 0$ then Left can jump to a position such that $p_R - p_L \geq 1$.*
(ii) *If $p_L > 0$ then Left can decrease p_L by 1 by placing a stone.*
(iii) *If $p_L > 0$ any Left jump increases $p_R - p_L$ by at most 1.*

Proof. If there are only stones to the right of the arrow then $p_L = 0$ and Left can jump and win immediately. Therefore, suppose that there are k stones immediately to the right of the arrow before the first empty square. Let m be the sum before these k stones are included in the count.

1: If $p_L = 0$ then $0 < m < k$ and specifically, $k > 1$. Choose k' even such that $k - 1 \leq k' \leq k$. Therefore $k' > 0$ and $\lfloor (k'-1)/2 \rfloor + 1 = \lfloor k'/2 \rfloor$. Then Left jumping over k' stones changes p_L from 0 to at most $\lfloor (k'-1)/2 \rfloor + 1$, i.e. at most $\lfloor k'/2 \rfloor$. Now p_R has changed either from $\lfloor m'/2 \rfloor + 1$ to $\lfloor (m'+k')/2 \rfloor + 1$ for some $m' > 0$ or from 0 to $\lfloor k'/2 \rfloor + 1$. In both cases, the new position has $p_L < p_R$.

2: If $p_L > 0$ then $m \geq k$. Placing a stone on the empty square immediately to the right of this group changes p_L from 1 to 0 if $m = k$ or $m = k + 1$; and if $m \geq k + 2$ then p_L changes from $\lfloor \frac{m-k}{2} \rfloor + 1$ to $\lfloor \frac{m-k}{2} \rfloor$.

3: Suppose that $p_L = \lfloor i/2 \rfloor + 1 > 0$ for some $i \geq 0$. Let Left jump forward over j stones. The Left potential changes from $\lfloor i/2 \rfloor + 1$ to $\lfloor (i+j)/2 \rfloor + 1$. The Right potential changes from either $\lfloor l/2 \rfloor + 1$ to $\lfloor (l+j)/2 \rfloor + 1$ for some $l \geq 0$

or from 0 to $\lfloor j/2 \rfloor + 1$. In all cases, the change in the potential is between $\lfloor j/2 \rfloor$ and $\lfloor j/2 \rfloor + 1$. Therefore $p_R - p_L$ changes by at most 1.

Note that if Left jumps backwards over j stones and $p_R > 0$ the same analysis shows $p_R - p_L$ changes by at most 1. However, if $p_R = 0$ with the Right sum being zero for the last q stones then the Right potential does not change if $j < q - 1$ and changes to $\lfloor (j-q)/2 \rfloor + 1$ if $q - 1 \le j$ whereas the Left potential has changed from $\lfloor i/2 \rfloor + 1$ to $\lfloor (i+j)/2 \rfloor + 1$ for some $i \ge 0$. Hence the change in the Right potential is not greater than $\lfloor j/2 \rfloor + 1$ and that of the Left is at least $\lfloor j/2 \rfloor$, so again the change in $p_R - p_L$ is less than or equal to 1. □

Proof of Theorem 1. We show that Left cannot lose by giving an explicit strategy and then show that the game is finite and hence Left must win.

We assume first that $p_L < p_R$ and it is Right to move. Note that since $p_R > 0$, Right cannot win in one move by jumping on or over the goal line.

If Right places a stone to the left of the arrow then she decreases her potential by at most 1 and does not affect Left's potential. If she places a stone to the right of the arrow then she does not affect p_R but may decrease p_L. If Right jumps forward or backwards then $p_L - p_R$ increases by at most 1. In all cases, it is then Left's turn with $p_L \le p_R$.

Suppose that $p_L = 0$ and that there are k stones in the string adjacent to the arrow. Left jumps over k' stones where $k' = k$ or $k' = k - 1$ and k' is even. With this k' we have $\lfloor (k' - 1)/2 \rfloor + 1 = \lfloor k'/2 \rfloor$. Then, by Lemma 2.1, $p_L < p_R$ and it is now Right's turn.

If $p_L > 0$ then Left places a stone at the end of the adjacent string of stones and p_L is reduced by at least 1 and therefore by Lemma 2.2, $p_L < p_R$ and it is now Right's turn.

Now suppose that it is Left's move with $p_L \le p_R$. If $p_L = 0$ then by Lemma 2.1 Left can jump to a position with $p_L - p_R \ge 1$. If $p_L > 0$ then by Lemma 2.2 Left can place a stone to decrease p_L by 1 without changing p_R. In both cases, it is Right's turn with $p_L < p_R$. It follows that if it is Left's move and $p_R \ge p_L$ or Right's move with $p_R > p_L$ then Right cannot win — it is never her move in a position with $p_R = 0$ so she never has a move that allows her to jump to the goal line. We now have to prove that the game finishes after a finite number of moves. Suppose that it does not. Let A be the rightmost square, on the Oddish board, that is changed infinitely often and suppose that the game has progressed far enough so that in the continuation no moves to the right of A are played.

The next time that the arrow passes over A to the right, it cannot be that Right has just jumped backwards past A since then Left's strategy has him placing a stone to the right of A. If Left has just jumped (immediately to the right of A) then Right must place a stone to the left and then Left would place a stone to the right of A, again a contradiction.

Therefore the game is finite and therefore there is a winner. Since the winner is not Right, it must be Left. $\qquad\square$

3. 1-dimensional Phutball Revisited

When we attempted to extend our analysis to unrestricted 1-dimensional phutball, we were able to develop a generalized potential function which *almost* works. Instead of the full solution we had hoped for, a small hole in the proof led us to discover the position shown earlier in which the only winning move is a backwards jump. The very existence of such a position implies that a fundamentally different approach is required to fully analyze the general game. Nonetheless, the generalized potential function does provide some insight into the game, so we include it here for completeness.

The Left *potential function*, P_L, takes into account only the ball and all the squares to the right of the ball. The following are the rules for the calculation.

1. Transformation: Replace the ball with the symbol immediately to the right, and for each odd group of contiguous empty squares replace the rightmost with a stone.
2. Summing: In this new configuration, compute the left sum by starting with a sum of 0 at the rightmost edge and, moving left, up to and including the square with the ball:

   add 1 if the next square is empty; and

   subtract 1 if the next square is a stone, *except if the sum is 0 in which case the sum remains 0*;
3. Calculation: The sum up to but not including the ball's square we denote by *psum* and the total sum we denote by *bsum*. So the Left potential can be calculated by

$$P_L = \lfloor bsum/4 \rfloor + (1 - \delta_p^b)$$

where δ_p^b is the Kronecker delta function which is 1 if $bsum = psum$ (and this can only happen if they both equal 0) and it equals 0 otherwise. The Right potential, P_R, is calculated similarly.

The analysis of when this potential function works and when it breaks down is similar to the analysis of the oddish potential function and is left as an exercise for the reader.

References

[1] E. R. Berlekamp, J. H. Conway, R. K. Guy, *Winning Ways for your mathematical plays*, New York, Academic Press, 1982.

[2] E. D. Demaine, M. L. Demaine, D. Eppstein, "Phutball endgames are hard, in *More games of no chance*, edited by R. Nowakowski, Math. Sci. Res. Publ. **42**, Cambridge Univ. Press, Cambridge, 2002.

J. P. GROSSMAN
DEPARTMENT OF ELECTRIC ENGINEERING AND COMPUTER SCIENCE
MASSACHUSETTS INSTITUTE OF TECHNOLOGY
CAMBRIDGE, MA 02139
UNITED STATES
  jpg@ai.mit.edu

RICHARD J. NOWAKOWSKI
DEPARTMENT OF MATHEMATICS AND STATISTICS
DALHOUSIE UNIVERSITY
HALIFAX, NS B3H 3J5
CANADA
  rjn@mathstat.dal.ca

More Games of No Chance
MSRI Publications
Volume **42**, 2002

A Symmetric Strategy
in Graph Avoidance Games

FRANK HARARY, WOLFGANG SLANY, AND OLEG VERBITSKY

ABSTRACT. In the graph avoidance game two players alternately color the edges of a graph G in red and in blue respectively. The player who first creates a monochromatic subgraph isomorphic to a forbidden graph F loses. A *symmetric strategy* of the second player ensures that, independently of the first player's strategy, the blue and the red subgraph are isomorphic after every round of the game. We address the class of those graphs G that admit a symmetric strategy for all F and discuss relevant graph-theoretic and complexity issues. We also show examples when, though a symmetric strategy on G generally does not exist, it is still available for a particular F.

1. Introduction

In a broad class of games that have been studied in the literature, two players, $\mathcal{A}$ and $\mathcal{B}$, alternately color the edges of a graph G in red and in blue respectively. In the achievement game the objective is to create a monochromatic subgraph isomorphic to a given graph F. In the avoidance game the objective is, on the contrary, to avoid creating such a subgraph. Both the achievement and the avoidance games have strong and weak versions. In the strong version $\mathcal{A}$ and $\mathcal{B}$ both have the same objective. In the weak version $\mathcal{B}$ just plays against $\mathcal{A}$, that is, tries either to prevent $\mathcal{A}$ from creating a copy of F in the achievement game or to force such creation in the avoidance game. The weak achievement game, known also as the Maker-Breaker game, is most studied [4; 1; 13]. Our paper is motivated by the strong avoidance game [7; 5] where monochromatic F-subgraphs of G are forbidden, and the player who first creates such a subgraph loses.

Slany's research was partly supported by Austrian Science Foundation grant Z29-INF. This work was done while Verbitsky was visiting the Institut für Informationssysteme at the Technische Universität Wien, supported by a Lise Meitner Fellowship of the Austrian Science Foundation (FWF grant M 532).

The instance of a strong avoidance game with complete graphs $G = K_6$ and $F = K_3$ is well known under the name SIM [15]. Since for any edge bicoloring of K_6 there is a monochromatic K_3, a draw in this case is impossible. It is proven in [12] that a winning strategy in SIM is available for $\mathcal{B}$. A few other results for small graphs are known [7]. Note that, in contrast to the weak achievement games, if $\mathcal{B}$ has a winning strategy in the avoidance game on G with forbidden F and if G is a subgraph of G', then it is not necessary that $\mathcal{B}$ also has a winning strategy on G' with forbidden F. Recognition of a winner seems generally to be a nontrivial task both from the combinatorial and the complexity-theoretic point of view (for complexity issues see, e.g., [16]).

In this paper we introduce the notion of a *symmetric strategy*[1] for $\mathcal{B}$. We say that $\mathcal{B}$ follows a symmetric strategy on G if after every move of $\mathcal{B}$ the blue and the red subgraphs are isomorphic, regardless of $\mathcal{A}$'s strategy. As easily seen, if $\mathcal{B}$ plays so, he at least does not lose in the avoidance game on G with any forbidden F. There is a similarity with the *mirror-image strategy* of $\mathcal{A}$ in the achievement game [2]. However, the latter strategy is used on two disjoint copies of the complete graph, and therefore in our case things are much more complicated.

We address the class $\mathcal{C}_{\mathrm{sym}}$ of those graphs G on which a symmetric strategy for $\mathcal{B}$ exists. We observe that $\mathcal{C}_{\mathrm{sym}}$ contains all graphs having an involutory automorphism without fixed edges. This subclass of $\mathcal{C}_{\mathrm{sym}}$, denoted by $\mathcal{C}_{\mathrm{auto}}$, includes even paths and cycles, bipartite complete graphs $K_{s,t}$ with s or t even, cubes, and the Platonic graphs except the tetrahedron. We therefore obtain a lot of instances of the avoidance game with a winning strategy for $\mathcal{B}$. More instances can be obtained based on closure properties of $\mathcal{C}_{\mathrm{auto}}$ that we check with respect to a few basic graph operations.

Nevertheless, recognizing a suitable automorphism and, therefore, using the corresponding symmetric strategy is not easy. Based on a related result of Lubiw [11], we show that deciding membership in $\mathcal{C}_{\mathrm{auto}}$ is NP-complete.

We then focus on games on complete graphs. We show that K_n is not in $\mathcal{C}_{\mathrm{sym}}$ for all $n \geq 4$. Moreover, for an arbitrary strategy of $\mathcal{B}$, $\mathcal{A}$ is able to violate the isomorphism between the red and the blue subgraphs in at most $n - 1$ moves. Nevertheless, we consider the avoidance game on K_n with forbidden P_2, a path of length 2, and point out a simple symmetric strategy making $\mathcal{B}$ the winner. This shows an example of a graph G for which, while a symmetric strategy in the avoidance game does not exist in general, it does exist for a particular forbidden F.

The paper is organized as follows. Section 2 contains the precise definitions. In Section 3 we compile the membership list for $\mathcal{C}_{\mathrm{sym}}$ and $\mathcal{C}_{\mathrm{auto}}$. In Section 4 we investigate the closure properties of $\mathcal{C}_{\mathrm{sym}}$ and $\mathcal{C}_{\mathrm{auto}}$ with respect to various graph products. In Section 5 we prove the NP-completeness of $\mathcal{C}_{\mathrm{auto}}$. Section 6 analyses the avoidance game on K_n with forbidden P_2.

---

[1]Note that this term has been used also in other game-theoretic situations (see, e.g., [14]).

2. Definitions

We deal with two-person positional games of the following kind. Two players, $\mathcal{A}$ and $\mathcal{B}$, alternately color the edges of a graph G in red and in blue respectively. Player $\mathcal{A}$ starts the game. In a *move* a player colors an edge that was so-far uncolored. The i-th *round* consists of the i-th move of $\mathcal{A}$ and the i-th move of $\mathcal{B}$. Let a_i (resp. b_i) denote an edge colored by $\mathcal{A}$ (resp. $\mathcal{B}$) in the i-th round.

A *strategy* for a player determines the edge to be colored by him at every round of the game. Formally, let ε denote the empty sequence. A strategy of $\mathcal{A}$ is a function S_1 that maps every possibly empty sequence of distinct edges $e_1, \ldots, e_i$ into an edge different from $e_1, \ldots$, and e_i and from $S_1(\varepsilon), S_1(e_1), S_1(e_1, e_2), \ldots$, and $S_1(e_1, \ldots, e_{i-1})$. A strategy of $\mathcal{B}$ is a function S_2 that maps every nonempty sequence of distinct edges $e_1, \ldots, e_i$ into an edge different from $e_1, \ldots$, and e_i and from $S_2(e_1), S_2(e_1, e_2), \ldots$, and $S_2(e_1, \ldots, e_{i-1})$. If $\mathcal{A}$ follows a strategy S_1 and $\mathcal{B}$ follows a strategy S_2, then $a_i = S_1(b_1, \ldots, b_{i-1})$ and $b_i = S_2(a_1, \ldots, a_i)$.

Let $A_i = \{a_1, \ldots, a_i\}$ (resp. $B_i = \{b_1, \ldots, b_i\}$) consist of the red (resp. blue) edges colored up to the i-th round. A *symmetric strategy of $\mathcal{B}$ on G* ensures that, regardless of $\mathcal{A}$'s strategy, the subgraphs A_i and B_i are isomorphic for every $i \leq m/2$, where m is the number of edges of G.

The class of all graphs G on which $\mathcal{B}$ has a symmetric strategy will be denoted by $\mathcal{C}_{\mathrm{sym}}$.

Suppose that we are given graphs G and F and that F is a subgraph of G. The *avoidance game on G with a forbidden subgraph F* or, shortly, the game AVOID(G, F) is played as described above with the following ending condition: The player who first creates a monochromatic subgraph of G isomorphic to F loses.

Note that a symmetric strategy of $\mathcal{B}$ on G is nonlosing for $\mathcal{B}$ in AVOID(G, F), for every forbidden F. Really, the assumption that $\mathcal{B}$ creates a monochromatic copy of F implies that such a copy is already created by $\mathcal{A}$ earlier in the same round.

3. Automorphism-based strategy

Recall that the *order* of a graph is its number of vertices and the *size* of a graph is its number of edges. Given a graph G, we denote its vertex set by $V(G)$ and its edge set by $E(G)$. An *automorphism* of a graph G is a permutation of $V(G)$ that preserves the vertex adjacency. Recall that the *order* of a permutation is the minimal k such that the k-fold composition of the permutation is the identity permutation. In particular, a permutation of order 2, also called an *involution*, coincides with its inversion. We call an automorphism of order 2 *involutory*.

The symmetric strategy can be realized if a graph G has an involutory automorphism that moves every edge. More precisely, an automorphism $\phi : V(G) \to V(G)$ determines a permutation $\phi' : E(G) \to E(G)$ by $\phi'(\{u, v\}) = \{\phi(u), \phi(v)\}$.

We assume that ϕ is involutory and ϕ' has no fixed element. In this case, whenever $\mathcal{A}$ chooses an edge e, $\mathcal{B}$ chooses the edge $\phi'(e)$. This strategy of $\mathcal{B}$ is well defined because $E(G)$ is partitioned into 2-subsets of the form $\{e, \phi'(e)\}$. This strategy is really symmetric because after completion of every round ϕ induces an isomorphism between the red and the blue subgraphs. We will call such a strategy *automorphism-based*.

Definition 3.1. $\mathcal{C}_{\text{auto}}$ is a subclass of $\mathcal{C}_{\text{sym}}$ consisting of all those graphs G on which $\mathcal{B}$ has an automorphism-based symmetric strategy.

We now list some examples of graphs in $\mathcal{C}_{\text{auto}}$.

Example 3.2 (Graphs in $\mathcal{C}_{\text{auto}}$). (i) P_n, a path of length n, if n is even.
(ii) C_n, a cycle of length n, if n is even.
(iii) Four Platonic graphs excluding the tetrahedron.
(iv) Cubes of any dimension.[2]
(v) Antipodal graphs (in the sense of [3]) of size more than 1. Those are connected graphs such that for every vertex v, there is a unique vertex $\bar{v}$ of maximum distance from v. The correspondence $\phi(v) = \bar{v}$ is an automorphism [9]. As easily seen, it is involutory and has no fixed edge. The class of antipodal graphs includes the graphs from the three preceding items.
(vi) $K_{s,t}$, a bipartite graph whose classes have s and t vertices, if st is even.
(vii) $K_{s,t} - e$, that is, $K_{s,t}$ with an edge deleted, provided st is odd.
(viii) K_n, a complete graph on n vertices, with a matching of size $\lfloor n/2 \rfloor$ deleted. Note that in this and the preceding examples, for all choices of edges to be deleted, the result of deletion is the same up to an isomorphism.

It turns out that a symmetric strategy is not necessarily automorphism-based.

Theorem 3.3. $\mathcal{C}_{\text{auto}}$ *is a proper subclass of* $\mathcal{C}_{\text{sym}}$.

Below is a list of a few separating examples.

Example 3.4 (Graphs in $\mathcal{C}_{\text{sym}} \setminus \mathcal{C}_{\text{auto}}$). (i) A triangle with one more edge attached (the first graph in Figure 1). This is the only connected separating example of even size we know. In particular, none of the connected graphs of size 6 is in $\mathcal{C}_{\text{sym}} \setminus \mathcal{C}_{\text{auto}}$. Note that the definition of $\mathcal{C}_{\text{sym}}$ does not exclude graphs of odd size, as given in the further examples.
(ii) The graphs of size 5 shown in Figure 1.
(iii) Paths P_1, P_3, and P_5.
(iv) Cycles C_3, C_5, and C_7.
(v) Stars $K_{1,n}$, if n is odd.

Note that in spite of items 4 and 5, P_7 and C_9 are not in $\mathcal{C}_{\text{sym}}$.

---

[2]More generally, cubes are a particular case of grids, i.e., Cartesian products of paths. The central symmetry of a grid moves each edge unless exactly one of the factors is an odd path.

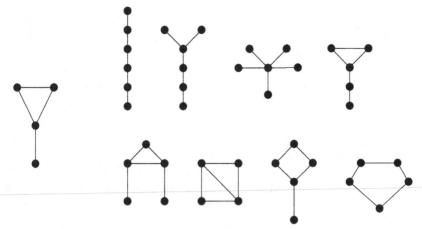

Figure 1. Graphs of size 4 and 5 that are in $\mathcal{C}_{\text{sym}}$ but not in $\mathcal{C}_{\text{auto}}$.

Question 3.5. How much larger is $\mathcal{C}_{\text{sym}}$ than $\mathcal{C}_{\text{auto}}$? Are there other connected separating examples than those listed above?

4. Closure properties of $\mathcal{C}_{\text{auto}}$

We now recall a few operations on graphs. Given two graphs G_1 and G_2, we define a product graph on the vertex set $V(G_1) \times V(G_2)$ in three ways. Two vertices (u_1, u_2) and (v_1, v_2) are adjacent in the *Cartesian product* $G_1 \times G_2$ if either $u_1 = v_1$ and $\{u_2, v_2\} \in E(G_2)$ or $u_2 = v_2$ and $\{u_1, v_1\} \in E(G_1)$; in the *lexicographic product* $G_1[G_2]$ if either $\{u_1, v_1\} \in E(G_1)$ or $u_1 = v_1$ and $\{u_2, v_2\} \in E(G_2)$; in the *categorical product* $G_1 \cdot G_2$ if $\{u_1, v_1\} \in E(G_1)$ and $\{u_2, v_2\} \in E(G_2)$.

If the vertex sets of G_1 and G_2 are disjoint, we define the *sum* (or *disjoint union*) $G_1 + G_2$ to be the graph with vertex set $V(G_1) \cup V(G_2)$ and edge set $E(G_1) \cup E(G_2)$.

Using these graph operations, from Example 3.2 one can obtain more examples of graphs in $\mathcal{C}_{\text{sym}}$. Note that the class of antipodal graphs itself is closed with respect to the Cartesian product [9].

Theorem 4.1. (i) $\mathcal{C}_{\text{auto}}$ *is closed with respect to the sum and with respect to the Cartesian, the lexicographic, and the categorical products.*
(ii) *Moreover, $\mathcal{C}_{\text{auto}}$ is an ideal with respect to the categorical product, that is, if G is in $\mathcal{C}_{\text{auto}}$ and H is arbitrary, then both $G \cdot H$ and $H \cdot G$ are in $\mathcal{C}_{\text{auto}}$.*

Proof. For the sum the claim 1 is obvious. Consider three auxiliary product notions. Given two graphs G_1 and G_2, we define product graphs $G_1 \otimes_1 G_2$, $G_1 \otimes_2 G_2$, and $G_1 \otimes_3 G_2$ each on the vertex set $V(G_1) \times V(G_2)$. Two vertices (u_1, u_2) and (v_1, v_2) are adjacent in $G_1 \otimes_1 G_2$ if $\{u_1, v_1\} \in E(G_1)$ and $u_2 = v_2$; in $G_1 \otimes_2 G_2$ if $\{u_1, v_1\} \in E(G_1)$ and $u_2 \neq v_2$; and in $G_1 \otimes_3 G_2$ if $u_1 = v_1$ and $\{u_2, v_2\} \in E(G_2)$.

Given two permutations, ϕ_1 of $V(G_1)$ and ϕ_2 of $V(G_2)$, we define a permutation ψ of $V(G_1) \times V(G_2)$ by $\psi(u_1, u_2) = (\phi_1(u_1), \phi_2(u_2))$. If both ϕ_1 and ϕ_2 are involutory, so is ψ. If ϕ_1 and ϕ_2 are automorphisms of G_1 and G_2 respectively, then ψ is an automorphism of each $G_1 \otimes_i G_2$, $i = 1, 2, 3$. Finally, it is not hard to see that if both ϕ_1 and ϕ_2 move all edges, so does ψ in each $G_1 \otimes_i G_2$, $i = 1, 2, 3$.

Notice now that $E(G_1 \otimes_1 G_2)$, $E(G_1 \otimes_2 G_2)$, and $E(G_1 \otimes_3 G_2)$ are pairwise disjoint. Notice also that $E(G_1 \times G_2) = E(G_1 \otimes_1 G_2) \cup E(G_1 \otimes_3 G_2)$ and $E(G_1[G_2]) = E(G_1 \otimes_1 G_2) \cup E(G_1 \otimes_2 G_2) \cup E(G_1 \otimes_3 G_2)$. It follows that if ϕ_1 and ϕ_2 are fixed-edge-free involutory automorphisms of G_1 and G_2 respectively, then ψ is a fixed-edge-free involutory automorphism of both $G_1 \times G_2$ and $G_1[G_2]$. Thus, $\mathcal{C}_{\text{auto}}$ is closed with respect to the Cartesian and the lexicographic products.

To prove the claim 2, let $G \in \mathcal{C}_{\text{auto}}$, ϕ be a fixed-edge-free involutory automorphism of G, and H be an arbitrary graph. Define a permutation ψ of $V(G \cdot H)$ by $\psi(u, v) = (\phi(u), v)$. It is not hard to see that ψ is a fixed-edge-free involutory automorphism of $G \cdot H$. Thus, $G \cdot H \in \mathcal{C}_{\text{auto}}$. The same is true for $H \cdot G$ because $G \cdot H$ and $H \cdot G$ are isomorphic. $\qquad\square$

Example 4.2 ($\mathcal{C}_{\text{sym}}$ is not closed with respect to the Cartesian, the lexicographic, and the categorical products). Denote the first graph in Example 3.4 by $K_3 + e$. The following product graphs are not in $\mathcal{C}_{\text{sym}}$: $(K_3 + e) \times P_2$, $P_2[K_3 + e]$, and $(K_3 + e) \cdot (K_3 + e)$. To show this, for each of these graphs we will describe a strategy allowing $\mathcal{A}$ to destroy an isomorphism between the red and the blue subgraphs, regardless of $\mathcal{B}$'s strategy.

$(K_3 + e) \times P_2$ has a unique vertex v of the maximum degree 5, and v is connected to the two vertices v_1 and v_2 of degree 4 that are connected to each other. In the first move of a symmetry-breaking strategy, $\mathcal{A}$ chooses the edge $\{v, v_1\}$. If $\mathcal{B}$ chooses an edge not incident to v, $\mathcal{A}$ creates a star $K_{1,5}$ and thus breaks $\mathcal{B}$'s symmetric strategy. If $\mathcal{B}$ chooses an edge incident to v but not $\{v, v_2\}$, $\mathcal{A}$ chooses $\{v, v_2\}$ and breaks $\mathcal{B}$'s symmetric strategy by creating a triangle K_3 in the third move. Assume therefore that in the second round $\mathcal{B}$ chooses $\{v, v_2\}$. In the next moves $\mathcal{A}$ creates a star with center v. If $\mathcal{B}$ tries to create a star with the same center, the symmetry eventually breaks because $\mathcal{A}$ can create a $K_{1,3}$ while $\mathcal{B}$ can create at most a $K_{1,2}$. Assume therefore that in the first four rounds $\mathcal{A}$ creates a $K_{1,4}$ with center v and $\mathcal{B}$ creates a $K_{1,4}$ with center v_2 (see Figure 2). In the rounds 5–8 $\mathcal{A}$ attaches a new edge to every leaf of the red star. Player $\mathcal{B}$'s symmetric strategy is broken because he cannot attach any edge to v.

$P_2[K_3 + e]$ consists of three copies of $K_3 + e$ on the vertex sets $\{u_1, u_2, u_3, u_4\}$, $\{v_1, v_2, v_3, v_4\}$, and $\{w_1, w_2, w_3, w_4\}$, and of 32 edges $\{v_i, u_j\}$ and $\{v_i, w_j\}$ for all $1 \le i, j \le 4$ (see Figure 3). The vertex v_1 has the maximum degree 11, v_2 and v_3 have degree 10, v_4 has degree 9, and all other vertices have degree at most 7. In the first move of a symmetry-breaking strategy $\mathcal{A}$ chooses the edge $\{v_1, v_2\}$. If $\mathcal{B}$ in response does not choose $\{v_1, v_3\}$, $\mathcal{A}$ does it and breaks $\mathcal{B}$'s

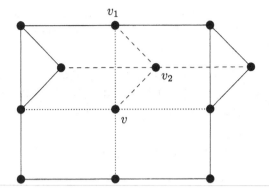

Figure 2. First four rounds of $\mathcal{A}$'s symmetry-breaking strategy on $(K_3 + e) \times P_2$ ($\mathcal{A}$'s edges dotted, $\mathcal{B}$'s edges dashed, uncolored edges continuous).

symmetric strategy by creating a star with center v_1. If $\mathcal{B}$ chooses $\{v_1, v_3\}$, in the second move $\mathcal{A}$ chooses $\{v_1, u_4\}$. If $\mathcal{B}$ then chooses an edge going out of v_1, $\mathcal{A}$ breaks $\mathcal{B}$'s symmetric strategy by creating a $K_{1,6}$. Assume therefore that in the second move $\mathcal{B}$ chooses an edge $\{v_3, x\}$. If $x = v_2$ or $x = u_4$, $\mathcal{A}$ chooses $\{u_4, v_2\}$ and breaks $\mathcal{B}$'s symmetric strategy by creating a triangle K_3. Assume therefore that x is another vertex (for example, $x = u_1$ as in Figure 3). In the third move $\mathcal{A}$ chooses $\{v_2, v_3\}$. If $\mathcal{B}$ chooses $\{x, v_2\}$, $\mathcal{A}$ chooses $\{u_4, v_3\}$ and breaks $\mathcal{B}$'s symmetric strategy by creating a quadrilateral C_4. Otherwise, in the next moves $\mathcal{A}$ creates a star $K_{1,10}$ with center v_2. Player $\mathcal{B}$'s symmetric strategy eventually is broken because he can create at most a $K_{1,9}$ with center v_1 or v_3 or at most a $K_{1,7}$ with center x.

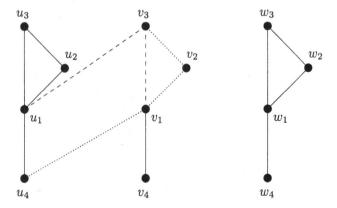

Figure 3. First three moves of $\mathcal{A}$'s symmetry-breaking strategy on $P_2[K_3 + e]$ ($\mathcal{A}$'s edges dotted, $\mathcal{B}$'s edges dashed, uncolored edges continuous, uncolored edges $\{v_i, u_j\}$, $\{v_i, w_j\}$ not shown).

$(K_3 + e) \cdot (K_3 + e)$ has a unique vertex v of the maximum degree 9, whereas all other vertices have degree at most 6. A symmetry-breaking strategy of $\mathcal{A}$ consists in creating a star with center v.

Remark 4.3. $\mathcal{C}_{\mathrm{sym}}$ is not closed with respect to the sum because, for example, it does not contain $K_3 + P_3$. Nevertheless, if G_1 and G_2 are in $\mathcal{C}_{\mathrm{sym}}$ and both have even size, $G_1 + G_2$ is easily seen to be in $\mathcal{C}_{\mathrm{sym}}$.

5. Complexity of $\mathcal{C}_{\mathrm{auto}}$

Though the graph classes listed in Example 3.2 have efficient membership tests, in general the existence of an involutory automorphism without fixed edges is not easy to determine.

Theorem 5.1. *Deciding membership of a given graph G in the class $\mathcal{C}_{\mathrm{auto}}$ is NP-complete.*

Proof. Consider the related problem ORDER 2 FIXED-POINT-FREE AUTOMOR-PHISM whose NP-completeness was proven in [11]. This is the problem of recognition if a given graph has an involutory automorphism without fixed *vertices*. We describe a polynomial time reduction R from ORDER 2 FIXED-POINT-FREE AUTOMORPHISM to $\mathcal{C}_{\mathrm{auto}}$.

Given a graph G, we perform two operations:

Step 1. Split every edge into two adjacent edges by inserting a new vertex, i.e., form the subdivision graph $S(G)$ (see [6, p. 80]).

Step 2. Attach a 3-star by a leaf at every non-isolated vertex of $S(G)$ which also was in G.

As a result we obtain $R(G)$ (see an example in Figure 4). We have to prove that G has an involutory automorphism without fixed vertices if and only if $R(G)$ has an involutory automorphism without fixed edges.

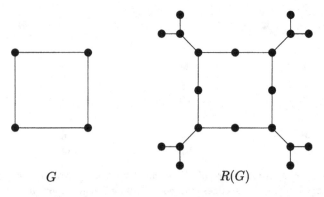

$$G \qquad\qquad R(G)$$

Figure 4. An example of the reduction.

Every involutory automorphism of G without fixed vertices determines an involutory automorphism of $R(G)$ that, thanks to the new vertices, has no fixed edge. On the other hand, consider an arbitrary automorphism ψ of $R(G)$. Since ψ maps the set of vertices of degree 1 in $R(G)$ onto itself, ψ maps every 3-star added in Step 2 into another such 3-star (or itself) and therefore it maps $V(G)$ onto itself. Suppose that u and v are two vertices adjacent in G and let z be the vertex inserted between u and v in Step 1. Then $\psi(z)$ is adjacent in $R(G)$ with both $\psi(u)$ and $\psi(v)$. As easily seen, $\psi(z)$ can appear in $R(G)$ only in Step 1 and therefore $\psi(u)$ and $\psi(v)$ are adjacent in G. This proves that ψ induces an automorphism of G. The latter is involutory if so is ψ. Finally, if ψ has no fixed edge, then every 3-star added in Step 2 is mapped to a different such 3-star and consequently the induced automorphism of G has no fixed vertex.   □

Theorem 5.1 implies that, despite the combinatorial simplicity of an automorphism-based strategy, realizing this strategy by $\mathcal{B}$ on $G \in \mathcal{C}_{\text{auto}}$ requires him to be at least NP powerful. The reason is that an automorphism-based strategy subsumes finding an involutory fixed-edge-free automorphism of any given $G \in \mathcal{C}_{\text{auto}}$, whereas this problem is at least as hard as testing membership in $\mathcal{C}_{\text{auto}}$.

Given the order or the size, there are natural ways of efficiently generating a graph in $\mathcal{C}_{\text{auto}}$ with respect to a certain probability distribution. Theorem 5.1 together with such a generating procedure has two imaginable applications in "real-life" situations.

Negative scenario. Player $\mathcal{B}$ secretly generates $G \in \mathcal{C}_{\text{auto}}$ and makes an offer to $\mathcal{A}$ to choose F at his discretion and play the game AVOID(G, F). If $\mathcal{A}$ accepts, then $\mathcal{B}$, who knows a suitable automorphism of G, follows the automorphism-based strategy and at least does not lose. $\mathcal{A}$ is not able to observe that $G \in \mathcal{C}_{\text{auto}}$, unless he can efficiently solve NP.[3]

Positive scenario. Player $\mathcal{A}$ insists that before the game an impartial third person, hidden from $\mathcal{B}$, permutes at random the vertices of G. Then applying the automorphism-based strategy in the worst case becomes for Player $\mathcal{B}$ as hard as testing isomorphism of graphs. More precisely, Player $\mathcal{B}$ faces the following search problem.

PAR (PERMUTED AUTOMORPHISM RECONSTRUCTION)
Input: G, H, and β, where G and H are isomorphic graphs in $\mathcal{C}_{\text{auto}}$, and β is a fixed-edge-free involutory automorphism of H.
Find: α, a fixed-edge-free involutory automorphism of G.

We relate this problem to GI, the GRAPH ISOMORPHISM problem, that is, given two graphs G_0 and G_1, to recognize if they are isomorphic. We use the notion of the Turing reducibility extended in a natural way over search problems. We

---

[3]We assume here that $\mathcal{A}$ fails to decide if $G \in \mathcal{C}_{\text{auto}}$ at least for some G. We could claim this failure for *most* G if $\mathcal{C}_{\text{auto}}$ would be proven to be complete for the average case [10].

say that two problems are polynomial-time equivalent if they are reducible one to another by polynomial-time Turing reductions.

Theorem 5.2. *The problems* PAR *and* GI *are polynomial-time equivalent.*

Proof. We use the well-known fact that the decision problem GI is polynomial-time equivalent with the search problem of finding an isomorphism between two given graphs [8, Section 1.2].

A *reduction from* PAR *to* GI. We describe a simple algorithm solving PAR under the assumption that we are able to construct a graph isomorphism. Given an input (G, H, β) of PAR, let π be an isomorphism from G to H. As easily seen, computing the composition $\alpha = \pi^{-1}\beta\pi$ gives us a solution of PAR.

A *reduction from* GI *to* PAR. We will describe a reduction to PAR from the problem of constructing an isomorphism between two graphs G_0 and G_1 of the same size. We assume that both G_0 and G_1 are connected and their size is odd. To ensure the odd size, one can just add an isolated edge to both of the graphs. To ensure the connectedness, one can replace the graphs with their complements. To not lose the odd size, the latter operation should be applied only for graphs of order n such that $n(n-1)/2$ is even; hence adding one or two isolated vertices may be required beforehand. If we find an isomorphism between the modified graphs, an isomorphism between the original graphs is easily reconstructed.

We form the triple (G, H, β) by setting $G = G_0 + G_1$, $H = G_0 + G_0$, and taking β to be the identity map between the two copies of G_0. If G_0 and G_1 are isomorphic, this is a legitimate instance of PAR. By the connectedness of G_0 and G_1, if $\alpha : V(G) \to V(G)$ is a solution of PAR on this instance, it either acts within the connected components $V(G_0)$ and $V(G_1)$ independently or maps $V(G_0)$ to $V(G_1)$ and vice versa. The first possibility actually cannot happen because the size of G_0 and G_1 is odd and hence α cannot be at the same time involutory and fixed-edge-free. Thus α is an isomorphism between G_0 and G_1.   □

Question 5.3. Is deciding membership in $\mathcal{C}_{\text{sym}}$ NP-hard? A priori we can say only that $\mathcal{C}_{\text{sym}}$ is in PSPACE. Of course, if the difference $\mathcal{C}_{\text{sym}} \setminus \mathcal{C}_{\text{auto}}$ is decidable in polynomial time, then NP-completeness of $\mathcal{C}_{\text{sym}}$ would follow from Theorem 5.1.

6. Game AVOID(K_n, P_2)

Games on complete graphs are particularly interesting. Notice first of all that in this case a symmetric strategy is not available.

Theorem 6.1. $K_n \notin \mathcal{C}_{\text{sym}}$ *for* $n \geq 4$.

Proof. We describe a strategy of $\mathcal{A}$ that violates the isomorphism between the red and the blue subgraphs at latest in the $(n-1)$-th round. In the first two rounds $\mathcal{A}$ chooses two adjacent edges ensuring that at least one of them is adjacent also to the first edge chosen by $\mathcal{B}$. Thus, after the second round the game can be in one of five positions depicted in Figure 5.

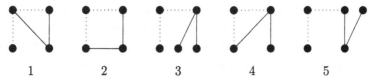

Figure 5. First two rounds of $\mathcal{A}$'s symmetry-breaking strategy ($\mathcal{A}$'s edges dotted, $\mathcal{B}$'s edges continuous).

In positions 1 and 2 $\mathcal{A}$ creates a triangle, which is impossible for $\mathcal{B}$. In positions 3, 4, and 5 $\mathcal{A}$ creates an $(n-1)$-star, while $\mathcal{B}$ is able to create at most an $(n-2)$-star (in position 5 $\mathcal{A}$ first of all chooses the uncolored edge connecting two vertices of degree 2).   □

Let us define $\mathcal{C}_{\mathrm{II}}$ to be the class of all graphs G such that, for all F, $\mathcal{B}$ has a nonlosing strategy in the game $\mathrm{AVOID}(G, F)$. Clearly, $\mathcal{C}_{\mathrm{II}}$ contains $\mathcal{C}_{\mathrm{sym}}$. It is easy to check that K_4 is in $\mathcal{C}_{\mathrm{II}}$, and therefore $\mathcal{C}_{\mathrm{sym}}$ is a proper subclass of $\mathcal{C}_{\mathrm{II}}$.

It is an interesting question if $K_n \in \mathcal{C}_{\mathrm{II}}$ for all n. We examine the case of a forbidden subgraph $F = P_2$, a path of length 2. For all $n > 2$, we describe an efficient winning strategy for $\mathcal{B}$ in $\mathrm{AVOID}(K_n, P_2)$. Somewhat surprisingly, this strategy, in contrast to Theorem 6.1, proves to be symmetric in a weaker sense.

More precisely, we say that a strategy of $\mathcal{B}$ is *symmetric in AVOID(G, F)* if, independently of $\mathcal{A}$'s strategy, the red and the blue subgraphs are isomorphic after every move of $\mathcal{B}$ in the game. Let us stress the difference with the notion of a symmetric strategy on G we used so far. While a strategy symmetric on G guarantees the isomorphism until G is completely colored (except one edge if G has odd size), a strategy symmetric in $\mathrm{AVOID}(G, F)$ guarantees the isomorphism only as long as $\mathcal{A}$ does not lose in $\mathrm{AVOID}(G, F)$.

Theorem 6.2. *Player $\mathcal{B}$ has a symmetric strategy in the game $AVOID(K_n, P_2)$.*

Proof. Let A_i (resp. B_i) denote the set of the edges chosen by $\mathcal{A}$ (resp. $\mathcal{B}$) in the first i rounds. The strategy of $\mathcal{B}$ is, as long as A_i is a matching, to choose an edge so that the subgraph of K_n with edge set $A_i \cup B_i$ is a path. The only case when this is impossible is that n is even and $i = n/2$. Then $\mathcal{B}$ chooses the edge that makes $A_i \cup B_i$ a Hamiltonian cycle (see Figure 6).   □

Figure 6. A game (K_6, P_2) ($\mathcal{A}$'s edges dotted, $\mathcal{B}$'s edges continuous).

Question 6.3. What is the complexity of deciding, given G, whether or not $\mathcal{B}$ has a winning strategy in $\text{AVOID}(G, P_2)$?

It is worth noting that in [16], PSPACE-completeness of the winner recognition in the avoidance game with precoloring is proven even for a fixed forbidden graph F, namely for two triangles with a common vertex called the "bowtie graph". Notice also that $\text{AVOID}(G, P_2)$ has an equivalent vertex-coloring version: the players color the vertices of the line graph $L(G)$ and the loser is the one who creates two adjacent vertices of the same color.

Question 6.4. Does K_n belong to $\mathcal{C}_{II}$? In particular, does $\mathcal{B}$ have winning strategies in $\text{AVOID}(K_n, K_{1,3})$, $\text{AVOID}(K_n, P_3)$, and $\text{AVOID}(K_n, K_3)$ for large enough n?

References

[1] J. Beck. Van der Waerden and Ramsey type games. *Combinatorica* 1:103–116 (1981).

[2] E. R. Berlekamp, J. H. Conway, R. K. Guy. *Winning ways for your mathematical plays*. Academic Press, New York (1982).

[3] A. Berman, A. Kotzig, G. Sabidussi. Antipodal graphs of diameter 4 and extremal girth. In: *Contemporary methods in graph theory*, R. Bodendiek (ed.), BI-Wiss.-Verl., Mannheim, pp. 137–150 (1990).

[4] P. Erdős, J. L. Selfridge. On a combinatorial game. *J. Combin. Theory A* 14:298–301 (1973).

[5] M. Erickson, F. Harary. Generalized Ramsey theory XV: Achievement and avoidance games for bipartite graphs. *Graph theory*, Proc. 1st Southeast Asian Colloq., Singapore 1983, Lect. Notes Math. 1073:212–216 (1984).

[6] F. Harary. *Graph theory*. Addison-Wesley, Reading MA (1969).

[7] F. Harary. Achievement and avoidance games for graphs. *Ann. Discrete Math.* 13:111–119 (1982).

[8] J. Köbler, U. Schöning, and J. Torán. *The Graph Isomorphism problem: its structural complexity*. Birkhäuser (1993).

[9] A. Kotzig On centrally symmetric graphs. (in Russian) *Czech. Math. J.* 18(93):606–615 (1968).

[10] L. A. Levin. Average case complete problems. *SIAM J. Comput.* 15:285–286 (1986).

[11] A. Lubiw. Some NP-complete problems similar to graph isomorphism. *SIAM J. Comput.* 10:11–21 (1981).

[12] E. Mead, A. Rosa, C. Huang. The game of SIM: A winning strategy for the second player. *Math. Mag.* 47:243–247 (1974).

[13] A. Pekeč. A winning strategy for the Ramsey graph game. *Comb. Probab. Comput.* 5(3):267–276 (1996).

[14] A. G. Robinson, A. J. Goldman. The Set Coincidence Game: Complexity, Attainability, and Symmetric Strategies. *J. Comput. Syst. Sci.* 39(3):376–387 (1989).

[15] G. J. Simmons. The game of SIM. *J. Recreational Mathematics*, 2(2):66 (1969).

[16] W. Slany. Endgame problems of Sim-like graph Ramsey avoidance games are PSPACE-complete. *Theoretical Computer Science*, to appear.

FRANK HARARY
COMPUTER SCIENCE DEPARTMENT
NEW MEXICO STATE UNIVERSITY
LAS CRUCES, NM 88003
UNITED STATES
fnh@crl.nmsu.edu

WOLFGANG SLANY
INSTITUT FÜR INFORMATIONSSYSTEME
TECHNISCHE UNIVERSITÄT WIEN
FAVORITENSTR. 9
A-1040 WIEN
AUSTRIA
wsi@dbai.tuwien.ac.at

OLEG VERBITSKY
DEPARTMENT OF MECHANICS AND MATHEMATICS
LVIV UNIVERSITY
UNIVERSYTETSKA 1
79000 LVIV
UKRAINE
oleg@ov.litech.net

More Games of No Chance
MSRI Publications
Volume **42**, 2002

A Simple FSM-Based Proof
of the Additive Periodicity
of the Sprague-Grundy Function
of Wythoff's Game

HOWARD A. LANDMAN

ABSTRACT. Dress et al.[4] recently proved the additive periodicity of rows of the Sprague–Grundy function[7,5] of a class of Nim-like games including Wythoff's Game[8]. This implies that the function for each row minus its saltus is periodic and can be computed by a finite state machine.

The proof presented here, which was developed independently and in ignorance of the above result, proceeds in exactly the opposite direction by explicitly constructing a finite state machine with $o(n^2)$ bits of state to compute $\mathcal{H}(m, n) = \mathcal{G}(m, n) - m + 2n$, for all m and fixed n. From this, the periodicity of $\mathcal{H}$ and the additive periodicity of $\mathcal{G}$ follow trivially.

1. Introduction

Note: The original lecture on this material is available on streaming video at http://www.msri.org/publications/ln/msri/2000/gametheory/landman/1/.

Wythoff's Game (also called Wythoff's Nim or simply Wyt) is an impartial 2-player game played with 2 piles of counters. Each player may, on their turn, remove any number of counters from either pile, or remove the same number of counters from both piles. The player removing the last counter wins. The set of winning positions (those for which the Sprague–Grundy value $\mathcal{G}(m, n)$ equals 0) is well-known[3,1]. However, no direct formula is known for computing the Sprague–Grundy value $\mathcal{G}(m, n)$ of an arbitrary position with m counters in one pile and n in the other; it appears necessary to compute all the $\mathcal{G}(i, j)$-values for smaller games ($i \leq m$, $j \leq n$, $i + j < m + n$) first.

It is clearly impossible to compute the n-th row of $\mathcal{G}$ (that is, the sequence $G_n(m) = \mathcal{G}(m, n)$) directly with a finite state machine. One reason is that the values grow without bound, and thus the number of bits needed to represent the value will eventually exceed the capacity of any FSM. It also appears at first that the FSM would have to remember an ever-growing number of values which have already been used. The following lemmas provide bounds on $\mathcal{G}$ and allow us to overcome these obstacles.

2. Bounds on $\mathcal{G}(m, n)$

Lemma 1. $\mathcal{G}(m, n) \leq m + n$.

Proof: by induction on $m + n$. First observe that $\mathcal{G}(0,0) = 0 \leq 0 + 0$. Assume that, for all i, j such that $i + j \leq m + n$, we have $\mathcal{G}(i, j) \leq i + j$. When calculating $\mathcal{G}(m, n)$, the set A of values to be excluded contains only values which are $< m + n$. So $\mathcal{G}(m, n) = \text{mex}(A) \leq \max(A) + 1 \leq (m + n - 1) + 1 = m + n$.

Lemma 2. $\mathcal{G}(m, n) \geq m - 2n$.

Proof: Assume $g = \mathcal{G}(m, n) < m - 2n$. This implies that g did not appear as any $\mathcal{G}(k, n)$ for $k < m$ (m times in all). Now there are only three reasons why g would not appear for any particular k. Either $\mathcal{G}(k, n) < g$, which can only happen at most g times, or g cannot appear because some $\mathcal{G}(k, j)$ for $j < n$ equals g, or g cannot appear because some $\mathcal{G}(k - i, n - i)$ for $0 \leq i \leq \min(k, n)$ equals g. But since g can appear at most once in each row, the second and third reasons can only happen at most n times each. So the total number of times g does not occur must be no more than $g + 2n < (m - 2n) + 2n = m$, which is a contradiction.

We note in passing that every non-negative integer g must eventually appear in each row, that is, for all g and n there exists m such that $g = \mathcal{G}(m, n)$. Not only must each g appear, but the above lemmas bound its position m: $g - n \leq m \leq g + 2n$. Since we also have (by definition of mex) that no integer can appear more than once, each row is a permutation of the non-negative integers.

3. Main Theorem

If we now define $\mathcal{H}(m, n) = \mathcal{G}(m, n) - m + 2n$, we can see from the above lemmas that $0 \leq \mathcal{H}(m, n) \leq 3n$ for all m. (Other saltus-adjusted functions besides $\mathcal{H}$ would also work for what follows, but this $\mathcal{H}$ seems one of the more natural choices.) If we know $\mathcal{H}$ we can compute $\mathcal{G}(m, n) = \mathcal{H}(m, n) + m - 2n$ quite easily. $\mathcal{G}$ is additively periodic iff $\mathcal{H}$ is periodic. And the values of $\mathcal{H}$ do not grow without bound, so the first obstacle is obviously overcome. What is less obvious but nonetheless true is that the second obstacle is also overcome.

Define the Left, Slanting, and Down sets: Let $L(m, n)$ be the set of integers which appear to the left of $\mathcal{G}(m, n)$, i.e. $L(m, n) = \{\mathcal{G}(m - k, n) : 1 \leq k \leq m\}$. Similarly let $S(m, n) = \{\mathcal{G}(m - k, n - k) : 1 \leq k \leq \min(m, n)\}$ and $D(m, n) = \mathcal{G}(m, n - k) : 1 \leq k \leq n$. L contains the $\mathcal{G}$-values corresponding to moves which remove counters from the pile of m, D to those which remove conters from the pile of n, and S to those which remove counters from both piles. We can calculate

$\mathcal{G}$ or $\mathcal{H}$ from L, S, and D, since $A(m,n) = L(m,n) \cup S(m,n) \cup D(m,n)$ is the set of elements to be excluded when calculating $\mathcal{G}$, that is, $\mathcal{G} = \text{mex}(A)$. And both S and D are bounded in size; they each never have more than n elements. Unfortunately $L(m,n)$ grows arbitrarily large as m increases, so it cannot be directly held in a FSM. We need to find a more clever and compact way to represent these sets.

Let $L'(m,n) = ((m-2n)\ldots(m+n)) - L(m,n)$, which, by lemmas 1 and 2, is the set of all numbers $\leq m+n$ which are *not* in L. $L'(m,n)$ can be represented as a bit-array of $3n+1$ bits for all m, indicating which of the numbers from $m-2n$ to $m+n$ it contains. We similarly construct $S'(m,n)$ and $D'(m,n)$, which have the same size. By definition we have $\mathcal{G}(m,n) = \text{mex}(L(m,n) \cup S(m,n) \cup D(m,n)) = \min(L'(m,n) \cap S'(m,n) \cap D'(m,n))$.

To compute $\mathcal{H}(m,n)$ for all m, it is sufficient to keep track of the sets $L'(m,0)$, $L'(m,1)$, ..., $L'(m,n-1)$, $L'(m,n)$, $S'(m,0)$, $S'(m,1)$, ..., $S'(m,n-1)$, $S(m,n)$, and $D'(m,n)$. ($D'(m,0)$, $D'(m,1)$, ..., $D'(m,n-1)$ are all subsets of $D'(m,n)$ and do not need to be stored separately if the bit-array representation is used.) This gives $o(n)$ sets which are each $o(n)$ bits, for a total number of bits which is $o(n^2)$. Only the sets $L'(m,n)$, $S'(m,n)$, and $D'(m,n)$ are needed directly to compute $\mathcal{H}(m,n)$ and $L'(m+1,n)$, but the other sets are needed to compute e.g. $S'(m+1,n)$ and $D'(m+1,n)$.

For example, to calculate $L'(m+1,k)$, we take $L'(m,k)$, unset the bit corresponding to $\mathcal{H}(m,k)$, shift over 1, and set the end bit corresponding to $m+k+1$. This is guaranteed never to shift a set bit off the other end (due to lemma 2), so the number of set bits in $L'(m,k)$ is constant (equal to $k+1$) as m varies. Similarly, $S'(m+1,k)$ can be calculated from $S'(m,k-1)$ and $\mathcal{H}(m,k)$. The $D'(m+1,k)$ can be computed "bottom up" in sequence starting from $k = 0$.

In addition to the above storage requirements, a program to calculate $\mathcal{H}$ by this method might also need one or more variables whose value is bounded by n (for example, to do the counting from $k = 0$ to n while computing D'). However, a finite number of these suffice, and each of them requires only $o(\log(n))$ bits. Therefore the n-th row can be computed by a finite state machine with $b = o(n^2)$ total bits of state.

This result tells us immediately that $\mathcal{H}(m,n)$ is eventually periodic for fixed n, since after no more than 2^b steps it must reenter a state previously visited and thereafter loop. Therefore $G_n(m)$ must be additively periodic.

4. Conclusion

The key point in the above proof was the finiteness of storage requirements. It was critical to show that the storage required did not depend on m, which increases, but rather only on n, which is constant as m varies. The precise amount of storage, or even its being $o(n^2)$, was not really important.

From the computational point of view, it is much simpler to prove periodicity of a bounded function than additive periodicity of an unbounded function, even though the two may be logically equivalent. This is because periodicity is equivalent to computability by a finite state machine; while additive periodicity requires a machine with unbounded storage, placing it in a different and harder computational complexity class.

The technique presented here should be generalizable to many other impartial games, such as the various proposed extensions of Wythoff's Game to more than 2 piles[2,6].

5. Bibliography

[1] U. Blass and A. S. Fraenkel [1990], The Sprague–Grundy function of Wythoff's game, *Theoret. Comput. Sci.* (Math Games) **75**, 311–333.

[2] I. G. Connell [1959], A generalization of Wythoff's game, *Canad. Math. Bull.* **2**, 181–190.

[3] H. S. M. Coxeter [1953], The golden section, phyllotaxis and Wythoff's game, *Scripta Math.* **19**, 135–143.

[4] A. Dress, A. Flammenkamp and N. Pink [1999], Additive periodicity of the Sprague–Grundy function of certain Nim games, *Adv. in Appl. Math.* **22**, 249–270.

[5] P. M. Grundy [1964], Mathematics and Games, *Eureka* **27**, 9–11, originally published: *ibid.* **2** (1939), 6–8.

[6] A. S. Fraenkel and I. Borosh [1973], A generalization of Wythoff's game, *J. Combin. Theory* (Ser. A) **15**, 175–191.

[7] R. Sprague [1935–36], Über mathematische Kampfspiele, *Tôhoku Math. J.* **41**, 438–444.

[8] W. A. Wythoff [1907], A modification of the game of Nim, *Nieuw Arch. Wisk.* **7**, 199–202.

HOWARD A. LANDMAN
FORT COLLINS, CO 80521
howard@riverrock.org

Puzzles and Life

More Games of No Chance
MSRI Publications
Volume **42**, 2002

The Complexity of Clickomania

THERESE C. BIEDL, ERIK D. DEMAINE, MARTIN L. DEMAINE,
RUDOLF FLEISCHER, LARS JACOBSEN, AND J. IAN MUNRO

ABSTRACT. We study a popular puzzle game known variously as Clickomania and Same Game. Basically, a rectangular grid of blocks is initially
colored with some number of colors, and the player repeatedly removes a
chosen connected monochromatic group of at least two square blocks, and
any blocks above it fall down. We show that one-column puzzles can be
solved, i.e., the maximum possible number of blocks can be removed, in
linear time for two colors, and in polynomial time for an arbitrary number
of colors. On the other hand, deciding whether a puzzle is solvable (all
blocks can be removed) is NP-complete for two columns and five colors, or
five columns and three colors.

1. Introduction

Clickomania is a one-player game (puzzle) with the following rules. The board
is a rectangular grid. Initially the board is full of square blocks each colored
one of k colors. A *group* is a maximal connected monochromatic polyomino;
algorithmically, start with each block as its own group, then repeatedly combine
groups of the same color that are adjacent along an edge. At any step, the
player can select (*click*) any group of size at least two. This causes those blocks
to disappear, and any blocks stacked above them fall straight down as far as they
can (the *settling* process). Thus, in particular, there is never an internal hole.
There is an additional twist on the rules: if an entire column becomes empty
of blocks, then this column is "removed," bringing the two sides closer to each
other (the *column shifting* process).

The basic goal of the game is to remove all of the blocks, or to remove as
many blocks as possible. Formally, the basic decision question is whether a
given puzzle is *solvable*: can all blocks of the puzzle be removed? More generally, the algorithmic problem is to find the maximum number of blocks that

Biedl and Munro were partially supported by NSERC. This work was done while Demaine,
Demaine, Fleischer, and Jacobsen were at the University of Waterloo.

can be removed from a given puzzle. We call these problems the *decision* and *optimization* versions of Clickomania.

There are several parameters that influence the complexity of Clickomania. One obvious parameter is the number of colors. For example, the problem is trivial if there is only one color, or every block is a different color. It is natural to ask whether there is some visible difference, in terms of complexity, between a constant number of colors and an arbitary number of colors, or between one constant number of colors and another. We give a partial answer by proving that even for just three colors, the problem is NP-complete. The complexity for two colors remains open.

Other parameters to vary are the number of rows and the number of columns in the rectangular grid. A natural question is whether enforcing one of these dimensions to be constant changes the complexity of the problem. We show that even for just two columns, the problem is NP-complete, whereas for one column (or equivalently, one row), the problem is solvable in polynomial time. It remains open precisely how the number of rows affects the complexity.

1.1. History. The origins of Clickomania seem unknown. We were introduced to the game by Bernie Cosell [1], who suggested analyzing the strategy involved in the game. In a followup email, Henry Baker suggested the idea of looking at a small constant number of colors. In another followup email, Michael Kleber pointed out that the game is also known under the title "Same Game."

Clickomania! is implemented by Matthias Schuessler in a freeware program for Windows, available from http://www.clickomania.ch/click/. On the same web page, you can find versions for the Macintosh, Java, and the Palm Pilot. There is even a "solver" for the Windows version, which appears to be based on a constant-depth lookahead heuristic.

1.2. Outline. The rest of this paper is outlined as follows. Section 2 describes several polynomial-time algorithms for the one-column case. Section 3 proves that the decision version of Clickomania is NP-complete for 5 colors and 2 columns. Section 4 gives the much more difficult NP-completeness proof for 2 colors and 5 columns. We conclude in Section 5 with a discussion of two-player variations and other open problems.

2. One Column in Polynomial Time

In this section we describe polynomial-time algorithms for the decision version and optimization version of one-column Clickomania (or equivalently, one-row Clickomania). In this context, a group with more than 2 blocks is equivalent to a group with just 2 blocks, so in time linear in the number of blocks we can reduce the problem to have size linear in the number of groups, n.

First, in Section 2.1, we show how to reduce the optimization version to the decision version by adding a factor of $O(n^2)$. Second, in Section 2.2, we give a general algorithm for the decision question running in $O(kn^3)$ where k is the number of colors, based on a context-free-grammer formulation. Finally, in Section 2.3, we improve this result to $O(n)$ time for $k = 2$ colors, using a combinatorial characterization of solvable puzzles for this case.

2.1. Reducing Optimization to Decision. If a puzzle is solvable, the optimization version is equivalent to the decision version (assuming that the algorithm for the decision version exhibits a valid solution, which our algorithms do). If a puzzle is not solvable, then there are some groups that are never removed. If we knew one of the groups that is not removed, we would split the problem into two subproblems, which would be independent subpuzzles of the original puzzle.

Thus, we can apply a dynamic-programming approach. Each subprogram is a consecutive subpuzzle of the puzzle. We start with the solvable cases, found by the decision algorithm. We then build up a solution to a larger puzzle by choosing an arbitrary group not to remove, adding up the scores of the two resulting subproblems, and maximizing over all choices for the group not to remove. If the decision version can be solved in $d(n, k)$ time, then this solution to the optimization version runs in $O(n^2 d(n, k) + n^3)$ time. It is easy to see that $d(n, k) = \Omega(n)$, thus proving

Lemma 1. *If the decision version of one-column Clickomania can be solved in $d(n, k)$ time, then the optimization version can be solved in $O(n^2 d(n, k))$ time.*

2.2. A General One-Column Solver. In this section we show that one-column Clickomania reduces to parsing context-free languages. Because strings are normally written left-to-right and not top-down, we speak about one-row Clickomania in this subsection, which is equivalent to one-column Clickomania. We can write a one-row k-color Clickomania puzzle as a word over the alphabet $\Sigma = \{c_1, \ldots, c_k\}$. Such words and Clickomania puzzles are in one-to-one correspondence, so we use them interchangably.

Now consider the context-free grammar

$$G : \ S \rightarrow \Lambda \mid SS \mid$$
$$c_i S c_i \mid c_i S c_i S c_i \quad \forall i \in \{1, 2, \ldots, k\}$$

We claim that a word can be parsed by this grammar precisely if it is solvable.

Theorem 2. *The context-free language $L(G)$ is exactly the language of solvable one-row Clickomania puzzles.*

Any solution to a Clickomania puzzle can be described by a sequence of moves (clicks), $m_1, m_2, \ldots m_s$, such that after removing m_s no blocks remain. We call a solution *internal* if the leftmost and rightmost blocks are removed in the last two moves (or the last move, if they have the same color). Note that in an internal solution we can choose whether to remove the leftmost or the rightmost block in the last move.

Lemma 3. *Every solvable one-row Clickomania puzzle has an internal solution.*

Proof. Let $m_1, \ldots, m_{b-1}, m_b, m_{b+1}, \ldots, m_s$ be a solution to a one-row Clickomania puzzle, and suppose that the leftmost block is removed in move m_b. Because move m_b removes the leftmost group, it cannot form new clickable groups. The sequence $m_1, \ldots, m_{b-1}, m_{b+1}, \ldots, m_s$ is then a solution to the same puzzle except perhaps for the group containing the leftmost block. If the leftmost block is removed in this subsequence, continue discarding moves from the sequence until the remaining subsequence removes all but the group containing the leftmost block. Now the puzzle can be solved by adding one more move, which removes the last group containing the leftmost block. Applying the same argument to the rightmost block proves the lemma.   □

We prove Theorem 2 in two parts:

Lemma 4. *If $w \in L(G)$, then w is solvable.*

Proof. Because $w \in L(G)$, there is a derivation $S \Rightarrow^* w$. The proof is by induction on the length n of this derivation. In the base case, $n = 1$, we have $w = \Lambda$, which is clearly solvable. Assume all strings derived in at most $n-1$ steps are solvable, for some $n \geq 2$. Now consider the first step in a n-step derivation. Because $n \geq 2$, the first production cannot be $S \rightarrow \Lambda$. So there are three cases.

- $S \Rightarrow SS \Rightarrow^* w$:

  In this case $w = xy$, such that $S \Rightarrow^* x$ and $S \Rightarrow^* y$ both in at most $n - 1$ steps. By the induction hypothesis, x and y are solvable. By Lemma 3, there are internal solutions for x and y, where the rightmost block of x and the leftmost block of y are removed last, respectively. Doing these two moves at the very end, we can now arbitrarily merge the two move sequences for x and y, removing all blocks of w.

- $S \Rightarrow c_i S c_i \Rightarrow^* w$:

  In this case $w = c_i x c_i$, such that $S \Rightarrow^* x$ in at most $n - 1$ steps. By the induction hypothesis, x is solvable. By Lemma 3, there is an internal solutions for x; if either the leftmost or rightmost block of x has color i, it can be chosen to be removed in the last move. Therefore, the solution for x followed by removing the remaining $c_i c_i$ (if it still exists) is a solution to w.

- $S \Rightarrow c_i S c_i S c_i \Rightarrow^* w$:

  This case is analogous to the previous case.   □

Lemma 5. *If $w \in \Sigma^*$ is solvable, then $w \in L(G)$.*

Proof. Suppose $w \in \Sigma^*$ be solvable. We will prove that $w \in L(G)$ by induction on $|w|$. The base case, $|w| = 0$ follows since $\Lambda \in L(G)$. Assume all solvable strings of length at most $n - 1$ are in $L(G)$, for some $n \geq 1$. Consider the case $|w| = n$.

Since w is solvable, there is a first move in a solution to w, let's say removing a group c_i^m for $m \geq 2$. Thus, $w = x c_i^m y$. Now, neither the last symbol of x nor the first symbol of y can be c_i. Let $w' = xy$. Since $|w'| \leq |w| - 2 = n - 2$, and w' is solvable, w' is in $L(G)$ by the induction hypothesis.

Observe that $c_i^m \in L(G)$ by one of the derivations:

$$S \Rightarrow^{(m-3)/2} c_i^{(m-3)/2} S c_i^{(m-3)/2} \Rightarrow c_i^{(m-3)/2} S c_i S c_i^{(m-3)/2} \Rightarrow^2 c_i^m$$

if m is odd, or

$$S \Rightarrow^{m/2} c_i^{m/2} S c_i^{m/2} \Rightarrow c_i^m$$

if m is even. Thus, if $x = \Lambda$, w can be derived as $S \Rightarrow SS \Rightarrow^* c_i^m S \Rightarrow^* c_i^m y = w$. Analogously for $y = \Lambda$. It remains to consider the case $x, y \neq \Lambda$.

Consider the first step in a derivation for w'. There are three cases.

- $S \Rightarrow SS \Rightarrow^* uS \Rightarrow^* uv = w'$:

  We can assume that $u, v \neq \Lambda$, otherwise we consider the derivation of w' in which this first step is skipped. By Lemma 4, u and v are both solvable. Consider the substring c_i^m of w that was removed in the first move. Either $w = u_1 c_i^m u_2 v$ (u_2 possibly empty) or $w = u v_1 c_i^m v_2$ (v_1 possibly empty). Without loss of generality, we assume the former case, i.e., $u = u_1 u_2$. Then $u' = u_1 c_i^m u_2$ is solvable because u is solvable and m was maximal. Since $v \neq \Lambda$, it follows that $|u'| < |w|$, and by the induction hypothesis, $u' \in L(G)$. Hence $S \Rightarrow SS \Rightarrow^* u'S \Rightarrow^* u'v = w$ is a derivation of w and $w \in L(G)$.

- $S \Rightarrow c_j S c_j \Rightarrow^* c_j u c_j = w'$:

  Since $x, y \neq \Lambda$, it must be the case that $w = c_j u_1 c_i^m u_2 c_j$, where $u = u_1 u_2$. By Lemma 4, u is solvable, hence so is $u' = u_1 c_i^m u_2$ because m was maximal. Moreover, $|u'| = |w| - 2$ and thus $u' \in L(G)$ by the induction hypothesis and $S \Rightarrow c_j S c_j \Rightarrow^* c_j u' c_j = w \in L(G)$.

- $S \Rightarrow c_j S c_j S c_j \Rightarrow^* c_j u c_j v c_j = w'$:

  Since $x, y \neq \Lambda$, either $w = c_j u_1 c_i^m u_2 c_j v c_j$ and $u = u_1 u_2$, or $w = c_j u c_j v_1 c_i^m v_2 c_j$ and $u = v_1 v_2$. Without loss of generality, assume $w = c_j u_1 c_i^m u_2 c_j v c_j$. Analogously to the previous case, $u' = u_1 c_i^m u_2 \in L(G)$, hence $S \Rightarrow c_j S c_j S c_j \Rightarrow^* c_j u' c_j v c_j = w \in L(G)$. $\square$

Thus, deciding if a one-row Clickomania puzzle is solvable reduces to deciding if the string w corresponding to the Clickomania puzzle is in $L(G)$. Since deciding $w \in L(G)$ is in P, so is deciding if a one-row Clickomania is solvable. This completes the proof of Theorem 2. In particular, we can obtain a polynomial-time algorithm for one-row Clickomania by applying standard parsing algorithms for context-free grammars.

Corollary 6. *We can decide in $O(kn^3)$ time whether a one-row (or one-column) k-color Clickomania puzzle is solvable.*

Proof. The context-free grammar can be converted into a grammar in Chomsky normal form of size $O(k)$ and with $O(1)$ nonterminals. The algorithm in [4, Theorem 7.14, pp. 240–241] runs in time $O(n^3)$ times the number of nonterminals plus the number of productions, which is $O(k)$. $\square$

Applying Lemma 1, we obtain

Corollary 7. *One-row (or one-column) k-color Clickomania can be solved in $O(kn^5)$ time.*

2.3. A Linear-Time Algorithm for Two Colors. In this section, we show how to decide solvability of a one-column two-color Clickomania puzzle in linear time. To do so, we give necessary and sufficient combinatorial conditions for a puzzle to be solvable. As it turns out, these conditions are very different depending on whether the number of groups in the puzzle is even or odd, with the odd case being the easier one.

We assume throughout the section that the groups are named $g_1, \ldots, g_n$. A group with just one block is called a *singleton*, and a group with at least two blocks in it is called a *nonsingleton*.

The characterization is based on the following simple notion. A *checkerboard* is a maximal-length sequence of consecutive groups each of size one. For a checkerboard C, $|C|$ denotes the number of singletons it contains. The following lemma formalizes the intuition that if a puzzle has a checkerboard longer than around half the total number of groups, then the puzzle is unsolvable.

Lemma 8. *Consider a solvable one-column two-color Clickomania puzzle with n groups, and let C be the longest checkerboard in this puzzle.*

(i) *If C is at an end of the puzzle, then $|C| \leq \frac{n-1}{2}$.*
(ii) *If C is strictly interior to the puzzle, then $|C| \leq \frac{n-2}{2}$.*

Proof. (i) Each group g of the checkerboard C must be removed. This is only possible if g is merged with some other group of the same color not in C, so there are at least $|C|$ groups outside of C. These groups must be separated from C by at least one extra group. Therefore, $n \geq 2|C| + 1$ or $|C| \leq \frac{n-1}{2}$.
(ii) Analogously, if C is not at one end of the puzzle, then there are two extra groups at either end of C. Therefore, $n \geq 2|C| + 2$ or $|C| \leq \frac{n-2}{2}$.   □

2.3.1. An Odd Number of Groups. The condition in Lemma 8 is also sufficient if the number of groups is odd (but not if the number of groups is even). The idea is to focus on the *median* group, which has index $m = \frac{n+1}{2}$. This is motivated by the following fact:

Lemma 9. *If the median group has size at least two, then the puzzle is solvable.*

Proof. Clicking on the median group removes that piece and merges its two neighbors into the new median group (it has two neighbors because n is odd). Therefore, the resulting puzzle again has a median group with size at least two, and the process repeats. In the end, we solve the puzzle.   □

Theorem 10. *A one-column two-color Clickomania puzzle with an odd number of groups, n, is solvable if and only if*

- *the length of the longest checkerboard is at most* $(n-3)/2$; *or*
- *the length of the longest checkerboard is exactly* $(n-1)/2$, *and the checkerboard occurs at an end of the puzzle.*

Proof. If the puzzle contains a checkerboard of length at least $m = \frac{n+1}{2}$, then it is unsolvable by Lemma 8. If the median has size at least two, then we are also done by Lemma 9, so we may assume that the median is a singleton. Thus there must be a nonsingleton somewhere to the left of the median that is not the leftmost group, and there must be a nonsingleton to the right of the median that is not the rightmost group. Also, there are two such nonsingletons with at most $\frac{n-2}{2}$ other groups between them.

Clicking on any one of these nonsingletons destroys two groups (the clicked-on group disappears, and its two neighbors merge). The new median moved one group right [left] of the old one if we clicked on the nonsingleton left [right] of the median. The two neighbors of the clicked nonsingleton merge into a new nonsingleton, and this new nonsingleton is one closer to the other nonsingleton than before. Therefore, we can continue applying this procedure until the median becomes a nonsingleton and then apply Lemma 9. Note that if one of the two nonsingletons ever reaches the end of the sequence then the other singleton must be the median. ☐

Note that there is a linear-time algorithm implicit in the proof of the previous lemma, so we obtain the following corollary.

Corollary 11. *One-column two-color Clickomania with n groups can be decided in time $O(n)$ if n is odd. If the problem is solvable, a solution can also be found in time $O(n)$.*

2.3.2. An Even Number of Groups. The characterization in the even case reduces to the odd case, by showing that a solvable even puzzle can be split into two solvable odd puzzles.

Theorem 12. *A one-column two-color Clickomania puzzle, $g_1, \ldots, g_n$, with n even is solvable if and only if there is an odd index i such that $g_1, \ldots, g_i$ and $g_{i+1}, \ldots, g_n$ are solvable puzzles.*

Proof. Sufficiency is a straightforward application of Lemma 3. First solve the instance $g_1, \ldots, g_i$ so that all groups but g_i disappear and g_i becomes a nonsingleton. Then solve instance $g_{i+1}, \ldots, g_n$ so that all groups but g_{i+1} disappear and g_{i+1} becomes a nonsingleton. These two solutions can be executed independently because g_i and g_{i+1} form a "barrier." Then g_i and g_{i+1} can be clicked to solve the puzzle.

For necessity, assume that $m_1, \ldots, m_l$ is a sequence of clicks that solves the instance. One of these clicks, say m_j, removes the blocks of group g_1. (Note that this group might well have been merged with other groups before, but we are interested in the click that actually removes the blocks.) Let i be maximal such that the blocks of group g_i are also removed during click m_j.

Clearly i is odd, since groups g_1 and g_i have the same color and we have only two colors. It remains to show that the instances $g_1, \ldots, g_i$ and $g_{i+1}, \ldots, g_n$ are solvable.

The clicks $m_1, \ldots, m_{j-1}$ can be distinguished into two kinds: those that affect blocks to the left of g_i, and those that affect blocks to the right of g_i. (Since g_i is not removed before m_j, a click cannot be of both kinds.)

Consider those clicks that affect blocks to the left of g_i, and apply the exact same sequence of clicks to instance $g_1, \ldots, g_i$. Since m_j removes g_1 and g_i at once, these clicks must have removed all blocks $g_2, \ldots, g_{i-1}$. They also merged g_1 and g_i, so that this group becomes a nonsingleton. One last click onto g_i hence gives a solution to instance $g_1, \ldots, g_i$.

Consider those clicks before m_j that affect blocks to the right of g_i. None of these clicks can merge g_i with a block g_k, $k > i$, since this would contradict the definition of i. Hence it does not matter whether we execute these clicks before or after m_j, as they have no effect on g_i or the blocks to the left of it.

If we took these clicks to the right of g_i, and combine them with the clicks after m_j (note that at this time, block g_i and everything to the left of it is gone), we obtain a solution to the instance $g_{i+1}, \ldots, g_n$. This proves the theorem. $\qquad\Box$

Using this theorem, it is possible to decide in linear time whether an even instance of one-column two-color Clickomania is solvable, though the algorithm is not as straightforward as in the odd case. The idea is to proceed in two scans of the input. In the first scan, in forward order, we determine for each odd index i whether $g_1, \ldots, g_i$ is solvable. We will explain below how to do this in amortized constant time. In the second scan, in backward order, we determine for each odd index i whether $g_{i+1}, \ldots, g_n$ is solvable. If any index appears in both scans, then we have a solution, otherwise there is none.

So all that remains to show is how to determine whether $g_1, \ldots, g_i$ is solvable in amortized constant time. (The procedure is similar for the reverse scan.) Assume that we are considering group g_i, $i = 1, \ldots, n$. Throughout the scan we maintain three indices, j, k and l. We use j and k to denote the current longest checkerboard from g_j to g_k. Index l is the minimal index such that $g_l, \ldots, g_i$ is a checkerboard. We initialize $i = j = k = l = 0$.

When considering group g_i, we first update l. If g_i is a singleton, then l is unchanged. Otherwise, $l = i + 1$. Next, we update j and k, by verifying whether $i - l > k - j$, and if so, setting $j = l$ and $k = i$. Clearly, this takes constant time.

For odd i, we now need to verify whether the instance $g_1, \ldots, g_i$ is solvable. This holds if $(k+1) - j \le (i-3)/2$, since then the longest checkerboard is short enough. If $(k+1) - j \ge (i+1)/2$, then the instance is not solvable. The only case that requires a little bit of extra work is $(k+1) - j = (i-1)/2$, since we then must verify whether the longest checkerboard is at the beginning or the end. This, however, is easy. If the longest checkerboard has length $(i-1)/2$ and is at the beginning or the end, then the median group of the instance $g_1, \ldots, g_i$, i.e., $g_{(i-1)/2}$ must be a nonsingleton. If the longest checkerboard is not at the beginning or the end, then the median group is a singleton. This can be tested in constant time. Hence we can test in amortized constant time whether the instance $g_1, \ldots, g_i$ is solvable.

Corollary 13. *One-column two-color Clickomania with n groups can be decided in time $O(n)$ if n is even. If the problem is solvable, a solution can also be found in time $O(n)$.*

---

3. Hardness for 5 Colors and 2 Columns

Theorem 14. *Deciding whether a Clickomania puzzle can be solved is NP-complete, even if we have only two columns and five colors.*

It is relatively easy to reduce two-column six-color Clickomania from the weakly NP-hard set-partition problem: given a set of integers, can it be partitioned into two subsets with equal sum? Unfortunately this does not prove NP-hardness of Clickomania, because the reduction would represent the integers in unary (as a collection of blocks). But the partition problem is only NP-hard for integers that are superpolynomial in size, so this reduction would not have polynomial size. (Set partition is solvable in pseudo-polynomial time, i.e., time polynomial in the sum of the integers [3].)

Thus we reduce from the 3-partition problem, which is strongly NP-hard [2; 3].

3-Partition Problem. Given a multiset $A = \{a_1, \ldots, a_n\}$ of $n = 3m$ positive integers bounded by a fixed polynomial in n, with the property that $\sum_{i=1}^{n} a_i = tm$, is there a partition of A into subsets $S_1, \ldots, S_m$ such that $\sum_{a \in S_i} a = t$ for all i?

Such a partition is called a *3-partition*. The problem is NP-hard in the case that $t/3 \leq a_i \leq 2t/3$ for all i. This implies that a 3-partition satisfies $|S_i| = 3$ for all i, which explains the name.

The construction has two columns; refer to Figure 1. The left column encodes the sets $S_1, \ldots, S_m$ (or more precisely, the sets $U_j = S_1 \cup \ldots \cup S_j$ for $j = 1, \ldots, m-1$, which is equivalent). The right column encodes the elements $a_1, \ldots, a_{3m}$, as well as containing separators and blocks to match the sets.

Essentially, the idea is that in order to remove the singleton that encodes set U_j, we must remove three blocks that encode elements in A, and these elements exactly sum to t, hence form the set S_j.

The precise construction is as follows. The left column consists, from bottom to top, of the following:

- $3m$ squares, alternately black and white

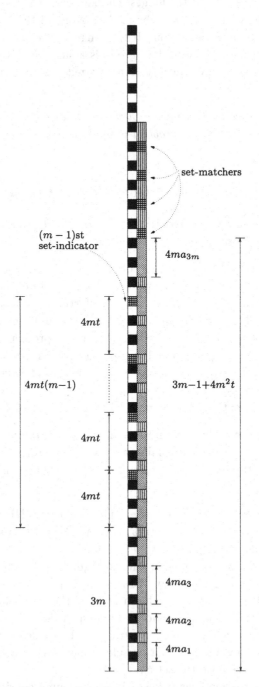

Figure 1. Overall construction, not to scale.

- $m - 1$ sections for the $m - 1$ sets $U_1, \ldots, U_{m-1}$, numbered from bottom to top. The section for U_j consists of $4mt - 1$ black and white squares, follows by one "red" square (indicated hashed in Figure 1). This red square is called the jth *set-indicator*.

  The black and white squares are colored alternatingly black and white, even across a set-indicator. That is, if the last square below a set-indicator is white, then the first one above it is black and vice versa.
- Another long stretch of alternating black and white squares. There are exactly as many black and white squares above the last set-indicator as there were below, and they are arranged in such a way that if we removed all set-indicators, the whole left column could collapse to nothing.

The right column contains at the bottom the elements in A, and at the top squares to remove the set-indicators. More precisely, the right column consists, from bottom to top, of the following:

- $3m$ sections for each element in A. The section for a_i consists of 1 "blue" square (indicated with vertical lines in Figure 1) and $4ma_i$ "green" squares (indicated with diagonals in Figure 1). Element a_1 does not have a separator.

  The blue squares are called *separators*, while the green squares are the one that encode the actual elements.
- $m - 1$ sections for each set. These consist of three squares each, one red and two blue. The red squares will also be called *set-matchers*, while the blue squares will again be called *separators*.

The total height of the construction is bounded by $8m^2t + 6m$, which is polynomial in the input. And it is not difficult to see that solutions to the puzzle correspond uniquely to solutions to the 3-partition problem.

4. Hardness for 3 Colors and 5 Columns

Theorem 15. *Deciding whether a Clickomania puzzle can be solved is NP-complete, even if we have only five columns and three colors.*

The proof is by reduction from 3-SAT. We now give the construction.

Let $F = C_1 \wedge \cdots \wedge C_m$ be a formula in conjunctive normal form with variables $x_1, \ldots, x_n$. We will construct a 5-column Clickomania puzzle using three colors, white, gray, and black, where the two leftmost columns, the *v-columns*, represent the variables, and the three rightmost columns, the *c-columns*, represent the clauses (see Figure 2(a)). Most of the board is white, and gray blocks are only used in the c-columns. In particular, a single gray block sits on top of the fourth column, and another white block on top of the gray block. We will show that this gray block can be removed together with another single gray block in the rightmost column if and only if there is a satisfying assignment for F.

All clauses occupy a rectangle CB of height h_{CB}. Each variable x_i occupies a rectangle V_i of height h_v. The variable groups are slightly larger than CB,

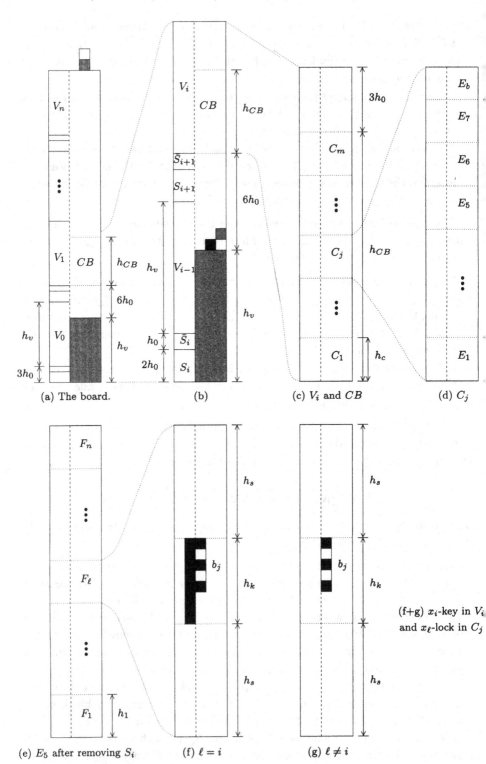

Figure 2. The Clickomania puzzle. The white area is not drawn to scale.

namely $h_v = h_{CB} + 3h_0$. The lowest group V_0 represents a dummy variable x_0 with no function other than elevating x_1 to the height of CB. The total height of the construction is therefore approximately $(n+1) \cdot (h_v + 3h_0)$.

For all i, there are two *sliding groups* S_{i+1} and $\bar{S}_{i+1}$ of size $2h_0$ and h_0, respectively, underneath V_i; their function will be explained later. The variable groups and the sliding groups are separated by single black rows which always count for the height of the group below. The variable groups contain some more black blocks in the second column to be explained later.

CB sits above a gray rectangle of height h_v at the bottom of the c-columns, a white row with a black block in the middle, a white row with a gray block to the right, and a white rectangle of height $6h_0 - 2$. Figure 2(b) shows the board after we have removed $V_0, \ldots, V_{i-2}$ from the board, i.e., assigned a value to the first $i - 1$ variables.

CB and V_i are divided into m chunks of height h_c, one for each clause (see Figure 2(c)). Note that V_i is larger than CB, so it also has a completely white rectangle on top of these m chunks. Each clause contains three *locks*, corresponding to its literals, each variable having a different lock (we distinguish between different locks by their position within the clause, otherwise the locks are indistinguishable). Each variable group V_i on the other hand contains matching x_i-*keys* which can be used to open a lock, thus satisfying the clause. After we have unlocked all clauses containing x_i we can slide V_i down by removing the white area of V_{i-1} which is now near the bottom of the v-columns. Thus we can satisfy clauses using all variables, one after the other.

Variables can appear as positive or negative literals, and we must prevent x_i-keys from opening both positive and negative locks. Either all x_i-keys must be used to open only x_i-locks (this corresponds to the assignment $x_i = 1$), or they are used to open only $\bar{x}_i$-locks (this corresponds to the assignment $x_i = 0$). To achieve this we use the sliding groups S_i and $\bar{S}_i$. Initially, a clause containing literal x_i has its x_i-lock $2h_0$ rows below the x_i-key; if it contains the literal $\bar{x}_i$ then the x_i-lock is h_0 rows below the x_i-key; and if it does not contain the variable x_i there is no x_i-lock. So before we can use any x_i-key we must slide down V_i by either h_0 (by removing $\bar{S}_i$) or by $2h_0$ (by removing S_i). Removing both S_i and $\bar{S}_i$ slides V_i down by $3h_0$ which again makes the keys useless, so either $x_i = 0$ in all clauses or $x_i = 1$.

To prevent removal of the large gray rectangle at the bottom of the c-columns prematurely, we divide each clause into seven chunks $E_1, \ldots, E_7$ of height h_0 each and a barrier group E_b (see Figure 2(d)). The locks for positive literals are located in E_5, and the locks for negative literals are in E_6. The keys are located in E_7. As said before, we can slide them down by either h_0 (i.e., $x = 0$), or by $2h_0$ (i.e., $x = 1$). The empty chunks $E_1, \ldots, E_5$ are needed to prevent misuse of keys by sliding them down more than $2h_0$.

We only describe E_5, the construction of E_6 is similar (see Figure 2(e)). To keep the drawings simple we assume that the v-columns have been slid down by $2h_0$, i.e., the chunk E_7 in the v-columns is now chunk E_5. E_5 is divided into n rectangle $F_1, \ldots, F_n$ of height h_1, one for each variable. In V_i, only F_i contains an x_i-key which is a black rectangle of height h_k in the second column (see Figure 2(f)), surrounded on both sides by white space of height h_s. In the c-columns, rectangle F_ℓ contains an x_ℓ-lock if and only if the literal x_ℓ appears in the clause. The lock is an alternating sequence of black and white blocks, where the topmost black block is aligned with the topmost black block of the x_ℓ-key (see Figure 2(f) and (g)). The number of black blocks in a lock varies between clauses, we denote it by b_j for clause C_j, and is independent of the variable x_i. Let $B_j = b_1 + \cdots + b_j$.

The barrier of clause C_j is located in the chunk E_b of that clause (see Figure 3). It is a single black block in column 4. There is another single black block in column 3, the *bomb*, B_j rows above the barrier. The rest of E_b is white. As long as the large white area exists, the only way to remove a barrier is to slide down a bomb to the same height as the barrier.

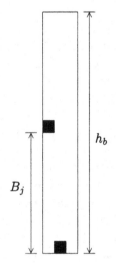

Figure 3. A barrier in E_b

With some effort one can show that this board can be solved if and only if the given formula has a satisfying assignment.

5. Conclusion

One intriguing direction for further research is *two-player Clickomania*, a combinatorial game suggested to us by Richard Nowakowski. In the impartial version of the game, the initial position is an arbitrary Clickomania puzzle, and the players take turns clicking on groups with at least t wo blocks; the last player to move wins. In the partizan version of the game, the initial position is a t wo-color Clickomania puzzle, and each player is assigned a color. Players take turns clicking on groups of their color with at least two blocks, and the last player to move wins.

Several interesting questions arise from these games. For example, what is the complexity of determining the game-theoretic value of an initial position? What is the complexity of the simpler problem of determining the outcome (winner) of a given game? These games are likely harder than the corresponding puzzles (i.e., at least NP-hard), although they are more closely tied to how many *moves* can be made in a given puzzle, instead of how many *blocks* can be removed as we have analyzed here. The games are obviously in PSPACE, and it would seem natural that they are PSPACE-complete.

Probably the more interesting direction to pursue is tractability of special cases. For example, this paper has shown polynomial solvability of one-column Clickomonia puzzles, both for the decision and optimization problems. Can this be extended to one-column games? Can both the outcome and the game-theoretic value of the game be computed in polynomial time? Even these problems seem to have an intricate structure, although we conjecture the answers are yes.

In addition, several open problems remain about one-player Clickomania:

1. What is the complexity of Clickomania with two colors?
2. What is the complexity of Clickomania with two rows? $O(1)$ rows?
3. What is the precise complexity of Clickomania with one column? Can any context-free-grammar parsing problem be converted into an equivalent Clickomania puzzle? Alternatively, can we construct an LR(k) grammar?
4. In some implementations, there is a scored version of the puzzle in which removing a group of size n results in $(n - 2)^2$ points, and the goal is to maximize score. What is the complexity of this problem? (This ignores that there is usually a large bonus for removing all blocks, which as we have shown is NP-complete to decide.)

Acknowledgment

This work was initiated during the University of Waterloo algorithmic open problem session held on June 19, 2000. We thank the attendees of that meeting for helpful discussions: Jonathan Buss, Eowyn Čenek, Yashar Ganjali, and Paul Nijjar (in addition to the authors).

References

[1] Bernie Cosell. Clickomania. Email on math-fun mailing list, June 2000.

[2] M. R. Garey and D. S. Johnson. Complexity results for multiprocessor scheduling under resource constraints. *SIAM Journal on Computing*, 4(4):397–411, 1975.

[3] Michael R. Garey and David S. Johnson. *Computers and Intractability: A Guide to the Theory of NP-Completeness*. W. H. Freeman & Co., 1979.

[4]   Michael Sipser. *Introduction to the Theory of Computation.* PWS Publishing Company, Boston, 1997.

THERESE C. BIEDL
DEPARTMENT OF COMPUTER SCIENCE
UNIVERSITY OF WATERLOO
WATERLOO, ONTARIO N2L 3G1
CANADA
    biedl@uwaterloo.ca

ERIK D. DEMAINE
MIT LABORATORY FOR COMPUTER SCIENCE
200 TECHNOLOGY SQUARE
CAMBRIDGE, MASSACHUSETTS 02139
UNITED STATES
    edemaine@mit.edu

MARTIN L. DEMAINE
MIT LABORATORY FOR COMPUTER SCIENCE
200 TECHNOLOGY SQUARE
CAMBRIDGE, MASSACHUSETTS 02139
UNITED STATES
    mdemaine@mit.edu

RUDOLF FLEISCHER
DEPARTMENT OF COMPUTER SCIENCE
THE HONG KONG UNIVERSITY OF SCIENCE AND TECHNOLOGY
CLEAR WATER BAY, KOWLOON
HONG KONG
    rudolf@cs.ust.hk

LARS JACOBSEN
DEPARTMENT OF MATHEMATICS AND COMPUTER SCIENCE
UNIVERSITY OF SOUTHERN DENMARK
CAMPUSVEJ 55
DK-5230 ODENSE M
DENMARK
    eljay@imada.sdu.dk

J. IAN MUNRO
DEPARTMENT OF COMPUTER SCIENCE
UNIVERSITY OF WATERLOO
WATERLOO, ONTARIO N2L 3G1
CANADA
    imunro@uwaterloo.ca

More Games of No Chance
MSRI Publications
Volume **42**, 2002

Coin-Moving Puzzles

ERIK D. DEMAINE, MARTIN L. DEMAINE, AND
HELENA A. VERRILL

ABSTRACT. We introduce a new family of one-player games, involving the
movement of coins from one configuration to another. Moves are restricted
so that a coin can be placed only in a position that is adjacent to at least
t wo other coins. The goal of this paper is to specify exactly which of these
games are solvable. By introducing the notion of a constant number of extra
coins, we give tight theorems c haracterizing solvable puzzles on the square
grid and equilateral-triangle grid. These existence results are supplemented
by polynomial-time algorithms for finding a solution.

1. Introduction

Consider a configuration of coins such as the one on the left of Figure 1. The
player is allowed to move any coin to a position that is determined rigidly by
incidences to other coins. In other words, a coin can be moved to any position
adjacent to at least t wo other coins. The puzzle or 1-player game is to reach
the configuration on the right of Figure 1 by a sequence of such moves. This
particular puzzle is most interesting when each move is restricted to *slide* a coin
in the plane without overlapping other coins.

Figure 1. Re-arrange the rhombus into the circle using three slides, such that
each coin is slid to a position adjacent to t wo other coins.

This puzzle is described in Gardner's Mathematical Games article on Penny
Puzzles [7], in *Winning Ways* [1], in *Tokyo Puzzles* [6], in *Moscow Puzzles* [8],
and in *The Penguin Book of Curious and Interesting Puzzles* [11]. Langman [9]
shows all 24 ways to solve the puzzle in three moves. Another classic puzzle of

this sort [2; 6; 7; 11] is shown in Figure 2. A final classic puzzle that originally motivated our work is shown in Figure 3; its source is unknown. Other related puzzles are presented by Dudeney [5], Fujimura [6], and Brooke [4].

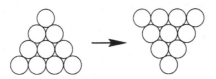

Figure 2. Turn the pyramid upside-down in three moves, such that each coin is moved to a position adjacent to two other coins.

Figure 3. Re-arrange the pyramid into a line in seven moves, such that each coin is moved to a position adjacent to two other coins.

The preceding puzzles always move the centers of coins to vertices of the equilateral-triangle grid. Another type of puzzle is to move coins on the square grid, which appears less often in the literature but has significantly more structure and can be more difficult. The only published example we are aware of is given by Langman [10], which is also described by Brooke [4], Bolt [3], and Wells [11]; see Figure 4. The first puzzle (H → O) is solvable on the square grid, and the second puzzle (O → H) can only be solved by a combination of the two grids.

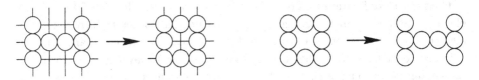

Figure 4. Re-arrange the H into the O in four moves while staying on the square grid (and always moving adjacent to two other coins), and return to the H in six moves using both the equilateral-triangle and square grids.

In this paper we study generalizations of these types of puzzles, in which coins are moved on some grid to positions adjacent to at least two other coins. Specifically, we address the basic algorithmic problem: is it possible to solve a puzzle with given start and finish configurations, and if so, find a sequence of moves. Surprisingly, we show that this problem has a polynomial-time solution in many cases. Our goal in this pursuit is to gain a better understanding of what is possible by these motions, and as a result to design new and interesting puzzles. For example, one puzzle we have designed is shown in Figure 5. We recommend the reader try this difficult puzzle before reading Section 5.3.1 which shows how to solve it. Figures 6–9 show a few of the other puzzles we have designed. The last t wo puzzles involve labeled coins.

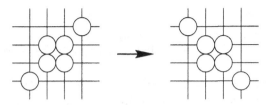

Figure 5. A difficult puzzle on the square grid. The optimal solution uses 18 moves, each of which places a coin adjacent to t wo others.

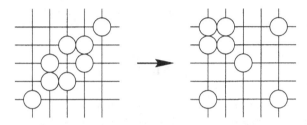

Figure 6. Another puzzle on the square grid. The optimal solution uses 24 moves, each of which places a coin adjacent to t wo others.

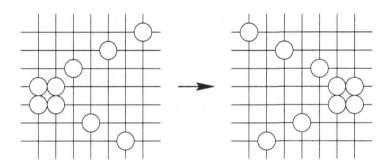

Figure 7. Another puzzle on the square grid with the same rules.

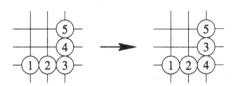

Figure 8. A puzzle on the square grid involving labeled coins. Solvable in eleven moves, each of which places a coin adjacent to t wo others; see Figure 31.

This paper studies t wo grids in particular: the equilateral-triangle grid, and the square grid. It turns out that the triangular grid has a relatively simple

Figure 9. A puzzle on the equilateral-triangle grid involving labeled coins. Solvable in eight moves, each of which places a coin adjacent to two others.

structure, and nearly all puzzles are solvable. An exact, efficient characterization of solvable puzzles is presented in Section 3. The square grid has a more complicated structure, requiring us to introduce the notion of "extra coins" to give a partial characterization of solvable puzzles. This result is described in Section 5 after some general tools for analysis are developed in Section 4.

Before we begin, the next section defines a general graph model of the puzzles under consideration.

2. Model

We begin by defining "token-moving" and "coin-moving" puzzles and related concepts. The *tokens* form a finite multiset T. We normally think of tokens as unlabeled, modeled by all elements of T being equal, but another possibility is to color tokens into more than one equivalence class (as in Figure 9). A *board* is any simple undirected graph $G = (P, E)$, possibly infinite, whose vertices are called *positions*. A *configuration* is a placement of the tokens onto distinct positions on the board, i.e., a one-to-one mapping $C : T \to P$. We will often associate a configuration C with its image, that is, the set of positions *occupied* by tokens.

A *move* from a configuration C changes the position of a single token t to an unoccupied position p, resulting in a new configuration. This move is denoted $t \mapsto p$, and the resulting configuration is denoted $C / t \mapsto p$. We stress that moves are not required to "slide" the token while avoiding other tokens (like the puzzle in Figure 1); the token can be picked up and placed in any unoccupied position.

The *configuration space* (or *game graph*) is the directed graph whose vertices are configurations and whose edges correspond to feasible moves. A typical *token-moving puzzle* asks for a sequence of moves to reach one configuration from another, i.e., for a path between two vertices in the configuration space, subject to some constraints. A *coin-moving puzzle* is a geometric instance of a token-moving puzzle, in which tokens are represented by *coins*—constant-radius disks in the plane, and constant-radius hyperballs in general—and the board is some lattice in the same dimension. If a token-moving or coin-moving puzzle with source configuration A and destination configuration B is solvable, we say that A can be *re-arranged* into B, and that B is *reachable* from A. This is equivalent to the existence of a directed path from A to B in the configuration space.

This paper addresses the natural question of what puzzles are solvable, subject to the following constraint on moves which makes the problem interesting. A move $t \mapsto p$ is *d-adjacent* if the new position p is adjacent to at least d tokens other than the moved token t. (Throughout, *adjacency* refers to the board graph G.) This constraint is particularly meaningful for d-dimensional coin-moving puzzles, because then a move is easy to "perform exactly" without any underlying lattice: the new position p is determined rigidly by the d coin adjacencies (sphere tangencies).

The *d-adjacency configuration space* is the subgraph of the configuration space in which moves are restricted to be d-adjacent. Studying connectivity in this graph is equivalent to studying solvable puzzles; for example, if the graph is strongly connected, then all puzzles are solvable.

Here we explore solvable puzzles on two boards, the equilateral-triangle grid and the square grid. Because these puzzles are two-dimensional, in the context of this paper we call a move *valid* if it is 2-adjacent, and a position a *valid destination* if it is unoccupied and adjacent to at least two occupied positions. Thus a valid move involves transferring a token from some source position to a valid destination position. When the context is clear, we will refer to a valid move just by "move." A move is *reversible* if the source position is also a valid destination.

3. Triangular Grid

This section studies the equilateral-triangle grid, where most puzzles are solvable. To state our result, we need a simple definition. Associated with any configuration is the subgraph of the board induced by the occupied positions. In particular, a *connected component* of a configuration is a connected component in this induced subgraph.

Theorem 1. *On the triangular grid with the 2-adjacency restriction and unlabeled coins, configuration A can be re-arranged into a different configuration B precisely if A has a valid move, the number of coins in A and B match, and at least one of four conditions holds:*

(i) *B contains three coins that are mutually adjacent (a triangle).*
(ii) *B has a connected component with at least four coins.*
(iii) *B has a connected component with at least three coins and another connected component with at least two coins.*
(iv) *There is a single move from A to B.*

The same result holds for labeled coins, except when there are exactly three coins in the puzzle, in which case the labelings and movements are controlled by the vertex 3-coloring of the triangular grid.

Furthermore, there is a polynomial-time algorithm to find a re-arrangement from A to B if one exists. Specifically, let n denote the number of coins and *d denote the maximum distance between two coins in A or B. Then a solution with $O(nd)$ moves can be found in $O(nd)$ time.*

The rest of this section is devoted to the proof of this theorem. We begin in the next subsection by proving necessity of the conditions: if a puzzle is solvable, then one of the conditions holds. Then in the following subsection we prove sufficiency of the conditions.

3.1. Necessity. Of course, it is necessary for A to have a valid move and for A and B to have the same number of coins. Necessity of at least one of the four conditions is also not difficult to show, because Conditions 1–3 are so broad, encompassing most possibilities for configuration B.

Suppose that a solvable puzzle does not satisfy any of Conditions 1–3, as in Figure 10. We prove that it must satisfy Condition 4, by considering play backwards from the goal configuration B. Specifically, a *reverse move* takes a coin currently adjacent to at least t wo others, and moves it to any other location. Because the puzzle is solvable, some coin in configuration B must be reverse-movable, i.e., must have at least t wo coins adjacent to it. Thus, some connected component of B has at least three coins. Because Condition 2 does not hold, this connected component has exactly three coins. Because Condition 1 does not hold, these three coins are not connected in a triangle. Because Condition 3 does not hold, every other component has exactly one coin.

Hence, one component of B is a path of exactly three coins, say c_1, c_2, c_3, and every other component of B has exactly one coin, as in the left of Figure 10. Certainly at this moment c_2 is the only reverse-movable coin. We claim that after a sequence of reverse moves, c_2 will continue to be the only reverse-movable coin. If we removed c_2, then every coin would be adjacent to no others. Thus, if we reverse move c_2 somewhere, then every other coin would be adjacent to at most one other (c_2). Hence, it remains that only c_2 can be reverse moved.

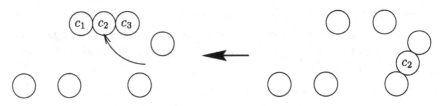

Figure 10. Reverse-moving a configuration B that does not satisfy any of Conditions 1–3.

Therefore, if we can reach A from B via reverse moves, we can do so in a single reverse move of c_2 directly to where it occurs in A. Thus Condition 4 holds, as desired.

3.2. Sufficiency. Next we prove the more difficult direction: provided one of Conditions 1–3 hold, there is a re-arrangement from A to B. (This fact is obvious when Condition 4 holds.) All three cases will follow a common outline: we first form a triangle (Section 3.2.1), then maneuver this triangle (Section 3.2.2) to transport all other coins (Section 3.2.3), and finally we place the three triangle coins appropriately depending on the case (Section 3.2.4).

3.2.1. Getting Started. It is quite simple to make some triangle of coins. By assumption, there is a valid move from configuration A. The destination of this move can have t wo basic forms, as shown in Figure 11. Either the move forms a triangle, as desired, or the move forms a path of three coins. In the latter case, if there is not a triangle already with a different triple of coins, a triangle can be formed by one more move as shown in the right of the figure.

Figure 11. Two types of valid destinations for a coin c. In the latter case, we show a move to form a triangle.

This triangle T_0 suffices for unlabeled coin puzzles. However, for labeled coin puzzles, we cannot use just any three coins in the triangle; we need a particular three, depending on B. For example, if B satisfies Condition 1, then the coins forming the triangle in B are the coins we would like in the triangle for maneuvering. To achieve this, we "bootstrap" the triangle T_0 formed above, using this triangle with the incorrect coins to form another triangle with the correct coins. Specifically, if we desire a triangle using coins t_1, t_2, and t_2, then we move each coin in the difference $\{t_1, t_2, t_3\} - T_0$ to be adjacent to appropriate coins in T_0. There are three cases, shown in Figure 12, depending on how many coins are in the difference. If ever we attempt to move a coin to an already occupied destination, we first move the coin located at that destination to any other valid destination.

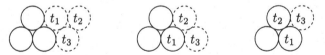

Figure 12. The three cases of building a triangle $\{t_1, t_2, t_3\}$ out of an existing triangle, depending on how many coins the t wo triangles share. From left to right, zero, one, and t wo coins of overlap.

3.2.2. Triangle Maneuvering. Consider a triangle of coins t_1, t_2, and t_3. The possible positions of this triangle on the triangular grid are in one-to-one correspondence with their centers, which are vertices of the dual hexagonal grid. Moving one coin (say t_1) to be adjacent to and on the other side of the others (t_2 and t_3) corresponds to moving the center of the triangle to one of the three neighboring centers on the hexagonal grid. Thus, without any other coins on the board, the triangle can be moved to any position by following a path in the hexagonal grid.

This approach can be modified to apply when there are additional obstacle coins; see Figure 13 for an example. Conceptually we always move one of the triangle coins, say t_i, in order to move the center of the triangle to an adjacent vertex of the hexagonal grid. But if the move of t_i is impossible because the destination is already occupied by another coin g_i, then in fact we do not make any move. There will be a triangle in the desired position now, but it will not consist of the usual three coins (t_1, t_2, and t_3); instead, t_i will be replaced by the "ghost coin" g_i. Such a triangle suffices for our purposes of transportation described in Section 3.2.3. One final detail is how the ghost coins behave: if we later need to move a ghost coin g_i, we instead move the original (unmoved) coin t_i. Thus ghost coins are never moved; only t_1, t_2, and t_3 are moved during triangle maneuvering (even if coins are labeled).

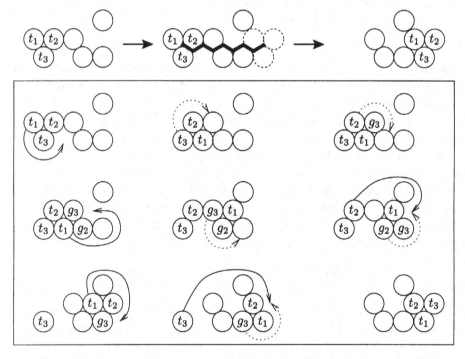

Figure 13. An example of triangle maneuvering. Dotted arrows denote conceptual moves, and solid arrows denote actual moves.

3.2.3. Transportation. Triangle maneuvering makes it easy to *transport* any other coin to any desired location. Specifically, suppose we want to move coin $c \notin \{t_1, t_2, t_3\}$ to destination position d. If d is already occupied by another coin c', we first move c' to an arbitrary valid destination; there is at least one because the triangle can be maneuvered. Now we maneuver the triangle so that the (potentially ghost) triangle has t wo coins adjacent to d, so that the third coin is not on d, and so that the triangle does not overlap c. This is easily arranged by examining the location of c and setting the destination of the triangle appropriately. For example, if c is within distance t wo of d, then there are four positions for the triangle that are adjacent to d and do not overlap c; otherwise, the triangle can be placed in any of the six positions adjacent to d. Finally, because d is now a valid destination—it is adjacent to t wo coins in the triangle—we can move c to d.

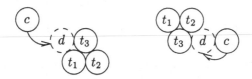

Figure 14. Transporting coin c to destination d using triangle $\{t_1, t_2, t_3\}$. In both cases, we choose the location of the triangle so that it does not overlap c.

By the properties of triangle maneuvering, this transportation process even preserves coin labels: the only actual coins moved are t_1, t_2, t_3, c, and possibly a coin at d. But any coin at position d must not have already been in its desired position, because d is c's desired position. Thus, applying the transportation process to every coin except t_1, t_2, and t_3 places all coins except these three in their desired locations.

3.2.4. Finale. Once transportation is complete, it only remains to place the triangle coins t_1, t_2, and t_3 in their desired locations. By the bootstrapping in Section 3.2.1, we are able to choose the unplaced coins $\{t_1, t_2, t_3\}$ however we like. This property will be exploited differently in the three cases.

Property 1. If there is a triangle in B, then we choose these three coins as the unplaced coins t_1, t_2, and t_3, and use them to transport all other coins. Then we maneuver the triangle $\{t_1, t_2, t_3\}$ exactly where it appears in B. Because all other coins have been moved to their proper location, in this position the triangle will not have any ghost coins.

However, it may be that the coins $\{t_1, t_2, t_3\}$ are labeled incorrectly among themselves, compared to B. Assuming there are more than three coins in the puzzle, this problem can be repaired as follows. We maneuver the triangle so that it does not overlap any other coins but is adjacent to at least one coin c; for example, there is such a position for the triangle just outside the smallest enclosing hexagon of the other coins. Refer to Figure 15. Now t wo coins of the triangle, say t_1 and t_2, are adjacent to three other coins each: each other, t_3, and c. Thus we can move t_1 to any other valid destination, and then move t_2 or t_3 to replace it. Afterwards we can move t_1 to take the place of t_2 or t_3, whichever moved. This procedure swaps t_1 and either t_2 or t_3. By suitable application, we can achieve any permutation of $\{t_1, t_2, t_3\}$, and thereby achieve the desired labeling of the triangle.

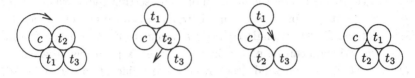

Figure 15. Swapping coins t_1 and t_2 in a triangle, using an adjacent coin c.

Property 2 but not Property 1. Refer to Figure 16. If there is not a triangle in B, but there is a connected component of B with at least four coins, then there is a path in B of length four, (p_1, p_2, p_3, p_4). From B we reverse move p_2 so that it is adjacent to p_3 and p_4. If this position is already occupied by a coin c, we first reverse move c to any other unoccupied position. Now p_2, p_3, and p_4 are mutually adjacent, so we have a new destination configuration B' with Property 1. As described above, we can re-arrange A into B'. Then we undo our reverse moves: move p_2 back adjacent to p_1 and p_3, and move c back adjacent to p_3 and p_4. This procedure re-arranges A into B.

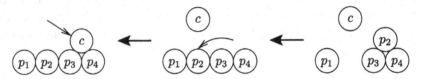

Figure 16. Reverse moving a configuration B with Property 2 into a configuration with Property 1.

Property 3 but not Property 1. This case is similar to the previous one; refer to Figure 17. There must be a path in B of length three, (p_1, p_2, p_3), as well as a pair of adjacent coins, (q_1, q_2), in different connected components of B. If both positions adjacent to both q_1 and q_2 are already occupied, we first reverse move one such coin (call it c) to an arbitrary unoccupied position. This frees up a position adjacent to q_1 and q_2, to which we reverse move p_2. Now $\{q_1, q_2, p_2\}$ form a triangle, so Property 1 holds, and we can reach this new configuration B' from A. Then we undo our reverse moves: move p_2 back adjacent to p_1 and p_3, and move c back adjacent to q_1 and q_2. This procedure re-arranges A into B.

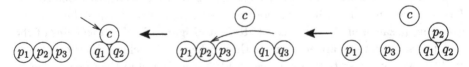

Figure 17. Reverse moving a configuration B with Property 3 into a configuration with Property 1.

This concludes the proof of Theorem 1.

4. General Tools

In this section we develop some general lemmas about token-moving puzzles. Although we only use these tools for the square grid, in Section 5, they apply to arbitrary boards and may be of more general use.

4.1. Picking Up and Dropping Tokens. First we observe that additional tokens cannot "get in the way":

Lemma 1. *If a token-moving puzzle is solvable, then it remains solvable if we add an additional token with an unspecified destination, provided tokens are unlabeled. This result also holds if all moves must be reversible.*

Proof. A move can be blocked by an extra token e at position p because p is occupied and hence an invalid destination. But if ever we encounter such a move of a token t to position p, we can just ignore the move, and swap the roles of e and t: treat e as the moved version of t, and treat t as an extra token replacing e. Thus, any sequence of moves in the original puzzle can be emulated by an equivalent sequence of moves in the augmented puzzle. We are not introducing any new moves, only removing existing moves, so all moves remain reversible if they were originally. □

This proof leads to a technique for emulating a more powerful model for solving puzzles. In addition to moving coins as in the normal model, we can conceptually *pick up* (remove) a token, and later *drop* (add) it onto any valid destination. At any moment we can have any number of tokens picked up. While a token t is conceptually picked up, we emulate any moves to its actual position p as in the proof of Lemma 1: if we attempt to move another token t' onto position p, we instead reverse the roles of t and t'. To drop a token onto a desired position p, we simply move the actual token to position p if it is not there already.

Of course, this process may permute the tokens. Nonetheless we will find this approach useful for puzzles with labeled tokens.

One might instead consider the emulation method used implicitly in Section 3.2.2 for triangular maneuvering: move original coins instead of ghost coins. This approach has the advantage that it preserves the labels of the coins. Unfortunately, the approach makes it difficult to preserve reversibility as in Lemma 1, and so is insufficient for our purposes here.

4.2. Span. The *span* of a configuration C is defined recursively as follows. Let $d_1, \ldots, d_m$ be the set of valid destinations for moves in C. If $m = 0$, the span of C is just C itself. Otherwise, it is the span of another configuration C', defined to be C with additional tokens at positions $d_1, \ldots, d_m$. If this process never terminates, the span is defined to be the limit, which exists because it is a countable union of finite sets.

Figure 18. In this example, the span is the smallest rectangle enclosing the configuration.

The span of a configuration lists all the positions we could hope to reach, or more precisely, the positions we could reach if we had an unlimited number of extra tokens that we could drop. In particular, we have the following:

Lemma 2. *If configuration A can be re-arranged into configuration B, then* span $A \supseteq B$ *and thus* span $A \supseteq$ span B.

In other words, valid moves can never cause the span of the current configuration to increase. Thus the most connected we could hope the configuration space to be is the converse of Lemma 2: for every pair of configurations with span $A \supseteq$ span B, A can be re-arranged into B. In words, we want that every configuration A can be re-arranged into any configuration B with the *same or smaller span*.

We call a configuration *span-minimal* if the removal of any of its tokens reduces the span. Span-minimal configurations are essentially the "skeleta" that keep configurations with the same span reachable. One general property of span-minimal configurations is the following:

Lemma 3. *If a configuration is span-minimal, any move will reduce the span.*

Proof. Suppose to the contrary that there is a move $t \mapsto p$ that does not reduce the span of a span-minimal configuration C. In particular, p must be a valid destination position in the subconfiguration $C - t$, because t does not count in the d-adjacency restriction. Hence, adding a new token at position p to the configuration $C - t$ has no effect on the span of C. But this two-step process of removing token t and adding a token at position p is equivalent to moving t to p, so span$(C/t \mapsto p) =$ span$(C - t)$. But we assumed that span$(C/t \mapsto p) =$ span C, and hence span $C =$ span$(C - t)$, contradicting that C is span-minimal. $\square$

Under the 2-adjacency restriction, a *chain* is a sequence of tokens with the property that the distance (in the board graph G) between two successive tokens is at most 2. We will use chains as basic "units" for creating a desired span.

Notice that the notion of span is useless for the already analyzed triangular grid: provided there is a valid move, the span of any configuration is the entire grid. Thus, for the triangular grid, a configuration is span-minimal precisely if it has no valid moves. For the square grid, however, the notion of span and span minimality is crucial.

4.3. Extra Tokens. As described in the previous section, we can only re-arrange configurations into configurations with the same or smaller span. Unfortunately, the converse is not true. Indeed, the key problem situations are span-minimal configurations; by Lemma 3, such configurations immediately lose span when we try to move them. Hence, any t wo distinct span-minimal configurations with the same span cannot reach each other. An example on the square grid is that the t wo opposite diagonals of a square are unreachable from each other, as shown in Figure 19.

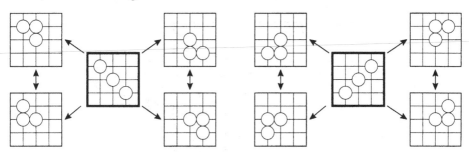

Figure 19. Subgraph of configuration space reachable from full-span configurations (outlined in bold) with no extra coins.

Thus we explore the notion of *extra tokens*, a set of tokens whose removal does not reduce the span of the configuration. Lemma 2 and Figure 19 shows that we need at least one extra token. In fact, the t wo opposite diagonals on the square grid shown in Figure 19 are difficult to reach from each other; as shown in Figure 20, even one extra token is insufficient. What is surprising is that a small number of extra tokens seem to be generally sufficient to make the configuration space strongly connected. We prove this for the square grid in the next section.

5. Square Grid

This section analyzes coin-moving puzzles on the square grid, using the tools from the previous section. In particular, we show that with just t wo extra coins, we can reach essentially every configuration on the square grid with the same or smaller span. The only restriction is that the extra coins can only be destined for positions that are adjacent to at least t wo other coins.

Theorem 2. *On the square grid with the 2-adjacency restriction and unlabeled coins, configuration A can be re-arranged into configuration B if there are coins e_1 and e_2 such that $\operatorname{span}(A - \{e_1, e_2\}) \supseteq \operatorname{span}(B - \{e_1, e_2\})$ and each e_i is adjacent to two other coins in B (excluding e_1 or e_2). Furthermore, there is an algorithm to find such a re-arranging sequence using $O(n^3)$ moves and $O(n^3)$ time, where n is the number of coins.*

We prove this theorem by showing that every configuration (in particular, A and B) can be brought to a canonical configuration with the same span via a

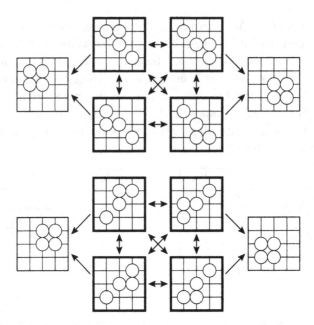

Figure 20. Subgraph of configuration space reachable from full-span configurations (outlined in bold) with one extra token.

sequence of (mostly) reversible moves. As a consequence, we can move from any configuration A to any other B by routing through this canonical configuration.

Our proof uses the model of picking up and dropping coins, which can be emulated as described in Section 4.1. However, we must be careful how we pick up and drop coins, so that the resulting moves are reversible. For example, initially we pick up the extra coins e_1 and e_2, and then drop them temporarily wherever needed. For re-arranging the source configuration A into the canonical configuration, this step may not result in reversible moves, but fortunately this is not necessary in this case. For re-arranging the destination configuration B into the canonical configuration, however, reversibility is crucial, and is guaranteed by the condition in the theorem of each e_i being adjacent to at least t wo other coins.

5.1. Basics. We begin with some preliminary lemmas. A *rectangle* is the full collection of coins between t wo x coordinates and t wo y coordinates. The *half-perimeter* of a rectangle is the number of distinct x coordinates plus the number of distinct y coordinates over all coins in the rectangle. The *distance* between t wo sets of coins is the minimum distance between t wo coins from different sets.

Lemma 4. *For the square grid, the span of any configuration is a disjoint union of (finite) rectangles with pairwise distances at least 3.*

Lemma 5. *For each rectangle (connected component) of the span, say with half-perimeter h, there must be at least $\lceil h/2 \rceil$ coins within that rectangle in the configuration.*

The following beautiful proof of this lemma has been distributed among several people, but its precise origin is unknown. We first heard it from Martin Farach-Colton, who heard it from Peter Winkler, who heard it from Pete Gabor Zoltan, who learned of it through the Russian magazine *Kvant* (around 1985–1987).

Proof. Consider how the (full) perimeter changes as we compute the span of the coins within the rectangle. Initially we have n coins, say, so the perimeter is at most $4n$. Each coin that we add while computing the span satisfies the 2-adjacency restriction, so the perimeter never increases. In the end we must have a rectangle with perimeter $2h$. Hence $4n \geq 2h$, i.e., $n \geq h/2$, and because n is integral, $n \geq \lceil h/2 \rceil$. $\qquad\square$

5.2. Canonical Configuration.

Observe that a chain has span equal to its smallest enclosing rectangle. We define an L to be a particular kind of chain, starting and ending at opposite corners of the rectangular span, and arranged along two edges of this rectangle, with the property that it has the minimum number of coins. See Figure 21 for examples. More precisely, if the half-perimeter of the rectangle (along which the L is arranged) is $2k$, then there must be precisely k coins, every consecutive pair at distance exactly two from each other. And if the half-perimeter is $2k + 1$, then there must be precisely $k + 1$ coins, every consecutive pair at distance exactly two from each other, except the last pair which are distance one from each other. In general, for half-perimeter h, an L has $\lceil h/2 \rceil$ coins.

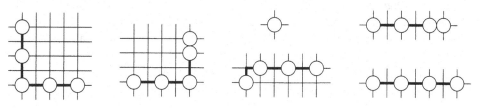

Figure 21. Examples of L's.

While L's can have any orientation, the *canonical L* is oriented like the letter L, starting at the top-left corner, continuing past the lower-left corner, and ending at the bottom-right corner.

Given a configuration, or more precisely, given its span and the number of coins in each connected component of the span, we define the *canonical configuration* as follows. Refer to Figure 22 for examples. Within each connected component (rectangle) of the span, say with half-perimeter h, we arrange the first $\lceil h/2 \rceil$ coins into the canonical L. (Lemma 5 implies that there are at least this many coins to place.) Any additional coins are placed one at a time, in the leftmost bottommost unoccupied position.

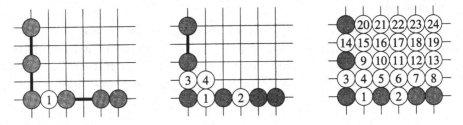

Figure 22. Examples of the canonical configuration of k coins within a rectangular span. (Left) One coin in addition to the canonical L. (Middle) Four coins in addition to the canonical L. (Right) All 24 additional coins.

This definition of the canonical configuration is fairly arbitrary, but it has the useful property that each successive position for an additional coin is a valid destination, given the previously placed additional coins. This allows us to focus on forming the canonical L, and then picking up all additional coins and dropping them in the order shown on the right of Figure 22.

5.3. Canonicalizing Algorithm. The main part of proving Theorem 2 is to show an algorithm for converting any configuration into the corresponding canonical configuration, using a sequence of (mostly) reversible moves. We will apply induction (or, equivalently, recursion) on the number of coins. That is, we assume that any configuration with fewer coins can be re-arranged into its canonical configuration.

For now, we assume that there are no extra coins in addition to e_1 and e_2. For if there were such a coin, we could immediately pick it up. Then we have a simpler configuration: it has one fewer coin. Thus we can apply the induction hypothesis, and re-arrange the remaining coins into their canonical configuration. Finally we must drop the previously picked-up coin in the appropriate location. This aspect is somewhat trickier than it may seem: if we are not careful, we may make an irreversible move. We delay this issue to Section 5.4.

The overall outline of the algorithm is as follows:

(i) Initialize the set of L's to be one for each coin.
(ii) Until the configuration is canonical:

    (i) Pick two L's whose bounding rectangles are distance at most two from each other.

    (ii) Re-orient the L's so that the L's themselves are distance at most three from each other.

    (iii) Merge the two L's.

Normally, each iteration of Step 2 decreases the number of L's by one, so the algorithm would terminate in at most n iterations. However, at any time we may find an extra coin in addition to e_1 and e_2, and pick it up. Fortunately, this operation can only split one L into at most t wo L's. Thus we can charge the cost of creating an extra L to the event of picking up an extra coin, which can happen at most n times. Hence, the total number of iterations of Step 2 is $O(n)$.

In the following t wo sections, we describe how Steps 2(b) and 2(c) can be done in $O(n^2)$ moves each. These bounds result in a total of $O(n^3)$ moves. The running time of the algorithms will be proportional to the number of moves.

5.3.1. Re-orienting L's. There are eight possible orientations for an L, depending at which corner it starts, and whether it hugs the top edge or bottom edge of the rectangular span. We will only be concerned with four different types of orientations, depending on whether it looks like the letter L rotated 0, 90°, 180°, or 270°. In other words, we are not concerned with the parity issue of which corner might have t wo adjacent coins.

It is relatively easy to *flip* an L about a diagonal, using t wo extra coins. Figure 23 shows how to do this in a constant number of moves for an L consisting of three coins. Figure 24 shows how to use these subroutines to flip an L of arbitrary size. Basically, we use the flips of three-coin L's to "bubble" the kink in the L up to the top, repeatedly until it is all the way right. The total number of moves is $O(n^2)$, and they can easily be computed in $O(n^2)$ time.

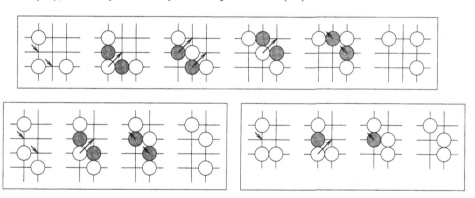

Figure 23. Flipping an L consisting of 3 coins. Extra coins are shaded.

The more difficult re-orientation to perform is a *rotation* of an L by $\pm 90°$. Perhaps one of the most surprising results of this paper is that this operation is possible with t wo extra coins. One way to do it for a square span, shown in Figure 25, is to convert the L into a diagonal, and then convert more and more of the diagonal into a rotated L. This is the basis for our "diagonal-flipping" puzzle in Figure 5.

A simpler way to argue that L's can be rotated is shown in Figure 26. Assume without loss of generality that the initial orientation is the canonical L. First we apply induction to the subconfiguration of all coins except the top-left coin.

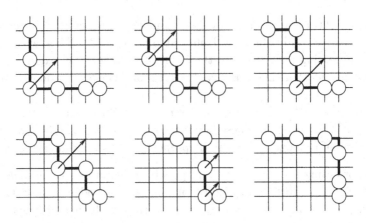

Figure 24. Flipping a general L, using the subroutines in Figure 23.

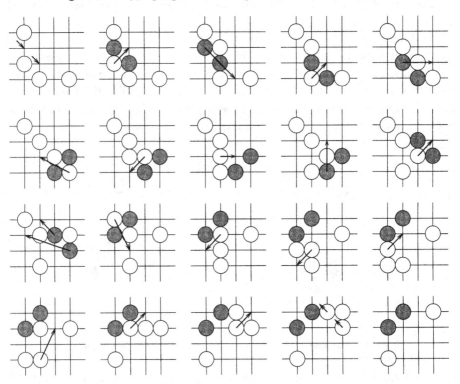

Figure 25. One method for rotating an L with a square span. Although this example places the extra coins in the final configuration, this is not necessary.

Thus all rows except the third row contain at most one coin each, assuming the L consists of at least three rows. Now we apply local operations in 3×3 or 3×2 rectangles (similar to Figure 23) to move the top-left coin to the far right. (We cannot perform this left-to-right motion in one step using induction, because there may be only three rows, and hence all coins may be involved in this motion.) Finally we flip the L in the top three rows, as described above, thereby obtaining the desired result. Again the number of moves and computation time are both $O(n^2)$. Note that the same approach of repeated local operations applies when the L consists of only t wo rows.

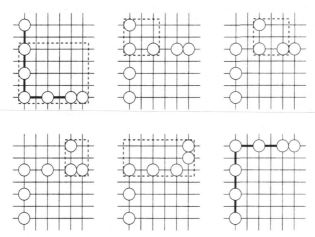

Figure 26. A general method for rotating an L. The first step is to apply induction, and the remaining steps apply subroutines similar to Figure 23.

5.3.2. Merging L's. Consider t wo L's L_1 and L_2 whose bounding rectangles R_1 and R_2 are distance at most t wo from each other. Equivalently, consider t wo L's such that span($L_1 \cup L_2$) has a single connected component. This section describes how to merge L_1 and L_2 into a single L. This step is the most complicated part of the algorithm, not because it is difficult in any one case, but because there are many cases involved.

First suppose that the rectangles R_1 and R_2 overlap. We claim that one of the L's, say L_1, can be re-oriented so that one of its coins is contained in the other L's bounding rectangle, R_2. This coin is therefore in the span of the L_2, and hence redundant, so as described above we can apply induction and finish the entire canonicalization process.

To prove the claim, there are three cases; see Figure 27. If a corner of one of the bounding rectangles, say R_1, is in the other bounding rectangle, R_2, then we can re-orient L_1 so that one of its end coins is at that corner of R_1 and hence in R_2; see Figure 27(a)). Otherwise, we have rectangles that form a kind of "thick plus sign" (Figure 27(b–c)); we distinguish the t wo rectangles as according to whether they form the *horizontal stroke* or *vertical stroke* of the plus sign. If the vertical stroke has width at least t wo (Figure 27(b)), then that rectangle already contains a coin of the other L, because that L cannot have t wo empty columns. Similarly, if the horizontal stroke has height at least t wo, then that rectangle already contains a coin of the other L, because that L cannot have t wo empty rows. Finally, if both strokes are of unit thickness, and there is not already a coin in their single-position intersection (Figure 27(c)), then we can splice and redefine the L's, so that one L is formed by the top half of the vertical stroke and the left half of the horizontal stroke, and the other L is formed by the bottom half of the vertical stroke and the right half of the horizontal stroke, and then we have the first case in which the bounding rectangles share a corner.

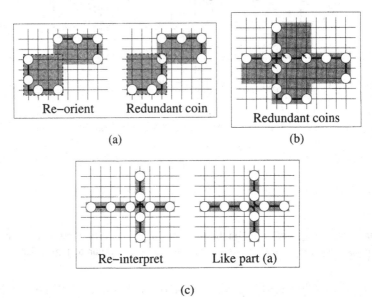

(a)   (b)

(c)

Figure 27. Merging two L's with overlapping bounding rectangles (shaded). (a) The corner of one L is contained in the other L's bounding rectangle. (b) A "thick plus sign" in which at least one stroke has thickness more than 1. (c) A plus sign in which both strokes have thickness 1.

Now suppose that the rectangles R_1 and R_2 do not overlap. Hence, either they share no x coordinates or they share no y coordinates. Assume by symmetry that R_1 and R_2 share no x coordinates. Assume again by symmetry that R_1 is to the left of R_2. A *leg* is a horizontal or vertical segment/edge of an L. Re-orient L_1 so that its vertical leg is on the right side, and re-orient L_2 so that its vertical leg is on the left side. Now L_1 and L_2 have distance at most three from each other; the distance may be as much as three because of parity.

We consider merging L_1 with each leg of L_2 one at a time. In other words, we merge L_1 with the nearest leg of L_2 within distance three of L_1, then we merge the result with the other leg of L_2. The second leg can be treated in the same way as the first leg, by induction. Thus there are two cases: either the first leg is the horizontal leg of L_2, or it is the vertical leg of L_2. We first show how the latter case reduces to the former case.

If the vertical leg of L_2 is the first leg, it can have only one coin within the y range of R_1 (by the assumption that extra coins are picked up). We can add this coin separately, as if it were a short horizontal leg of its own L. This leaves a portion of the vertical leg of L_2 outside of the y range of R_2. Thus what remain of R_1 and R_2 do not share any y coordinates, so we can rotate the picture 90 degrees and return to a horizontal problem.

This argument reduces merging two L's to at most three merges between an L and a horizontal leg. Still several cases remain, as illustrated in Figure 28. Case 1 is when the horizontal leg is aligned with a coin in the L. Case 2 is when they are out of alignment. Case 3 is a special case occurring at the corner of the L, where the horizontal legs are aligned but distances are higher than in Case 1. The above three cases are subdivided into subcases (a) and (b), depending on how close the horizontal leg is to the L. Finally, Case 4 is when the L and horizontal leg do not share x or y coordinates.

By the procedures in Figure 28, in all cases, the merging can be done in $O(1)$ flips and rotations of L's, $O(1)$ leapfrogs, and $O(n)$ shuffles. In total, $O(n^2)$ moves are required to merge an L and a horizontal leg, or equivalently to merge two L's.

5.4. Final Sweep. Thus far we have shown how to reversibly re-arrange a configuration (A or B) into the canonical configuration, using two extra coins. However, during this process, we may have picked up extra coins, and now need to drop them appropriately. In reality, these coins sit in arbitrary locations on the board. For re-arranging the source configuration A into the canonical configuration, the moves need not be reversible, so we can simply drop the extra coins in the canonical order, as in Figure 22. For re-arranging the destination configuration B into the canonical configuration, we need to effectively drop these coins by a sequence of reverse moves.

More directly, starting from the canonical configuration, we need to show how to distribute the extra coins to arbitrary locations on the board. We can achieve this effect by making a complete sweep over the board. More precisely, we flip the L as in Section 5.3.1, which has the effect of passing over every position on the board with the operations shown in Figure 23. During this process, we will pass over the extra coins; at this point we treat them as if they were picked up, applying the emulation in Section 4.1. Then we flip the L back to its original orientation. On the way back, whenever we apply an operation in Figure 23 and pass over the desired destination d for one of the extra coins, we move the extra coin to d while there are at least two adjacent coins from the L. By monotonicity of the flipping process, this extra coin will not be passed over later by the flip, so once an extra coin is placed in its desired location, it remains there.

5.5. Reducing Span. Now that we know any configuration can be brought to the corresponding canonical configuration with a sequence of (mostly) reversible moves, it follows immediately that any configuration can be re-arranged into any configuration with the same span. More generally, if we are given configurations A and B satisfying span $A \supseteq$ span B, we can first pick up all coins in $A -$ span B, then reversibly re-arrange both configurations into the same canonical

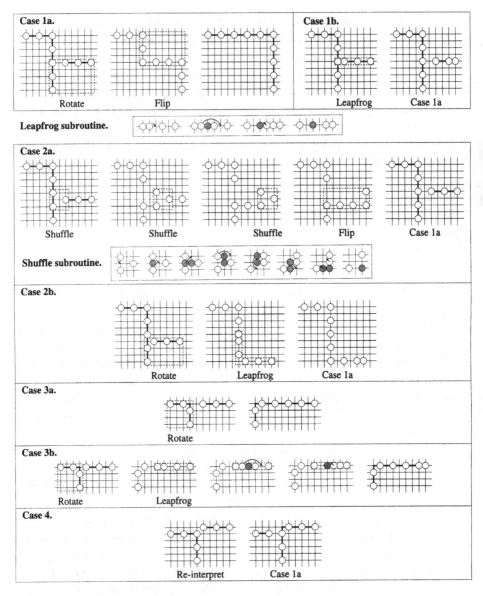

Figure 28. Merging an L and the horizontal leg of a nearby L.

configuration. Putting these two sequence of moves together, we obtain a re-arrangement from A − span B to B with some coins missing. Then we simply drop the previously picked up coins in the appropriate positions to create B.

Note that these moves need not be reversible, because we are only concerned with the direction from A to B. Indeed, the moves cannot be made reversible, because the span cannot increase (Lemma 2).

This concludes the proof of Theorem 2.

5.6. Lower Bound. The bound on the number of moves in Theorem 2 is in fact tight:

Theorem 3. *The "V to diagonal" puzzle in Figure 29 requires $\Theta(n^3)$ moves to solve.*

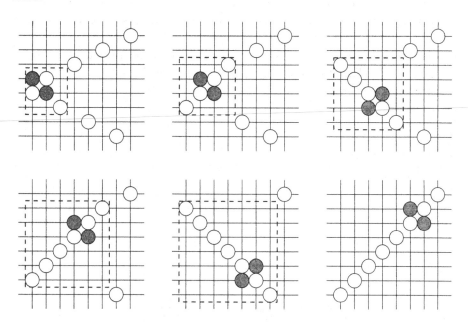

Figure 29. Re-arranging the V-shape in the upper left into the diagonal in the lower right requires repeated rotations of diagonals as in Figure 5 (or repeated rotations of L's).

Proof. We claim that re-arranging the V shape into a diagonal effectively requires repeated "diagonal flipping." At any time, only one component of coins can be actively manipulated (drawn with dotted lines in the figure); all other coins are isolated from movement. Thus we must repeatedly re-arrange the active component so that it can reach the nearest isolated coin. More specifically, we must re-arrange the active component into a chain starting at the corner of the bounding rectangle that is near the isolated coin, and ending at the opposite corner of the bounding rectangle of the active component. These two corners alternate for each isolated coin we pick up, and that is the sense in which we must "flip a diagonal." It is fairly easy to see that each diagonal flipping of a chain with k coins takes $\Omega(k^2)$ time. In total, the puzzle requires $\Theta(\sum_{k=1}^{n} k^2) = \Theta(n^3)$ moves. □

This theorem is the motivation for the puzzle in Figure 7.

5.7. Labeled Coins. We conjecture that Theorem 2 holds even when coins are labeled, subject to a few constraints. The idea is that permutation of the coins is relatively easy once we reach the canonical configuration. Examples of methods for swapping coins within one L are shown in Figure 30. The top figure shows how to swap a pair of coins when the canonical configuration is nothing more than a canonical L. The middle figure shows how to perform the same swap when there are four additional coins. Note that swapping the corner coin 3 works in exactly the same way; indeed, this method works whenever the coins to be swapped have t wo other coins adjacent to them, and there is another valid destination. The bottom figure shows how to swap one of the end coins, which is more difficult. This last method begins with moving the bend of the L toward the end coin, and then works locally on the coins $1, 2, 3$.

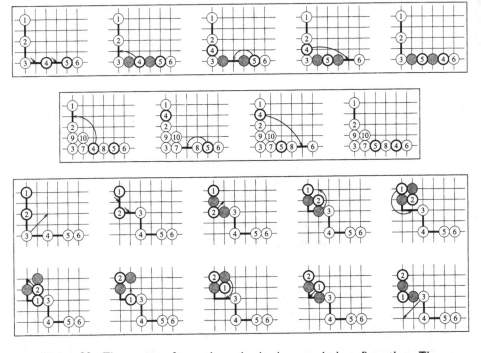

Figure 30. Three cases of swapping coins in the canonical configuration. The coins to be swapped have a thick outline.

One obvious constraint for these methods is that if there are no valid moves, permutation is impossible. Also, if the bounding rectangle of an L has width or height 1, then the t wo end coins of the L cannot be moved. Subject to these constraints, Figure 30 proves that the coins in a single connected component of the span, other than the extra coins e_1 and e_2, can be permuted arbitrarily.

It only remains to show that a coin can be swapped with e_1 or e_2, which implies that coins between different connected components of the span can be swapped. We have not proved this in general yet, but one illustrating example is the puzzle in Figure 8, whose solution is shown in Figure 31. The idea is that coins 2 and 4 are e_1 and e_2, and so we succeed in swapping e_2 with coin 3.

A slight generalization of this approach may complete a solution to the labeled coins.

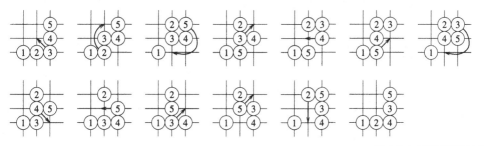

Figure 31. Solution to the puzzle in Figure 8.

5.8. Fewer Extra Coins. We have shown that the configuration space is essentially strongly connected provided there is a pair of extra coins, i.e., the removal of these t wo coins does not reduce the span. This section summarizes what we know about configurations without this property.

If we have a span-minimal configuration with no extra coins, Lemma 3 tells us that every move decreases the span. With an overhead of a factor of n^2, we can simply try all possible moves, in each case obtaining a configuration with smaller span, which furthermore must have an extra coin (the moved coin). Now we only need to recursively check these configurations.

Unfortunately, the situation is trickier with one extra coin. The key difficulty is that multiple coins could individually be considered extra, but no pair of coins is extra. In other words, there may be t wo coins such that removing either one does not reduce the span, but removing both of them reduces the span. Two simple examples are shown in Figure 32.

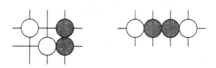

Figure 32. The shaded coins are individually extra, but do not form a pair of extra coins suitable for Theorem 2.

This difficulty makes "one" extra coin surprisingly powerful. For example, using one extra coin, an L with odd parity can be flipped, although it cannot be rotated, and an L with even parity cannot be flipped or rotated. In Figure 33 we exploit this property to make an interesting solvable puzzle initially with no pair of extra coins; it takes significant work before a pair of extra coins appears.

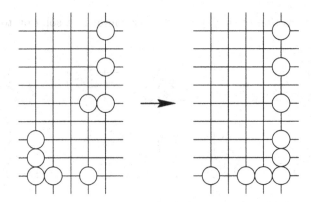

Figure 33. A puzzle on the square grid with no initial pair of extra coins.

6. Conclusion

We have begun the study of deciding solvability of coin-moving puzzles and more generally token-moving puzzles. We gave an exact characterization of solvable puzzles with labeled coins on the equilateral-triangle grid. By introducing the notion of a constant number of extra coins, we have given a tight theorem characterizing solvable puzzles on the square grid. Specifically, we have shown that any configuration can be re-arranged into any configuration with the same or smaller span using t wo extra coins, and that this is best possible in general. The number of moves is also best possible in the worst case.

Several open questions remain:

(i) What is the complexity of solving a puzzle using the fewest moves?
(ii) How do our results change if moves are forced to be *slides* that avoid other coins? We conjecture that Theorem 1 still holds for unlabeled coins.
(iii) Can we extend our results on the square grid to the hypercube lattice in any dimension?
(iv) Can we combine Theorems 1 and 2 to deal with a mix of the square and equilateral-triangle lattice, like the second puzzle in Figure 4?
(v) Can we prove similar results for general graphs?

Acknowledgments

We thank J. P. Grossman for writing a program to find optimal solutions to the puzzles in Figures 5 and 6.

References

[1]   Elwyn R. Berlekamp, John H. Conway, and Richard K. Guy. A solitaire-like puzzle and some coin-sliding problems. In *Winning Ways*, volume 2, pages 755–756. Academic Press, London, 1982.

[2] Brian Bolt. Invert the triangle. In *The Amazing Mathematical Amusement Arcade*, amusement 53, page 30. Cambridge University Press, Cambridge, 1984.

[3] Brian Bolt. A two touching transformation. In *Mathematical Cavalcade*, puzzle 20, page 10. Cambridge University Press, Cambridge, 1991.

[4] Maxey Brooke. *Fun for the Money*. Charles Scriber's Sons, New York, 1963. Reprinted as *Coin Games and Puzzles* by Dover Publications, 1973.

[5] Henry Ernest Dudeney. "The four pennies" and "The six pennies". In *536 Puzzles & Curious Problems*, problems 382–383, page 138. Charles Scribner's Sons, New York, 1967.

[6] Kobon Fujimura. "Coin pyramids," "Four pennies," "Six pennies," and "Five coins". In *The Tokyo Puzzles*, puzzles 23 and 25–27, pages 29–33. Charles Scribner's Sons, New York, 1978.

[7] Martin Gardner. Penny puzzles. In *Mathematical Carnival*, chapter 2, pages 12–26. Alfred A. Knopf, New York, 1975.

[8] Boris A. Kordemsky. A ring of disks. In *The Moscow Puzzles*, problem 117, page 47. Charles Scribner's Sons, New York, 1972.

[9] Harry Langman. Curiosa 261: A disc puzzle. *Scripta Mathematica*, 17(1–2):144, March–June 1951.

[10] Harry Langman. Curiosa 342: Easy but not obvious. *Scripta Mathematica*, 19(4):242, December 1953.

[11] David Wells. "Six pennies," "OH-HO," and "Inverted triangle". In *The Penguin Book of Curious and Interesting Puzzles*, puzzle 305, 375, and 376, pages 101–102 and 125. Penguin Books, 1992.

ERIK D. DEMAINE
MIT LABORATORY FOR COMPUTER SCIENCE
200 TECHNOLOGY SQUARE
CAMBRIDGE, MA 02139
UNITED STATES
edemaine@mit.edu

MARTIN L. DEMAINE
MIT LABORATORY FOR COMPUTER SCIENCE
200 TECHNOLOGY SQUARE
CAMBRIDGE, MA 02139
UNITED STATES
mdemaine@mit.edu

HELENA A. VERRILL
INSTITUT FOR MATEMATISKE FAG
UNIVERSITETSPARKEN 5
DK-2100 KØBENHAVN
DENMARK
verrill@math.ku.dk
http://hverrill.net

More Games of No Chance
MSRI Publications
Volume 42, 2002

Searching for Spaceships

DAVID EPPSTEIN

ABSTRACT. We describe software that searches for spaceships in Conway's Game of Life and related two-dimensional cellular automata. Our program searches through a state space related to the de Bruijn graph of the automaton, using a method that combines features of breadth first and iterative deepening search, and includes fast bit-parallel graph reachability and path enumeration algorithms for finding the successors of each state. Successful results include a new $2c/7$ spaceship in Life, found by searching a space with 2^{126} states.

1. Introduction

John Conway's Game of Life has fascinated and inspired many enthusiasts, due to the emergence of complex behavior from a very simple system. One of the many interesting phenomena in Life is the existence of gliders and spaceships: small patterns that move across space. When describing gliders, spaceships, and other early discoveries in Life, Martin Gardner wrote (in 1970) that spaceships "are extremely hard to find" [10]. Very small spaceships can be found by human experimentation, but finding larger ones requires more sophisticated methods. Can computer software aid in this search? The answer is yes – we describe here a program, gfind, that can quickly find large low-period spaceships in Life and many related cellular automata.

Among the interesting new patterns found by gfind are the "weekender" $2c/7$ spaceship in Conway's Life (Figure 1, right), the "dragon" $c/6$ Life spaceship found by Paul Tooke (Figure 1, left), and a $c/7$ spaceship in the Diamoeba rule (Figure 2, top). The middle section of the Diamoeba spaceship simulates a simple one-dimensional parity automaton and can be extended to arbitrary lengths. David Bell discovered that two back-to-back copies of these spaceships form a pattern that fills space with live cells (Figure 2, bottom). The existence of infinite-growth patterns in Diamoeba had previously been posed as an open problem (with a $50 bounty) by Dean Hickerson in August 1993, and was later

Figure 1. The "dragon" (left) and "weekender" (right) spaceships in Conway's Life (B3/S23). The dragon moves right one step every six generations (speed $c/6$) while the weekender moves right two steps every seven generations (speed $2c/7$).

Figure 2. A $c/7$ spaceship in the "Diamoeba" rule (B35678/S5678), top, and a spacefilling pattern formed by two back-to-back spaceships.

included in a list of related open problems by Gravner and Griffeath [11]. Our program has also found new spaceships in well known rules such as HighLife and Day&Night as well as in thousands of unnamed rules.

As well as providing a useful tool for discovering cellular automaton behavior, our work may be of interest for its use of state space search techniques. Recently, Buckingham and Callahan [8] wrote "So far, computers have primarily been used to speed up the design process and fill gaps left by a manual search. Much potential remains for increasing the level of automation, suggesting that Life may merit more attention from the computer search community." Spaceship searching provides a search problem with characteristics intriguingly different from standard test cases such as the 15-puzzle or computer chess, including a large state space that fluctuates in width instead of growing exponentially at each level, a tendency for many branches of the search to lead to dead ends, and the lack of any kind of admissable estimate for the distance to a goal state. Therefore, the search community may benefit from more attention to Life.

The software described here, a database of spaceships in Life-like automata, and several programs for related computations can be found online at http://www.ics.uci.edu/~eppstein/ca/.

Figure 3. Large $c/60$ spaceship in rule B36/S035678, found by brute force search.

2. A Brief History of Spaceship Searching

According to Berlekamp et al. [5], the $c/4$ diagonal glider in Life was first discovered by simulating the evolution of the R-pentomino, one of only 21 connected patterns of at most five live cells. This number is small enough that the selection of patterns was likely performed by hand. Gliders can also be seen in the evolution of random initial conditions in Life as well as other automata such as B3/S13 [14], but this technique often fails to work in other automata due to the lack of large enough regions of dead cells for the spaceships to fly through. Soon after the discovery of the glider, Life's three small $c/2$ orthogonal spaceships were also discovered.

Probably the first automatic search method developed to look for interesting patterns in Life and other cellular automata, and the method most commonly programmed, is a brute force search that tests patterns of bounded size, patterns with a bounded number of live cells, or patterns formed out of a small number of known building blocks. These tests might be exhaustive (trying all possible patterns) or they might perform a sequence of trials on randomly chosen small patterns. Such methods have found many interesting oscillators and other patterns in Life, and Bob Wainwright collected a large list of small spaceships found in this way for many other cellular automaton rules [19]. Currently, it is possible to try all patterns that fit within rectangles of up to 7×8 cells (assuming symmetric initial conditions), and this sort of exhaustive search can sometimes find spaceships as large as 12×15 (Figure 3). However, brute force methods have not been able to find spaceships in Life with speeds other than $c/2$ and $c/4$.

The first use of more sophisticated search techniques came in 1989, when Dean Hickerson wrote a backtracking search program which he called LS. For each generation of each cell inside a fixed rectangle, LS stored one of three states: unknown, live, or dead. LS then attempted to set the state of each unknown cell by examining neighboring cells in the next and previous generations. If no unknown cell's state could be determined, the program performed a depth first branching step in which it tried both possible states for one of the cells. Using this program, Hickerson discovered many patterns including Life's $c/3$, $c/4$, and $2c/5$ orthogonal spaceships. Hartmut Holzwart used a similar program to find many variant $c/2$ and $c/3$ spaceships in Life, and related patterns including the "spacefiller" in which four $c/2$ spaceships stretch the corners of a growing

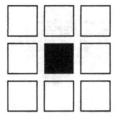

Figure 4. The eight neighbors in the Moore neighborhood of a cell.

diamond shaped still life. David Bell reimplemented this method in portable C, and added a fourth "don't care" state; his program, lifesrc, is available at http://www.canb.auug.org.au/~dbell/programs/lifesrc-3.7.tar.gz.

In 1996, Tim Coe discovered another Life spaceship, moving orthogonally at speed $c/5$, using a program he called knight in the hope that it could also find "knightships" such as those in Figure 5. Knight used breadth first search (with a fixed amount of depth-first lookahead per node to reduce the space needs of BFS) on a representation of the problem based on *de Bruijn graphs* (described in Section 4). The search took 38 cpu-weeks of time on a combination of Pentium Pro 133 and Hypersparc processors. The new search program we describe here can be viewed as using a similar state space with improved search algorithms and fast implementation techniques. Another recent program by Keith Amling also uses a state space very similar to Coe's, with a depth first search algorithm.

Other techniques for finding cellular automaton patterns include complementation of nondeterministic finite automata, by Jean Hardouin-Duparc [12, 13]; strong connectivity analysis of de Bruijn graphs, by Harold McIntosh [16,17]; randomized hill-climbing methods for minimizing the number of cells with incorrect evolution, by Paul Callahan (http://www.radicaleye.com/lifepage/stilledit.html); a backtracking search for still life backgrounds such that an initial perturbation remains bounded in size as it evolves, by Dean Hickerson; Gröbner basis methods, by John Aspinall; and a formulation of the search problem as an integer program, attempted by Richard Schroeppel and later applied with more success by Robert Bosch [7]. However to our knowledge none of these techniques has been used to find new spaceships.

3. Notation and Classification of Patterns

We consider here only *outer totalistic* rules, like Life, in which any cell is either "live" or "dead" and in which the state of any cell depends only on its previous state and on the total number of live neighbors among the eight adjacent cells of the Moore neighborhood (Figure 4). For some results on spaceships in cellular automata with larger neighborhoods, see Evans' thesis [9]. Outer totalistic rules are described with a string of the form $Bx_1x_2\ldots/Sy_1y_2\ldots$ where the x_i are digits listing the number of neighbors required for a cell to be born (change from dead

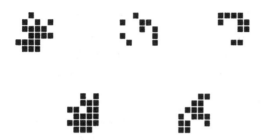

Figure 5. Slope 2 and slope 3/2 spaceships. Left to right: (a) B356/S02456, 2c/11, slope 2. (b) B3/S01367, c/13, slope 2. (c) B36/S01347, 2c/23, slope 2. (d) B34578/S358, 2c/25, slope 2. (e) B345/S126, 3c/23, slope 3/2.

Figure 6. Long narrow c/6 spaceship in Day&Night (B3678/S34678).

Figure 7. 9c/28 spaceship formed from 10 R-pentomino puffers in B37/S23.

to live) while the y_i list the number of neighbors required for a cell to survive (stay in the live state after already being live). For instance, Conway's Life is described in this notation as B3/S23.

A *spaceship* is a pattern which repeats itself after some number p of generations, in a different position from where it started. We call p the *period* of the spaceship. If the pattern moves x units horizontally and y units vertically every p steps, we say that it has slope y/x and speed $\max(|x|, |y|)c/p$, where c denotes the maximum speed at which information can propagate in the automaton (the so-called *speed of light*). Most known spaceships move orthogonally or diagonally, and many have an axis of symmetry parallel to the direction of motion. Others, such as the glider and small $c/2$ spaceships in Life, have *glide-reflect symmetry*: a mirror image of the original pattern appears in generations $p/2$, $3p/2$, etc.; spaceships with this type of symmetry must also move orthogonally or diagonally. However, a few asymmetric spaceships move along lines of slope 2 or even $3/2$ (Figure 5). According to Berlekamp et al. [5], there exist spaceships in Life that move with any given rational slope, but the argument for the existence of such spaceships is not very explicit and would lead to extremely large patterns.

Related types of patterns include oscillators (patterns that repeat in the same position), still lifes (oscillators with period 1), puffers (patterns which repeat some distance away after a fixed period, leaving behind a trail of discrete patterns such as oscillators, still lifes, and spaceships), rakes (spaceship puffers), guns (oscillators which send out a moving trail of discrete patterns such as spaceships or rakes), wickstretchers (patterns which leave behind one or more connected stable or oscillating regions), replicators (patterns which produce multiple copies of themselves), breeders (patterns which fill a quadratically-growing area of space with discrete patterns, for instance replicator puffers or rake guns), and spacefillers (patterns which fill a quadratically-growing area with one or more connected patterns). See Paul Callahan's Life Pattern Catalog (http://www.radicaleye.com/lifepage/patterns/contents.html) for examples of many of these types of patterns in Life, or http://www.ics.uci.edu/~eppstein/ca/replicators/ for a number of replicator-based patterns in other rules.

We can distinguish among several classes of spaceships, according to the methods that work best for finding them.

- Spaceships with small size but possibly high period can be found by brute force search. The patterns depicted in Figures 3 and 5 fall into this class, as do Life's glider and $c/2$ spaceships.

- Spaceships in which the period and one dimension are small, while the other dimension may be large, can be found by search algorithms similar to the ones described in this paper. The new Life spaceships in Figure 1 fall into this class. For a more extreme example of a low period ship which is long but narrow, see Figure 6. "Small" is a relative term—our search program has found $c/2$ spaceships with minimum dimension as high as 42 (Figure 10) as well as narrower spaceships with period as high as nine.

- Sometimes small non-spaceship objects, such as puffers, wickstretchers, or replicators, can be combined by human engineering into a spaceship. For in-

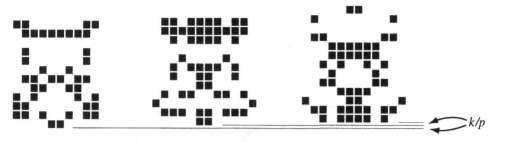

Figure 8. The three phases of the $c/3$ "turtle" Life spaceship, shifted by $1/3$ cell per phase.

stance, in the near-Life rule B37/S23, the R-pentomino pattern acts as an (unstable) puffer. Figure 7 depicts a $9c/28$ spaceship in which a row of six pentominoes stabilize each other while leaving behind a trail of still lifes, which are cleaned up by a second row of four pentominoes. Occasionally, spaceships found by our search software will appear to have this sort of structure (Figure 10). The argument for the existence of Life spaceships with any rational slope would also lead to patterns of this type. Several such spaceships have been constructed by Dean Hickerson, notably the $c/12$ diagonal "Cordership" in Conway's Life, discovered by him in 1991. Hickerson's web page http://www.math.ucdavis.edu/~dean/RLE/slowships.html has more examples of structured ships.

- The remaining spaceships have large size, large period, and little internal structure. We believe such spaceships should exist, but none are known and we know of no effective method for finding them.

4. State Space

Due to the way our search is structured, we need to arrange the rows from all phases of the pattern we are searching for into a single sequence. We now show how to do this in a way that falls out naturally from the motion of the spaceship.

Suppose we are searching for a spaceship that moves k units down every p generations. For simplicity of exposition, we will assume that $\gcd(k, p) = 1$. We can then think of the spaceship we are searching for as living in a cellular automaton modified by shifting the grid upward k/p units per generation. In this modified automaton, the shifting of the grid exactly offsets the motion of the spaceship, so the ship acts like an oscillator instead of like a moving pattern. We illustrate this shifted grid with the turtle, a $c/3$ spaceship in Conway's Life (Figure 8).

Because of the shifted grid, and the assumption that $\gcd(k, p) = 1$, each row of each phase of the pattern exists at a distinct vertical position. We form a

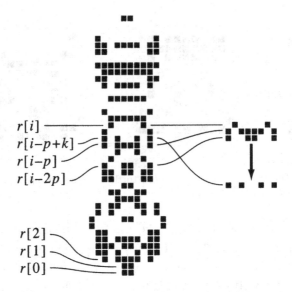

$$r[i] - \\ r[i-p+k] - \\ r[i-p] - \\ r[i-2p] -$$

$$r[2] - \\ r[1] - \\ r[0] -$$

Figure 9. Merged sequence of rows from all three phases of the turtle, illustrating equation (∗).

doubly-infinite sequence

$$\ldots r[-2], r[-1], r[0], r[1], r[2] \ldots$$

of rows, by taking each row from each phase in order by the rows' vertical positions (Figure 9). For any i, the three rows $r[i - 2p], r[i - p], r[i]$ form a contiguous height-three strip in a single phase of the pattern, and we can apply the cellular automaton rule within this strip to calculate

$$r[i - p + k] = \text{evolve}(r[i - 2p], r[i - p], r[i]). \qquad (\ast)$$

Conversely, any doubly-infinite sequence of rows, in which equation (∗) is satisfied for all i, and in which there are only finitely many live cells, corresponds to a spaceship or sequence of spaceships. Further, any finite sequence of rows can be extended to such a doubly infinite sequence if the first and last $2p$ rows of the finite sequence contain only dead cells.

Diagonal spaceships, such as Life's glider, can be handled in this framework by modifying equation (∗) to shift rows i, $i-p$, and $i-2p$ with respect to each other before performing the evolution rule. We can also handle glide-reflect spaceships such as Life's small $c/2$ spaceships, by modifying equation (∗) to reverse the order of the cells in row $i - p + k$ (when k is odd) or in the two rows $i - p$ and $i - p + k$ (when p is odd). Note that, by the assumption that $\gcd(k, p) = 1$, at least one of k and p will be odd. In these cases, p should be considered as the half-period of the spaceship, the generation at which a flipped copy of the original pattern appears. Searches for which $\gcd(k, p) > 1$ can be handled by adjusting the indices in equation $(*)$ depending on the phase to which row $r[i]$ belongs.

Our state space, then, consists of finite sequences of rows, such that equation $(*)$ is satisfied whenever all four rows in the equation belong to the finite sequence. The initial state for our search will be a sequence of $2p$ rows, all of which contain only dead cells. If our search discovers another state in which the last $2p$ rows also contain only dead cells, it outputs the pattern formed by every pth row of the state as a spaceship.

As in many game playing programs, we use a *transposition table* to detect equivalent states, avoid repeatedly searching the same states, and stop searching in a finite amount of time even when the state space may be infinite. We would like to define two states as being equivalent if, whenever one of them can be completed to form a spaceship, the same completion works for the other state as well. However, this notion of equivalence seems too difficult to compute (it would require us to be able to detect states that can be completed to spaceships, but if we could do that then much of our search could be avoided). So, we use a simpler sufficient condition: two states are equivalent if their last $2p$ rows are identical. If our transposition table detects two equivalent states, the longer of the two is eliminated, since it can not possibly lead to the shortest spaceship for that rule and period.

We can form a finite directed graph, the *de Bruijn graph* [16,17], by forming a vertex for each equivalence class of states in our state space, and an edge between two vertices whenever a state in one equivalence class can be extended to form a state in the other class. The size of the de Bruijn graph provides a rough guide to the complexity of a spaceship search problem. If we are searching for patterns with width w, the number of vertices in this de Bruijn graph is 2^{2pw}. For instance, in the search for the weekender spaceship, the effective width was nine (due to an assumption of bilateral symmetry), so the de Bruijn graph contained 2^{126} vertices. Fortunately, most of the vertices in this graph were unreachable from the start state.

Coe's search program `knight` uses a similar state space formed by sequences of pattern rows, but only uses the rows from a single phase of the spaceship. In place of our equation $(*)$, he evolves subsequences of $2p+1$ rows for p generations and tests that the middle row of the result matches the corresponding row of the subsequence. As with our state space, one can form a (different) de Bruijn graph by forming equivalence classes according to the last $2p$ rows of a state. Coe's approach has some advantages; for instance, it can find patterns in which some phases exceed the basic search width (as occurred in Coe's $c/5$ spaceship). However, it does not seem to allow the fast neighbor-finding techniques we describe in Section 7.

5. Search Strategies

There are many standard algorithms for searching state spaces [20], however each has some drawbacks in our application:

- Depth first search requires setting an arbitrary depth limit to avoid infinite recursion. The patterns it finds may be much longer than necessary, and the search may spend a long time exploring deep regions of the state space before reaching a spaceship. Further, DFS does not make effective use of the large amounts of memory available on modern computers.
- Breadth first search is very effective for small searches, but quickly runs out of memory for larger searches, even when large amounts of memory are available.
- Depth first iterative deepening [15] has been proposed as a method of achieving the fast search times of breadth first search within limited space. However, our state space often does not have the exponential growth required for iterative deepening to be efficient; rather, as the search progresses from level to level the number of states in the search frontier can fluctuate up and down, and typically has a particularly large bulge in the earlier levels of the search. The overall depth of the search (and hence the number of deepening iterations) can often be as large as several hundred. For these reasons, iterative deepening can be much slower than breadth first search. Further, the transposition table used to detect equivalent states does not work as well with depth first as with breadth first search: to save space, we represent this table as a collection of pointers to states, rather than explicitly listing the $2p$ rows needed to determine equivalence, so when searching depth first we can only detect repetitions within the current search path. Finally, depth first search does not give us much information about the speed at which the search is progressing, which we can use to narrow the row width when the search becomes too slow.
- Other techniques such as the A* algorithm, recursive best first search, and space-bounded best first search, depend on information unavailable in our problem, such as varying edge weights or admissable estimates of the distance to a solution.

Therefore, we developed a new search algorithm that combines the best features of breadth first and iterative deepening search, and that takes advantage of the fact that, in our search problem, many branches of the search eventually lead only to dead ends. Our method resembles the MREC algorithm of Sen and Bagchi [18], in that we perform deepening searches from the breadth first search frontier, however unlike MREC we use the deepening stages to prune the search tree, allowing additional breadth first searching.

Our search algorithm begins by performing a standard breadth first search. We represent each state as a single row together with a pointer to its predecessor, so the search must maintain the entire breadth first search tree. By default, we allocate storage for 2^{22} nodes in the tree, which is adequate for small searches yet well within the memory limitations of most computers.

On larger searches (longer than a minute or so), the breadth first search will eventually run out of space. When this happens, we suspend the breadth first search and perform a round of depth first search, starting at each node of the current breadth first search queue. This depth first search has a depth limit which is normally δ levels beyond the current search frontier, for a small value δ that we set in our implementation to equal the period of the pattern we are searching for. (Setting δ to a fixed small constant would likely work as well.) However, if a previous round of depth first searching reached a level past the current search frontier, we instead limit the new depth first search round to δ levels beyond the previous round's limit.

When we perform a depth first search from a BFS queue node, one of three things can happen. First, we may discover a spaceship; in that case we terminate the entire search. Second, we may reach the depth limit; in that case we terminate the depth first search and move on to the next BFS queue node. Third, the depth first search may finish normally, without finding any spaceships or deep nodes. In this case, we know that the root of the search leads only to dead ends, and we remove it from the breadth first search queue.

After we have performed this depth first search for all nodes of the queue, we compact the remaining nodes and continue with the previously suspended breadth first search. Generally, only a small fraction of the previous breadth first search tree remains after the compaction, leaving plenty of space for the next round of breadth first search.

There are two common modes of behavior for this searching algorithm, depending on how many levels of breadth first searching occur between successive depth first rounds. If more than δ levels occur between each depth first search round, then each depth first search is limited to only δ levels beyond the breadth first frontier, and the sets of nodes searched by successive depth first rounds are completely disjoint from each other. In this case, each node is searched at most twice (once by the breadth first and once by the depth first parts of our search), so we only incur a constant factor slowdown over the more memory-intensive pure breadth first search. In the second mode of behavior, successive depth first search rounds begin from frontiers that are fewer than δ levels apart. If this happens, the ith round of depth first search will be limited to depth $i \cdot \delta$, so the search resembles a form of iterative deepening. Unlike iterative deepening, however, the breadth first frontier always makes some progress, permanently removing nodes from the actively searched part of the state space. Further, the early termination of depth-first searches when they reach deep nodes allows our algorithm to avoid searching large portions of the state space that pure iterative deepening would have to examine.

On typical large searches, the amount of deepening can be comparable to the level of the breadth-first search frontier. For instance, in the search for the weekender, the final depth first search round occurred when the frontier had reached level 90, and this round searched from each frontier node to an additional 97 levels. We allow the user to supply a maximum value for the deepening amount; if this maximum is reached, the state space is pruned by reducing the row width by one cell and the depth first search limit reverts back to δ.

There is some possibility that a spaceship found in one of the depth first searches may be longer than the optimum, but this has not been a problem in practice. Even this small amount of suboptimality could be averted by moving on to the next node instead of terminating the search when the depth first phase of the algorithm discovers a spaceship.

6. Lookahead

Suppose that our search reaches state

$$S = r[0], r[1], \ldots r[i-1].$$

The natural set of neighboring states to consider would be all sequences of rows

$$r[0], r[1], \ldots r[i-1], r[i]$$

where the first i rows match state S and we try all choices of row $r[i]$ that satisfy equation $(*)$.

However, it is likely that some of these choices will result in inconsistent states for which equation $(*)$ can not be satisfied the next time the new row $r[i]$ is involved in the equation. Since that next time will not occur until we choose row $r[i+p-k]$, any work performed in the intermediate levels of the search between these two rows could be wasted. To avoid this problem, when making choices for row $r[i]$, we simultaneously search for pairs of rows $r[i]$ and $r[i+p-k]$ satisfying both equation $(*)$ and its shifted form

$$r[i] = \text{evolve}(r[i-p-k], r[i-k], r[i+p-k]). \tag{L}$$

Note that $r[i-p-k]$ and $[i-k]$ are both already present in state S and so do not need to be searched for. We use as the set of successors to row S the sequences of rows $r[0], r[1], \ldots r[i-1], r[i]$ such that equation $(*)$ is true and equation (L) has a solution.

One could extend this idea further, and (as well as searching for rows $r[i]$ and $r[i+p-k]$) search for two additional rows $r[i+p-2k]$ and $r[i+2p-2k]$ such that the double lookahead equations

$$\begin{aligned}
r[i-k] &= \text{evolve}(r[i-p-2k], r[i-2k], r[i+p-2k]) \quad \text{and} \\
r[i+p-k] &= \text{evolve}(r[i-2k], r[i+p-2k], r[i+2p-2k])
\end{aligned} \tag{LL}$$

have a solution. However this double lookahead technique would greatly increase the cost of searching for the successor states of S and provide only diminished

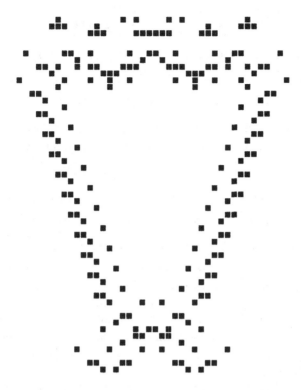

Figure 10. 42×53 $c/2$ spaceship in rule B27/S0.

returns. In our implementation, we use a much cheaper approximation to this technique: for every three consecutive cells of the row $r[i + p - k]$ being searched for as part of our single lookahead technique, we test that there could exist five consecutive cells in rows $r[i + p - 2k]$ and $r[i + 2p - 2k]$ satisfying equations (LL) for those three cells. The advantage of this test over the full search for rows $r[i + p - 2k]$ and $r[i + 2p - 2k]$ is that it can be performed with simple table lookup techniques, described in the next section.

For the special case $p = 2$, the double lookahead technique described above does not work as well, because $r[i + p - 2k] = r[i]$. Instead, we perform a reachability computation on a de Bruijn graph similar to the one used for our state space, five cells wide with don't-care boundary conditions, and test whether triples of consecutive cells from the new rows $r[i]$ and $r[i + p - k]$ correspond to a de Bruijn graph vertex that can reach a terminal state. This technique varies considerably in effectiveness, depending on the rule: for Life, 18.5% of the 65536 possible patterns are pruned as being unable to reach a terminal state, and for B27/S0 (Figure 10) the number is 68.3%, but for many rules it can be 0%.

7. Fast Neighbor-Finding Algorithm

To complete our search algorithm, we need to describe how to find the successors of each state. That is, given a state

$$S = r[0], r[1], \ldots r[i-1]$$

we wish to find all possible rows $r[i]$ such that (1) $r[i]$ satisfies equation (∗), (2) there exists another row $r[i + p - k]$ such that rows $r[i]$ and $r[i + p - k]$ satisfy equation (L), and (3) every three consecutive cells of $r[i + p - k]$ can be part of a solution to equations (LL).

Note that (because of our approximation to equations (LL)) all these constraints involve only triples of adjacent cells in unknown rows. That is, equation (∗) can be phrased as saying that every three consecutive cells of $r[i]$, $r[i - p]$, and $r[i - 2p]$ form a 3×3 square such that the result of the evolution rule in the center of the square is correct; and equation (L) can be phrased similarly. For this reason, we need a representation of the pairs of rows $r[i], r[i + p - k]$ in which we can access these triples of adjacent cells.

Such a representation is provided by the graph depicted in Figure 11. This graph (which like our state space can be viewed as a kind of de Bruijn graph) is formed by a sequence of columns of vertices. Each column contains 16 vertices, representing the 16 ways of forming a 2×2 block of cells. We connect a vertex in one column with a vertex in the next column whenever the two blocks overlap to form a single 2×3 block of cells; that is, an edge exists whenever the right half of the left 2×2 block matches the left half of the right 2×2 block.

If we have any two rows of cells, one placed above the other, we can form a path in this graph in which the vertices correspond to 2×2 blocks drawn from the pair of rows. Conversely, the fact that adjacent blocks are required to match implies that any path in this graph corresponds to a pair of rows.

Since each triple of adjacent cells from the two rows corresponds to an edge in this graph, the constraints of equations (∗), (L), and (LL) can be handled simply by removing the edges from the graph that correspond to triples not satisfying those constraints. In this constrained subgraph, any path corresponds to a pair of rows satisfying all the constraints. Thus, our problem has been reduced to finding the constrained subgraph, and searching for paths through it between appropriately chosen start and end terminals. The choice of which vertices to use as terminals depends on the symmetry type of the pattern we are searching for.

This search can be understood as being separated into three stages, although our actual implementation interleaves the first two of these.

In the first stage, we find the edges of the graph corresponding to blocks of cells satisfying the given constraints. We represent the set of 64 possible edges between each pair of columns in the graph (as shown in Figure 11) as a 64-bit quantity, where a bit is nonzero if an edge exists and zero otherwise. The set

Figure 11. Graph formed by placing each of 16 2 × 2 blocks of cells in each column, and connecting a block in one column to a block in the next column whenever the two blocks overlap to form a 2 × 3 block. Paths in the graph represent pairs of rows $r[i], r[i + p - k]$; by removing edges from the graph we can constrain triples of adjacent cells from these rows.

of edges corresponding to blocks satisfying equation (∗) can be found by a table lookup with an index that encodes the values of three consecutive cells of $r[i-p]$ and $r[i - 2p]$ together with one cell of $r[i - p + k]$. Similarly, the set of edges corresponding to blocks satisfying equation (L) can be found by a table lookup with an index that encodes the values of three consecutive cells of $r[i - k]$ and $r[i - p - k]$. In our implementation, we combine these two constraints into a single table lookup. Finally, the set of edges corresponding to blocks satisfying equations (LL) can be found by a table lookup with an index that encodes the values of five consecutive cells of $r[i - 2k]$ and $r[i - p - 2k]$ together with three consecutive cells of $r[i - k]$. The sets coming from equations (∗), (L), and (LL) are combined with a simple bitwise Boolean and operation. The various tables used by this stage depend on the cellular automaton rule, and are precomputed based on that rule before we do any searching.

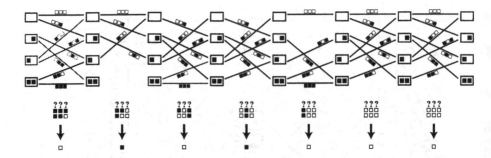

Figure 12. Simplified example (without lookahead) of graph representing equation (∗) for the rows from Fig. 9. Gray regions represent portions of the graph not reachable from the two start vertices.

In the second stage, we compute a 16-bit quantity for each column of vertices, representing the set of vertices in that column that can be reached from the start terminal. This set can be computed from the set of reachable vertices in the previous column by a small number of shift and mask operations involving the sets of edges computed in the previous stage.

In the third stage, we wish to find the actual rows $r[i]$ that correspond to the paths in the constrained graph. We perform a backtracking search for these rows, starting with the end terminal of the graph. At each step, we maintain a 16-bit quantity, representing the set of vertices in the current column of the graph that could reach the end terminal by a path matching the current partial row. To find the possible extensions of a row, we find the predecessors of this set of vertices in the graph by a small number of shift and mask operations (resembling those of the previous stage) and separate these predecessors into two subsets, those for which the appropriate cell of $r[i]$ is alive or dead. We then continue recursively in one or both of these subsets, depending on whichever of the two has a nonempty intersection with the set of vertices in the same column that can reach the start vertex.

Because of the reachability computation in the second stage, the third stage never reaches a dead end in which a partial row can not be extended. Therefore, this algorithm spends a total amount of time at most proportional to the width of the rows (in the first two stages) plus the width times the number of successor states (in the third stage). The time for the overall search algorithm is bounded by the width times the number of states reached.

A simplified example of this graph representation of our problem is depicted in Figure 12, which shows the graph formed by the rows from the Life turtle pattern depicted in Figure 9. To reduce the complexity of the figure, we have only incorporated equation (∗), and not the two lookahead equations. Therefore, the vertices in each column represent the states of two adjacent cells in row $r[i]$ only, instead of also involving row $r[i + p - k]$, and there are four vertices per column instead of sixteen. Each edge represents the state of three adjacent cells of $r[i]$, and connects vertices corresponding to the left two and right two of these cells; we have marked each edge with the corresponding cell states.

Due to the symmetry of the turtle, the effective search width is six, so we need to enforce equation $(*)$ for six cells of row $r[i-p+k]$. Six of the seven columns of edges in the graph correspond to these cells; the seventh represents a cell outside the search width which must remain blank to prevent the pattern from growing beyond its assigned bounds. Below each of these seven columns, we have shown the cells from previous rows $r[i-p]$, $r[i-2p]$, and $r[i-p+k]$ which determine the set of edges in the column and which are concatenated together to form the table index used to look up this set of edges.

The starting vertices in this example are the top and bottom left ones in the graph, which represent the possible states for the center two cells of $r[i]$ that preserve the pattern's symmetry. The destination vertex is the upper right one; it represents the state of two cells in row $r[i]$ beyond the given search width, so both must be blank. Each path from a start vertex to the destination vertex represents a possible choice for the cells in row $r[i]$ that would lead to the correct evolution of row $r[i-p+k]$ according to the rules of Conway's life. There are 13 such paths in the graph shown. The reachability information computed in the second stage of the algorithm is depicted by marking unreachable vertices and edges in gray; in this example, as well as the asymmetric states in the first column, there is one more unreachable vertex.

For this simplified example, a list of all 13 paths in this graph could be found in the third stage by a recursive depth first search from the destination vertex backwards, searching only the black edges and vertices. Thus even this simplified representation reduces the number of row states that need to be considered from 64 to 13, and automatically selects only those states for which the evolution rule leads to the correct outcome. The presence of lookahead complicates the third stage in our actual program since multiple paths can correspond to the same value of row $r[i]$; the recursive search procedure described above finds each such value exactly once.

8. Conclusions

To summarize, we have described an algorithm that finds spaceships in outer totalistic Moore neighborhood cellular automata, by a hybrid breadth first iterative deepening search algorithm in a state space formed by partial sequences of pattern rows. The algorithm represents the successors of each state by paths in a regularly structured graph with roughly $16w$ vertices; this graph is constructed by performing table lookups to quickly find the sets of edges representing the constraints of the cellular automaton evolution rule and of our lookahead formulations. We use this graph to find a state's successors by performing a 16-way bit-parallel reachability algorithm in this graph, followed by a recursive backtracking stage that uses the reachability information to avoid dead ends.

Figure 13. Large period-114 $c/3$ blinker puffer in Life, found by David Bell, Jason Summers, and Stephen Silver using a combination of automatic and human-guided searching.

Empirically, the algorithm works well, and is able to find large new patterns in many cellular automaton rules.

This work raises a number of interesting research questions, beyond the obvious one of what further improvements are possible to our search program:

- The algorithm described here, and the two search algorithms previously used by Hickerson and Coe, use three different state spaces. Do these spaces really lead to different asymptotic search performance, or are they merely three different ways of looking at the same thing? Can one make any theoretical arguments for why one space might work better than another?

- Is it possible to explain the observed fluctuations in size of the levels of the state space? A hint of an explanation for the fact that fluctuations exist comes from the idea that we typically run as narrow as possible a search as we can to find a given period spaceship. Increasing the width seems to increase the branching factor, and perhaps we should expect to find spaceships as soon as the state space develops infinite branches, at which point the branching factor will likely be quite close to one. However this rough idea depends on the unexplained assumption that the start of a spaceship is harder to find than the tail, and it does not explain other features of the state space size such as a large bulge near the early levels of the search.

- We have mentioned in Section 4 that one can use the size of the de Bruijn graph as a rough guide to the complexity of the search. However, our searches typically examine far fewer nodes than are present in the de Bruijn graph. Further, there seem to be other factors that can influence the search complexity; for instance, increasing k seems to decrease the running time, so that e.g. a width-nine search for a $c/7$ ship in Life would likely take much more time than the search for the weekender. Can we find a better formula for predicting search run time?

- What if anything can one say about the computational complexity of spaceship searching? Is the problem of determining whether a given outer totalistic rule has a spaceship of a given speed or period even decidable?

- David Bell and others have had success finding large spaceships with "arms" (Figure 13) by examining partial results from an automatic search, placing "don't care" cells at appropriate connection points, and then doing secondary searches for arms that can complete the pattern from each connection point. To what extent can this human-guided search procedure be automated?

- What other types of patterns can be found by our search techniques? For instance, one possibility would be a search for predecessors of a given pattern. The rows of each predecessor satisfy a consistency condition similar to equation ($*$), and it would not be difficult to incorporate a lookahead technique similar to the one we use for spaceship searching. However due to the fixed depth of the search it seems that depth first search would be a more appropriate strategy than the breadth first techniques we are using in our spaceship searches.

- Carter Bays has had some success using brute force methods to find small spaceships in various three-dimensional rules [1, 2, 4], and rules on the triangular planar lattice [3]. How well can other search methods such as the ones described here work for these types of automata?

- To what areas other than cellular automata can our search techniques be applied? A possible candidate is in document improvement, where Bern, Goldberg, and others [6] have been developing algorithms that look for a single high-resolution image that best matches a given set of low-resolution samples of the same character or image. Our graph based techniques could be used to replace a local optimization technique that changes a single pixel at a time, with a technique that finds an optimal assignment to an entire row of pixels, or to any other linear sequence such as the set of pixels around the boundary of an object.

Acknowledgements

Thanks go to Matthew Cook, Nick Gotts, Dean Hickerson, Harold McIntosh, Gabriel Nivasch, and Bob Wainwright for helpful comments on drafts of this paper; to Noam Elkies, Rich Korf, and Jason Summers for useful suggestions regarding search algorithms; to Keith Amling, David Bell, and Tim Coe, for making available the source codes of their spaceship searching programs; and to Richard Schroeppel and the members of the Life mailing list for their support and encouragement of cellular automaton research.

References

[1] C. Bays. Candidates for the game of Life in three dimensions. *Complex Systems* 1(3):373–400, June 1987.

[2] C. Bays. The discovery of a new glider for the game of three-dimensional Life. *Complex Systems* 4(6):599–602, December 1990.

[3] C. Bays. Cellular automata in the triangular tessellation. *Complex Systems* 8(2):127–150, April 1994.

[4] C. Bays. Further notes on the game of three-dimensional Life. *Complex Systems* 8(1):67–73, February 1994.

[5] E. R. Berlekamp, J. H. Conway, and R. K. Guy. What is Life? *Winning Ways For Your Mathematical Plays*, vol. 2, chapter 25, pp. 817–850. Academic Press, 1982.

[6] M. Bern and D. Goldberg. Scanner-model-based document image improvement. *Proc. 7th IEEE Int. Conf. Image Processing*, vol. 2, pp. 582-585, 2000.

[7] R. A. Bosch. Integer programming and Conway's game of Life. *SIAM Review* 41(3):594–604, 1999, http://epubs.siam.org/sam-bin/dbq/article/33825.

[8] D. J. Buckingham and P. B. Callahan. Tight bounds on periodic cell configurations in Life. *Experimental Mathematics* 7(3):221–241, 1998, http://www.expmath.com/restricted/7/7.3/callahan.ps.gz.

[9] K. M. Evans. *Larger Than Life: it's so nonlinear.* Ph.D. thesis, Univ. of Wisconsin, Madison, 1996, http://www.csun.edu/~kme52026/thesis.html.

[10] M. Gardner. The Game of Life, part I. *Wheels, Life, and Other Mathematical Amusements*, chapter 20, pp. 214–225. W. H. Freeman, 1983.

[11] J. Gravner and D. Griffeath. Cellular automaton growth on $\mathbb{Z}^2$: theorems, examples, and problems. *Advances in Applied Mathematics* 21(2):241–304, August 1998, http://psoup.math.wisc.edu/extras/r1shapes/r1shapes.html.

[12] J. Hardouin-Duparc. À la recherche du paradis perdu. *Publ. Math. Univ. Bordeaux Année* (4):51–89, 1972/73.

[13] J. Hardouin-Duparc. Paradis terrestre dans l'automate cellulaire de Conway. *Rev. Française Automat. Informat. Recherche Opérationnelle Sér. Rouge* 8(R-3):64–71, 1974.

[14] J.-C. Heudin. A new candidate rule for the game of two-dimensional Life. *Complex Systems* 10(5):367–381, October 1996.

[15] R. E. Korf. Depth-first iterative-deepening: an optimal admissable tree search. *Artificial Intelligence* 27:97–109, 1985.

[16] H. V. McIntosh. A zoo of Life forms, http://delta.cs.cinvestav.mx/~mcintosh/comun/zool/zoo.pdf. Manuscript, October 1988.

[17] H. V. McIntosh. Life's still lifes, http://delta.cs.cinvestav.mx/~mcintosh/comun/still/still.pdf. Manuscript, September 1988.

[18] A. K. Sen and A. Bagchi. Fast recursive formulations of best-first search that allow controlled use of memory. *Proc. 11th Int. Joint Conf. Artificial Intelligence*, vol. 1, pp. 297–302, 1989.

[19] R. T. Wainwright. Some characteristics of the smallest reported Life and alienlife spaceships. Manuscript, November 1994.

[20] W. Zhang. *State-Space Search: Algorithms, Complexity, Extensions, and Applications.* Springer Verlag, 1999.

DAVID EPPSTEIN
DEPTARTMENT OF INFORMATION AND COMPUTER SCIENCE
UNIVERSITY OF CALIFORNIA
IRVINE, CA 92697-3425
UNITED STATES
  eppstein@ics.uci.edu

Surveys

More Games of No Chance
MSRI Publications
Volume **42**, 2002

Unsolved Problems in Combinatorial Games

RICHARD K. GUY AND RICHARD J. NOWAKOWSKI

We have retained the numbering from the list of unsolved problems given on pp. 183–189 of AMS *Proc. Sympos. Appl. Math.* **43**(1991), called PSAM **43** below, and on pp. 475–491 of this volume's predecessor, *Games of No Chance*, hereafter referred to as GONC. This list also contains more detail about some of the games mentioned below. References in brackets, e.g., Ferguson [1974], are listed in Fraenkel's Bibliography later in this book; WW refers to

Elwyn Berlekamp, John Conway and Richard Guy, *Winning Ways for your Mathematical Plays*, Academic Press, 1982. A.K.Peters, 2000.

and references in parentheses, e.g., Kraitchik (1941), are at the end of this article.

1. Subtraction games are known to be periodic. Investigate the relationship between the subtraction set and the length and structure of the period. The same question can be asked about *partizan* subtraction games, in which each player is assigned an individual subtraction set. See Fraenkel and Kotzig [1987].

See also Subtraction Games in WW, 83–86, 487–498 and in the Impartial Games article in GONC. A move in the game $S(s_1, s_2, s_3, \ldots)$ is to take a number of beans from a heap, provided that number is a member of the **subtraction-set**, $\{s_1, s_2, s_3, \ldots\}$. Analysis of such a game and of many other heap games is conveniently recorded by a **nim-sequence**,

$$n_0 n_1 n_2 n_3 \ldots,$$

meaning that the nim-value of a heap of h beans is n_h, $h = 0, 1, 2, \ldots$, i.e., that the value of a heap of h beans in this particular game is the **nimber** $*n_h$. To avoid having to print stars, we say that the nim-value of a position is n, meaning that its value is the nimber $*n$.

For examples see Table 2 in § **4** on p. 67 of the Impartial Games paper in GONC.

In subtraction games the nim-values 0 and 1 are remarkably related by **Ferguson's Pairing Property** [Ferguson [1974]; WW, 86, 422]: if s_1 is the least member of the subtraction-set, then

$$\mathcal{G}(n) = 1 \qquad \text{just if} \qquad \mathcal{G}(n - s_1) = 0.$$

Here and later "$\mathcal{G}(n) = v$" means that the nim-value of a heap of n beans is v.

It would now seem feasible to give the complete analysis for games whose subtraction sets have just three members, but the detail has so far eluded those who have looked at the problem.

2. Are all finite **octal games** ultimately periodic? Resolve any number of outstanding particular cases, e.g., ·**6** (Officers), ·**06**, ·**14**, ·**36**, ·**64**, ·**74**, ·**76**, ·**004**, ·**005**, ·**006**, ·**007** (One-dimensional tic–tac–toe, Treblecross), ·**016**, ·**106**, ·**114**, ·**135**, ·**136**, ·**142**, ·**143**, ·**146**, ·**162**, ·**163**, ·**172**, ·**324**, ·**336**, ·**342**, ·**362**, ·**371**, ·**374**, ·**404**, ·**414**, ·**416**, ·**444**, ·**564**, ·**604**, ·**606**, ·**744**, ·**764**, ·**774**, ·**776** and **Grundy's Game** (split a heap into two unequal heaps), which has been analyzed, mainly by Dan Hoey, as far as heaps of 5×2^{32} beans.

A similar unsolved game is John Conway's **Couples-Are-Forever** where a move is to split any heap except a heap of two. The first 50 million nim-values haven't displayed any periodicity. See Caines et al. [1999].

Explain the structure of the periods of games known to be periodic.

[If the binary expansion of the kth code digit in the game with code

$$d_0 \cdot d_1 d_2 d_3 \ldots$$

is

$$\mathbf{d}_k = 2^{a_k} + 2^{b_k} + 2^{c_k} + \cdots,$$

where $0 \le a_k < b_k < c_k < \cdots$, then it is legal to remove k beans from a heap, provided that the rest of the heap is left in exactly a_k or b_k or c_k or ... nonempty heaps. See WW, 81–115. Some specimen games are exhibited in Table 3 of §**5** of the Impartial Games paper in GONC.]

In GONC, p. 476, we listed ·**644**, but its period, 442, had been found by Richard Austin in his thesis [1976].

Gangolli and Plambeck [1989] established the ultimate periodicity of four octal games which were previously unknown: ·**16** has period 149459 (a prime!), the last exceptional value being $\mathcal{G}(105350) = 16$. The game ·**56** has period 144 and last exceptional value $\mathcal{G}(326639) = 26$. The games ·**127** and ·**376** each have period 4 (with cycles of values 4, 7, 2, 1 and 17, 33, 16, 32 respectively) and last exceptional values $\mathcal{G}(46577) = 11$ and $\mathcal{G}(2268247) = 42$.

Achim Flammenkamp has recently settled ·**454**: it has the remarkable period and preperiod of 60620715 and 160949018, in spite of only $\mathcal{G}(124) = 17$ for the last sparse value and 41 for the largest nim-value, and even more recently has determined that ·**104** has period and preperiod 11770282 and 197769598 but no sparse space. For information on the current status of each of these games, see Flammenkamp's web page at http://www.uni-bielefeld.de/~achim/octal.html.

In Problem 38 in *Discrete Math.*, **44**(1983) 331–334 Fraenkel raises questions concerning the computational complexity of octal games. In Problem 39, he and Kotzig define **partizan octal games** in which distinct octals are assigned to the two players. In Problem 40, Fraenkel introduces **poset games**, played on a partially ordered set of heaps, each player in turn selecting a heap and removing a positive number of beans from this heap and all heaps which are above it in the poset ordering. Compare Problem **23** below.

3. Examine some **hexadecimal games**.

[**Hexadecimal games** are those with code digits d_k in the interval from **0** to **F** ($= 15$), so that there are options splitting a heap into three heaps. See WW, 116–117.]

Such games may be arithmetically periodic. That is, the nim-values belong to a finite set of arithmetic progressions with the same common difference. The number of progressions is the period and their common difference is called the **saltus**. Sam Howse has calculated the first 1500 nim-values for each of the 1-, 2- and 3-digit games. Richard Austin's theorem 6.8 in his 1976 thesis suffices to confirm the (ultimate) arithmetic periodicity of several of these games.

For example ·**XY**, where **X** and **Y** are each **A, B, E** or **F** and ·**E8**, ·**E9**, ·**EC** and ·**ED** are each equivalent to Nim.

·**0A**, ·**0B**, ·**0E**, ·**0F**, ·**1A**, ·**1B**, ·**48**, ·**4A**, ·**4C**, ·**4E**, ·**82**, ·**8A**, ·**8E** and ·**CZ**, where **Z** is any even digit, are equivalent to Duplicate Nim, while ·**0C**, ·**80**, ·**84**, ·**88**, and ·**8C** are like Triplicate Nim.

Some games displayed ordinary periodicity; ·**A2**, ·**A3**, ·**A6**, ·**A7**, ·**B2**, ·**B3**, ·**B7** have period 4, and ·**81**, ·**85**, ·**A0**, ·**A1**, ·**A4**, ·**A5**, ·**B0**, ·**B1**, ·**B5**, ·**D0**, ·**F0**, ·**F1** are all essentially She-Loves-Me-She-Loves-Me-Not.

·**9E**, ·**9F**, ·**BC**, ·**C9**, ·**CB**, ·**CD** and ·**CF** have (apparent ultimate) period 3 and saltus 2; ·**89**, ·**8D**, ·**A8**, ·**A9**, ·**AC**, ·**AD** each have period 4 and saltus 2, while ·**8B**, ·**8F** and ·**9B** have period 7 and saltus 4.

More interesting specimens are ·**28** = ·**29**, which have period 53 and saltus 16, the only exceptional value being $\mathcal{G}(0) = 0$; ·**9C**, which has period 36, preperiod 28 and saltus 16; and ·**F6** with period 43 and saltus 32, but its apparent preperiod of 604 and failure to satisfy one of the conditions of the theorem prevent us from verifying the ultimate periodicity.

The above accounts for nearly half of the two-digit genuinely hexadecimal (i.e., containing at least one **8**) games. There remain almost a hundred for which a pattern has yet to be established.

Kenyon's Game, ·**3F**, had been the only example found whose saltus of 3 countered the conjecture of Guy and Smith that it should always be a power of two. But Nowakowski has now shown that ·**3F3** has period 10 and saltus 5; ·**209**, ·**228** have period 9 with saltus 3; and ·**608** has period 6 and saltus 3. Further examples whose saltus is not a power of two may be ·**338**, probably with period 17 and saltus 6 and several, probably isomorphic, with period 9 and saltus 3.

The game ·**9** has not so far yielded its complete analysis, but, as far as analyzed, i.e. to 12000, exhibits a remarkable fractal-like set of nim-values. See Austin, Howse and Nowakowski (2002).

4. Extend the analysis of **Domineering**.

[Left and Right take turns to place dominoes on a checker-board. Left orients her dominoes North-South and Right orients his East-West. Each domino exactly covers two squares of the board and no two dominoes overlap. A player unable to play loses.]

See Berlekamp [1988] and the second edition of WW, 138–142, where some new values are given. For example David Wolfe and Dan Calistrate have found the values (to within '-ish', i.e., infinitesimally shifted) of 4×8, 5×6 and 6×6 boards. Lachmann, Moore and Rapaport (this volume) discover who wins on rectangular, toroidal and cylindrical boards of widths 2, 3, 5 and 7, but do not find their values.

Berlekamp asks, as a hard problem, to characterize all hot Domineering positions to within "ish". As a possibly easier problem he asks for a Domineering position with a new temperature, i.e., one not occurring in Table 1 on GONC, p. 477.

5. Analyze positions in the game of **Go**.

Compare Berlekamp [1988], his book with Wolfe [1994], and continuing discoveries, discussed in GONC and the present volume, which also contains Spight's analysis of an enriched environment Go game and Takizawa's rogue Ko positions.

6. Go-Moku. Solved by Allis, Herik and Huntjens [1996].

7. Complete the analysis of impartial **Eatcakes** (WW, 269, 271, 276–277).

[Eatcakes is an example of a **join** or **selective compound** of games. Each player plays in **all** the component games. It is played with a number of rectangles, $m_i \times n_i$; a move is to remove a strip $m_i \times 1$ or $1 \times n_i$ from each rectangle, either splitting it into two rectangles, or reducing the length or breadth by one. Winner removes the last strip.]

For fixed breadth the remoteness becomes constant when the length is sufficiently large. But 'sufficiently large' seems to be an increasing function of the breadth and doesn't, in the hand calculations already made, settle into any clear pattern. Perhaps computer calculations will reveal something.

8. Complete the analysis of **Hotcakes** (WW, 279–282).

[Also played with integer-sided rectangles, but as a **union** or **selective compound** in which each player moves in **some** of the components. Left cuts as many rectangles vertically along an integer line as she wishes, and then rotates one from each pair of resulting rectangles through a right angle. Right cuts as many rectangles as he wishes, horizontally into pairs of integer-sided rectangles and rotates one rectangle from each pair through a right angle. The **tolls** for rectangles with one dimension small are understood, but much remains to be discovered.]

9. Develop a **misère theory** for unions of partizan games.

[In a union of two or more games, you move in as many component games as you wish. In misère play, the last player *loses*.]

10. Extend the analysis of **Squares Off** (WW, 299).

[Played with heaps of beans. A move is to take a perfect square (> 1) number of beans from any number of heaps. Heaps of 0, 1, 2 or 3 cannot be further reduced. A move leaving a heap of 0 is an overriding win for the player making it. A move leaving 1 is an overriding win for Right, and one leaving 2 is an overriding win for Left. A move leaving 3 doesn't end the game unless all other heaps are of size 3, in which case the last player wins.]

11. Extend the analysis of **Top Entails** (WW, 376–377).

[Played with stacks of coins. Either split a stack into two smaller ones, or remove the top coin from a stack. In the latter case your opponent's move must use the same stack. Last player wins. Don't leave a stack of 1 on the board, since your opponent must take it and win, since it's now your turn to move in an empty stack!]

We are unable to report any advance on Julian West's discovery of loony positions at 2403 coins, 2505 coins, and 33,243 coins. The authors of *Winning Ways* did not know of a loony stack of more than 3 coins. These results are typical of the apparently quite unpredictable nature of combinatorial games, even when they have quite simple rules.

12. Extend the analysis of **All Square** (WW, 385).

[This game involves **complimenting moves** after which the same player has an extra **bonus move**. Note that this happens in Dots-and-Boxes when a box is completed. All Square is played with heaps of beans. A move splits a heap into two smaller ones. If both heap sizes are perfect squares, the player must move again: if he can't he loses!]

13. Extend the misère analysis of various octal games, e.g., **Officers, Dawson's Chess**, ..., and of **Grundy's Game** (WW, 411–421).

William L. Sibert made a breakthrough by completing the analysis of misère Kayles; see the post-script to Sibert and Conway [1992]. Plambeck [1992] has used their method to analyze a few other games, but there's a wide open field here. Recently, Allemang (2001) has extended the content of his 1984 thesis to include the complete analysis of ·26, ·53, ·54, ·72, ·75 and 4·7. (See also http://spdcc.com:8431/summary.html.)

We can ask the same question for Hexadecimal games (see Problem **3**).

14. Moebius, when played on 18 coins has a remarkable pattern. Is there any trace of pattern for larger numbers of coins? Can any estimate be made for the rate of growth of the nim-values?

[See Coin-turning games in WW, 432–435; and Vera Pless's lecture and the Impartial Games lecture in PSAM **43**. Moebius is played with a row of coins. A move turns 1, 2, 3, 4 or 5 coins, of which the rightmost must go from heads to tails (to make sure the game satisfies the Ending Condition). The winner is the player who makes all coins tails.]

15. Mogul has an even more striking pattern when played on 24 coins, which has some echoes when played on 40, 56 or 64 coins. Thereafter, is there complete chaos?

[See references for Problem **14**. A move turns 1, 2, . . . , 7 coins.]

16. Find an analysis of **Antonim** with four or more coins (WW, 459–462).

[Played with coins on a strip of squares. A move moves a coin from one square to a smaller-numbered square. Only one coin to a square, except that square zero can have any number of coins. It is known that (a, b, c) is a $\mathcal{P}$-position in Antonim just if $(a + 1, b + 1, c + 1)$ is a $\mathcal{P}$-position in Nim, but for more than 3 coins much remains to be discovered.]

17. Extend the analysis of **Kotzig's Nim** (WW, 481–483). Is the game eventually periodic in terms of the length of the circle for every finite move set? Analyze the misère version of Kotzig's Nim.

[Players alternately place coins on a circular strip, at most one coin on a square. Each coin must be placed m squares clockwise from the previously placed coin, provided m is in the given **move set**, and provided the square is not already occupied. The complete analysis is known only for a few small move sets.]

See Fraenkel, Jaffray, Kotzig and Sabidussi [1995].

18. Obtain asymptotic estimates for the proportions of $\mathcal{N}$-, $\mathcal{O}$- and $\mathcal{P}$-positions in Epstein's **Put-or-Take-a-Square** game (WW, 484–486).

[Played with one heap of beans. At each turn there are just two options, to take away or add the largest perfect square number of beans that there is in the heap. 5 is a $\mathcal{P}$-position, because 5 ± 4 are both squares; 2 and 3 are $\mathcal{O}$-positions, a win for neither player, since the best play is to go from one to the other, and not to 1 or 4 which are $\mathcal{N}$-positions.]

19. Simon Norton's game of **Tribulations** is similar to Epstein's game, but squares are replaced by triangular numbers. Norton conjectures that there are no $\mathcal{O}$-positions, and that the $\mathcal{N}$-positions outnumber the $\mathcal{P}$-positions in golden ratio. True up to 5000 beans.

Investigate other put-or-take games. If the largest number of form $2^k - 1$ is put or taken, we have yet another disguise for She-Loves-Me-She-Loves-Me-Not, with the remoteness given by the binary representation of the number of beans. For Fibulations and Tribulations, see WW 501–503. If the largest number used is of form $T_n + 1$, where T_n is a triangular number, the $\mathcal{P}$-positions are the multiples of 3.

20. Complete the analysis of **D.U.D.E.N.E.Y**

[Played with a single heap of beans. Either player may take any number of beans from 1 to Y, except that the immediately previous move mustn't be repeated. When you can't move you lose. Analysis easy for Y even, and known (WW, 487–489) for 53/64 of the odd values of Y.]

Marc Wallace and Alan Jaffray made a little progress here, but is the situation one in which there is always a small fraction of cases remaining, no matter how far the analysis is pursued?

21. Schuhstrings is the same as D.U.D.E.N.E.Y, except that a deduction of zero is also allowed, but cannot be immediately repeated (WW, 489–490).

22. Analyze **Nim** in which you are not allowed to repeat a move. There are at least five possible forms (assume that b beans have been taken from heap H):

medium local: b beans may not be taken from heap H until some other move is made in heap H.

short local: b beans may not be taken from heap H on the next move.

long local: b beans may never again be taken from heap H.

short global: b beans may not be taken from any heap on the next move.

long global: b beans may never again be taken from any heap.

23. Burning-the-Candle-at-Both-Ends. John Conway and Aviezri Fraenkel ask us to analyze Nim played with a row of heaps. A move may only be made in the leftmost or in the rightmost heap. When a heap becomes empty, then its neighbor becomes the end heap.

Albert and Nowakowski [2001] have solved the impartial and partizan versions. But there is also **Hub-and-Spoke Nim**, proposed by Fraenkel. One heap is the hub and the others are arranged in rows forming spokes radiating from the hub.

There are several versions:

(a) beans may be taken only from a heap at the end of a spoke;

(b) beans may also be taken from the hub;

(c) beans may be taken from the hub only when all the heaps in a spoke are exhausted;

(d) beans may be taken from the hub only when just one spoke remains;

(e) in versions (b), (c) and (d), when the hub is exhausted, beans may be taken from a heap at either end of any remaining spoke; i.e. the game becomes the sum of a number of games of Burning-the-Candle-at-Both-Ends.

Albert notes that Hub-and-Spoke Nim can be generalized to playing on a forest, i.e., a graph each of whose components is a tree. The most natural variant is that beans may only be taken from a leaf (valence 1) or isolated vertex (valence 0).

24. Continue the analysis of **The Princess and the Roses** (WW, 490–494).

[Played with heaps of beans. Take one bean, or two beans, one from each of two different heaps. The rules seem trivially simple, but the analysis takes on remarkable ramifications.]

25. Extend the analysis of the Conway-Paterson game of **Sprouts** in either the normal or misère form. (WW, 564–568).

[A move joins two spots, or a spot to itself by a curve which doesn't meet any other spot or previously drawn curve. When a curve is drawn, a new spot must be placed on it. The valence of any spot must not exceed three.]

Applegate, Jacobson and Sleator [1999] have pushed the normal analysis to 11 initial spots and the misère analysis to 9.

number of spots	1	2	3	4	5	6	7	8	9	10	11
normal play	$\mathcal{P}$	$\mathcal{P}$	$\mathcal{N}$	$\mathcal{N}$	$\mathcal{N}$	$\mathcal{P}$	$\mathcal{P}$	$\mathcal{P}$	$\mathcal{N}$	$\mathcal{N}$	$\mathcal{N}$
misère play	$\mathcal{N}$	$\mathcal{P}$	$\mathcal{P}$	$\mathcal{P}$	$\mathcal{N}$	$\mathcal{N}$	$\mathcal{P}$	$\mathcal{P}$	$\mathcal{P}$		

where $\mathcal{P}$ and $\mathcal{N}$ denote previous-player and next-player winners. There is a temptation to conjecture that the patterns continue.

26. Extend the analysis of **Sylver Coinage** (WW, 575–597).

[Players alternately name different positive integers, but may not name a number which is the sum of previously named ones, with repetitions allowed. Whoever names 1 loses. See Section 3 of Richard Nowakowski's chapter in PSAM **43**.]

27. Extend the analysis of **Chomp** (WW, 598–599).

[Players alternately name divisors of N, which may not be multiples of previously named numbers. Whoever names 1 loses. David Gale offers a prize of US$100.00 for the first complete analysis of 3D-Chomp, i.e., where N has three distinct prime divisors, raised to arbitrarily high powers.]

Doron Zeilberger (www.ics.uci.edu/~eppstein/cgt) has analyzed Chomp for $N = 2^2 3^n$ up to $n = 114$. For an excursion into infinite Chomp, see Huddleston and Shurman [2001] in this volume.

28. Extend Úlehla's or Berlekamp's analysis of **von Neumann's game** from directed forests to directed acyclic graphs (WW, 570–572; Úlehla [1980]).

[von Neumann's game, or Hackendot, is played on one or more rooted trees. The roots induce a direction, towards the root, on each edge. A move is to delete a node, together with all nodes on the path to the root, and all edges incident with those nodes. Any remaining subtrees are rooted by the nodes that were adjacent to deleted nodes.]

Since Chomp and the superset game (Gale and Neyman [1982]) can be described in terms of directed acyclic graphs but not by directed forests, a partial analysis of such an extension of von Neumann's game could throw some light on these two unsolved games. Fraenkel and Harary [1989] discuss a similar game, but with the directions determined by shortest distances. They find winning strategies for trees in normal play, circuits in normal and misère play; and for complete graphs with rays of equal length in normal play.

29. Prove that Black doesn't have a forced win in **Chess**.

Andrew Buchanan has recently emailed that he has examined some simpler (sub-)problems in which the moves 1. e4, e5 are made followed by either a Bishop move by each player, or a Queen move by each player. He claims that at most six of each of these sets of positions can be wins for Black.

30. A **King and Rook v. King** problem. Played on a quarter-infinite board, with initial position WKa1, WRb2 and BKc3. Can White win? If so, in how few moves? It may be better to ask, "what is the smallest board (if any) that White can win on if Black is given a win if he walks off the North or East edges of the board?" Is the answer 9×11? In an earlier edition of this paper I attributed this problem to Simon Norton, but it was proposed as a kriegsspiel problem, with unspecified position of the WK, and with W to win with probability 1, by Lloyd Shapley around 1960.

31. David Gale's version of **Lion and Man**. L and M are confined to the non-negative quadrant of the plane. They move alternately a distance of at most one unit. For which initial positions can L catch M?

David Gale's Lion-and-Man has been solved by Jiří Sgall [2001].

Variation. Replace quadrant by wedge-shaped region.

32. Gale's Vingt-et-un. Cards numbered 1 through 10 are laid on the table. L chooses a card. Then R chooses cards until his total of chosen cards exceeds the card chosen by L. Then L chooses until her cumulative total exceeds that of R, etc. The first player to get 21 wins. Who is it?

[As posed here it is not clear if the object is to get 21 exactly or 21-or-more. Jeffery Magnoli, a student of Julian West, thought that the latter rule was the more interesting and found a first-player win in 6-card Onze and in 8-card Dixsept.]

33. Subset Take-away. Given a finite set, players alternately choose proper subsets subject to the rule that once a subset has been chosen, none of *its* subsets may be chosen subsequently by either player. Last player wins.

[David Gale conjectures that it's a second player win—this is true for sets of less than six elements.]

34. Eggleton and Fraenkel ask for a theory of **Cannibal Games** or an analysis of special families of positions. They are played on an arbitrary finite digraph. Place any numbers of "cannibals" on any vertices. A move is to select a cannibal and move it along a directed edge to a neighboring vertex. If this is occupied, the incoming cannibal eats the whole population (**Greedy Cannibals**) or just one cannibal (**Polite Cannibals**). A player unable to move loses. Draws are possible. A partizan version can be played with cannibals of two colors, each eating only the opposite color.

35. Welter's Game on an arbitrary digraph. Place a number of monochromatic tokens on distinct vertices of a directed acyclic graph. A token may be moved to any *unoccupied* immediate follower. Last player wins. Make a dictionary of $\mathcal{P}$-positions and formulate a winning strategy for other positions. See Kahane and Fraenkel [1987] and Kahane and Ryba [2001].

36. Restricted Positrons and Electrons. Fraenkel places a number of Positrons (Pink tokens) and Electrons (Ebony tokens) on distinct vertices of a Welter strip. Any particle can be moved by either player leftward to any square u provided that u is either unoccupied or occupied by a particle of the opposite type. In the latter case, of course, both particles become annihilated (i.e., they are removed from the strip), as physicists tell us positrons and electrons do. Play ends when the excess particles of one type over the other are jammed in the lowest positions of the strip. Last player wins. Formulate a winning strategy for those positions where one exists. Note that if the particles are of one type only, this is Welter's Game. As a strategy is known for Misère Welter [WW, 480–481] it may not be unreasonable to ask for a misère analysis as well. See Problem 47, *Discrete Math.*, **46** (1983) 215–216.

37. General Positrons and Electrons. As Problem **36** but played on an arbitrary digraph. Last player wins.

38. Fulves's Merger. Start with heaps of 1, 2, 3, 4, 5, 6 and 7 beans. Two players alternately transfer any number of beans from one heap to another, except that beans may not be transferred from a larger to a smaller heap. The player who makes all the heaps *even* in size is the winner.

The total number of beans remains constant, and is even (28 in this case, though one is interested in even numbers in general: a similar game can be played in which the total number is odd and the object is to make all the heaps odd in size).

No progress has been reported for the general game.

39. Sowing or **Mancala Games.** Kraitchik (1941, p. 282) describes Ruma, which he attributes to Mr. Punga. Bell and Cornelius (1988, pp. 22–38) list Congklak, from Indonesia; Mankal'ah L'ib Al-Ghashim (the game of the unlearned); Leab El-Akil (the game of the wise or intelligent); Wari; Kalah, from Sumeria; Gabata, from Ethiopia; Kiarabu, from Africa; as well as some solitaire games based on a similar idea. Botermans et al. (1989, pp. 174–179) describe Mefuhva from Malawi and Kiuthi from Kenya. Many of these games go back for thousands of years, but several should be susceptible to present day methods of analysis. See Jeff Erickson's article in GONC.

Conway starts with a **line** of heaps of beans. A typical move is to take (some of) a heap of size N and do something with it that depends on the game and on N. He regards the nicest move as what he calls the **African move** in which **all** the beans in a heap are picked up and 'sowed' onto successive heaps, **and** subject to the condition that the last bean must land on a nonempty heap. Beans are sowed to the right if you are Left, to the left if you are Right, or either way if you're playing impartially.

In the partizan version, the position 1 (a single bean) has value 0, of course; the position $1.1 = \{0.2 \,|\, 2.0\}$ has value $\{0 \,|\, 0\} = *$; and so does

$$1.1.1 = \{1.0.2, 2.1 \,|\, 1.2, 2.0.1\},$$

since $2.1 = \{\ \,|\, 3.0\}$, $3.0 = \{\ \,|\ \} = 0$ and $1.0.2 = \{\ \,|\, 2.1\}$, so that '3' has value 0, 2.1 has value -1, 1.0.2 value -2, 1.1.1 value $\{-2, -1 \,|\, 1, 2\} = 0$, 1.1.1.1 value 0, and 1.1.1.1.1 value $\pm\frac{1}{2}$.

Recent papers on mancala-type games are Björner and Lovász [1992], Broline and Loeb [1995] and Yeh Yeong-Nam [1995].

40. Chess. Noam Elkies has found endgames with values 0, 1, $\frac{1}{2}$, $*$, $*k$ for many k, $\uparrow$, $\uparrow *$, $\Uparrow *$, etc.; see his papers in GONC and this volume. See also Problems **29**, **30** and **45**.

41. Sequential compounds of games have been studied by Stromquist and Ullman. They mention a more general compound. Let $(P, <)$ be a finite poset and for each $x \in P$ let G_x be a game. Consider a game $G(P)$ played as follows. Moves are allowed in any single component G_x provided that no legal moves remain in any component G_y with $y > x$. A player unable to move loses. The sequential compound is the special case when $(P, <)$ is a chain (or linear order). The **sum** or disjunctive compound is the case where $(P, <)$ is an antichain. They have no coherent theory of games $G(P)$ for arbitrary posets. They list some more specific problems which may be more tractable. Compare Problem **23** above.

42. Beanstalk and **Beans-Don't-Talk** are games invented by John Isbell and by John Conway. See Guy [1986]. Beanstalk is played between Jack and the Giant. The Giant chooses a positive integer, n_0. Then J. and G. play alternately $n_1, n_2, n_3, \ldots$ according to the rule $n_{i+1} = n_i/2$ if n_i is even, $= 3n_i \pm 1$ if n_i is odd; i.e. if n_i is even, there's only one option, while if n_i is odd there are just two. The winner is the person moving to 1. If the Giant chooses an odd number > 1, can Jack always win? Not by using the Greedy Strategy (always descend when it's safe to do so) as this can lead to cycles (draws).

In Beans-Don't-Talk, the move is from n to $(3n \pm 1)/2^*$ where 2^* is the highest power of two dividing the numerator; the winner is still the person moving to 1. Are there any drawn positions? There are certainly drawn **plays**, e.g., 7 (5) 7 (5) ..., but 5 is an $\mathcal{N}$-position because there is the immediate winning option $(5 \times 3 + 1)/2^4 = 1$, and 7 is a $\mathcal{P}$-position since the other option $(7 \times 3 + 1)/2 = 11$ is met by $(11 \times 3 - 1)/2^5 = 1$. What we want to know is: *are there any $\mathcal{O}$-positions* (positions of infinite remoteness)?

[For remoteness see Chapter 9 of WW. There are several unanswered questions about the remotenesses of positions in these two games. Remoteness may also be the best tool we have for Problems **18** and **19** above.]

43. Inverting Hackenbush. John Conway turns Blue-Red Hackenbush, played on finite strings of edges, into a hot game by amending the move to 'remove an edge of your color and everything thus disconnected from the ground, and then turn the remaining string upside-down and replant it'. The analysis replaces the 'number tree' (WW, p. 25) by a similar tree, but with the smaller binary fractions replaced by increasingly hot games. The game can be generalized to play on trees: a move which prunes the tree at a vertex V includes replanting the tree with V as its root.

44. Konane. See the paper by Ernst and Berlekamp in GONC. There is much to be discovered about this fascinating and eminently playable game, which exhibits the values 0, $*$, $*2$, $\uparrow$, 2^{-n}, and many other infinitesimals and also hot values of arbitrarily high temperature. Chan and Tsai, in this volume, give some values for $1 \times n$ boards.

45. Elwyn Berlekamp asks for the **habitat** of $*2$. [$*2 = \{0, *|0, *\}$.] It does not occur in Blockbusting, Hackenbush, Col or Go. It does occur in Konane and 6×6 Chess. What about Chilled Go, Domineering and 8×8 Chess? Elkies (see this volume) has modified and generalized the game Dawson's Chess to give games, on suitably large boards, of value $*k$ for many large k and conjectures that such values exist for all k.

46. There are various ways of playing **two-dimensional Nim**. One form is discussed on p. 313 of WW. Another is proposed by Berman, Fraenkel and Kahane in Problem 41, *Discrete Math.*, **45** (1983) 137–138. Start with a rectangular array of heaps of beans. At each move a row or column is selected and a positive number of beans is taken from some of the heaps in that row or column; see Fremlin [1973]. Ferguson's [2001] variant has the move as choosing a number and subtracting that number from all members of a row or column. He finds the outcomes for 2×2 matrices in both the impartial and partizan versions. Another variant is where beans may be taken only from contiguous heaps. Other variants are played on triangular or hexagonal boards; a special case of this last is Piet Hein's Nimbi, solved by Fraenkel and Herda [1980].

47. Many results are known concerning tiling rectangles with **polyominoes**. One can extend such problems to disconnected polyominoes. E.g., in GONC we asked if a rectangle can be tiled by

        ⊓⊔   ⊓⊔   or by   ⊓⊔   ⊓⊔   □

If so, what are the rectangles of least size that can be so tiled?

Juha Saukkola showed that the former will not tile a rectangle, but that the latter can tile a 12×15 rectangle. Since then Joe Devincentis, Erich Friedman, Patrick Hamlyn, Mike Reid and others have examined numerous other cases. See http://www.stetson.edu/~efriedma/mathmagic/0299.html.

There is an obvious generalization of Domineering (see Problem **4** above) to a two-player game in which the players alternately place polyominoes of given shape and orientation on a rectangular or other board.

48. Find all words which can be reduced to 1 peg in 1-dimensional **Peg Solitaire**. E.g., 1, 011, 110, 1101, 110101, $1(10)^k1$. Here 1 represents a peg and 0 an empty space. A move is for a peg to jump over an adjacent peg into an empty adjacent space, and remove the jumped-over peg. E.g., $1101 \rightarrow 0011 \rightarrow 0100$. Georg Gunther, Bert Hartnell and Richard Nowakowski found that for an $n \times 1$ board with one empty space, n must be even and the space next but one to the end. If the board is cyclic, the condition is simply n even. Christopher Moore and David Eppstein, indicate, in this volume, that this problem has been solved many times but does not seem to have been published. They coin the term Duotaire for one-dimensional peg solitaire played as a two-player game. They give some decomposition theorems and conjecture that arbitrarily high nim-values occur. J. P. Grossman notes that the position

$$\bigcirc\bigcirc \bullet (\bullet \bigcirc\bigcirc \bullet \bigcirc \bullet)^{20601} \bigcirc$$

i.e. a strip of 123610 squares, of which the first, second and last squares are occupied, together with the $(6n+5)$-th, $(6n+6)$-th and $(6n+8)$-th, for $0 \le n \le 20600$, has nim-value 197.

49. Elwyn Berlekamp asks if there is a game which has
    1. simple, playable rules,
    2. an intricate explicit solution, and
    3. is provably NP or harder.

Is Phutball (WW, 688–691) such a game? See Demaine, Demaine and Eppstein and also Grossman and Nowakowski in this volume. Compare Problem **57** below.

50. John Selfridge asks: is **Four-File** a draw? Four-File is played on a chessboard with the chess pieces in their usual starting positions, but only on the a-, c-, e- and g-files. I.e. a rook, a bishop, a king, a knight and four pawns on each side. The moves are normal chess moves except that play takes place only on these four files. I.e., each move ends on one of the files a, c, e or g; pawns cannot capture and there is no castling. The aim is to checkmate your opponent's king.

51. Elwyn Berlekamp asks for a complete theory of closed $1 \times n$ **Dots-and-Boxes**. I.e., with starting position

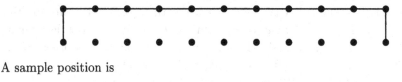

A sample position is

See WW, Chapter 16 and Berlekamp's book, *The Dots-and-Boxes Game* [2000]. Are there more nimber decomposition theorems? Compile a datebase of nim-values.

Nowakowski and Ottaway conjecture that closed $1 \times n$ Dots-and-Triangles (a row of n dots on top and n+1 on bottom with the top and sides already having lines) is a first player win except for $n = 2$.

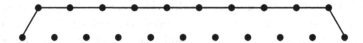

52. How does one play **sums** of games with varied overheating operators? Berlekamp notes that overheating operators provide a very concise way of expressing closed-form solutions to many games, and David Moews observes that monotonicity and linearity depend on the parameters and the domain. Find a simple, elegant way of relating the operator parameters to the game. See WW, pp. 163–175, Berlekamp [1988], Berlekamp and Wolfe [1994] and Calistrate's paper in GONC.

53. N-heap Wythoff Game. Aviezri Fraenkel asks some questions and makes some conjectures. The set of all integers $\geq m$ is denoted by $Z_{\geq m}$ and $\oplus$ denotes Nim addition. For any subset $S \subset Z_{\geq 0}$, $S \neq Z_{\geq} 0$, let $\operatorname{mex} S = \min(Z_{\geq} 0 \setminus S) =$ least nonnegative integer not in S.

Define an N-heap Wythoff game as follows: Given $N \geq 2$ heaps of finitely many tokens, whose sizes are $p_1, \ldots, p_N$. The moves are to take any positive number of tokens from a *single* heap or to take $(a_1, \ldots, a_N) \in Z_{\geq 0}^N$ from *all* the heaps — a_i from the i-th heap — subject to the conditions: (i) $a_i > 0$ for some i, (ii) $a_i \leq p_i$ for all i, (iii) $a_1 \oplus \ldots \oplus a_N = 0$. The player making the last move wins and the opponent loses. Note that the classical Wythoff game is the case $N = 2$.

For $N = 3$, denote by (A_n, B_n, C_n) the P-positions of the game, with $A_n \leq B_n \leq C_n$. We conjecture that

For every fixed $A_k = k \in Z_{\geq 1}$ there exists an integer $m = m(k) \in Z_{\geq 1}$ such that $B_n = \operatorname{mex}\left(\{B_i, C_i : i < n\} \cup T\right)$, $C_n = B_n + n$ for all $n \geq m$, where T is a (small) set of integers which depends only on k.

For example, for $k = 1$ we have $T = \{2\}$; and it seems that $m = 23$. A related conjecture is that

For every fixed $A_k = k \in Z_{\geq 1}$ there exist integers $a = a(k)$, $j = j(k)$, $m = m(k) \in Z_{\geq 1}$ with $j < a$, such that $B_n \in \{\lfloor n\phi \rfloor - (a + j), \lfloor n\phi \rfloor - (a + j - 1), \ldots, \lfloor n\phi \rfloor - (a - j + 1), \lfloor n\phi \rfloor - (a - j)\}$, $C_n = B_n + n$ for all $n \geq m$, where $\phi = (1 + \sqrt{5})/2$ (the golden section).

This appears to hold for $a = 4$, $j = 1$, $m = 64$ (perhaps a somewhat smaller value of m will do) when $k = 1$. Is perhaps $j = 1$ for all $k \geq 1$?

See also Coxeter [1953], Fraenkel and Ozery [1998] and Fraenkel and Zusman [2001].

54. Fox and Geese. Jonathan Welton notices that the conclusion of Chapter 20 of WW, namely that the value of Fox and Geese is $1 + 1/\textbf{on}$, is incorrect. He believes that he can show that the geese can win with the fox having $1 + 1/32$ passes, and probably the actual value is still higher. What is the correct value?

[Fox and Geese is played on an ordinary checkerboard, the geese being four white checkers, moving diagonally forward, starting on squares a1, c1, e1, g1; while the fox is a black checker king moving in any diagonal direction, starting on d8. There is no capturing: the geese try to encircle the fox; the fox endeavors to break through.]

55. Amazons was invented by the Argentinian Walter Zamkauskas in 1988. It is played on a 10×10 board. Each player has four amazons. The white amazons are initially on a4, d1, g1, j4 and the black ones are on a7, d10, g10, j7. White moves first. Each move consists of two mandatory parts. First, an amazon moves just like a chess queen. After an amazon has moved she shoots a burning arrow, which also moves like a chess queen. The square where the arrow lands is burnt and is blocked for the rest of the game; neither an amazon nor an arrow can move to or over that square, nor to or over a square occupied by another amazon. There are no captures in Amazons. Nor are there draws: the aim is to control territory: the winner is the last player to complete a move.

Analyses of smaller boards with fewer amazons have been made. For example in Solving 5×5 Amazons (2001 preprint), Martin Müller shows that 5×5 Amazons (with amazons on a2, b1, d1, e2 and on a4, b5, d5, e4) is a first player win. Berlekamp [2000,2001] investigates sums of $2 \times n$ Amazon games. See also the papers of Müller and Tegos and of Snatzke in this volume.

56. Are there any draws in **Beggar-my-Neighbor**?

[Two players deal single cards in turn onto a common stack. If a court card (J, Q, K, A) is dealt, the next player must cover it with respectively 1, 2, 3, 4 cards. If one of these is a court card, the obligation to cover reverts to the previous player. If they are not court cards, the previous player acquires the stack, which he inverts and places beneath his own hand, and starts dealing again. A player loses if she is unable to play.]

This problem reappears periodically. It was one of Conway's 'anti-Hilbert problems' about 40 years ago, but must have suggested itself to players of the game over the several centuries of its existence.

Marc Paulhus [1999] exhibited some cycles with small decks, and used a computer to show that there were no cycles when the game is played with a half-deck, although the addition or subtraction of two non-court cards produced cycles. Michael Kleber found an arrangement of two 26-card hands which required the dealing of 5790 cards before a winner was declared.

57. Aviezri Fraenkel describes a game as **succinct** if its input size is logarithmic. Thus Nim is succinct, because its input size is the sum of the logarithms of its heap sizes. It has a polynomial time winning strategy, yet the loser can make length of play exponentially long. (A trivial example: two heaps of the same size, where Player I keeps removing a single token from one heap, which has to be matched by Player II taking a single token from the other heap.)

(a) Is there a nonsuccinct game with a polynomial winning strategy in which play can be made to last exponentially long?

(b) Node Kayles, on a general graph, was proved to be Pspace-complete by Schaefer [1978]. Its succinct form, the octal game ·**137**, is polynomial. Is there a game which has a polynomial strategy on a general graph, but its succinct form is at least NP-hard?

[Node Kayles is played on a graph. A move is to place a counter on an unoccupied node that is not adjacent to any occupied node. Equivalently, to delete a node and all its neighbors. The game ·**137** is Dawson's Chess, i.e. Node

Kayles played on a path, and occurs in the analysis of several other games, notably Dots-and-Boxes. See WW, 92, 251, 466, 470, 532, 552.]

58. The one-dimensional version of **Clobber** is played on a $1 \times n$ strip of squares where there are blue and red pieces alternating, one to a square. Right moves the red pieces and Left the blue. A piece moves to an adjacent square but only if the square is occupied by an opposing piece. This piece is then removed from the board, i.e. it has been clobbered. Albert, Grossman and Nowakowski conjecture that $1 \times n$ Clobber is a first player win for $n \geq 13$. They also show that played on an arbitrary graph with one blue piece and the rest red, deciding the value of the game is NP-complete.

[Clobber is a special case of partizan Polite Cannibals (see Problem **34**) in which moves may only be made to occupied nodes.]

Bibliography

[those not listed here may be found in Fraenkel's Bibliography]

Michael H. Albert, J. P. Grossman, Richard J. Nowakowski, Clobber, preprint (2002). [Problem **58**]

Dean Thomas Allemang, Generalized genus sequences for misère octal games, *Internat. J. Game Theory*, (submitted, 2001 preprint). [Problem **13**]

Richard B. Austin, Sam Howse and Richard J. Nowakowski, Hexadecimal games, *Integers (Electron. J. Number Theory)* (submitted 2002). [Problem **3**]

Robbie Bell and Michael Cornelius, *Board Games Round the World*, Cambridge Univ. Press, 1988. [Problem **39**]

Jack Botermans, Tony Burrett, Pieter van Delft and Carla van Splunteren, *The World of Games*, Facts on File, New York and Oxford, 1989. [Problem **39**]

Maurice Kraitchik, *Mathematical Recreations*, Geo. Allen and Unwin, 1941; 2nd ed. Dover, 1953; p. 282. [Problem **39**]

RICHARD K. GUY
MATHEMATICS AND STATISTICS DEPARTMENT
UNIVERSITY OF CALGARY
2500 UNIVERSITY AVENUE
ALBERTA, T2N 1N4
CANADA
   rkg@cpsc.ucalgary.ca

RICHARD J. NOWAKOWSKI
DEPARTMENT OF MATHEMATICS
DALHOUSIE UNIVERSITY
HALIFAX, NS
CANADA B3H 3J5
   rjn@mathstat.dal.ca

More Games of No Chance
MSRI Publications
Volume **42**, 2002

Combinatorial Games: Selected Bibliography With A Succinct Gourmet Introduction

AVIEZRI S. FRAENKEL

1. What are Combinatorial Games?

Roughly speaking, the family of *combinatorial games* consists of two-player games with perfect information (no hidden information as in some card games), no chance moves (no dice) and outcome restricted to (lose, win), (tie, tie) and (draw, draw) for the two players who move alternately. Tie is an end position such as in tic-tac-toe, where no player wins, whereas draw is a dynamic tie: any position from which a player has a nonlosing move, but cannot force a win. Both the easy game of Nim and the seemingly difficult chess are examples of combinatorial games. And so is Go. The shorter terminology *game, games* is used below to designate combinatorial games.

2. Why are Games Intriguing and Tempting?

Amusing oneself with games may sound like a frivolous occupation. But the fact is that the bulk of interesting and natural mathematical problems that are hardest in complexity classes beyond *NP*, such as *P*space, *E*xptime and *E*xpspace, are two-player games; occasionally even one-player games (puzzles) or even zero-player games (Conway's "Life"). Some of the reasons for the high complexity of two-player games are outlined in the next section. Before that we note that in addition to a natural appeal of the subject, there are applications or connections to various areas, including complexity, logic, graph and matroid theory, networks, error-correcting codes, surreal numbers, on-line algorithms and biology.

But when the chips are down, it is this "natural appeal" that compels both amateurs and professionals to become addicted to the subject. What is the essence of this appeal? Perhaps the urge to play games is rooted in our primal beastly instincts; the desire to corner, torture, or at least dominate our peers. An intellectually refined version of these dark desires, well hidden under the façade of a passion for local, national or international tournaments or for scientific research, is the consuming strive "to beat them all", to be more clever than the most clever, in short — to create the tools to *Math-master* them all in hot *comb*inatorial *comb*at! Reaching this goal is particularly satisfying and sweet in the context of combinatorial games, in view of their inherent high complexity.

To further explore the nature of games, we consider, informally, two subclasses.

(i) Games People Play (*PlayGames*): games that are challenging to the point that people will purchase them and play them.

(ii) Games Mathematicians Play (*MathGames*): games that are challenging to a mathematician or other scientist to play with and ponder about, but not necessarily to "the man in the street".

Examples of PlayGames are chess, go, hex, reversi; of MathGames: Nim-type games, Wythoff games, annihilation games, octal games.

Some "rule of thumb" properties, which seem to hold for the majority of PlayGames and MathGames are listed below.

I. Complexity. PlayGames tend to be computationally intractable, whereas MathGames may have more accessible strategies, though many of them are also computationally intractable. But most games still live in *Wonderland*: we are wondering about their as yet unknown complexity. Those whose complexity is known, unless polynomial, are normally at least Pspace-hard, often Exptime-complete, sometimes Expspace-complete. Roughly speaking, NP-hardness is a necessary but not a sufficient condition for being a PlayGame!

Incidentally, combinatorial games offer an opportunity to study higher complexity classes not normally encountered in existential decision problems, which usually lie in *NP*.

II. Boardfeel. For Nim-type games and other MathGames, a player without prior knowledge of the strategy has no inkling whether any given position is "strong" or "weak" for a player. Even two positions before total defeat, the player sustaining it may be in the dark about the outcome, which will stump him. The player has no *boardfeel*. (Most MathGames, including Nim-type games, can be played, equivalently, on a board.)

Paradoxically, the intractable PlayGames have a boardfeel: None of us may know an exact strategy from a midgame position of chess, but even a novice gets some feel who of the two players is in a stronger position, merely by looking at the board. In the boardfeel sense, simple games are complex and complex games are simple! Also this property doesn't seem to have an analog in the realm of decision problems.

The boardfeel is the main ingredient which makes PlayGames interesting to play. Its lack causes MathGames not to be played by "the man in the street".

III. **Math Appeal.** PlayGames, in addition to being interesting to play, also have considerable mathematical appeal. This has been exposed recently by the theory of partizan games established by Conway and applied to endgames of Go by Berlekamp. On the other hand, MathGames have their own special combinatorial appeal, of a somewhat different flavor. They appeal to and are created by mathematicians of various disciplines, who find special intellectual challenges in analyzing them. As Peter Winkler called a subset of them: "games people don't play". We might also call them, in a more positive vein, "games mathematicians play". Both classes of games have applications to areas outside game theory. Examples: surreal numbers (PlayGames), error correcting codes (MathGames). Furthermore, they provide enlightenment through bewilderment, as David Wolfe and Tom Rodgers put it.

IV. **Distribution.** There are only relatively few interesting PlayGames around. It seems to be hard to invent a PlayGame that catches the masses. In contrast, MathGames abound. They appeal to a large subclass of mathematicians and other scientists, who cherish producing them and pondering about them. The large proportion of MathGames-papers in games bibliographies reflects this phenomenon.

We conclude, inter alia, that for PlayGames, high complexity is desirable. Whereas in all respectable walks of life we strive towards solutions or at least approximate solutions which are polynomial, there are two less respectable human activities in which high complexity is appreciated. These are cryptography (covert warfare) and games (overt warfare). The desirability of high complexity in cryptography — at least for the encryptor! — is clear. We claim that it is also desirable for PlayGames.

It's no accident that games and cryptography team up: in both there are adversaries, who pit their wits against each other! But games are, in general, considerably harder than cryptography. For the latter, the claim that the designer of a cryptosystem has a safe system is expressed with two quantifiers only: ∃ a cryptosystem such that ∀ attacks on it, the cryptosystem remains unbroken. In contrast, games are normally associated with a large number of alternating quantifiers. See also the next section.

3. Why are Combinatorial Games Hard?

Existential decision problems, such as graph hamiltonicity and Traveling Salesperson (Is there a round tour through specified cities of cost $\leq C$?), involve a single existential quantifier ("Is there...?"). In mathematical terms an existential problem boils down to finding a path—sometimes even just verifying its existence—in a large "decision-tree" of all possibilities, that satisfies specified properties. The above two problems, as well as thousands of other interesting and important combinatorial-type problems are NP-*complete*. This means that they are *conditionally intractable*, i.e., the best way to solve them seems to require traversal of most if not all of the decision tree, whose size is exponential in the input size of the problem. No essentially better method is known to date at any rate, and, roughly speaking, if an efficient solution will ever be found for any NP-complete problem, then all NP-complete problems will be solvable efficiently.

The decision problem whether White can win if White moves first in a chess game, on the other hand, has the form: Is there a move of White such that for *every* move of Black there is a move of White such that for *every* move of Black there is a move of White ... such that White can win? Here we have a large number of alternating existential and universal quantifiers rather than a single existential one. We are looking for an entire subtree rather than just a path in the decision tree. Because of this, most nonpolynomial games are at least Pspace-hard. The problem for generalized chess on an $n \times n$ board, and even for a number of seemingly simpler MathGames, is, in fact, Exptime-complete, which is a *provable intractability*.

Put in simple language, in analyzing an instance of Traveling Salesperson, the problem itself is passive: it does not resist your attempt to attack it, yet it is difficult. In a game, in contrast, there is your opponent, who, at every step, attempts to foil your effort to win. It's similar to the difference between an autopsy and surgery. Einstein, contemplating the nature of physics said, "Der Allmächtige ist nicht boshaft; Er ist raffiniert" (The Almighty is not mean; He is sophisticated). NP-complete existential problems are perhaps sophisticated. But your opponent in a game can be very mean!

Another reason for the high complexity of games is associated with a most basic tool of a game : its *game-graph*. It is a directed graph G whose vertices are the positions of the game, and (u, v) is an edge if and only if there is a move from position u to position v. Since every combination of tokens in the given game is a *single* vertex in G, the latter has normally exponential size. This holds, in particular, for both Nim and chess. Analyzing a game means reasoning about its game-graph. We are thus faced with a problem that is *a priori* exponential, quite unlike many present-day interesting existential problems.

A fundamental notion is the *sum* (disjunctive compound) of games. A sum is a finite collection of disjoint games; often very basic, simple games. Each of the two players, at every turn, selects one of the games and makes a move in it. If the outcome is not a draw, the sum-game ends when there is no move left in any of the component games. If the outcome is not a tie either, then in *normal* play, the player first unable to move loses and the opponent wins. The outcome is reversed in *misère* play.

If a game decomposes into a *disjoint* sum of its components, either from the beginning (Nim) or after a while (domineering), the potential for its tractability increases despite the exponential size of the game graph. As Elwyn Berlekamp remarked, the situation is similar to that in other scientific endeavors, where we often attempt to decompose a given system into its functional components. This approach may yield improved insights into hardware, software or biological systems, human organizations, and abstract mathematical objects such as groups. In most cases, there are interesting issues concerning the interactions between subsystems and their neighbors.

If a game doesn't decompose into a sum of disjoint components, it is more likely to be intractable (Geography or Poset Games). Intermediate cases happen when the components are not quite fixed (which explains why misère play of sums of games is much harder than normal play) or not quite disjoint (Welter).

4. Breaking the Rules

As the experts know, some of the most exciting games are obtained by breaking some of the rules for combinatorial games, such as permitting a player to pass a bounded or unbounded number of times, i.e., relaxing the requirement that players play alternately; or permitting a number of players other than two.

But permitting a payoff function other than (0,1) for the outcome (lose, win) and a payoff of $(\frac{1}{2}, \frac{1}{2})$ for either (tie, tie) or (draw, draw) usually, but not always, leads to games that are not considered to be combinatorial games; or to borderline cases.

5. Why Is the Bibliography Vast?

In the realm of existential problems, such as sorting or Traveling Salesperson, most present-day interesting decision problems can be classified into tractable, conditionally intractable, and provably intractable ones. There are exceptions, to be sure, such as graph isomorphism and primality testing, whose complexity is still unknown. But the exceptions are few. In contrast, most games are still in Wonderland, as pointed out in §2(I) above. Only a few games have been classified into the complexity classes they belong to. Despite recent impressive progress, the tools for reducing Wonderland are still few and inadequate.

To give an example, many interesting games have a very succinct input size, so a polynomial strategy is often more difficult to come by (Richard Guy and Cedric Smith' octal games; Grundy's game). Succinctness and non-disjointness of games in a sum may be present simultaneously (Poset games). In general, the alternating quantifiers, and, to a smaller measure, "breaking the rules", add to the volume of Wonderland. We suspect that the large size of Wonderland, a fact of independent interest, is the main contributing factor to the bulk of the bibliography on games.

6. Why Isn't it Larger?

The bibliography below is a *partial* list of books and articles on combinatorial games and related material. It is partial not only because I constantly learn of additional relevant material I did not know about previously, but also because of certain self-imposed restrictions. The most important of these is that only papers with some original and nontrivial mathematical content are considered. This excludes most historical reviews of games and most, but not all, of the work on heuristic or artificial intelligence approaches to games, especially the large literature concerning computer chess. I have, however, included the compendium Levy [1988], which, with its 50 articles and extensive bibliography, can serve as a first guide to this world. Also some papers on chess-endgames and clever exhaustive computer searches of some games have been included.

On the other hand, papers on games that break some of the rules of combinatorial games are included liberally, as long as they are interesting and retain a combinatorial flavor. These are vague and hard to define criteria, yet combinatorialists usually recognize a combinatorial game when they see it. Besides, it is interesting to break also this rule sometimes! We have included some references to one-player games, e.g., towers of Hanoi, n-queen problems, 15-puzzle and peg-solitaire, but only few zero-player games (such as Life and games on "sand piles"). We have also included papers on various applications of games, especially when the connection to games is substantial or the application is interesting or important.

During 1990–2001, *Theoretical Computer Science* ran a special Mathematical Games Section whose main purpose was to publish papers on combinatorial games. TCS still solicits papers on games. In 2001, *INTEGERS—Electronic J. of Combinatorial Number Theory* has started a Combinatorial Games Section. Lately, *Internat. J. Game Theory* has begun an effort to publish more papers on combinatorial games. It remains to be seen whether any of these forums, or others, will become focal points for high-class research results in the field of combinatorial games.

7. The Dynamics of the Literature

The game bibliography below is very dynamic in nature. Previous versions have been circulated to colleagues, intermittently, since the early 1980's. Prior to every mailing updates were prepared, and usually also afterwards, as a result of the comments received from several correspondents. The listing can never be "complete". Thus also the present form of the bibliography is by no means complete.

Because of its dynamic nature, it is natural that the bibliography became a "Dynamic Survey" in the Dynamic Surveys (DS) section of the *Electronic Journal of Combinatorics* (ElJC) and *The World Combinatorics Exchange* (WCE).

The ElJC and WCE are on the World Wide Web (WWW), and the DS can be accessed at http://www.combinatorics.org/Surveys/index.html. The journal itself can be found at http://www.combinatorics.org/. There are mirrors at various locations. Furthermore, the European Mathematical Information Service (EMIS) mirrors this Journal, as do all of its mirror sites (currently over forty of them). See http://www.emis.de/tech/mirrors.html.

8. An Appeal

Hereby I ask the readers to continue sending to me corrections and comments; and inform me of significant omissions, remembering, however, that it is a *selected* bibliography. I prefer to get reprints, preprints or URLs, rather than only titles — whenever possible.

Material on games is mushrooming on the Web. The URLs can be located using a standard searcher, such as Google.

9. Idiosyncrasies

A year or so after the bibliography became available electronically, I stopped snailmailing hard copies to potential readers.

Most of the bibliographic entries refer to items written in English, though there is a sprinkling of Danish, Dutch, French, German, Japanese, Slovakian and Russian, as well as some English translations from Russian. The predominance of English may be due to certain prejudices, but it also reflects the fact that nowadays the *lingua franca* of science is English. In any case, I'm soliciting also papers in languages other than English, especially if accompanied by an abstract in English.

On the administrative side, Technical Reports, submitted papers and unpublished theses have normally been excluded; but some exceptions have been made. Abbreviations of book series and journal names follow the *Math Reviews* conventions. Another convention is that de Bruijn appears under D, not B; von Neumann under V, not N, McIntyre under M not I, etc.

Earlier versions of this bibliography have appeared, under the title "Selected bibliography on combinatorial games and some related material", as the master bibliography for the book *Combinatorial Games*, AMS Short Course Lecture Notes, Summer 1990, Ohio State University, Columbus, OH, *Proc. Symp. Appl. Math.* **43** (R. K. Guy, ed.), AMS 1991, pp. 191–226 with 400 items, and in the *Dynamic Surveys* section of the *Electronic J. of Combinatorics* in November 1994 with 542 items (updated there at odd times). It also appeared as the master bibliography in *Games of No Chance*, Proc. MSRI Workshop on Combinatorial Games, July, 1994, Berkeley, CA (R. J. Nowakowski, ed.), MSRI Publ. Vol. 29, Cambridge University Press, Cambridge, pp. 493–537, under the present title, containing 666 items. The current version constitutes a growth of 38%. Published in the palindromic year 2002, it contains the palindromic number 919 of references.

10. Acknowledgments

Many people have suggested additions to the bibliography, or contributed to it in other ways. Among those that contributed more than two or three items are: Akeo Adachi, Ingo Althöfer, Thomas Andreae, Eli Bachmupsky, Adriano Barlotti, József Beck, Claude Berge, Gerald E. Bergum, H. S. MacDonald Coxeter, Thomas S. Ferguson, James A. Flanigan, Fred Galvin, Martin Gardner, Alan J. Goldman, Solomon W. Golomb, Richard K. Guy, Shigeki Iwata, David S. Johnson, Victor Klee, Donald E. Knuth, Anton Kotzig, Jeff C. Lagarias, Michel Las Vergnas, Hendrik W. Lenstra, Hermann Loimer, F. Lockwood Morris, Richard J. Nowakowski, Judea Pearl, J. Michael Robson, David Singmaster, Wolfgang Slany, Cedric A. B. Smith, Rastislaw Telgársky, Yōhei Yamasaki and many others. Thanks to all and keep up the game! Special thanks are due to Ms. Sarah Fliegelmann and Mrs. Carol Weintraub, who maintained and updated the bibliography-file, expertly and skilfully, over several different TEX generations in the past (now I do this myself), to Silvio Levy, who edited and transformed it into LATEX2e in 1996, and to Wolfgang Slany, who transformed it into a BIBTeX file at the end of the previous millenium.

11. The Bibliography

1. B. Abramson and M. Yung [1989], Divide and conquer under global constraints: a solution to the n-queens problem, *J. Parallel Distrib. Comput.* **6**, 649–662.

2. A. Adachi, S. Iwata and T. Kasai [1981], Low level complexity for combinatorial games, *Proc. 13th Ann. ACM Symp. Theory of Computing (Milwaukee, WI, 1981)*, Assoc. Comput. Mach., New York, NY, pp. 228–237.

3. A. Adachi, S. Iwata and T. Kasai [1984], Some combinatorial game problems require $\Omega(n^k)$ time, *J. Assoc. Comput. Mach.* **31**, 361–376.

4. H. Adachi, H. Kamekawa and S. Iwata [1987], Shogi on $n \times n$ board is complete in exponential time, *Trans. IEICE* **J70-D**, 1843–1852 (in Jpanese).

5. W. Ahrens [1910], *Mathematische Unterhaltungen und Spiele*, Vol. I, Teubner, Leipzig, Zweite vermehrte und verbesserte Auflage.

6. M. Aigner [1995], Ulams Millionenspiel, *Math. Semesterber.* **42**, 71–80.

7. M. Aigner [1996], Searching with lies, *J. Combin. Theory* (Ser. A) **74**, 43–56.

8. M. Aigner and M. Fromme [1984], A game of cops and robbers, *Discrete Appl. Math.* **8**, 1–12.

9. M. Ajtai, L. Csirmaz and Z. Nagy [1979], On a generalization of the game Go-Moku I, *Studia Sci. Math. Hungar.* **14**, 209–226.

10. E. Akin and M. Davis [1985], Bulgarian solitaire, *Amer. Math. Monthly* **92**, 237–250.

11. M. H. Albert and R. J. Nowakowski [2001], The game of End-Nim, *Electronic J. Combin.* **8(2)**, #R1, 12pp., Volume in honor of Aviezri S. Fraenkel. http://www.combinatorics.org/

12. R. E. Allardice and A. Y. Fraser [1884], La tour d'Hanoï, *Proc. Edinburgh Math. Soc.* **2**, 50–53.

13. D. T. Allemang [1984], Machine computation with finite games, M.Sc. Thesis, Cambridge University.

14. D. T. Allemang [2002], Generalized genus sequences for misère octal games, *Intern. J. Game Theory,* to appear.

15. J. D. Allen [1989], A note on the computer solution of Connect-Four, *Heuristic Programming in Artificial Intelligence* 1: *The First Computer Olympiad* (D. N. L. Levy and D. F. Beal, eds.), Ellis Horwood, Chichester, England, pp. 134–135.

16. L. V. Allis [1994], Searching for solutions in games and artificial intelligence, Ph.D. Thesis, University of Limburg. ftp://ftp.cs.vu.nl/pub/victor/PhDthesis/thesis.ps.Z

17. L. V. Allis and P. N. A. Schoo [1992], Qubic solved again, *Heuristic Programming in Artificial Intelligence* 3: *The Third Computer Olympiad* (H. J. van den Herik and L. V. Allis, eds.), Ellis Horwood, Chichester, England, pp. 192–204.

18. L. V. Allis, H. J. van den Herik and M. P. H. Huntjens [1993], Go-Moku solved by new search techniques, *Proc. 1993 AAAI Fall Symp. on Games: Planning and Learning*, AAAI Press Tech. Report FS93–02, Menlo Park, CA, pp. 1–9.

19. J.-P. Allouche, D. Astoorian, J. Randall and J. Shallit [1994], Morphisms, squarefree strings, and the tower of Hanoi puzzle, *Amer. Math. Monthly* **101**, 651–658.

20. S. Alpern and A. Beck [1991], Hex games and twist maps on the annulus, *Amer. Math. Monthly* **98**, 803–811.

21. I. Althöfer [1988], Nim games with arbitrary periodic moving orders, *Internat. J. Game Theory* **17**, 165–175.

22. I. Althöfer [1988], On the complexity of searching game trees and other recursion trees, *J. Algorithms* **9**, 538–567.

23. I. Althöfer [1989], Generalized minimax algorithms are no better error correctors than minimax is itself, in: *Advances in Computer Chess* (D. F. Beal, ed.), Vol. 5, Elsevier, Amsterdam, pp. 265–282.

24. I. Althöfer and J. Bültermann [1995], Superlinear period lengths in some subtraction games, *Theoret. Comput. Sci.* (Math Games) **148**, 111–119.

25. M. Anderson and T. Feil [1998], Turning lights out with linear algebra, *Math. Mag.* **71**, 300–303.

26. M. Anderson and F. Harary [1987], Achievement and avoidance games for generating abelian groups, *Internat. J. Game Theory* **16**, 321–325.

27. R. Anderson, L. Lovász, P. Shor, J. Spencer, E. Tardós and S. Winograd [1989], Disks, balls and walls: analysis of a combinatorial game, *Amer. Math. Monthly* **96**, 481–493.

28. T. Andreae [1984], Note on a pursuit game played on graphs, *Discrete Appl. Math.* **9**, 111–115.

29. T. Andreae [1986], On a pursuit game played on graphs for which a minor is excluded, *J. Combin. Theory* (Ser. B) **41**, 37–47.

30. T. Andreae, F. Hartenstein and A. Wolter [1999], A two-person game on graphs where each player tries to encircle his opponent's men, *Theoret. Comput. Sci.* (Math Games) **215**, 305–323.

31. V. V. Anshelevich [2000], The Game of Hex: an automatic theorem proving approach to game programming, *Proc. 17-th National Conference on Artificial Intelligence (AAAI-2000)*, AAAI Press, Menlo Park, CA, pp. 189–194.

32. R. P. Anstee and M. Farber [1988], On bridged graphs and cop-win graphs, *J. Combin. Theory* (Ser. B) **44**, 22–28.

33. D. Applegate, G. Jacobson and D. Sleator [1999], Computer analysis of Sprouts, in: *The Mathemagician and Pied Puzzler*, honoring Martin Gardner; E. Berlekamp and T. Rodgers, eds., A K Peters, Natick, MA, pp. 199-201.

34. A. F. Archer [1999], A modern treatment of the 15 puzzle, *Amer. Math. Monthly* **106**, 793–799.

35. P. Arnold, ed. [1993], *The Book of Games*, Hamlyn, Chancellor Press.

36. A. A. Arratia-Quesada and I. A. Stewart [1997], Generalized Hex and logical characterizations of polynomial space, *Inform. Process. Lett.* **63**, 147–152.

37. M. Ascher [1987], Mu Torere: An analysis of a Maori game, *Math. Mag.* **60**, 90–100.

38. I. M. Asel'derova [1974], On a certain discrete pursuit game on graphs, *Cybernetics* **10**, 859–864, trans. of *Kibernetika* **10** (1974) 102–105.

39. J. A. Aslam and A. Dhagat [1993], On-line algorithms for 2-coloring hypergraphs via chip games, *Theoret. Comput. Sci.* (Math Games) **112**, 355–369.

40. M. D. Atkinson [1981], The cyclic towers of Hanoi, *Inform. Process. Lett.* **13**, 118–119.

41. J. M. Auger [1991], An infiltration game on k arcs, *Naval Res. Logistics* **38**, 511–529.

42. J. Auslander, A. T. Benjamin and D. S. Wilkerson [1993], Optimal leapfrog-ging, *Math. Mag.* **66**, 14–19.

43. R. Austin [1976], Impartial and partisan games, M.Sc. Thesis, Univ. of Calgary.

44. J. O. A. Ayeni and H. O. D. Longe [1985], Game people play: Ayo, *Internat. J. Game Theory* **14**, 207–218.

45. L. Babai and S. Moran [1988], Arthur–Merlin games: a randomized proof system, and a hierarchy of complexity classes, *J. Comput. System Sci.* **36**, 254–276.

46. R. J. R. Back and J. von Wright [1995], Games and winning strategies, *Inform. Process. Lett.* **53**, 165–172.

47. C. K. Bailey and M. E. Kidwell [1985], A king's tour of the chessboard, *Math. Mag.* **58**, 285–286.

48. W. W. R. Ball and H. S. M. Coxeter [1987], *Mathematical Recreations and Essays*, Dover, New York, NY, 13th edn.

49. B. Banaschewski and A. Pultr [1990/91], Tarski's fixpoint lemma and com-binatorial games, *Order* **7**, 375–386.

50. R. B. Banerji [1971], Similarities in games and their use in strategy con-struction, *Proc. Symp. Computers and Automata* (J. Fox, ed.), Polytechnic Press, Brooklyn, NY, pp. 337–357.

51. R. B. Banerji [1980], *Artificial Intelligence, A Theoretical Approach*, Else-vier, North-Holland, New York, NY.

52. R. B. Banerji and C. A. Dunning [1992], On misere games, *Cybernetics and Systems* **23**, 221–228.

53. R. B. Banerji and G. W. Ernst [1972], Strategy construction using homo-morphisms between games, *Artificial Intelligence* **3**, 223–249.

54. R. Bar Yehuda, T. Etzion and S. Moran [1993], Rotating-table games and derivatives of words, *Theoret. Comput. Sci.* (Math Games) **108**, 311–329.

55. I. Bárány [1979], On a class of balancing games, *J. Combin. Theory* (Ser. A) **26**, 115–126.

56. J. G. Baron [1974], The game of nim — a heuristic approach, *Math. Mag.* **47**, 23–28.

57. R. Barua and S. Ramakrishnan [1996], σ-game, σ^+-game and two-dimen-sional additive cellular automata, *Theoret. Comput. Sci.* (Math Games) **154**, 349–366.

58. V. J. D. Baston and F. A. Bostock [1985], A game locating a needle in a cirular haystack, *J. Optimization Theory and Applications* **47**, 383–391.

59. V. J. D. Baston and F. A. Bostock [1986], A game locating a needle in a square haystack, *J. Optimization Theory and Applications* **51**, 405–419.

60. V. J. D. Baston and F. A. Bostock [1987], Discrete hamstrung squad car games, *Internat. J. Game Theory* **16**, 253–261.

61. V. J. D. Baston and F. A. Bostock [1988], A simple cover-up game, *Amer. Math. Monthly* **95**, 850–854.

62. V. J. D. Baston and F. A. Bostock [1989], A one-dimensional helicopter-submarine game, *Naval Res. Logistics* **36**, 479–490.

63. V. J. D. Baston and F. A. Bostock [1993], Infinite deterministic graphical games, *SIAM J. Control Optim.* **31**, 1623–1629.

64. J. Baumgartner, F. Galvin, R. Laver and R. McKenzie [1975], Game theoretic versions of partition relations, in: *Colloquia Mathematica Societatis János Bolyai 10, Proc. Internat. Colloq. on Infinite and Finite Sets*, Vol. 1, Keszthely, Hungary, 1973 (A. Hajnal, R. Rado and V. T. Sós, eds.), North-Holland, pp. 131–135.

65. J. D. Beasley [1985], *The Ins & Outs of Peg Solitaire*, Oxford University Press, Oxford.

66. J. D. Beasley [1990], *The Mathematics of Games*, Oxford University Press, Oxford.

67. P. Beaver [1995], *Victorian Parlour Games*, Magna Books.

68. A. Beck [1969], Games, in: *Excursions into Mathematics* (A. Beck, M. N. Bleicher and D. W. Crowe, eds.), Worth Publ., Chap. 5, pp. 317–387.

69. J. Beck [1981], On positional games, *J. Combin. Theory* (Ser. A) **30**, 117–133.

70. J. Beck [1981], Van der Waerden and Ramsey type games, *Combinatorica* **1**, 103–116.

71. J. Beck [1982], On a generalization of Kaplansky's game, *Discrete Math.* **42**, 27–35.

72. J. Beck [1982], Remarks on positional games, I, *Acta Math. Acad. Sci. Hungar.* **40**(1–2), 65–71.

73. J. Beck [1983], Biased Ramsey type games, *Studia Sci. Math. Hung.* **18**, 287–292.

74. J. Beck [1985], Random graphs and positional games on the complete graph, *Ann. Discrete Math.* **28**, 7–13.

75. J. Beck [1993], Achievement games and the probabilistic method, in: *Combinatorics, Paul Erdős is Eighty*, Vol. 1, Bolyai Soc. Math. Stud., János Bolyai Math. Soc., Budapest, pp. 51–78.

76. J. Beck [1994], Deterministic graph games and a probabilistic intuition, *Combin. Probab. Comput.* **3**, 13–26.

77. J. Beck [1996], Foundations of positional games, *Random Structures Algorithms* **9**, 15–47, appeared first in: Proc. Seventh International Conference on Random Structures and Algorithms, Atlanta, GA, 1995.

78. J. Beck [1997], Games, randomness and algorithms, in: *The Mathematics of Paul Erdős* (R. L. Graham and J. Nešetřil, eds.), Vol. I, Springer, pp. 280–310.

79. J. Beck [1997], Graph games, *Proc. Int. Coll. Extremal Graph Theory*, Balatonlelle, Hungary.

80. J. Beck and L. Csirmaz [1982], Variations on a game, *J. Combin. Theory* (Ser. A) **33**, 297–315.

81. R. Beigel and W. I. Gasarch [1991], The mapmaker's dilemma, *Discrete Appl. Math.* **34**, 37–48.

82. R. C. Bell [1960, 1969], *Board and Table Games from Many Civilisations*, Vol. I & II, Oxford University Press, revised in 1979, Dover.

83. R. Bell [1988], *Board Games Round the World*, Cambridge University Press, Cambridge (Third Printing: 1993).

84. S. J. Benkoski, M. G. Monticino and J. R. Weisinger [1991], A survey of the search theory literature, *Naval Res. Logistics* **38**, 469–494.

85. G. Bennett [1994], Double dipping: the case of the missing binomial coefficient identities, *Theoret. Comput. Sci.* (Math Games) **123**, 351–375.

86. D. Berengut [1981], A random hopscotch problem or how to make Johnny read more, in: *The Mathematical Gardner* (D. A. Klarner, ed.), Wadsworth Internat., Belmont, CA, pp. 51–59.

87. B. Berezovskiy and A. Gnedin [1984], The best choice problem, *Akad. Nauk, USSR, Moscow* (in Russian) .

88. C. Berge [1976], Sur les jeux positionnels, *Cahiers du Centre Études Rech. Opér.* **18**, 91–107.

89. C. Berge [1977], Vers une théorie générale des jeux positionnels, in: *Mathematical Economics and Game Theory, Essays in Honor of Oskar Morgenstern*, Lecture Notes in Economics (R. Henn and O. Moeschlin, eds.), Vol. 141, Springer Verlag, Berlin, pp. 13–24.

90. C. Berge [1981], Some remarks about a Hex problem, in: *The Mathematical Gardner* (D. A. Klarner, ed.), Wadsworth Internat., Belmont, CA, pp. 25–27.

91. C. Berge [1985], *Graphs*, North-Holland, Amsterdam, Chap. 14.

92. C. Berge [1989], *Hypergraphs*, Elsevier (French: Gauthier Villars 1988), Chap. 4.

93. C. Berge [1996], Combinatorial games on a graph, *Discrete Math.* **151**, 59–65.

94. C. Berge and P. Duchet [1988], Perfect graphs and kernels, *Bull. Inst. Math. Acad. Sinica* **16**, 263–274.

95. C. Berge and P. Duchet [1990], Recent problems and results about kernels in directed graphs, *Discrete Math.* **86**, 27–31.

96. C. Berge and M. Las Vergnas [1976], Un nouveau jeu positionnel, le "Match-It", ou une construction dialectique des couplages parfaits, *Cahiers du Centre Études Rech. Opér.* **18**, 83–89.

97. E. R. Berlekamp [1972], Some recent results on the combinatorial game called Welter's Nim, *Proc. 6th Ann. Princeton Conf. Information Science and Systems*, pp. 203–204.

98. E. R. Berlekamp [1974], The Hackenbush number system for compresssion of numerical data, *Inform. and Control* **26**, 134–140.

99. E. R. Berlekamp [1988], Blockbusting and domineering, *J. Combin. Theory* (Ser. A) **49**, 67–116, an earlier version, entitled Introduction to blockbusting and domineering, appeared in: *The Lighter Side of Mathematics*, Proc. E. Strens Memorial Conf. on Recr. Math. and its History, Calgary, 1986, Spectrum Series (R. K. Guy and R. E. Woodrow, eds.), Math. Assoc. of America, Washington, DC, 1994, pp. 137–148.

100. E. Berlekamp [1990], Two-person, perfect-information games, in: *The Legacy of John von Neumann* (Hempstead NY, 1988), *Proc. Sympos. Pure Math.*, Vol. 50, Amer. Math. Soc., Providence, RI, pp. 275–287.

101. E. R. Berlekamp [1991], Introductory overview of mathematical Go endgames, in: *Combinatorial Games*, Proc. Symp. Appl. Math. (R. K. Guy, ed.), Vol. 43, Amer. Math. Soc., Providence, RI, pp. 73–100.

102. E. R. Berlekamp [1996], The economist's view of combinatorial games, in: *Games of No Chance*, Proc. MSRI Workshop on Combinatorial Games, July, 1994, Berkeley, CA, MSRI Publ. (R. J. Nowakowski, ed.), Vol. 29, Cambridge University Press, Cambridge, pp. 365–405.

103. E. R. Berlekamp [2000], Sums of 2 × N Amazons, in: *Institute of Mathematical Statistics Lecture Notes–Monograph Series* (F.T. Bruss and L.M. Le Cam, eds.), Vol. 35, Beechwood, Ohio: Institute of Mathematical Statistics, pp. 1–34, Papers in honor of Thomas S. Ferguson.

104. E. R. Berlekamp [2000], *The Dots-and-Boxes Game: Sophisticated Child's Play*, A K Peters, Natick, MA.

105. E. R. Berlekamp [2002], Four games for Gardner, in: *Puzzler's Tribute: a Feast for the Mind*, pp. 383–386, honoring Martin Gardner (D. Wolfe and T. Rodgers, eds.), A K Peters, Natick, MA.

106. E. R. Berlekamp, J. H. Conway and R. K. Guy [1982], *Winning Ways for your Mathematical Plays*, Vol. I & II, Academic Press, London, 2nd edition of vol. 1 (of four volumes), 2001, A K Peters, Natick, MA; translated into German: *Gewinnen, Strategien für Mathematische Spiele* by G. Seiffert, Foreword by K. Jacobs, M. Reményi and Seiffert, Friedr. Vieweg & Sohn, Braunschweig (four volumes), 1985.

107. E. R. Berlekamp and Y. Kim [1996], Where is the "Thousand-Dollar Ko?", in: *Games of No Chance*, Proc. MSRI Workshop on Combinatorial Games, July, 1994, Berkeley, CA, MSRI Publ. (R. J. Nowakowski, ed.), Vol. 29, Cambridge University Press, Cambridge, pp. 203–226.

108. E. Berlekamp and T. Rodgers, eds. [1999], *The Mathemagician and Pied Puzzler*, A K Peters, Natick, MA, A collection in tribute to Martin Gardner.

Papers from the Gathering for Gardner Meeting (G4G1) held in Atlanta, GA, January 1993.

109. E. Berlekamp and K. Scott [2002], Forcing your opponent to stay in control of a loony dotsandboxes endgame, in: *More Games of No Chance*, Proc. MSRI Workshop on Combinatorial Games, July, 2000, Berkeley, CA, MSRI Publ. (R. J. Nowakowski, ed.), Vol. 42, Cambridge University Press, Cambridge, pp. 317–330.

110. E. Berlekamp and D. Wolfe [1994], *Mathematical Go — Chilling Gets the Last Point*, A K Peters, Natick, MA.

111. P. Berloquin [1976], *100 Jeux de Table*, Flammarion, Paris.

112. P. Berloquin [1995], *100 Games of Logic*, Barnes & Noble.

113. P. Berloquin and D. Dugas (Illustrator) [1999], *100 Perceptual Puzzles*, Barnes & Noble.

114. N. L. Biggs [1999], Chip-firing and the critical group of a graph, *J. Algebr. Comb.* **9**, 25–45.

115. K. Binmore [1992], *Fun and Games: a Text on Game Theory*, D.C. Heath, Lexington.

116. J. Bitar and E. Goles [1992], Parallel chip firing games on graphs, *Theoret. Comput. Sci.* **92**, 291–300.

117. A. Björner and L. Lovász [1992], Chip-firing games on directed graphs, *J. Algebraic Combin.* **1**, 305–328.

118. A. Björner, L. Lovász and P. Chor [1991], Chip-firing games on graphs, *European J. Combin.* **12**, 283–291.

119. D. Blackwell and M. A. Girshick [1954], *Theory of Games and Statistical Decisions*, Wiley, New York, NY.

120. U. Blass and A. S. Fraenkel [1990], The Sprague–Grundy function for Wythoff's game, *Theoret. Comput. Sci.* (Math Games) **75**, 311–333.

121. U. Blass, A. S. Fraenkel and R. Guelman [1998], How far can Nim in disguise be stretched?, *J. Combin. Theory* (Ser. A) **84**, 145–156.

122. M. Blidia [1986], A parity digraph has a kernel, *Combinatorica* **6**, 23–27.

123. M. Blidia, P. Duchet, H. Jacob, F. Maffray and H. Meyniel [1999], Some operations preserving the existence of kernels, *Discrete Math.* **205**, 211–216.

124. M. Blidia, P. Duchet and F. Maffray [1993], On kernels in perfect graphs, *Combinatorica* **13**, 231–233.

125. J.-P. Bode and H. Harborth [1997], Hexagonal polyomino achievement, *Discrete Math.* **212**, 5–18.

126. J.-P. Bode and H. Harborth [1998], Achievement games on Platonic solids, *Bull. Inst. Combin. Appl.* **23**, 23–32.

127. H. L. Bodlaender [1991], On the complexity of some coloring games, *Internat. J. Found. Comput. Sci.* **2**, 133–147.

128. H. L. Bodlaender [1993], Complexity of path forming games, *Theoret. Comput. Sci.* (Math Games) **110**, 215–245.

129. H. L. Bodlaender [1993], Kayles on special classes of graphs—an application of Sprague-Grundy theory, in: *Graph-Theoretic Concepts in Computer Science* (Wiesbaden-Naurod, 1992), Lecture Notes in Comput. Sci., Vol. 657, Springer, Berlin, pp. 90–102.

130. H. L. Bodlaender and D. Kratsch [1992], The complexity of coloring games on perfect graphs, *Theoret. Comput. Sci.* (Math Games) **106**, 309–326.

131. K. D. Boklan [1984], The n-number game, *Fibonacci Quart.* **22**, 152–155.

132. B. Bollobás and T. Szabó [1998], The oriented cycle game, *Discrete Math.* **186**, 55–67.

133. D. L. Book [1998, Sept. 9-th], What the Hex, *The Washington Post* p. H02.

134. E. Borel [1921], La théorie du jeu et les équations integrales à noyau symmetrique gauche, *C. R. Acad. Sci. Paris* **173**, 1304–1308.

135. E. Boros and V. Gurevich [1996], Perfect graphs are kernel solvable, *Discrete Math.* **159**, 35–55.

136. E. Boros and V. Gurevich [1998], A corrected version of the Duchet kernel conjecture, *Discrete Math.* **179**, 231–233.

137. C. L. Bouton [1902], Nim, a game with a complete mathematical theory, *Ann. of Math.* **3**(2), 35–39.

138. J. Boyce [1981], A Kriegspiel endgame, in: *The Mathematical Gardner* (D. A. Klarner, ed.), Wadsworth Internat., Belmont, CA, pp. 28–36.

139. G. Brandreth [1981], *The Bumper Book of Indoor Games*, Victorama, Chancellor Press.

140. D. M. Breuker, J. W. H. M. Uiterwijk and H. J. van den Herik [2000], Solving 8 × 8 Domineering, *Theoret. Comput. Sci.* (Math Games) **230**, 195–206.

141. D. M. Broline and D. E. Loeb [1995], The combinatorics of Mancala-type games: Ayo, Tchoukaillon, and $1/\pi$, *UMAP J.* **16**(1), 21–36.

142. A. Brousseau [1976], Tower of Hanoi with more pegs, *J. Recr. Math.* **8**, 169–178.

143. C. Browne [2000], *HEX Strategy: Making the Right Connections*, A K Peters, Natick, MA.

144. R. A. Brualdi and V. S. Pless [1993], Greedy codes, *J. Combin. Theory* (Ser. A) **64**, 10–30.

145. A. A. Bruen and R. Dixon [1975], The n-queen problem, *Discrete Math.* **12**, 393–395.

146. P. Buneman and L. Levy [1980], The towers of Hanoi problem, *Inform. Process. Lett.* **10**, 243–244.

147. A. P. Burger, E. J. Cockayne and C. M. Mynhardt [1977], Domination and irredundance in the queen's graph, *Discrete Math.* **163**, 47–66.

148. D. W. Bushaw [1967], On the name and history of Nim, *Washington Math.* **11**, Oct. 1966. Reprinted in: *NY State Math. Teachers J.*, **17**, pp. 52–55.

149. P. J. Byrne and R. Hesse [1996], A Markov chain analysis of jai alai, *Math. Mag.* **69**, 279–283.

150. L. Cai and X. Zhu [2001], Game chromatic index of k-degenerate graphs, *J. Graph Theory* **36**, 144–155.

151. J.-Y. Cai, A. Condon and R. J. Lipton [1992], On games of incomplete information, *Theoret. Comput. Sci.* **103**, 25–38.

152. I. Caines, C. Gates, R. K. Guy and R. J. Nowakowski [1999], Periods in taking and splitting games, *Amer. Math. Monthly* **106**, 359–361.

153. D. Calistrate [1996], The reduced canonical form of a game, in: *Games of No Chance,* Proc. MSRI Workshop on Combinatorial Games, July, 1994, Berkeley, CA, MSRI Publ. (R. J. Nowakowski, ed.), Vol. 29, Cambridge University Press, Cambridge, pp. 409–416.

154. D. Calistrate, M. Paulhus and D. Wolfe [2002], On the lattice structure of finite games, in: *More Games of No Chance,* Proc. MSRI Workshop on Combinatorial Games, July, 2000, Berkeley, CA, MSRI Publ. (R. J. Nowakowski, ed.), Vol. 42, Cambridge University Press, Cambridge, pp. 25–30.

155. C. Cannings and J. Haigh [1992], Montreal solitaire, *J. Combin. Theory* (Ser. A) **60**, 50–66.

156. A. Chan and A. Tsai [2002], $1 \times n$ Konane: a summary of results, in: *More Games of No Chance,* Proc. MSRI Workshop on Combinatorial Games, July, 2000, Berkeley, CA, MSRI Publ. (R. J. Nowakowski, ed.), Vol. 42, Cambridge University Press, Cambridge, pp. 331–339.

157. A. K. Chandra, D. C. Kozen and L. J. Stockmeyer [1981], Alternation, *J. Assoc. Comput. Mach.* **28**, 114–133.

158. A. K. Chandra and L. J. Stockmeyer [1976], Alternation, *Proc. 17th Ann. Symp. Foundations of Computer Science* (Houston, TX, Oct. 1976), IEEE Computer Soc., Long Beach, CA, pp. 98–108.

159. S. M. Chase [1972], An implemented graph algorithm for winning Shannon switching games, *Commun. Assoc. Comput. Mach.* **15**, 253–256.

160. G. Chen, R. H. Schelp and W. E. Shreve [1997], A new game chromatic number, *European J. Combin.* **18**, 1–9.

161. B. S. Chlebus [1986], Domino-tiling games, *J. Comput. System Sci.* **32**, 374–392.

162. F. R. K. Chung [1989], Pebbling in hypercubes, *SIAM J. Disc. Math.* **2**, 467–472.

163. F. R. K. Chung, J. E. Cohen and R. L. Graham [1988], Pursuit-evasion games on graphs, *J. Graph Theory* **12**, 159–167.

164. F. Chung, R. Graham, J. Morrison and A. Odlyzko [1995], Pebbling a chessboard, *Amer. Math. Monthly* **102**, 113–123.

165. V. Chvátal [1973], On the computational complexity of finding a kernel, Report No. CRM-300, Centre de Recherches Mathématiques, Université de Montréal.

166. V. Chvátal [1981], Cheap, middling or dear, in: *The Mathematical Gardner* (D. A. Klarner, ed.), Wadsworth Internat., Belmont, CA, pp. 44–50.

167. V. Chvátal [1983], Mastermind, *Combinatorica* **3**, 325–329.

168. V. Chvátal and P. Erdős [1978], Biased positional games, *Ann. Discrete Math.* **2**, 221–229, Algorithmic Aspects of Combinatorics, (B. Alspach, P. Hell and D. J. Miller, eds.), Qualicum Beach, BC, Canada, 1976, North-Holland.

169. F. Cicalese, D. Mundici and U. Vaccaro [2002], Least adaptive optimal search with unreliable tests, *Theoret. Comput. Sci.* (Math Games) **270**, 877–893.

170. F. Cicalese and U. Vaccaro [2000], Optimal strategies against a liar, *Theoret. Comput. Sci.* (Math Games) **230**, 167–193.

171. C. Clark [1996], On achieving channels in a bipolar game, in: *African Americans in Mathematics* (Piscataway, NJ, 1996), DIMACS Ser. Discrete Math. Theoret. Comput. Sci., Vol. 34, Amer. Math. Soc., Providence, RI, pp. 23–27.

172. D. S. Clark [1986], Fibonacci numbers as expected values in a game of chance, *Fibonacci Quart.* **24**, 263–267.

173. N. E. Clarke and R. J. Nowakowski [2000], Cops, robber, and photo radar, *Ars Combin.* **56**, 97–103.

174. E. J. Cockayne [1990], Chessboard domination problems, *Discrete Math.* **86**, 13–20.

175. E. J. Cockayne and S. T. Hedetniemi [1986], On the diagonal queens domination problem, *J. Combin. Theory* (Ser. A) **42**, 137–139.

176. A. J. Cole and A. J. T. Davie [1969], A game based on the Euclidean algorithm and a winning strategy for it, *Math. Gaz.* **53**, 354–357.

177. D. B. Coleman [1978], Stretch: a geoboard game, *Math. Mag.* **51**, 49–54.

178. A. Condon [1989], *Computational Models of Games*, ACM Distinguished Dissertation, MIT Press, Cambridge, MA.

179. A. Condon [1991], Space-bounded probabilistic game automata, *J. Assoc. Comput. Mach.* **38**, 472–494.

180. A. Condon [1992], The complexity of Stochastic games, *Information and Computation* **96**, 203–224.

181. A. Condon [1993], On algorithms for simple stochastic games, in: *Advances in Computational Complexity Theory* (New Brunswick, NJ, 1990),

DIMACS Ser. Discrete Math. Theoret. Comput. Sci., Vol. 13, Amer. Math. Soc., Providence, RI, pp. 51–71.

182. A. Condon, J. Feigenbaum, C. Lund and P. Shor [1993], Probabilistically checkable debate systems and approximation algorithms for PSPACE-hard functions, *Proc. 25th Ann. ACM Symp. Theory of Computing*, Assoc. Comput. Mach., New York, NY, pp. 305–314.

183. A. Condon and R. E. Ladner [1988], Probabilistic game automata, *J. Comput. System Sci.* **36**, 452–489, preliminary version in: Proc. Structure in complexity theory (Berkeley, CA, 1986), Lecture Notes in Comput. Sci., Vol. 223, Springer, Berlin, pp. 144–162.

184. I. G. Connell [1959], A generalization of Wythoff's game, *Canad. Math. Bull.* **2**, 181–190.

185. J. H. Conway [1972], All numbers great and small, Res. Paper No. 149, Univ. of Calgary Math. Dept.

186. J. H. Conway [1976], *On Numbers and Games*, Academic Press, London, 2nd edition, 2001, A K Peters, Natick, MA; translated into German: *Über Zahlen und Spiele* by Brigitte Kunisch, Friedr. Vieweg & Sohn, Braunschweig, 1983.

187. J. H. Conway [1977], All games bright and beautiful, *Amer. Math. Monthly* **84**, 417–434.

188. J. H. Conway [1978], A gamut of game theories, *Math. Mag.* **51**, 5–12.

189. J. H. Conway [1978], Loopy Games, *Ann. Discrete Math.* **3**, 55–74, Proc. Symp. Advances in Graph Theory, Cambridge Combinatorial Conf. (B. Bollobás, ed.), Cambridge, May 1977.

190. J. H. Conway [1990], Integral lexicographic codes, *Discrete Math.* **83**, 219–235.

191. J. H. Conway [1991], More ways of combining games, in: *Combinatorial Games,* Proc. Symp. Appl. Math. (R. K. Guy, ed.), Vol. 43, Amer. Math. Soc., Providence, RI, pp. 57–71.

192. J. H. Conway [1991], Numbers and games, in: *Combinatorial Games,* Proc. Symp. Appl. Math. (R. K. Guy, ed.), Vol. 43, Amer. Math. Soc., Providence, RI, pp. 23–34.

193. J. H. Conway [1996], The angel problem, in: *Games of No Chance,* Proc. MSRI Workshop on Combinatorial Games, July, 1994, Berkeley, CA, MSRI Publ. (R. J. Nowakowski, ed.), Vol. 29, Cambridge University Press, Cambridge, pp. 3–12.

194. J. H. Conway [2002], Infinite games, in: *More Games of No Chance,* Proc. MSRI Workshop on Combinatorial Games, July, 2000, Berkeley, CA, MSRI Publ. (R. J. Nowakowski, ed.), Vol. 42, Cambridge University Press, Cambridge, pp. 31–36.

195. J. H. Conway and H. S. M. Coxeter [1973], Triangulated polygons and frieze patterns, *Math. Gaz.* **57**, 87–94; 175–183.

196. J. H. Conway and N. J. A. Sloane [1986], Lexicographic codes: error-correcting codes from game theory, *IEEE Trans. Inform. Theory* **IT-32**, 337–348.

197. M. L. Cook and L. E. Shader [1979], A strategy for the Ramsey game "Tritip", *Proc. 10th Southeast. Conf. Combinatorics, Graph Theory and Computing, Boca Raton*, Vol. 1 of *Congr. Numer. 23*, Utilitas Math., pp. 315–324.

198. M. Copper [1993], Graph theory and the game of sprouts, *Amer. Math. Monthly* **100**, 478–482.

199. M. Cornelius and A. Parr [1991], *What's Your Game?*, Cambridge University Press, Cambridge.

200. H. S. M. Coxeter [1953], The golden section, phyllotaxis and Wythoff's game, *Scripta Math.* **19**, 135–143.

201. J. W. Creely [1987], The length of a two-number game, *Fibonacci Quart.* **25**, 174–179.

202. J. W. Creely [1988], The length of a three-number game, *Fibonacci Quart.* **26**, 141–143.

203. H. T. Croft [1964], 'Lion and man': a postscript, *J. London Math. Soc.* **39**, 385–390.

204. D. W. Crowe [1956], The n-dimensional cube and the tower of Hanoi, *Amer. Math. Monthly* **63**, 29–30.

205. L. Csirmaz [1980], On a combinatorial game with an application to Go-Moku, *Discrete Math.* **29**, 19–23.

206. L. Csirmaz and Z. Nagy [1979], On a generalization of the game Go-Moku II, *Studia Sci. Math. Hung.* **14**, 461–469.

207. J. Culberson [1999], Sokoban is PSPACE complete, in: *Fun With Algorithms*, Vol. 4 of *Proceedings in Informatics*, Carleton Scientific, University of Waterloo, Waterloo, Ont., pp. 65–76, Conference took place on the island of Elba, June 1998.

208. P. Cull and E. F. Ecklund, Jr. [1982], On the towers of Hanoi and generalized towers of Hanoi problems, *Congr. Numer.* **35**, 229–238.

209. P. Cull and E. F. Ecklund, Jr. [1985], Towers of Hanoi and analysis of algorithms, *Amer. Math. Monthly* **92**, 407–420.

210. P. Cull and C. Gerety [1985], Is towers of Hanoi really hard?, *Congr. Numer.* **47**, 237–242.

211. J. Czyzowicz, D. Mundici and A. Pelc [1988], Solution of Ulam's problem on binary search with two lies, *J. Combin. Theory* (Ser. A) **49**, 384–388.

212. J. Czyzowicz, D. Mundici and A. Pelc [1989], Ulam's searching game with lies, *J. Combin. Theory* (Ser. A) **52**, 62–76.

213. G. Danaraj and V. Klee [1977], The connectedness game and the c-complexity of certain graphs, *SIAM J. Appl. Math.* **32**, 431–442.

214. C. Darby and R. Laver [1998], Countable length Ramsey games, *Set Theory: Techniques and Applications.* Proc. of the conferences, Curacao, Netherlands Antilles, June 26–30, 1995 and Barcelona, Spain, June 10–14, 1996 (C. A. Di Prisco et al., eds.), Kluwer, Dordrecht, pp. 41–46.

215. A. L. Davies [1970], Rotating the fifteen puzzle, *Math. Gaz.* **54**, 237–240.

216. M. Davis [1963], Infinite games of perfect information, *Ann. of Math. Stud., Princeton* **52**, 85–101.

217. R. W. Dawes [1992], Some pursuit-evasion problems on grids, *Inform. Process. Lett.* **43**, 241–247.

218. T. R. Dawson [1934], Problem 1603, *Fairy Chess Review* p. 94, Dec.

219. T. R. Dawson [1935], Caissa's Wild Roses, reprinted in: *Five Classics of Fairy Chess*, Dover, 1973.

220. N. G. de Bruijn [1972], A solitaire game and its relation to a finite field, *J. Recr. Math.* **5**, 133–137.

221. N. G. de Bruijn [1981], Pretzel Solitaire as a pastime for the lonely mathematician, in: *The Mathematical Gardner* (D. A. Klarner, ed.), Wadsworth Internat., Belmont, CA, pp. 16–24.

222. F. de Carteblanche [1970], The princess and the roses, *J. Recr. Math.* **3**, 238–239.

223. F. deCarte Blanche [1974], The roses and the princes, *J. Recr. Math.* **7**, 295–298.

224. A. P. DeLoach [1971], Some investigations into the game of SIM, *J. Recr. Math.* **4**, 36–41.

225. E. D. Demaine, M. L. Demaine and D. Eppstein [2002], Phutball Endgames are Hard, in: *More Games of No Chance,* Proc. MSRI Workshop on Combinatorial Games, July, 2000, Berkeley, CA, MSRI Publ. (R. J. Nowakowski, ed.), Vol. 42, Cambridge University Press, Cambridge, pp. 351–360.

226. E. D. Demaine, M. L. Demaine and H. A. Verrill [2002], Coin-moving puzzles, in: *More Games of No Chance,* Proc. MSRI Workshop on Combinatorial Games, July, 2000, Berkeley, CA, MSRI Publ. (R. J. Nowakowski, ed.), Vol. 42, Cambridge University Press, Cambridge, pp. 405–421.

227. H. de Parville [1884], La tour d'Hanoï et la question du Tonkin, *La Nature* **12**, 285–286.

228. C. Deppe [2000], Solution of Ulam's searching game with three lies or an optimal adaptive strategy for binary three-error-correcting codes, *Discrete Math.* **224**, 79–98.

229. B. Descartes [1953], Why are series musical?, *Eureka* **16**, 18–20, reprinted *ibid.* **27** (1964) 29–31.

230. A. K. Dewdney [1984 –], Computer Recreations, a column in Scientific American (since May, 1984).

231. A. Dhagat, P. Gács and P. Winkler [1992], On playing "twenty questions" with a liar, *Proceedings of the Third Annual ACM-SIAM Symposium on Discrete Algorithms*, (Orlando, FL, 1992), ACM, New York, pp. 16–22.

232. C. S. Dibley and W. D. Wallis [1981], The effect of starting position in jai-alai, *Congr. Numer.* **32**, 253–259, Proc. 12-th Southeastern Conference on Combinatorics, Graph Theory and Computing, Vol. I (Baton Rouge, LA, 1981).

233. C. G. Diderich [1995], Bibliography on minimax game theory, sequential and parallel algorithms.
http://diwww.epfl.ch/~diderich/bibliographies.html

234. R. Diestel and I. Leader [1994], Domination games on infinite graphs, *Theoret. Comput. Sci.* (Math Games) **132**, 337–345.

235. T. Dinski and X. Zhu [1999], A bound for the game chromatic number of graphs, *Discrete Math.* **196**, 109–115.

236. A. P. Domoryad [1964], *Mathematical Games and Pastimes*, Pergamon Press, Oxford, translated by H. Moss.

237. D. Dor and U. Zwick [1999], SOKOBAN and other motion planning problems, *Comput. Geom.* **13**, 215–228.

238. M. Dresher [1951], Games on strategy, *Math. Mag.* **25**, 93–99.

239. A. Dress, A. Flammenkamp and N. Pink [1999], Additive periodicity of the Sprague-Grundy function of certain Nim games, *Adv. in Appl. Math.* **22**, 249–270.

240. P. Duchet [1980], Graphes noyau-parfaits, *Ann. Discrete Math.* **9**, 93–101.

241. P. Duchet [1987], A sufficient condition for a digraph to be kernel-perfect, *J. Graph Theory* **11**, 81–85.

242. P. Duchet [1987], Parity graphs are kernel-M-solvable, *J. Combin. Theory* (Ser. B) **43**, 121–126.

243. P. Duchet and H. Meyniel [1981], A note on kernel-critical graphs, *Discrete Math.* **33**, 103–105.

244. P. Duchet and H. Meyniel [1983], Une généralisation du théorème de Richardson sur l'existence de noyaux dans le graphes orientés, *Discrete Math.* **43**, 21–27.

245. P. Duchet and H. Meyniel [1993], Kernels in directed graphs: a poison game, *Discrete Math.* **115**, 273–276.

246. H. E. Dudeney [1958], *The Canterbury Puzzles*, Mineola, NY, 4th edn.

247. H. E. Dudeney [1989], *Amusements in Mathematics*, reprinted by Dover, Mineola, NY.

248. N. Duvdevani and A. S. Fraenkel [1989], Properties of k-Welter's game, *Discrete Math.* **76**, 197–221.

249. J. Edmonds [1965], Lehman's switching game and a theorem of Tutte and Nash–Williams, *J. Res. Nat. Bur. Standards* **69B**, 73–77.

250. R. Ehrenborg and E. Steingrímsson [1996], Playing Nim on a simplicial complex, *Electronic J. Combin.* **3**(1), #R9, 33pp. http://www.combinatorics.org/

251. A. Ehrenfeucht and J. Mycielski [1979], Positional strategies for mean payoff games, *Internat. J. Game Theory* **8**, 109–113.

252. N. D. Elkies [1996], On numbers and endgames: combinatorial game theory in chess endgames, in: *Games of No Chance,* Proc. MSRI Workshop on Combinatorial Games, July, 1994, Berkeley, CA, MSRI Publ. (R. J. Nowakowski, ed.), Vol. 29, Cambridge University Press, Cambridge, pp. 135–150.

253. N. D. Elkies [2002], Higher nimbers in pawn endgames on large chessboards, in: *More Games of No Chance,* Proc. MSRI Workshop on Combinatorial Games, July, 2000, Berkeley, CA, MSRI Publ. (R. J. Nowakowski, ed.), Vol. 42, Cambridge University Press, Cambridge, pp. 61–78.

254. D. Engel [1972], DIM: three-dimensional Sim, *J. Recr. Math.* **5**, 274–275.

255. R. J. Epp and T. S. Ferguson [1980], A note on take-away games, *Fibonacci Quart.* **18**, 300–303.

256. R. A. Epstein [1977], *Theory of Gambling and Statistial Logic*, Academic Press, New York, NY.

257. M. C. Er [1982], A representation approach to the tower of Hanoi problem, *Comput. J.* **25**, 442–447.

258. M. C. Er [1983], An analysis of the generalized towers of Hanoi problem, *BIT* **23**, 429–435.

259. M. C. Er [1983], An iterative solution to the generalized towers of Hanoi problem, *BIT* **23**, 295–302.

260. M. C. Er [1984], A generalization of the cyclic towers of Hanoi, *Intern. J. Comput. Math.* **15**, 129–140.

261. M. C. Er [1984], The colour towers of Hanoi: a generalization, *Comput. J.* **27**, 80–82.

262. M. C. Er [1984], The cyclic towers of Hanoi: a representation approach, *Comput. J.* **27**, 171–175.

263. M. C. Er [1984], The generalized colour towers of Hanoi: an iterative algorithm, *Comput. J.* **27**, 278–282.

264. M. C. Er [1984], The generalized towers of Hanoi problem, *J. Inform. Optim. Sci.* **5**, 89–94.

265. M. C. Er [1985], The complexity of the generalized cyclic towers of Hanoi problem, *J. Algorithms* **6**, 351–358.

266. M. C. Er [1987], A general algorithm for finding a shortest path between two *n*-configurations, *Information Sciences* **42**, 137–141.

267. C. Erbas, S. Sarkeshik and M. M. Tanik [1992], Different perspectives of the N-queens problem, *Proc. ACM Computer Science Conf.*, Kansas City, MO.

268. C. Erbas and M. M. Tanik [1994], Parallel memory allocation and data alignment in SIMD machines, *Parallel Algorithms and Applications* **4**, 139–151, preliminary version appeared under the title: Storage schemes for parallel memory systems and the N-queens problem, in: Proc. 15th Ann. Energy Tech. Conf., Houston, TX, Amer. Soc. Mech. Eng., Vol. 43, 1992, pp. 115–120.

269. C. Erbas, M. M. Tanik and Z. Aliyazicioglu [1992], Linear conguence equations for the solutions of the N-queens problem, *Inform. Process. Lett.* **41**, 301–306.

270. P. Erdős and J. L. Selfridge [1973], On a combinatorial game, *J. Combin. Theory* (Ser. A) **14**, 298–301.

271. J. Erickson [1996], New toads and frogs results, in: *Games of No Chance*, Proc. MSRI Workshop on Combinatorial Games, July, 1994, Berkeley, CA, MSRI Publ. (R. J. Nowakowski, ed.), Vol. 29, Cambridge University Press, Cambridge, pp. 299–310.

272. J. Erickson [1996], Sowing games, in: *Games of No Chance*, Proc. MSRI Workshop on Combinatorial Games, July, 1994, Berkeley, CA, MSRI Publ. (R. J. Nowakowski, ed.), Vol. 29, Cambridge University Press, Cambridge, pp. 287–297.

273. M. Erickson and F. Harary [1983], Picasso animal achievement games, *Bull. Malaysian Math. Soc.* **6**, 37–44.

274. N. Eriksen, H. Eriksson and K. Eriksson [2000], Diagonal checker-jumping and Eulerian numbers for color-signed permutations, *Electron. J. Combin.* **7**, #R3, 11 pp.
http://www.combinatorics.org/

275. H. Eriksson [1995], Pebblings, *Electronic J. Combin.* **2**, #R7, 18pp.
http://www.combinatorics.org/

276. H. Eriksson and B. Lindström [1995], Twin jumping checkers in Z^d, *European J. Combin.* **16**, 153–157.

277. K. Eriksson [1991], No polynomial bound for the chip firing game on directed graphs, *Proc. Amer. Math. Soc.* **112**, 1203–1205.

278. K. Eriksson [1992], Convergence of Mozes' game of numbers, *Linear Algebra Appl.* **166**, 151–165.

279. K. Eriksson [1994], Node firing games on graphs, *Contemp. Math.* **178**, 117–127.

280. K. Eriksson [1994], Reachability is decidable in the numbers game, *Theoret. Comput. Sci.* (Math Games) **131**, 431–439.

281. K. Eriksson [1995], The numbers game and Coxeter groups, *Discrete Math.* **139**, 155–166.

282. K. Eriksson [1996], Chip-firing games on mutating graphs, *SIAM J. Discrete Math.* **9**, 118–128.

283. K. Eriksson [1996], Strong convergence and a game of numbers, *European J. Combin.* **17**, 379–390.

284. K. Eriksson [1996], Strong convergence and the polygon property of 1-player games, *Discrete Math.* **153**, 105–122, Proc. 5th Conf. on Formal Power Series and Algebraic Combinatorics (Florence 1993).

285. J. M. Ettinger [2000], A metric for positional games, *Theoret. Comput. Sci.* (Math Games) **230**, 207–219.

286. R. J. Evans [1974], A winning opening in reverse Hex, *J. Recr. Math.* **7**, 189–192.

287. R. J. Evans [1975–76], Some variants of Hex, *J. Recr. Math.* **8**, 120–122.

288. R. J. Evans [1979], Silverman's game on intervals, *Amer. Math. Monthly* **86**, 277–281.

289. R. J. Evans and G. A. Heuer [1992], Silverman's game on discrete sets, *Linear Algebra Appl.* **166**, 217–235.

290. S. Even and R. E. Tarjan [1976], A combinatorial problem which is complete in polynomial space, *J. Assoc. Comput. Mach.* **23**, 710–719, also appeared in Proc. 7th Ann. ACM Symp. Theory of Computing (Albuquerque, NM, 1975), Assoc. Comput. Mach., New York, NY, 1975, pp. 66–71.

291. G. Exoo [1980-81], A new way to play Ramsey games, *J. Recr. Math.* **13**(2), 111–113.

292. U. Faigle, W. Kern, H. Kierstead and W. T. Trotter [1993], On the game chromatic number of some classes of graphs, *Ars Combin.* **35**, 143–150.

293. U. Faigle, W. Kern and J. Kuipers [1998], Computing the nucleolus of min-cost spanning tree games is NP-hard, *Internat. J. Game Theory* **27**, 443–450.

294. E. Falkener [1961], *Games Ancient and Modern*, Dover, New York, NY. (Published previously by Longmans Green, 1897.).

295. B.-J. Falkowski and L. Schmitz [1986], A note on the queens' problem, *Inform. Process. Lett.* **23**, 39–46.

296. T. Feder [1990], Toetjes, *Amer. Math. Monthly* **97**, 785–794.

297. T. S. Ferguson [1974], On sums of graph games with last player losing, *Internat. J. Game Theory* **3**, 159–167.

298. T. S. Ferguson [1984], Misère annihilation games, *J. Combin. Theory* (Ser. A) **37**, 205–230.

299. T. S. Ferguson [1989], Who solved the secretary problem?, *Statistical Science* **4**, 282–296.

300. T. S. Ferguson [1992], Mate with bishop and knight in kriegspiel, *Theoret. Comput. Sci.* (Math Games) **96**, 389–403.

301. T. S. Ferguson [1998], Some chip transfer games, *Theoret. Comp. Sci.* (Math Games) **191**, 157–171.

302. T. S. Ferguson [2001], Another form of matrix Nim, *Electronic J. Combin.* **8(2)**, #R9, 9pp., Volume in honor of Aviezri S. Fraenkel. http://www.combinatorics.org/

303. A. S. Finbow and B. L. Hartnell [1983], A game related to covering by stars, *Ars Combinatoria* **16-A**, 189–198.

304. M. J. Fischer and R. N. Wright [1990], An application of game-theoretic techniques to cryptography, *Advances in Computational Complexity Theory* (New Brunswick, NJ, 1990), DIMACS Ser. Discrete Math. Theoret. Comput. Sci.,, Vol. 13, pp. 99–118.

305. P. C. Fishburn and N. J. A. Sloane [1989], The solution to Berlekamp's switching game, *Discrete Math.* **74**, 263–290.

306. D. C. Fisher and J. Ryan [1992], Optimal strategies for a generalized "scissors, paper, and stone" game, *Amer. Math. Monthly* **99**, 935–942.

307. D. C. Fisher and J. Ryan [1995], Probabilities within optimal strategies for tournament games, *Discrete Appl. Math.* **56**, 87–91.

308. D. C. Fisher and J. Ryan [1995], Tournament games and positive tournaments, *J. Graph Theory* **19**, 217–236.

309. G. W. Flake and E. B. Baum [2002], *Rush Hour* is PSPACE-complete, or "Why you should generously tip parking lot attendants", *Theoret. Comput. Sci.* (Math Games) **270**, 895–911.

310. A. Flammenkamp [1996], Lange Perioden in Subtraktions-Spielen, Ph.D. Thesis, University of Bielefeld.

311. J. A. Flanigan [1978], Generalized two-pile Fibonacci nim, *Fibonacci Quart.* **16**, 459–469.

312. J. A. Flanigan [1981], On the distribution of winning moves in random game trees, *Bull. Austr. Math. Soc.* **24**, 227–237.

313. J. A. Flanigan [1981], Selective sums of loopy partizan graph games, *Internat. J. Game Theory* **10**, 1–10.

314. J. A. Flanigan [1982], A complete analysis of black-white Hackendot, *Internat. J. Game Theory* **11**, 21–25.

315. J. A. Flanigan [1982], One-pile time and size dependent take-away games, *Fibonacci Quart.* **20**, 51–59.

316. J. A. Flanigan [1983], Slow joins of loopy games, *J. Combin. Theory* (Ser. A) **34**, 46–59.

317. J. O. Flynn [1973], Lion and man: the boundary constraint, *SIAM J. Control* **11**, 397–411.

318. J. O. Flynn [1974], Lion and man: the general case, *SIAM J. Control* **12**, 581–597.

319. J. O. Flynn [1974], Some results on max-min pursuit, *SIAM J. Control* **12**, 53–69.

320. F. V. Fomin [1998], Helicopter search problems, bandwidth and pathwidth, *Discrete Appl. Math.* **85**, 59–70.

321. F. V. Fomin [1999], Note on a helicopter search problem on graphs, *Discrete Appl. Math.* **95**, 241–249, Proc. Conf. on Optimal Discrete Structures and Algorithms — ODSA '97 (Rostock).

322. F. V. Fomin and N. N. Petrov [1996], Pursuit-evasion and search problems on graphs, *Congr. Numer.* **122**, 47–58, Proc. 27-th Southeastern Intern. Conf. on Combinatorics, Graph Theory and Computing (Baton Rouge, LA, 1996).

323. L. R. Foulds and D. G. Johnson [1984], An application of graph theory and integer programming: chessboard non-attacking puzzles, *Math. Mag.* **57**, 95–104.

324. A. S. Fraenkel [1974], Combinatorial games with an annihilation rule, in: *The Influence of Computing on Mathematical Research and Education*, Missoula MT, August 1973, Proc. Symp. Appl. Math., (J. P. LaSalle, ed.), Vol. 20, Amer. Math. Soc., Providence, RI, pp. 87–91.

325. A. S. Fraenkel [1977], The particles and antiparticles game, *Comput. Math. Appl.* **3**, 327–328.

326. A. S. Fraenkel [1980], From Nim to Go, *Ann. Discrete Math.* **6**, 137–156, Proc. Symp. on Combinatorial Mathematics, Combinatorial Designs and Their Applications (J. Srivastava, ed.), Colorado State Univ., Fort Collins, CO, June 1978.

327. A. S. Fraenkel [1981], Planar kernel and Grundy with $d \leq 3$, $d_{out} \leq 2$, $d_{in} \leq 2$ are NP-complete, *Discrete Appl. Math.* **3**, 257–262.

328. A. S. Fraenkel [1982], How to beat your Wythoff games' opponent on three fronts, *Amer. Math. Monthly* **89**, 353–361.

329. A. S. Fraenkel [1983], 15 Research problems on games, *Discrete Math.* in "Research Problems" section, Vols. **43-46**.

330. A. S. Fraenkel [1984], Wythoff games, continued fractions, cedar trees and Fibonacci searches, *Theoret. Comput. Sci.* **29**, 49–73, an earlier version appeared in Proc. 10th Internat. Colloq. on Automata, Languages and Programming (J. Diaz, ed.), Vol. 154, Barcelona, July 1983, Lecture Notes in Computer Science, Springer Verlag, Berlin, 1983, pp. 203–225,.

331. A. S. Fraenkel [1991], Complexity of games, in: *Combinatorial Games*, Proc. Symp. Appl. Math. (R. K. Guy, ed.), Vol. 43, Amer. Math. Soc., Providence, RI, pp. 111–153.

332. A. S. Fraenkel [1994], Even kernels, *Electronic J. Combinatorics* **1**, #R5, 13pp.
http://www.combinatorics.org/

333. A. S. Fraenkel [1994], Recreation and depth in combinatorial games, in: *The Lighter Side of Mathematics,* Proc. E. Strens Memorial Conf. on Recr. Math. and its History, Calgary, 1986, Spectrum Series (R. K. Guy and R. E. Woodrow, eds.), Math. Assoc. of America, Washington, DC, pp. 176–194.

334. A. S. Fraenkel [1996], Error-correcting codes derived from combinatorial games, in: *Games of No Chance,* Proc. MSRI Workshop on Combinatorial Games, July, 1994, Berkeley, CA, MSRI Publ. (R. J. Nowakowski, ed.), Vol. 29, Cambridge University Press, Cambridge, pp. 417–431.

335. A. S. Fraenkel [1996], Scenic trails ascending from sea-level Nim to alpine chess, in: *Games of No Chance,* Proc. MSRI Workshop on Combinatorial Games, July, 1994, Berkeley, CA, MSRI Publ. (R. J. Nowakowski, ed.), Vol. 29, Cambridge University Press, Cambridge, pp. 13–42.

336. A. S. Fraenkel [1997], Combinatorial game theory foundations applied to digraph kernels, *Electronic J. Combinatorics* **4**(2), #R10, 17pp., Wilf Festschrift.
http://www.combinatorics.org/

337. A. S. Fraenkel [1998], Heap games, numeration systems and sequences, *Annals of Combinatorics* **2**, 197–210, an earlier version appeared in: *Fun With Algorithms,* Vol. 4 of *Proceedings in Informatics* (E. Lodi, L. Pagli and N. Santoro, eds.), Carleton Scientific, University of Waterloo, Waterloo, Ont., pp. 99–113, 1999. Conference took place on the island of Elba, June 1998.

338. A. S. Fraenkel [1998], Multivision: an intractable impartial game with a linear winning strategy, *Amer. Math. Monthly* **105**, 923–928.

339. A. S. Fraenkel [2000], Recent results and questions in combinatorial game complexities, *Theoret. Comput. Sci.* **249**, 265–288, Conference version in: Proc. AWOCA98 — Ninth Australasian Workshop on Combinatorial Algorithms, C.S. Iliopoulos, ed., Perth, Western Australia, 27–30 July, 1998, special AWOCA98 issue, pp. 124-146.

340. A. S. Fraenkel [2000], Arrays, Numeration systems and Frankenstein games, *Theoret. Comput. Sci.* special "Fun With Algorithms" issue, to appear.

341. A. S. Fraenkel [2001], Virus versus mankind, *Proc. 2nd Intern. Conference on Computers and Games CG'2000* (T. Marsland and I. Frank, eds.), Vol. 2063, Hamamatsu, Japan, Oct. 2000, Lecture Notes in Computer Science, Springer, pp. 204–213.

342. A. S. Fraenkel [2002], Mathematical chats between two physicists, in: *Puzzler's Tribute: a Feast for the Mind,* honoring Martin Gardner (D. Wolfe and T. Rodgers, eds.), A K Peters, Natick, MA, pp. 383-386.

343. A. S. Fraenkel [2002], Two-player games on cellular automata, in: *More Games of No Chance,* Proc. MSRI Workshop on Combinatorial Games, July, 2000, Berkeley, CA, MSRI Publ. (R. J. Nowakowski, ed.), Vol. 42, Cambridge University Press, Cambridge, pp. 279–305.

344. A. S. Fraenkel and I. Borosh [1973], A generalization of Wythoff's game, *J. Combin. Theory* (Ser. A) **15**, 175–191.

345. A. S. Fraenkel, M. R. Garey, D. S. Johnson, T. Schaefer and Y. Yesha [1978], The complexity of checkers on an $n \times n$ board — preliminary report, *Proc. 19th Ann. Symp. Foundations of Computer Science* (Ann Arbor, MI, Oct. 1978), IEEE Computer Soc., Long Beach, CA, pp. 55–64.

346. A. S. Fraenkel and E. Goldschmidt [1987], Pspace-hardness of some combinatorial games, *J. Combin. Theory* (Ser. A) **46**, 21–38.

347. A. S. Fraenkel and F. Harary [1989], Geodetic contraction games on graphs, *Internat. J. Game Theory* **18**, 327–338.

348. A. S. Fraenkel and H. Herda [1980], Never rush to be first in playing Nimbi, *Math. Mag.* **53**, 21–26.

349. A. S. Fraenkel, A. Jaffray, A. Kotzig and G. Sabidussi [1995], Modular Nim, *Theoret. Comput. Sci.* (Math Games) **143**, 319–333.

350. A. S. Fraenkel and A. Kotzig [1987], Partizan octal games: partizan subtraction games, *Internat. J. Game Theory* **16**, 145–154.

351. A. S. Fraenkel and D. Lichtenstein [1981], Computing a perfect strategy for $n \times n$ chess requires time exponential in n, *J. Combin. Theory* (Ser. A) **31**, 199–214, preliminary version in Proc. 8th Internat. Colloq. Automata, Languages and Programming (S. Even and O. Kariv, eds.), Vol. 115, Acre, Israel, 1981, Lecture Notes in Computer Science, Springer Verlag, Berlin, pp. 278–293.

352. A. S. Fraenkel, M. Loebl and J. Nešetřil [1988], Epidemiography II. Games with a dozing yet winning player, *J. Combin. Theory* (Ser. A) **49**, 129–144.

353. A. S. Fraenkel and M. Lorberbom [1989], Epidemiography with various growth functions, *Discrete Appl. Math.* **25**, 53–71, special issue on Combinatorics and Complexity.

354. A. S. Fraenkel and M. Lorberbom [1991], Nimhoff games, *J. Combin. Theory* (Ser. A) **58**, 1–25.

355. A. S. Fraenkel and J. Nešetřil [1985], Epidemiography, *Pacific J. Math.* **118**, 369–381.

356. A. S. Fraenkel and M. Ozery [1998], Adjoining to Wythoff's game its *P*-positions as moves, *Theoret. Comput. Sci.* **205**, 283–296.

357. A. S. Fraenkel and Y. Perl [1975], Constructions in combinatorial games with cycles, *Coll. Math. Soc. János Bolyai* **10**, 667–699, Proc. Internat. Colloq. on Infinite and Finite Sets, Vol. 2 (A. Hajnal, R. Rado and V. T. Sós, eds.) Keszthely, Hungary, 1973, North-Holland.

358. A. S. Fraenkel and O. Rahat [2001], Infinite cyclic impartial games, *Theoret. Comput. Sci.* **252**, 13–22, special "Computers and Games" issue; first version appeared in Proc. 1st Intern. Conf. on Computer Games CG'98, Tsukuba, Japan, Nov. 1998, *Lecture Notes in Computer Science*, Vol. 1558, Springer, pp. 212-221, 1999.

359. A. S. Fraenkel and E. R. Scheinerman [1991], A deletion game on hypergraphs, *Discrete Appl. Math.* **30**, 155–162.

360. A. S. Fraenkel, E. R. Scheinerman and D. Ullman [1993], Undirected edge geography, *Theoret. Comput. Sci.* (Math Games) **112**, 371–381.

361. A. S. Fraenkel and S. Simonson [1993], Geography, *Theoret. Comput. Sci.* (Math Games) **110**, 197–214.

362. A. S. Fraenkel and U. Tassa [1975], Strategy for a class of games with dynamic ties, *Comput. Math. Appl.* **1**, 237–254.

363. A. S. Fraenkel and U. Tassa [1982], Strategies for compounds of partizan games, *Math. Proc. Camb. Phil. Soc.* **92**, 193–204.

364. A. S. Fraenkel, U. Tassa and Y. Yesha [1978], Three annihilation games, *Math. Mag.* **51**, 13–17, special issue on Recreational Math.

365. A. S. Fraenkel and Y. Yesha [1976], Theory of annihilation games, *Bull. Amer. Math. Soc.* **82**, 775–777.

366. A. S. Fraenkel and Y. Yesha [1979], Complexity of problems in games, graphs and algebraic equations, *Discrete Appl. Math.* **1**, 15–30.

367. A. S. Fraenkel and Y. Yesha [1982], Theory of annihilation games — I, *J. Combin. Theory* (Ser. B) **33**, 60–86.

368. A. S. Fraenkel and Y. Yesha [1986], The generalized Sprague–Grundy function and its invariance under certain mappings, *J. Combin. Theory* (Ser. A) **43**, 165–177.

369. A. S. Fraenkel and D. Zusman [2001], A new heap game, *Theoret. Comput. Sci.* **252**, 5–12, special "Computers and Games" issue; first version appeared in Proc. 1st Intern. Conf. on Computer Games CG'98, Tsukuba, Japan, Nov. 1998, *Lecture Notes in Computer Science*, Vol. 1558, Springer, pp. 205-211, 1999.

370. C. N. Frangakis [1981], A backtracking algorithm to generate all kernels of a directed graph, *Intern. J. Comput. Math.* **10**, 35–41.

371. P. Frankl [1987], Cops and robbers in graphs with large girth and Cayley graphs, *Discrete Appl. Math.* **17**, 301–305.

372. P. Frankl [1987], On a pursuit game on Cayley graphs, *Combinatorica* **7**, 67–70.

373. D. Fremlin [1973], Well-founded games, *Eureka* **36**, 33–37.

374. G. H. Fricke, S. M. Hedetniemi, S. T. Hedetniemi, A. A. McRae, C. K. Wallis, M. S. Jacobson, H. W. Martin and W. D. Weakley [1995], Combinatorial problems on chessboards: a brief survey, in: *Graph Theory, Combinatorics,*

and Applications: Proc. 7th Quadrennial Internat. Conf. on the Theory and Applications of Graphs (Y. Alavi and A. Schwenk, eds.), Vol. 1, Wiley, pp. 507–528.

375. W. W. Funkenbusch [1971], SIM as a game of chance, *J. Recr. Math.* 4(4), 297–298.

376. Z. Füredi and Á. Seress [1994], Maximal triangle-free graphs with restrictions on the degrees, *J. Graph Theory* 18, 11–24.

377. H. N. Gabow and H. H. Westermann [1992], Forests, frames, and games: algorithms for matroid sums and applications, *Algorithmica* 7, 465–497.

378. D. Gale [1974], A curious Nim-type game, *Amer. Math. Monthly* 81, 876–879.

379. D. Gale [1979], The game of Hex and the Brouwer fixed-point theorem, *Amer. Math. Monthly* 86, 818–827.

380. D. Gale [1986], Problem 1237 (line-drawing game), *Math. Mag.* 59, 111, solution by J. Hutchinson and S. Wagon, *ibid.* 60 (1987) 116.

381. D. Gale and A. Neyman [1982], Nim-type games, *Internat. J. Game Theory* 11, 17–20.

382. D. Gale and F. M. Stewart [1953], Infinite games with perfect information, *Ann. of Math. Stud.* (Contributions to the Theory of Games), Princeton 2(28), 245–266.

383. H. Galeana-Sánchez [1982], A counterexample to a conjecture of Meyniel on kernel-perfect graphs, *Discrete Math.* 41, 105–107.

384. H. Galeana-Sánchez [1986], A theorem about a conjecture of Meyniel on kernel-perfect graphs, *Discrete Math.* 59, 35–41.

385. H. Galeana-Sánchez [1995], B_1 and B_2-orientable graphs in kernel theory, *Discrete Math.* 143, 269–274.

386. H. Galeana-Sánchez [2000], Semikernels modulo F and kernels in digraphs, *Discrete Math.* 218, 61–71.

387. H. Galeana-Sánchez and V. Neuman-Lara [1984], On kernels and semikernels of digraphs, *Discrete Math.* 48, 67–76.

388. H. Galeana-Sánchez and V. Neuman-Lara [1994], New extensions of kernel perfect digraphs to kernel imperfect critical digraphs, *Graphs Combin.* 10, 329–336.

389. F. Galvin [1978], Indeterminacy of point-open games, *Bull. de l'Academie Polonaise des Sciences* (Math. astr. et phys.) 26, 445–449.

390. F. Galvin [1985], Stationary strategies in topological games, *Proc. Conf. on Infinitistic Mathematics* (Lyon, 1984), Publ. Dép. Math. Nouvelle Sér. B, 85-2, Univ. Claude-Bernard, Lyon, pp. 41–43.

391. F. Galvin [1990], Hypergraph games and the chromatic number, in: *A Tribute to Paul Erdős*, Cambridge Univ Press, Cambridge, pp. 201–206.

392. F. Galvin, T. Jech and M. Magidor [1978], An ideal game, *J. Symbolic Logic* **43**, 284–292.

393. F. Galvin and M. Scheepers [1992], A Ramseyan theorem and an infinite game, *J. Combin. Theory* (Ser. A) **59**, 125–129.

394. F. Galvin and R. Telgársky [1986], Stationary strategies in topological games, *Topology Appl.* **22**, 51–69.

395. A. Gangolli and T. Plambeck [1989], A note on periodicity in some octal games, *Internat. J. Game Theory* **18**, 311–320.

396. T. E. Gantner [1988], The game of Quatrainment, *Math. Mag.* **61**, 29–34.

397. M. Gardner [1956], *Mathematics, Magic and Mystery*, Dover, New York, NY.

398. M. Gardner [Jan. 1957–Dec. 1981], Mathematical Games, a column in Scientific American.

399. M. Gardner [1959], *Fads and Fallacies in the Name of Science*, Dover, NY.

400. M. Gardner [1959], *Logic Machines and Diagrams*, McGraw-Hill, NY.

401. M. Gardner [1959], *Mathematical Puzzles of Sam Loyd*, Dover, New York, NY.

402. M. Gardner [1960], *More Mathematical Puzzles of Sam Loyd*, Dover, New York, NY.

403. M. Gardner [1966], *More Mathematical Puzzles and Diversions*, Harmondsworth, Middlesex, England (Penguin Books), translated into German: *Mathematische Rätsel und Probleme*, Vieweg, Braunschweig, 1964.

404. M. Gardner, ed. [1967], *536 Puzzles and Curious Problems*, Scribner's, NY, reissue of H. E. Dudeney's *Modern Puzzles* and *Puzzles and Curious Problems*.

405. M. Gardner [1968], *Logic Machines, Diagrams and Boolean Algebra*, Dover, NY.

406. M. Gardner [1970], *Further Mathematical Diversions*, Allen and Unwin, London.

407. M. Gardner [1977], *Mathematical Magic Show*, Knopf, NY.

408. M. Gardner [1978], *Aha! Insight*, Freeman, New York, NY.

409. M. Gardner [1979], *Mathematical Circus*, Knopf, NY.

410. M. Gardner [1981], *Entertaining Science Experiments with Everyday Objects*, Dover, NY.

411. M. Gardner [1981], *Science Fiction Puzzle Tales*, Potter.

412. M. Gardner [1982], *Aha! Gotcha!*, Freeman, New York, NY.

413. M. Gardner [1983], *New Mathematical Diversions from Scientific American*, University of Chicago Press, Chicago, before that appeared in 1971, Simon and Schuster, New York, NY. First appeared in 1966.

414. M. Gardner [1983], *Order and Surprise*, Prometheus Books, Buffalo, NY.

415. M. Gardner [1983], *Wheels, Life and Other Mathematical Amusements*, Freeman, New York, NY.

416. M. Gardner [1984], *Codes, Ciphers and Secret Writing*, Dover, NY.

417. M. Gardner [1984], *Puzzles from Other Worlds*, Random House.

418. M. Gardner [1984], *The Magic Numbers of Dr. Matrix*, Prometheus.

419. M. Gardner [1984], *The Sixth Book of Mathematical Games*, Univ. of Chicago Press. First appeared in 1971, Freeman, New York, NY.

420. M. Gardner [1986], *Knotted Doughnuts and Other Mathematical Entertainments*, Freeman, New York, NY.

421. M. Gardner [1987], *The Second Scientific American Book of Mathematical Puzzles and Diversions*, University of Chicago Press, Chicago. First appeared in 1961, Simon and Schuster, NY.

422. M. Gardner [1988], *Hexaflexagons and Other Mathematical Diversions*, University of Chicago Press, Chicago, 1988. A first version appeared under the title *The Scientific American Book of Mathematical Puzzles and Diversions*, Simon & Schuster, 1959, NY.

423. M. Gardner [1988], *Perplexing Puzzles and Tantalizing Teasers*, Dover, NY.

424. M. Gardner [1988], *Riddles of the Sphinx*, Math. Assoc. of America, Washington, DC.

425. M. Gardner [1988], *Time Travel and Other Mathematical Bewilderments*, Freeman, New York, NY.

426. M. Gardner [1989], *How Not to Test a Psychic*, Prometheus Books, Buffalo, NY.

427. M. Gardner [1989], *Mathematical Carnival*, Math. Assoc. of America, Washington, DC. First appeared in 1975, Knopf, NY.

428. M. Gardner [1990], *The New Ambidextrous Universe*, Freeman, New York, NY. First appeared in 1964, Basic Books, then in 1969, New American Library.

429. M. Gardner [1991], *The Unexpected Hanging and Other Mathematical Diversions*, University of Chicago Press. First appeared in 1969, Simon and Schuster, NY, translated into German: *Logik Unterm Galgen*, Vieweg, Braunschweig, 1980.

430. M. Gardner [1992], *Fractal Music, Hypercards and More*, Freeman, New York, NY.

431. M. Gardner [1992], *On the Wild Side*, Prometheus Books, Buffalo, NY.

432. M. Gardner [1993], *The Book of Visual Illusions*, Dover, NY.

433. M. Gardner [1997], *Penrose Tiles to Trapdoor Ciphers*, The Math. Assoc. of America, Washington, DC. First appeared in 1989, Freeman, New York, NY. Freeman, New York, NY.

434. M. Gardner [1997], *The Last Recreations*, Copernicus, NY.

435. M. Gardner [2001], *A Gardner's Workout: Training the Mind and Entertaining the Spirit*, A K Peters, Natick, MA, in press.

436. M. R. Garey and D. S. Johnson [1979], *Computers and Intractability: A Guide to the Theory of NP-Completeness*, Freeman, San Francisco, Appendix A8: Games and Puzzles, pp. 254-258.

437. R. Gasser [1996], Solving nine men's Morris, in: *Games of No Chance*, Proc. MSRI Workshop on Combinatorial Games, July, 1994, Berkeley, CA, MSRI Publ. (R. J. Nowakowski, ed.), Vol. 29, Cambridge University Press, Cambridge, pp. 101-114.

438. B. Gerla [2000], Conditioning a state by Łukasiewicz event: a probabilistic approach to Ulam Games, *Theoret. Comput. Sci.* (Math Games) **230**, 149-166.

439. J. R. Gilbert, T. Lengauer and R. E. Tarjan [1980], The pebbling problem is complete in polynomial space, *SIAM J. Comput.* **9**, 513-524, preliminary version in Proc. 11th Ann. ACM Symp. Theory of Computing (Atlanta, GA, 1979), Assoc. Comput. Mach., New York, NY, pp. 237-248.

440. M. Ginsberg [2002], Alphabeta pruning under partial orders, in: *More Games of No Chance*, Proc. MSRI Workshop on Combinatorial Games, July, 2000, Berkeley, CA, MSRI Publ. (R. J. Nowakowski, ed.), Vol. 42, Cambridge University Press, Cambridge, pp. 37-48.

441. J. Ginsburg [1939], Gauss's arithmetization of the problem of 8 queens, *Scripta Math.* **5**, 63-66.

442. A. S. Goldstein and E. M. Reingold [1995], The complexity of pursuit on a graph, *Theoret. Comput. Sci.* (Math Games) **143**, 93-112.

443. E. Goles [1991], Sand piles, combinatorial games and cellular automata, *Math. Appl.* **64**, 101-121.

444. E. Goles and M. A. Kiwi [1993], Games on line graphs and sand piles, *Theoret. Comput. Sci.* (Math Games) **115**, 321-349 (0-player game).

445. E. Goles and M. Margenstern [1997], Universality of the chip-firing game, *Theoret. Comput. Sci.* (Math Games) **172**, 121-134.

446. S. W. Golomb [1966], A mathematical investigation of games of "take-away", *J. Combin. Theory* **1**, 443-458.

447. S. W. Golomb [1994], *Polyominoes: Puzzles, Patterns, Problems, and Packings*, Princeton University Press. Original edition: *Polyominoes*, Scribner's, NY, 1965; Allen and Unwin, London, 1965.

448. D. M. Gordon, R. W. Robinson and F. Harary [1994], Minimum degree games for graphs, *Discrete Math.* **128**, 151-163.

449. E. Grädel [1990], Domino games and complexity, *SIAM J. Comput.* **19**, 787-804.

450. S. B. Grantham [1985], Galvin's "racing pawns" game and a well-ordering of trees, *Memoirs Amer. Math. Soc.* **53**(316), 63 pp.

451. R. Greenlaw, H. J. Hoover and W. L. Ruzzo [1995], *Limits to Parallel Computation: P-completeness Theory*, Oxford University Press, New York.

452. J. P. Grossman and R. J. Nowakowski [2002], One-dimensional Phutball, in: *More Games of No Chance,* Proc. MSRI Workshop on Combinatorial Games, July, 2000, Berkeley, CA, MSRI Publ. (R. J. Nowakowski, ed.), Vol. 42, Cambridge University Press, Cambridge, pp. 361–377.

453. P. M. Grundy [1964], Mathematics and Games, *Eureka* **27**, 9–11, originally published: *ibid.* **2** (1939), 6–8.

454. P. M. Grundy, R. S. Scorer and C. A. B. Smith [1944], Some binary games, *Math. Gaz.* **28**, 96–103.

455. P. M. Grundy and C. A. B. Smith [1956], Disjunctive games with the last player losing, *Proc. Camb. Phil. Soc.* **52**, 527–533.

456. F. Grunfeld and R. C. Bell [1975], *Games of the World*, Holt, Rinehart and Winston.

457. C. D. Grupp [1976], *Brettspiele-Denkspiele*, Humboldt-Taschenbuchverlag, München.

458. D. J. Guan and X. Zhu [1999], Game chromatic number of outerplanar graphs, *J. Graph Theory* **30**, 67–70.

459. L. J. Guibas and A. M. Odlyzko [1981], String overlaps, pattern matching, and nontransitive games, *J. Combin. Theory* (Ser. A) **30**, 183–208.

460. S. Gunther [1874], Zur mathematischen Theorie des Schachbretts, *Arch. Math. Physik* **56**, 281–292.

461. R. K. Guy [1976], Packing $[1, n]$ with solutions of $ax + by = cz$; the unity of combinatorics, *Atti Conv. Lincei #17, Acad. Naz. Lincei, Tomo II*, Rome, pp. 173–179.

462. R. K. Guy [1976], Twenty questions concerning Conway's sylver coinage, *Amer. Math. Monthly* **83**, 634–637.

463. R. K. Guy [1977], Games are graphs, indeed they are trees, *Proc. 2nd Carib. Conf. Combin. and Comput.*, Letchworth Press, Barbados, pp. 6–18.

464. R. K. Guy [1977], She loves me, she loves me not; relatives of two games of Lenstra, Een Pak met een Korte Broek (papers presented to H. W. Lenstra), Mathematisch Centrum, Amsterdam.

465. R. K. Guy [1978], Partizan and impartial combinatorial games, *Colloq. Math. Soc. János Bolyai* **18**, 437–461, Proc. 5th Hungar. Conf. Combin. Vol. I (A. Hajnal and V. T. Sós, eds.), Keszthely, Hungary, 1976, North-Holland.

466. R. K. Guy [1979], Partizan Games, *Colloques Internationaux C. N. R. No. 260 — Problèmes Combinatoires et Théorie des Graphes*, pp. 199–205.

467. R. K. Guy [1981], Anyone for twopins?, in: *The Mathematical Gardner* (D. A. Klarner, ed.), Wadsworth Internat., Belmont, CA, pp. 2–15.

468. R. K. Guy [1983], Graphs and games, in: *Selected Topics in Graph Theory* (L. W. Beineke and R. J. Wilson, eds.), Vol. 2, Academic Press, London, pp. 269–295.

469. R. K. Guy [1986], John Isbell's game of beanstalk and John Conway's game of beans-don't-talk, *Math. Mag.* **59**, 259–269.

470. R. K. Guy [1989], *Fair Game*, COMAP Math. Exploration Series, Arlington, MA.

471. R. K. Guy [1990], A guessing game of Bill Sands, and Bernardo Recamán's Barranca, *Amer. Math. Monthly* **97**, 314–315.

472. R. K. Guy [1991], Mathematics from fun & fun from mathematics; an informal autobiographical history of combinatorial games, in: *Paul Halmos: Celebrating 50 Years of Mathematics* (J. H. Ewing and F. W. Gehring, eds.), Springer Verlag, New York, NY, pp. 287–295.

473. R. K. Guy [1995], Combinatorial games, in: *Handbook of Combinatorics*, (R. L. Graham, M. Grötschel and L. Lovász, eds.), Vol. II, North-Holland, Amsterdam, pp. 2117–2162.

474. R. K. Guy [1996], Impartial Games, in: *Games of No Chance*, Proc. MSRI Workshop on Combinatorial Games, July, 1994, Berkeley, CA, MSRI Publ. (R. J. Nowakowski, ed.), Vol. 29, Cambridge University Press, Cambridge, pp. 61–78, earlier version in: *Combinatorial Games*, Proc. Symp. Appl. Math. (R. K. Guy, ed.), Vol. 43, Amer. Math. Soc., Providence, RI, 1991, pp. 35–55.

475. R. K. Guy [1996], What is a game?, in: *Games of No Chance*, Proc. MSRI Workshop on Combinatorial Games, July, 1994, Berkeley, CA, MSRI Publ. (R. J. Nowakowski, ed.), Vol. 29, Cambridge University Press, Cambridge, pp. 43–60, earlier version in: *Combinatorial Games*, Proc. Symp. Appl. Math. (R. K. Guy, ed.), Vol. 43, Amer. Math. Soc., Providence, RI, 1991, pp. 1–21.

476. R. K. Guy [1996], Unsolved problems in combinatorial games, in: *Games of No Chance*, Proc. MSRI Workshop on Combinatorial Games, July, 1994, Berkeley, CA, MSRI Publ. (R. J. Nowakowski, ed.), Vol. 29, Cambridge University Press, Cambridge, pp. 475–491, update with 52 problems of earlier version with 37 problems, in: *Combinatorial Games*, Proc. Symp. Appl. Math. (R. K. Guy, ed.), Vol. 43, Amer. Math. Soc., Providence, RI, 1991, pp. 183–189.

477. R. K. Guy and R. J. Nowakowski [2002], Unsolved problems in combinatorial games, in: *More Games of No Chance*, Proc. MSRI Workshop on Combinatorial Games, July, 2000, Berkeley, CA, MSRI Publ. (R. J. Nowakowski, ed.), Vol. 42, Cambridge University Press, Cambridge, pp. 455–473.

478. R. K. Guy and C. A. B. Smith [1956], The *G*-values of various games, *Proc. Camb. Phil. Soc.* **52**, 514–526.

479. R. K. Guy and R. E. Woodrow, eds. [1994], *The Lighter Side of Mathematics,* Proc. E. Strens Memorial Conf. on Recr. Math. and its History, Calgary, 1986, Spectrum Series, Math. Assoc. Amer., Washington, DC.

480. W. Guzicki [1990], Ulam's searching game with two lies, *J. Combin. Theory* (Ser. A) **54**, 1–19.

481. A. Hajnal and Z. Nagy [1984], Ramsey games, *Trans. American Math. Soc.* **284**, 815–827.

482. D. R. Hale [1983], A variant of Nim and a function defined by Fibonacci representation, *Fibonacci Quart.* **21**, 139–142.

483. A. W. Hales and R. I. Jewett [1963], Regularity and positional games, *Trans. Amer. Math. Soc.* **106**, 222–229.

484. L. Halpenny and C. Smyth [1992], A classification of minimal standard-path 2×2 switching networks, *Theoret. Comput. Sci.* (Math Games) **102**, 329–354.

485. Y. O. Hamidoune [1987], On a pursuit game of Cayley digraphs, *Europ. J. Combin.* **8**, 289–295.

486. Y. O. Hamidoune and M. Las Vergnas [1985], The directed Shannon switching game and the one-way game, in: *Graph Theory and Its Applications to Algorithms and Computer Science* (Y. Alavi et al., eds.), Wiley, pp. 391–400.

487. Y. O. Hamidoune and M. Las Vergnas [1986], Directed switching games on graphs and matroids, *J. Combin. Theory* (Ser. B) **40**, 237–269.

488. Y. O. Hamidoune and M. Las Vergnas [1987], A solution to the box game, *Discrete Math.* **65**, 157–171.

489. Y. O. Hamidoune and M. Las Vergnas [1988], A solution to the misère Shannon switching game, *Discrete Math.* **72**, 163–166.

490. O. Hanner [1959], Mean play of sums of positional games, *Pacific J. Math.* **9**, 81–99.

491. F. Harary [1982], Achievement and avoidance games for graphs, *Ann. Discrete Math.* **13**, 111–120.

492. F. Harary [2002], Sum-free games, in: *Puzzler's Tribute: a Feast for the Mind,* pp. 395–398, honoring Martin Gardner (D. Wolfe and T. Rodgers, eds.), A K Peters, Natick, MA.

493. F. Harary and K. Plochinski [1987], On degree achievement and avoidance games for graphs, *Math. Mag.* **60**, 316–321.

494. F. Harary, W. Slany and O. Verbitsky [2002], A symmetric strategy in graph avoidance games, in: *More Games of No Chance,* Proc. MSRI Workshop on Combinatorial Games, July, 2000, Berkeley, CA, MSRI Publ. (R. J. Nowakowski, ed.), Vol. 42, Cambridge University Press, Cambridge, pp. 369–381.

495. H. Harborth and M. Seemann [1996], Snaky is an edge-to-edge loser, *Geombinatorics* **5**, 132–136.

496. H. Harborth and M. Seemann [1997], Snaky is a paving winner, *Bull. Inst. Combin. Appl.* **19**, 71–78.

497. P. J. Hayes [1977], A note on the towers of Hanoi problem, *Computer J.* **20**, 282–285.

498. O. Heden [1992], On the modular n-queen problem, *Discrete Math.* **102**, 155–161.

499. O. Heden [1993], Maximal partial spreads and the modular n-queen problem, *Discrete Math.* **120**, 75–91.

500. O. Heden [1995], Maximal partial spreads and the modular n-queen problem II, *Discrete Math.* **142**, 97–106.

501. P. Hein [1942], Polygon, *Politiken* (description of Hex in this Danish newspaper of Dec. 26) .

502. D. Hensley [1988], A winning strategy at Taxman, *Fibonacci Quart.* **26**, 262–270.

503. C. W. Henson [1970], Winning strategies for the ideal game, *Amer. Math. Monthly* **77**, 836–840.

504. R. I. Hess [1999], Puzzles from around the world, in: *The Mathemagician and Pied Puzzler*, honoring Martin Gardner; E. Berlekamp and T. Rodgers, eds., A K Peters, Natick, MA, pp. 53-84.

505. G. A. Heuer [1982], Odds versus evens in Silverman-type games, *Internat. J. Game Theory* **11**, 183–194.

506. G. A. Heuer [1989], Reduction of Silverman-like games to games on bounded sets, *Internat. J. Game Theory* **18**, 31–36.

507. G. A. Heuer [2001], Three-part partition games on rectangles, *Theoret. Comput. Sci.* (Math Games) **259**, 639–661.

508. G. A. Heuer and U. Leopold-Wildburger [1995], *Silverman's game, A special class of two-person zero-sum games,* with a foreword by Reinhard Selten, Vol. 424 of Lecture Notes in Economics and Mathematical Systems, Springer-Verlag, Berlin.

509. G. A. Heuer and W. D. Rieder [1988], Silverman games on disjoint discrete sets, *SIAM J. Disc. Math.* **1**, 485–525.

510. R. Hill and J. P. Karim [1992], Searching with lies: the Ulam problem, *Discrete Math.* **106/107**, 273–283.

511. T. P. Hill and U. Krengel [1991], Minimax-optimal stop rules and distributions in secretary problems, *Ann. Probab.* **19**, 342–353.

512. T. P. Hill and U. Krengel [1992], On the game of Googol, *Internat. J. Game Theory* **21**, 151–160.

513. P. G. Hinman [1972], Finite termination games with tie, *Israel J. Math.* **12**, 17–22.

514. A. M. Hinz [1989], An iterative algorithm for the tower of Hanoi with four pegs, *Computing* **42**, 133–140.

515. A. M. Hinz [1989], The tower of Hanoi, *Enseign. Math.* **35**, 289–321.

516. A. M. Hinz [1992], Pascal's triangle and the tower of Hanoi, *Amer. Math. Monthly* **99**, 538–544.

517. A. M. Hinz [1992], Shortest paths between regular states of the tower of Hanoi, *Inform. Sci.* **63**, 173–181.

518. S. Hitotumatu [1968], Some remarks on nim-like games, *Comment. Math. Univ. St. Paul* **17**, 85–98.

519. E. J. Hoffman, J. C. Loessi and R. C. Moore [1969], Construction for the solution of the n-queens problem, *Math. Mag.* **42**, 66–72.

520. J. C. Holladay [1957], Cartesian products of termination games, *Ann. of Math. Stud.* (Contributions to the Theory of Games), Princeton **3**(39), 189–200.

521. J. C. Holladay [1958], Matrix nim, *Amer. Math. Monthly* **65**, 107–109.

522. J. C. Holladay [1966], A note on the game of dots, *Amer. Math. Monthly* **73**, 717–720.

523. J. E. Hopcroft, J. T. Schwartz and M. Sharir [1984], On the complexity of motion planning for multiple independent objects: PSPACE-hardness of the "Warehouseman's problem", *J. Robot. Res.* **3**, 76–88.

524. E. Hordern [1986], *Sliding piece puzzles*, Oxford University Press, Oxford.

525. S. Huddleston and J. Shurman [2002], Transfinite Chomp, in: *More Games of No Chance,* Proc. MSRI Workshop on Combinatorial Games, July, 2000, Berkeley, CA, MSRI Publ. (R. J. Nowakowski, ed.), Vol. 42, Cambridge University Press, Cambridge, pp. 183–212.

526. K. Igusa [1985], Solution of the Bulgarian solitaire conjecture, *Math. Mag.* **58**, 259–271.

527. J. Isbell [1992], The Gordon game of a finite group, *Amer. Math. Monthly* **99**, 567–569.

528. O. Itzinger [1977], The South American game, *Math. Mag.* **50**, 17–21.

529. S. Iwata and T. Kasai [1994], The Othello game on an $n \times n$ board is PSPACE-complete, *Theoret. Comput. Sci.* (Math Games) **123**, 329–340.

530. T. A. Jenkyns and J. P. Mayberry [1980], The skeleton of an impartial game and the Nim-function of Moore's Nim_k, *Internat. J. Game Theory* **9**, 51–63.

531. T. R. Jensen and B. Toft [1995], *Graph Coloring Problems*, Wiley-Interscience Series in Discrete Mathematics and Optimization, Wiley, New York, NY.

532. D. S. Johnson [1983], The NP-Completeness Column: An Ongoing Guide, *J. Algorithms* **4**, 397–411 (9th quarterly column (games); column started in 1981).

533. W. W. Johnson [1879], Note on the "15" puzzle, *Amer. J. Math.* **2**, 397–399.

534. J. P. Jones [1982], Some undecidable determined games, *Internat. J. Game Theory* **11**, 63–70.

535. J. P. Jones and A. S. Fraenkel [1995], Complexities of winning strategies in diophantine games, *J. Complexity* **11**, 435–455.

536. M. Kac [1974], Hugo Steinhaus, a reminiscence and a tribute, *Amer. Math. Monthly* **81**, 572–581 (p. 577).

537. J. Kahane and A. S. Fraenkel [1987], k-Welter — a generalization of Welter's game, *J. Combin. Theory* (Ser. A) **46**, 1–20.

538. J. Kahane and A. J. Ryba [2001], The Hexad game, *Electronic J. Combin.* **8(2)**, #R11, 9pp., Volume in honor of Aviezri S. Fraenkel. http://www.combinatorics.org/

539. J. Kahn, J. C. Lagarias and H. S. Witsenhausen [1987], Single-suit two-person card play, *Internat. J. Game Theory* **16**, 291–320.

540. J. Kahn, J. C. Lagarias and H. S. Witsenhausen [1988], Single-suit two-person card play, II. Dominance, *Order* **5**, 45–60.

541. J. Kahn, J. C. Lagarias and H. S. Witsenhausen [1989], On Lasker's card game, in: *Differential Games and Applications,* Lecture Notes in Control and Information Sciences (T. S. Başar and P. Bernhard, eds.), Vol. 119, Springer Verlag, Berlin, pp. 1–8.

542. J. Kahn, J. C. Lagarias and H. S. Witsenhausen [1989], Single-suit two-person card play, III. The misère game, *SIAM J. Disc. Math.* **2**, 329–343.

543. L. Kalmár [1928], Zur Theorie der abstrakten Spiele, *Acta Sci. Math. Univ. Szeged* **4**, 65–85.

544. B. Kalyanasundram [1991], On the power of white pebbles, *Combinatorica* **11**, 157–171.

545. B. Kalyanasundram and G. Schnitger [1988], On the power of white pebbles, *Proc. 20th Ann. ACM Symp. Theory of Computing* (Chicago, IL, 1988), Assoc. Comput. Mach., New York, NY, pp. 258–266.

546. M. Kano [1983], Cycle games and cycle cut games, *Combinatorica* **3**, 201–206.

547. M. Kano [1993], An edge-removing game on a graph (a generalization of Nim and Kayles), in: *Optimal Combinatorial Structures on Discrete Mathematical Models* (in Japanese, Kyoto, 1992), Sūrikaisekikenkyūsho Kōkyūroku, pp. 82–90.

548. M. Kano [1996], Edge-removing games of star type, *Discrete Math.* **151**, 113–119.

549. M. Kano, T. Sasaki, H. Fujita and S. Hoshi [1993], Life games of Ibadai type, in: *Combinatorial Structure in Mathematical Models* (in Japanese, Kyoto, 1993), Sūrikaisekikenkyūsho Kōkyūroku, pp. 108–117.

550. R. M. Karp and Y. Zhang [1989], On parallel evaluation of game trees, *Proc. ACM Symp. Parallel Algorithms and Architectures*, pp. 409–420.

551. T. Kasai, A. Adachi and S. Iwata [1979], Classes of pebble games and complete problems, *SIAM J. Comput.* **8**, 574–586.

552. Y. Kawano [1996], Using similar positions to search game trees, in: *Games of No Chance*, Proc. MSRI Workshop on Combinatorial Games, July, 1994, Berkeley, CA, MSRI Publ. (R. J. Nowakowski, ed.), Vol. 29, Cambridge University Press, Cambridge, pp. 193–202.

553. R. Kaye [2000], Minesweeper is NP-complete, *Math. Intelligencer* **22**, 9–15.

554. J. C. Kenyon [1967], A Nim-like game with period 349, Res. Paper No. 13, Univ. of Calgary, Math. Dept.

555. J. C. Kenyon [1967], Nim-like games and the Sprague–Grundy theory, M.Sc. Thesis, Univ. of Calgary.

556. B. Kerst [1933], *Mathematische Spiele*, Reprinted by Dr. Martin Sändig oHG, Wiesbaden 1968.

557. H. A. Kierstead and W. T. Trotter [1994], Planar graph coloring with an uncooperative partner, *J. Graph Theory* **18**, 569–584.

558. H. A. Kierstead and W. T. Trotter [2001], Competitive colorings of oriented graphs, *Electronic J. Combin.* **8(2)**, #R12, 15pp., Volume in honor of Aviezri S. Fraenkel.
http://www.combinatorics.org/

559. Y. Kim [1996], New values in domineering, *Theoret. Comput. Sci.* (Math Games) **156**, 263–280.

560. H. Kinder [1973], Gewinnstrategien des Anziehenden in einigen Spielen auf Graphen, *Arch. Math.* **24**, 332–336.

561. M. A. Kiwi, R. Ndoundam, M. Tchuente and E. Goles [1994], No polynomial bound for the period of the parallel chip firing game on graphs, *Theoret. Comput. Sci.* **136**, 527–532.

562. D. A. Klarner, ed. [1998], *Mathematical recreations. A collection in honor of Martin Gardner*, Dover, Mineola, NY, Corrected reprint of *The Mathematical Gardner*, Wadsworth Internat., Belmont, CA, 1981.

563. M. M. Klawe [1985], A tight bound for black and white pebbles on the pyramids, *J. Assoc. Comput. Mach.* **32**, 218–228.

564. C. S. Klein and S. Minsker [1993], The super towers of Hanoi problem: large rings on small rings, *Discrete Math.* **114**, 283–295.

565. D. J. Kleitman and B. L. Rothschild [1972], A generalization of Kaplansky's game, *Discrete Math.* **2**, 173–178.

566. T. Kløve [1977], The modular n-queen problem, *Discrete Math.* **19**, 289–291.

567. T. Kløve [1981], The modular n-queen problem II, *Discrete Math.* **36**, 33–48.

568. M. Knor [1996], On Ramsey-type games for graphs, *Australas. J. Combin.* **14**, 199–206.

569. D. E. Knuth [1974], *Surreal Numbers*, Addison-Wesley, Reading, MA.

570. D. E. Knuth [1976], The computer as Master Mind, *J. Recr. Math.* **9**, 1–6.

571. D. E. Knuth [1977], Are toy problems useful?, *Popular Computing* **5**, 3–10.

572. D. E. Knuth [1993], *The Stanford Graph Base*: A Platform for Combinatorial Computing, ACM Press, New York.

573. P. Komjáth [1984], A simple strategy for the Ramsey-game, *Studia Sci. Math. Hung.* **19**, 231–232.

574. A. Kotzig [1946], O *k*-posunutiach (On *k*-translations; in Slovakian), *Časop. pro Pěst. Mat. a Fys.* **71**, 55–61, extended abstract in French, pp. 62–66.

575. G. Kowalewski [1930], *Alte und neue mathematische Spiele*, Reprinted by Dr. Martin Sändig oHG, Wiesbaden 1968.

576. K. Koyama and T. W. Lai [1993], An optimal Mastermind strategy, *J. Recr. Math.* **25**, 251–256.

577. M. Kraitchik [1953], *Mathematical Recreations*, Dover, New York, NY, 2nd edn.

578. B. Kummer [1980], *Spiele auf Graphen*, Internat. Series of Numerical Mathematics, Birkhäuser Verlag, Basel.

579. R. E. Ladner and J. K. Norman [1985], Solitaire automata, *J. Comput. System Sci.* **30**, 116–129.

580. J. C. Lagarias [1977], Discrete balancing games, *Bull. Inst. Math. Acad. Sinica* **5**, 363–373.

581. J. Lagarias and D. Sleator [1999], Who wins misère Hex?, in: *The Mathemagician and Pied Puzzler*, honoring Martin Gardner; E. Berlekamp and T. Rodgers, eds., A K Peters, Natick, MA, pp. 237-240.

582. S. P. Lalley [1988], A one-dimensional infiltration game, *Naval Res. Logistics* **35**, 441–446.

583. T. K. Lam [1997], Connected sprouts, *Amer. Math. Monthly* **104**, 116–119.

584. H. A. Landman [1996], Eyespace values in Go, in: *Games of No Chance*, Proc. MSRI Workshop on Combinatorial Games, July, 1994, Berkeley, CA, MSRI Publ. (R. J. Nowakowski, ed.), Vol. 29, Cambridge University Press, Cambridge, pp. 227–257.

585. L. Larson [1977], A theorem about primes proved on a chessboard, *Math. Mag.* **50**, 69–74.

586. E. Lasker [1931], *Brettspiele der Völker, Rätsel und mathematische Spiele*, Berlin.

587. I. Lavalée [1985], Note sur le problème des tours d'Hanoï, *Rev. Roumaine Math. Pures Appl.* **30**, 433–438.

588. E. L. Lawler and S. Sarkissian [1995], An algorithm for "Ulam's game" and its application to error correcting codes, *Inform. Process. Lett.* **56**, 89–93.

589. A. J. Lazarus, D. E. Loeb, J. G. Propp and D. Ullman [1996], Richman Games, in: *Games of No Chance*, Proc. MSRI Workshop on Combinatorial Games, July, 1994, Berkeley, CA, MSRI Publ. (R. J. Nowakowski, ed.), Vol. 29, Cambridge University Press, Cambridge, pp. 439–449.

590. A. J. Lazarus, D. E. Loeb, J. G. Propp, W. R. Stromquist and D. Ullman [1999], Combinatorial games under auction play, *Games Econom. Behav.* **27**, 229–264.

591. D. B. Leep and G. Myerson [1999], Marriage, magic and solitaire, *Amer. Math. Monthly* **106**, 419–429.

592. A. Lehman [1964], A solution to the Shannon switching game, *SIAM J. Appl. Math.* **12**, 687–725.

593. T. Lengauer and R. Tarjan [1980], The space complexity of pebble games on trees, *Inform. Process. Lett.* **10**, 184–188.

594. T. Lengauer and R. Tarjan [1982], Asymptotically tight bounds on time-space trade-offs in a pebble game, *J. Assoc. Comput. Mach.* **29**, 1087–1130.

595. H. W. Lenstra, Jr. [1977], On the algebraic closure of two, *Proc. Kon. Nederl. Akad. Wetensch.* (Ser. A) **80**, 389–396.

596. H. W. Lenstra, Jr. [1977/1978], Nim multiplication, Séminaire de Théorie des Nombres No. 11, Université de Bordeaux, France.

597. D. N. L. Levy, ed. [1988], *Computer Games* I, II, Springer-Verlag, New York, NY.

598. J. Lewin [1986], The lion and man problem revisited, *J. Optimization Theory and Applications* **49**, 411–430.

599. T. Lewis and S. Willard [1980], The rotating table, *Math. Mag.* **53**, 174–179.

600. S.-Y. R. Li [1974], Generalized impartial games, *Internat. J. Game Theory* **3**, 169–184.

601. S.-Y. R. Li [1976], Sums of Zuchswang games, *J. Combin. Theory* (Ser. A) **21**, 52–67.

602. S.-Y. R. Li [1977], *N*-person nim and *N*-person Moore's games, *Internat. J. Game Theory* **7**, 31–36.

603. D. Lichtenstein [1982], Planar formulae and their uses, *SIAM J. Comput.* **11**, 329–343.

604. D. Lichtenstein and M. Sipser [1980], Go is Polynomial-Space hard, *J. Assoc. Comput. Mach.* **27**, 393–401, earlier version appeared in Proc. 19th Ann. Symp. Foundations of Computer Science (Ann Arbor, MI, Oct. 1978), IEEE Computer Soc., Long Beach, CA, 1978, pp. 48–54.

605. H. Liebeck [1971], Some generalizations of the 14-15 puzzle, *Math. Mag.* **44**, 185–189.

606. D. E. Loeb [1995], How to win at Nim, *UMAP J.* **16**, 367–388.

607. D. E. Loeb [1996], Stable winning coalitions, in: *Games of No Chance*, Proc. MSRI Workshop on Combinatorial Games, July, 1994, Berkeley, CA,

MSRI Publ. (R. J. Nowakowski, ed.), Vol. 29, Cambridge University Press, Cambridge, pp. 451–471.

608. A. M. Lopez, Jr. [1991], A prolog Mastermind program, *J. Recr. Math.* **23**, 81–93.

609. C. M. López [1997], Chip firing and the Tutte polynomial, *Annals of Combinatorics* **1**, 253–259.

610. S. Loyd [1914], *Cyclopedia of Puzzles and Tricks*, Franklin Bigelow Corporation, Morningside Press, NY, reissued and edited by M. Gardner under the name *The Mathematical Puzzles of Sam Loyd* (two volumes), Dover, New York, NY, 1959.

611. X. Lu [1991], A matching game, *Discrete Math.* **94**, 199–207.

612. X. Lu [1992], A characterization on n-critical economical generalized tic-tac-toe games, *Discrete Math.* **110**, 197–203.

613. X. Lu [1992], Hamiltonian games, *J. Combin. Theory* (Ser. B) **55**, 18–32.

614. X. Lu [1995], A Hamiltonian game on $K_{n,n}$, *Discrete Math.* **142**, 185–191.

615. X.-M. Lu [1986], Towers of Hanoi graphs, *Intern. J. Comput. Math.* **19**, 23–38.

616. X.-M. Lu [1988], Towers of Hanoi problem with arbitrary $k \geq 3$ pegs, *Intern. J. Comput. Math.* **24**, 39–54.

617. X.-M. Lu [1989], An iterative solution for the 4-peg towers of Hanoi, *Comput. J.* **32**, 187–189.

618. É. Lucas [1960], *Récréations Mathématiques*, Vol. I – IV, A. Blanchard, Paris. Previous editions: Gauthier-Villars, Paris, 1891–1894.

619. É. Lucas [1974], *Introduction aux Récréations Mathématiques: L'Arithmétique Amusante*, reprinted by A. Blanchard, Paris. Originally published by A. Blanchard, Paris, 1895.

620. A. L. Ludington [1988], Length of the 7-number game, *Fibonacci Quart.* **26**, 195–204.

621. A. Ludington-Young [1990], Length of the n-number game, *Fibonacci Quart.* **28**, 259–265.

622. M. Maamoun and H. Meyniel [1987], On a game of policemen and robber, *Discrete Appl. Math.* **17**, 307–309.

623. P. A. MacMahon [1921], *New Mathematical Pastimes*, Cambridge University Press, Cambridge.

624. F. Maffray [1986], On kernels in i-triangulated graphs, *Discrete Math.* **61**, 247–251.

625. F. Maffray [1992], Kernels in perfect line-graphs, *J. Combin. Theory* (Ser. B) **55**, 1–8.

626. R. Mansfield [1996], Strategies for the Shannon switching game, *Amer. Math. Monthly* **103**, 250–252.

627. G. Martin [1991], *Polyominoes: Puzzles and Problems in Tiling*, Math. Assoc. America, Wasington, DC.

628. O. Martín-Sánchez and C. Pareja-Flores [2001], Two reflected analyses of lights out, *Math. Mag.* **74**, 295–304.

629. J. G. Mauldon [1978], Num, a variant of nim with no first player win, *Amer. Math. Monthly* **85**, 575–578.

630. D. P. McIntyre [1942], A new system for playing the game of nim, *Amer. Math. Monthly* **49**, 44–46.

631. R. McConville [1974], *The History of Board Games*, Creative Publications, Palo Alto, CA.

632. R. McNaughton [1993], Infinite games played on finite graphs, *Ann. Pure Appl. Logic* **65**, 149–184.

633. J. W. A. McWorter [1981], Kriegspiel Hex, *Math. Mag.* **54**, 85–86, solution to Problem 1084 posed by the author in Math. Mag. **52** (1979) 317.

634. E. Mead, A. Rosa and C. Huang [1974], The game of SIM: A winning strategy for the second player, *Math. Mag.* **47**, 243–247.

635. N. Megiddo, S. L. Hakimi, M. R. Garey, D. S. Johnson and C. H. Papadimitriou [1988], The complexity of searching a graph, *J. Assoc. Comput. Mach.* **35**, 18–44.

636. K. Mehlhorn, S. Näher and M. Rauch [1990], On the complexity of a game related to the dictionary problem, *SIAM J. Comput.* **19**, 902–906, earlier draft appeared in Proc. 30th Ann. Symp. Foundations of Computer Science, pp. 546–548.

637. N. S. Mendelsohn [1946], A psychological game, *Amer. Math. Monthly* **53**, 86–88.

638. C. G. Méndez [1981], On the law of large numbers, infinite games and category, *Amer. Math. Monthly* **88**, 40–42.

639. D. Mey [1994], Finite games for a predicate logic without contractions, *Theoret. Comput. Sci.* (Math Games) **123**, 341–349.

640. F. Meyer auf der Heide [1981], A comparison of two variations of a pebble game on graphs, *Theoret. Comput. Sci.* **13**, 315–322.

641. H. Meyniel and J.-P. Roudneff [1988], The vertex picking game and a variation of the game of dots and boxes, *Discrete Math.* **70**, 311–313.

642. D. Michie and I. Bratko [1987], Ideas on knowledge synthesis stemming from the KBBKN endgame, *Internat. Comp. Chess Assoc. J.* **10**, 3–10.

643. J. Milnor [1953], Sums of positional games, *Ann. of Math. Stud.* (Contributions to the Theory of Games, H. W. Kuhn and A. W. Tucker, eds.), Princeton **2**(28), 291–301.

644. P. Min Lin [1982], Principal partition of graphs and connectivity games, *J. Franklin Inst.* **314**, 203–210.

645. S. Minsker [1989], The towers of Hanoi rainbow problem: coloring the rings, *J. Algorithms* **10**, 1–19.

646. S. Minsker [1991], The towers of Antwerpen problem, *Inform. Process. Lett.* **38**, 107–111.

647. J. Missigman and R. Weida [2001], An easy solution to mini lights out, *Math. Mag.* **74**, 57–59.

648. D. Moews [1991], Sum of games born on days 2 and 3, *Theoret. Comput. Sci.* (Math Games) **91**, 119–128.

649. D. Moews [1992], Pebbling graphs, *J. Combin. Theory* (Ser. B) **55**, 244–252.

650. D. Moews [1996], Coin-sliding and Go, *Theoret. Comput. Sci.* (Math Games) **164**, 253–276.

651. D. Moews [1996], Infinitesimals and coin-sliding, in: *Games of No Chance,* Proc. MSRI Workshop on Combinatorial Games, July, 1994, Berkeley, CA, MSRI Publ. (R. J. Nowakowski, ed.), Vol. 29, Cambridge University Press, Cambridge, pp. 315–327.

652. D. Moews [1996], Loopy games and Go, in: *Games of No Chance,* Proc. MSRI Workshop on Combinatorial Games, July, 1994, Berkeley, CA, MSRI Publ. (R. J. Nowakowski, ed.), Vol. 29, Cambridge University Press, Cambridge, pp. 259–272.

653. D. Moews [2002], The abstract structure of the group of games, in: *More Games of No Chance,* Proc. MSRI Workshop on Combinatorial Games, July, 2000, Berkeley, CA, MSRI Publ. (R. J. Nowakowski, ed.), Vol. 42, Cambridge University Press, Cambridge, pp. 49–57.

654. C. Moore and D. Eppstein [2002], One-dimensional peg solitaire, and duotaire, in: *More Games of No Chance,* Proc. MSRI Workshop on Combinatorial Games, July, 2000, Berkeley, CA, MSRI Publ. (R. J. Nowakowski, ed.), Vol. 42, Cambridge University Press, Cambridge, pp. 341–350.

655. E. H. Moore [1909–1910], A generalization of the game called nim, *Ann. of Math.* (Ser. 2) **11**, 93–94.

656. A. H. Moorehead and G. Mott-Smith [1963], *Hoyle's Rules of Games,* Signet, New American Library.

657. F. L. Morris [1981], Playing disjunctive sums is polynomial space complete, *Internat. J. Game Theory* **10**, 195–205.

658. M. Morse and G. A. Hedlund [1944], Unending chess, symbolic dynamics and a problem in semigroups, *Duke Math. J.* **11**, 1–7.

659. M. Müller and R. Gasser [1996], Experiments in computer Go endgames, in: *Games of No Chance,* Proc. MSRI Workshop on Combinatorial Games, July, 1994, Berkeley, CA, MSRI Publ. (R. J. Nowakowski, ed.), Vol. 29, Cambridge University Press, Cambridge, pp. 273–284.

660. M. Müller and T. Tegos [2002], Experiments in computer Amazons, in: *More Games of No Chance*, Proc. MSRI Workshop on Combinatorial Games, July, 2000, Berkeley, CA, MSRI Publ. (R. J. Nowakowski, ed.), Vol. 42, Cambridge University Press, Cambridge, pp. 245–262.

661. D. Mundici and T. Trombetta [1997], Optimal comparison strategies in Ulam's searching game with two errors, *Theoret. Comput. Sci.* (Math Games) **182**, 217–232.

662. H. J. R. Murray [1952], *A History of Board Games Other Than Chess*, Oxford University Press.

663. B. Nadel [1990], Representation selection for constraint satisfaction: a case study using n-queens, *IEEE Expert* pp. 16–23.

664. A. Napier [1970], A new game in town, Empire Mag., Denver Post, May 2.

665. A. Negro and M. Sereno [1992], Solution of Ulam's problem on binary search with three lies, *J. Combin. Theory* (Ser. A) **59**, 149–154.

666. J. Nešetřil and E. Sopena [2001], On the oriented game chromatic number, *Electronic J. Combin.* **8(2)**, #R14, 13pp., Volume in honor of Aviezri S. Fraenkel.
http://www.combinatorics.org/

667. J. Nešetřil and R. Thomas [1987], Well quasi ordering, long games and combinatorial study of undecidability, in: *Logic and Combinatorics, Contemp. Math.* (S. G. Simpson, ed.), Vol. 65, Amer. Math. Soc., Providence, RI, pp. 281–293.

668. S. Neufeld and R. J. Nowakowski [1993], A vertex-to-vertex pursuit game played with disjoint sets of edges, in: *Finite and Infinite Combinatorics in Sets and Logic* (N. W. Sauer et al., eds.), Kluwer, Dordrecht, pp. 299–312.

669. S. Neufeld and R. J. Nowakowski [1998], A game of cops and robbers played on products of graphs, *Discrete Math.* **186**, 253–268.

670. I. Niven [1988], Coding theory applied to a problem of Ulam, *Math. Mag.* **61**, 275–281.

671. R. J. Nowakowski [1991], ..., Welter's game, Sylver coinage, dots-and-boxes, ..., in: *Combinatorial Games*, Proc. Symp. Appl. Math. (R. K. Guy, ed.), Vol. 43, Amer. Math. Soc., Providence, RI, pp. 155–182.

672. R. J. Nowakowski [2001], The game of End-Nim, *Electronic J. Combinatorics* **8**, in press.

673. R. J. Nowakowski and D. G. Poole [1996], Geography played on products of directed cycles, in: *Games of No Chance*, Proc. MSRI Workshop on Combinatorial Games, July, 1994, Berkeley, CA, MSRI Publ. (R. J. Nowakowski, ed.), Vol. 29, Cambridge University Press, Cambridge, pp. 329–337.

674. R. Nowakowski and P. Winkler [1983], Vertex-to-vertex pursuit in a graph, *Discrete Math.* **43**, 235–239.

675. S. P. Nudelman [1995], The modular n-queens problem in higher dimensions, *Discrete Math.* **146**, 159–167.

676. T. H. O'Beirne [1984], *Puzzles and Paradoxes*, Dover, New York, NY. (Appeared previously by Oxford University Press, London, 1965.).

677. G. L. O'Brian [1978-79], The graph of positions in the game of SIM, *J. Recr. Math.* **11**, 3–9.

678. H. K. Orman [1996], Pentominoes: a first player win, in: *Games of No Chance*, Proc. MSRI Workshop on Combinatorial Games, July, 1994, Berkeley, CA, MSRI Publ. (R. J. Nowakowski, ed.), Vol. 29, Cambridge University Press, Cambridge, pp. 339–344.

679. J. Pach [1992], On the game of Misery, *Studia Sci. Math. Hungar.* **27**, 355–360.

680. E. W. Packel [1987], The algorithm designer versus nature: a game-theoretic approach to information-based complexity, *J. Complexity* **3**, 244–257.

681. C. H. Papadimitriou [1985], Games against nature, *J. Comput. System Sci.* **31**, 288–301.

682. C. H. Papadimitriou [1994], *Computational Complexity*, Addison-Wesley, Chapter 19: Polynomial Space.

683. C. H. Papadimitriou, P. Raghavan, M. Sudan and H. Tamaki [1994], Motion planning on a graph (extended abstract), *Proc. 35-th Annual IEEE Symp. on Foundations of Computer Science*, Santa Fe, NM, pp. 511–520.

684. A. Papaioannou [1982], A Hamiltonian game, *Ann. Discrete Math.* **13**, 171–178.

685. T. Pappas [1994], *The Magic of Mathematics*, Wide World, San Carlos.

686. D. Parlett [1999], *The Oxford History of Board Games*, Oxford University Press.

687. T. D. Parsons [1978], Pursuit-evasion in a graph, in: *Theory and Applications of Graphs* (Y. Alavi and D. R. Lick, eds.), Springer-Verlag, pp. 426–441.

688. T. D. Parsons [1978], The search number of a connected graph, *Proc. 9th South-Eastern Conf. on Combinatorics, Graph Theory, and Computing,*, pp. 549–554.

689. O. Patashnik [1980], Qubic: $4 \times 4 \times 4$ Tic-Tac-Toe, *Math. Mag.* **53**, 202–216.

690. J. L. Paul [1978], Tic-Tac-Toe in n dimensions, *Math. Mag.* **51**, 45–49.

691. W. J. Paul, E. J. Prauss and R. Reischuk [1980], On alternation, *Acta Informatica* **14**, 243–255.

692. W. J. Paul and R. Reischuk [1980], On alternation, II, *Acta Informatica* **14**, 391–403.

693. W. J. Paul, R. E. Tarjan and J. R. Celoni [1976/77], Space bounds for a game on graphs, *Math. Systems Theory* **10**, 239–251, correction ibid. **11** (1977/78), 85. First appeared in Eighth Annual ACM Symposium on

Theory of Computing (Hershey, Pa., 1976), Assoc. Comput. Mach., New York, NY, 1976, pp 149–160.

694. M. M. Paulhus [1999], Beggar my neighbour, *Amer. Math. Monthly* **106**, 162–165.

695. J. Pearl [1980], Asymptotic properties of minimax trees and game-searching procedures, *Artificial Intelligence* **14**, 113–138.

696. J. Pearl [1984], *Heuristics: Intelligent Search Strategies for Computer Problem Solving*, Addison-Wesley, Reading, MA.

697. A. Pekeč [1996], A winning strategy for the Ramsey graph game, *Combin. Probab. Comput.* **5**, 267–276.

698. A. Pelc [1987], Solution of Ulam's problem on searching with a lie, *J. Combin. Theory* (Ser. A) **44**, 129–140.

699. A. Pelc [1988], Prefix search with a lie, *J. Combin. Theory* (Ser. A) **48**, 165–173.

700. A. Pelc [1989], Detecting errors in searching games, *J. Combin. Theory* (Ser. A) **51**, 43–54.

701. A. Pelc [2002], Searching games with errors—fifty years of coping with liars, *Theoret. Comput. Sci.* (Math Games) **270**, 71–109.

702. D. H. Pelletier [1987], Merlin's magic square, *Amer. Math. Monthly* **94**, 143–150.

703. H. Peng and C. H. Yan [1998], Balancing game with a buffer, *Adv. in Appl. Math.* **21**, 193–204.

704. G. L. Peterson [1979], Press-Ups is Pspace-complete, Dept. of Computer Science, The University of Rochester, Rochester New York, 14627, unpublished manuscript.

705. G. L. Peterson and J. H. Reif [1979], Multiple-person alternation, *Proc. 20th Ann. Symp. Foundations Computer Science* (San Juan, Puerto Rico, Oct. 1979), IEEE Computer Soc., Long Beach, CA, pp. 348–363.

706. C. Pickover [2002], The fractal society, in: *Puzzler's Tribute: a Feast for the Mind*, pp. 377–381, honoring Martin Gardner (D. Wolfe and T. Rodgers, eds.), A K Peters, Natick, MA.

707. N. Pippenger [1980], Pebbling, *Proc. 5th IBM Symp. Math. Foundations of Computer Science*, IBM, Japan, 19 pp.

708. N. Pippenger [1982], Advances in pebbling, *Proc. 9th Internat. Colloq. Automata, Languages and Programming,* Lecture Notes in Computer Science (M. Nielson and E. M. Schmidt, eds.), Vol. 140, Springer Verlag, New York, NY, pp. 407–417.

709. T. E. Plambeck [1992], Daisies, Kayles, and the Sibert–Conway decomposition in misère octal games, *Theoret. Comput. Sci.* (Math Games) **96**, 361–388.

710. V. Pless [1991], Games and codes, in: *Combinatorial Games*, Proc. Symp. Appl. Math. (R. K. Guy, ed.), Vol. 43, Amer. Math. Soc., Providence, RI, pp. 101–110.

711. A. Pluhár [2002], The accelerated k-in-a-row game, *Theoret. Comput. Sci.* (Math Games) **270**, 865–875.

712. R. Polizzi and F. Schaefer [1991], *Spin Again*, Chronicle Books.

713. G. Pólya [1921], *Über die "doppelt-periodischen" Lösungen des n-Damen-Problems, in: W. Ahrens, Mathematische Unterhaltungen und Spiele, Vol. 1*, B.G. Teubner, Leipzig, also appeared in vol. IV of Pólya's Collected Works, G.-C Rota ed., pp. 237-247.

714. D. Poole [1992], The bottleneck towers of Hanoi problem, *J. Recr. Math.* **24**, 203–207.

715. D. G. Poole [1994], The towers and triangles of Professor Claus (or, Pascal knows Hanoi), *Math. Mag.* **67**, 323–344.

716. D. Pritchard [1994], *The Family Book of Games*, Sceptre Books, Time-Life Books, Brockhampton Press.

717. J. Propp [1994], A new take-away game, in: *The Lighter Side of Mathematics*, Proc. E. Strens Memorial Conf. on Recr. Math. and its History, Calgary, 1986, Spectrum Series (R. K. Guy and R. E. Woodrow, eds.), Math. Assoc. of America, Washington, DC, pp. 212–221.

718. J. Propp [1996], About David Richman (Prologue to the paper by J. D. Lazarus et al.), in: *Games of No Chance*, Proc. MSRI Workshop on Combinatorial Games, July, 1994, Berkeley, CA, MSRI Publ. (R. J. Nowakowski, ed.), Vol. 29, Cambridge University Press, Cambridge, p. 437.

719. J. Propp [2000], Three-player impartial games, *Theoret. Comput. Sci.* (Math Games) **233**, 263–278.

720. J. Propp and D. Ullman [1992], On the cookie game, *Internat. J. Game Theory* **20**, 313–324.

721. P. Pudlák and J. Sgall [1997], An upper bound for a communication game related to time-space tradeoffs, in: *The Mathematics of Paul Erdős* (R. L. Graham and J. Nešetřil, eds.), Vol. I, Springer, Berlin, pp. 393–399.

722. A. Pultr and F. L. Morris [1984], Prohibiting repetitions makes playing games substantially harder, *Internat. J. Game Theory* **13**, 27–40.

723. A. Pultr and J. Úlehla [1985], Remarks on strategies in combinatorial games, *Discrete Appl. Math.* **12**, 165–173.

724. A. Quilliot [1982], Discrete pursuit games, *Proc. 13th Conf. on Graphs and Combinatorics*, Boca Raton, FL.

725. A. Quilliot [1985], A short note about pursuit games played on a graph with a given genus, *J. Combin. Theory* (Ser. B) **38**, 89–92.

726. M. O. Rabin [1957], Effective computability of winning strategies, *Ann. of Math. Stud.* (Contributions to the Theory of Games), Princeton **3**(39), 147–157.

727. M. O. Rabin [1976], Probabilistic algorithms, *Proc. Symp. on New Directions and Recent Results in Algorithms and Complexity* (J. F. Traub, ed.), Carnegie-Mellon, Academic Press, New York, NY, pp. 21–39.

728. D. Ratner and M. Warmuth [1990], The $(n^2 - 1)$-puzzle and related relocation problems, *J. Symbolic Comput.* **10**, 111–137.

729. B. Ravikumar and K. B. Lakshmanan [1984], Coping with known patterns of lies in a search game, *Theoret. Comput. Sci.* **33**, 85–94.

730. N. Reading [1999], Nim-regularity of graphs, *Electronic J. Combinatorics* **6**, #R11, 8pp.
http://www.combinatorics.org/

731. M. Reichling [1987], A simplified solution of the N queens' problem, *Inform. Process. Lett.* **25**, 253–255.

732. J. H. Reif [1984], The complexity of two-player games of incomplete information, *J. Comput. System Sci.* **29**, 274–301, earlier draft entitled Universal games of incomplete information, appeared in Proc. 11th Ann. ACM Symp. Theory of Computing (Atlanta, GA, 1979), Assoc. Comput. Mach., New York, NY, pp. 288–308.

733. S. Reisch [1980], Gobang ist PSPACE-vollständig, *Acta Informatica* **13**, 59–66.

734. S. Reisch [1981], Hex ist PSPACE-vollständig, *Acta Informatica* **15**, 167–191.

735. C. S. ReVelle and K. E. Rosing [2000], Defendens imperium romanum [Defending the Roman Empire]: a classical problem in military strategy, *Amer. Math. Monthly* **107**, 585–594.

736. M. Richardson [1953], Extension theorems for solutions of irreflexive relations, *Proc. Nat. Acad. Sci. USA* **39**, 649.

737. M. Richardson [1953], Solutions of irreflexive relations, *Ann. of Math.* **58**, 573–590.

738. R. D. Ringeisen [1974], Isolation, a game on a graph, *Math. Mag.* **47**, 132–138.

739. R. L. Rivest, A. R. Meyer, D. J. Kleitman, K. Winklman and J. Spencer [1980], Coping with errors in binary search procedures, *J. Comput. System Sci.* **20**, 396–404.

740. I. Rivin, I. Vardi and P. Zimmermann [1994], The n-queens problem, *Amer. Math. Monthly* **101**, 629–638.

741. I. Rivin and R. Zabih [1992], A dynamic programming solution to the N-queens problem, *Inform. Process. Lett.* **41**, 253–256.

742. E. Robertson and I. Munro [1978], NP-completeness, puzzles and games, *Utilitas Math.* **13**, 99–116.

743. A. G. Robinson and A. J. Goldman [1989], The set coincidence game: complexity, attainability, and symmetric strategies, *J. Comput. System Sci.* **39**, 376–387.

744. A. G. Robinson and A. J. Goldman [1990], On Ringeisen's isolation game, *Discrete Math.* **80**, 297–312.

745. A. G. Robinson and A. J. Goldman [1990], On the set coincidence game, *Discrete Math.* **84**, 261–283.

746. A. G. Robinson and A. J. Goldman [1991], On Ringeisen's isolation game, II, *Discrete Math.* **90**, 153–167.

747. A. G. Robinson and A. J. Goldman [1993], The isolation game for regular graphs, *Discrete Math.* **112**, 173–184.

748. J. M. Robson [1983], The complexity of Go, *Proc. Information Processing 83* (R. E. A. Mason, ed.), Elsevier, Amsterdam, pp. 413–417.

749. J. M. Robson [1984], Combinatorial games with exponential space complete decision problems, *Proc. 11th Symp. Math. Foundations of Computer Science*, Praha, Czechoslovakia, Lecture Notes in Computer Science (M. P. Chytie and V. Koubek, eds.), Vol. 176, Springer, Berlin, pp. 498–506.

750. J. M. Robson [1984], *N* by *N* checkers is Exptime complete, *SIAM J. Comput.* **13**, 252–267.

751. J. M. Robson [1985], Alternation with restrictions on looping, *Inform. and Control* **67**, 2–11.

752. E. Y. Rodin [1989], A pursuit-evasion bibliography – version 2, *Comput. Math. Appl.* **18**, 245–250.

753. J. S. Rohl [1983], A faster lexicographical *n*-queens algorithm, *Inform. Process. Lett.* **17**, 231–233.

754. I. Roizen and J. Pearl [1983], A minimax algorithm better than alpha-beta? Yes and no, *Artificial Intelligence* **21**, 199–220.

755. I. Rosenholtz [1993], Solving some variations on a variant of Tic-Tac-Toe using invariant subsets, *J. Recr. Math.* **25**, 128–135.

756. A. S. C. Ross [1953], The name of the game of Nim, Note 2334, *Math. Gaz.* **37**, 119–120.

757. A. E. Roth [1978], A note concerning asymmetric games on graphs, *Naval Res. Logistics* **25**, 365–367.

758. A. E. Roth [1978], Two-person games on graphs, *J. Combin. Theory* (Ser. B) **24**, 238–241.

759. T. Roth [1974], The tower of Brahma revisited, *J. Recr. Math.* **7**, 116–119.

760. E. M. Rounds and S. S. Yau [1974], A winning strategy for SIM, *J. Recr. Math.* **7**, 193–202.

761. W. L. Ruzzo [1980], Tree-size bounded alternation, *J. Comput. Systems Sci.* **21**, 218–235.

762. S. Sackson [1946], *A Gamut of Games*, Random House.

763. M. Saks and A. Wigderson [1986], Probabilistic Boolean decision trees and the complexity of evaluating game trees, *Proc. 27th Ann. Symp. Foundations of Computer Science* (Toronto, Ont., Canada), IEEE Computer Soc., Washington, DC, pp. 29–38.

764. U. K. Sarkar [2000], On the design of a constructive algorithm to solve the multi-peg towers of Hanoi problem, *Theoret. Comput. Sci.* (Math Games) **237**, 407–421.

765. M. Sato [1972], Grundy functions and linear games, *Publ. Res. Inst. Math. Sciences,* Kyoto Univ. **7**, 645–658.

766. W. L. Schaaf [1955, 1970, 1973, 1978], *A Bibliography of Recreational Mathematics*, Vol. I – IV, Nat'l. Council of Teachers of Mathematics, Reston, VA.

767. T. J. Schaefer [1976], Complexity of decision problems based on finite two-person perfect information games, *Proc. 8th Ann. ACM Symp. Theory of Computing* (Hershey, PA, 1976), Assoc. Comput. Mach., New York, NY, pp. 41–49.

768. T. J. Schaefer [1978], On the complexity of some two-person perfect-information games, *J. Comput. System Sci.* **16**, 185–225.

769. J. Schaeffer and R. Lake [1996], Solving the game of checkers, in: *Games of No Chance*, Proc. MSRI Workshop on Combinatorial Games, July, 1994, Berkeley, CA, MSRI Publ. (R. J. Nowakowski, ed.), Vol. 29, Cambridge University Press, Cambridge, pp. 119–133. http://www.msri.org/publications/books/Book29/files/sch% aeffer.ps.gz

770. M. Scheepers [1994], Variations on a game of Gale (II): Markov strategies, *Theoret. Comput. Sci.* (Math Games) **129**, 385–396.

771. G. Schmidt and T. Ströhlein [1985], On kernels of graphs and solutions of games: a synopsis based on relations and fixpoints, *SIAM J. Alg. Disc. Math.* **6**, 54–65.

772. R. W. Schmittberger [1992], *New Rules for Classic Games*, Wiley, New York.

773. G. Schrage [1985], A two-dimensional generalization of Grundy's game, *Fibonacci Quart.* **23**, 325–329.

774. H. Schubert [1953], *Mathematische Mussestunden*, De Gruyter, Berlin, neubearbeitet von F. Fitting, Elfte Auflage.

775. F. Schuh [1952], Spel van delers (The game of divisors), *Nieuw Tijdschrift voor Wiskunde* **39**, 299–304.

776. F. Schuh [1968], *The Master Book of Mathematical Recreations*, Dover, New York, NY, translated by F. Göbel, edited by T. H. O'Beirne.

777. B. L. Schwartz [1971], Some extensions of Nim, *Math. Mag.* **44**, 252–257.

778. B. L. Schwartz, ed. [1979], *Mathematical solitaires and games*, Baywood Publishing Company, Farmingdale, NY, pp. 37–81.

779. A. J. Schwenk [1970], Take-away games, *Fibonacci Quart.* **8**, 225–234.

780. A. J. Schwenk [2000], What is the correct way to seed a knockout tournament?, *Amer. Math. Monthly* **107**, 140–150.

781. R. S. Scorer, P. M. Grundy and C. A. B. Smith [1944], Some binary games, *Math. Gaz.* **28**, 96–103.

782. Á. Seress [1992], On Hajnal's triangle-free game, *Graphs Combin.* **8**, 75–79.

783. Á. Seress and T. Szabó [1999], On Erdös' Eulerian trail game, *Graphs Combin.* **15**, 233–237.

784. J. Sgall [2001], Solution of David Gale's lion and man problem, *Theoret. Comput. Sci.* (Math Games) **259**, 663–670.

785. L. E. Shader [1978], Another strategy for SIM, *Math. Mag.* **51**, 60–64.

786. L. E. Shader and M. L. Cook [1980], A winning strategy for the second player in the game Tri-tip, *Proc. Tenth S.E. Conference on Computing, Combinatorics and Graph Theory*, Utilitas, Winnipeg.

787. A. S. Shaki [1979], Algebraic solutions of partizan games with cycles, *Math. Proc. Camb. Phil. Soc.* **85**, 227–246.

788. A. Shamir, R. L. Rivest and L. M. Adleman [1981], Mental Poker, in: *The Mathematical Gardner* (D. A. Klarner, ed.), Wadsworth Internat., Belmont, CA, pp. 37–43.

789. A. Shen [2000], Lights out, *Math. Intelligencer* **22**, 20–21.

790. R. Sheppard and J. Wilkinson [1995], *Strategy Games: A Collection Of 50 Games & Puzzles To Stimulate Mathematical Thinking*, Parkwest Publications.

791. G. J. Sherman [1978], A child's game with permutations, *Math. Mag.* **51**, 67–68.

792. W. L. Sibert and J. H. Conway [1992], Mathematical Kayles, *Internat. J. Game Theory* **20**, 237–246.

793. R. Silber [1976], A Fibonacci property of Wythoff pairs, *Fibonacci Quart.* **14**, 380–384.

794. R. Silber [1977], Wythoff's Nim and Fibonacci representations, *Fibonacci Quart.* **15**, 85–88.

795. J.-N. O. Silva [1993], Some game bounds depending on birthdays, *Portugaliae Math.* **3**, 353–358.

796. R. Silver [1967], The group of automorphisms of the game of 3-dimensional ticktacktoe, *Amer. Math. Monthly* **74**, 247–254.

797. D. L. Silverman [1971], *Your Move*, McGraw-Hill.

798. G. J. Simmons [1969], The game of SIM, *J. Recr. Math.* **2**, 193–202.

799. D. Singmaster [1981], Almost all games are first person games, *Eureka* **41**, 33–37.

800. D. Singmaster [1982], Almost all partizan games are first person and almost all impartial games are maximal, *J. Combin. Inform. System Sci.* **7**, 270–274.

801. D. Singmaster [1999], Some diophantine recreations, in: *The Mathemagician and Pied Puzzler*, honoring Martin Gardner; E. Berlekamp and T. Rodgers, eds., A K Peters, Natick, MA, pp. 219-235.

802. C. A. B. Smith [1966], Graphs and composite games, *J. Combin. Theory* **1**, 51–81, reprinted in slightly modified form in: *A Seminar on Graph Theory* (F. Harary, ed.), Holt, Rinehart and Winston, New York, NY, 1967.

803. C. A. B. Smith [1968], Compound games with counters, *J. Recr. Math.* **1**, 67–77.

804. C. A. B. Smith [1971], Simple game theory and its applications, *Bull. Inst. Math. Appl.* **7**, 352–357.

805. D. E. Smith and C. C. Eaton [1911], Rithmomachia, the great medieval number game, *Amer. Math. Montly* **18**, 73–80.

806. R. G. Snatzke [2002], Exhaustive search in Amazons, in: *More Games of No Chance,* Proc. MSRI Workshop on Combinatorial Games, July, 2000, Berkeley, CA, MSRI Publ. (R. J. Nowakowski, ed.), Vol. 42, Cambridge University Press, Cambridge, pp. 261–278.

807. R. Sosic and J. Gu [1990], A polynomial time algorithm for the n-queens problem, *SIGART* **1**, 7–11.

808. J. Spencer [1977], Balancing games, *J. Combin. Theory* (Ser. B) **23**, 68–74.

809. J. Spencer [1984], Guess a number with lying, *Math. Mag.* **57**, 105–108.

810. J. Spencer [1986], Balancing vectors in the max norm, *Combinatorica* **6**, 55–65.

811. J. Spencer [1991], Threshold spectra via the Ehrenfeucht game, *Discrete Appl. Math.* **30**, 235–252.

812. J. Spencer [1992], Ulam's searching game with a fixed number of lies, *Theoret. Comput. Sci.* (Math Games) **95**, 307–321.

813. J. Spencer [1994], Randomization, derandomization and antirandomization: three games, *Theoret. Comput. Sci.* (Math Games) **131**, 415–429.

814. W. L. Spight [2001], Extended thermography for multile kos in go, *Theoret. Comput. Sci.* **252**, 23–43, special "Computers and Games" issue; first version appeared in Proc. 1st Intern. Conf. on Computer Games CG'98, Tsukuba, Japan, Nov. 1998, *Lecture Notes in Computer Science*, Vol. 1558, Springer, pp. 232-251, 1999.

815. E. L. Spitznagel, Jr. [1967], A new look at the fifteen puzzle, *Math. Mag.* **40**, 171–174.

816. E. L. Spitznagel, Jr. [1973], Properties of a game based on Euclid's algorithm, *Math. Mag.* **46**, 87–92.

817. R. Sprague [1935–36], Über mathematische Kampfspiele, *Tôhoku Math. J.* **41**, 438–444.

818. R. Sprague [1937], Über zwei Abarten von Nim, *Tôhoku Math. J.* **43**, 351–359.

819. R. Sprague [1947], Bemerkungen über eine spezielle Abelsche Gruppe, *Math. Z.* **51**, 82–84.

820. R. Sprague [1961], *Unterhaltsame Mathematik*, Vieweg and Sohn, Braunschweig, Paperback reprint, translation by T. H. O'Beirne: *Recreations in Mathematics*, Blackie, 1963.

821. H. Steinhaus [1960], Definitions for a theory of games and pursuit, *Naval Res. Logistics* **7**, 105–108.

822. V. N. Stepanenko [1975], Grundy games under conditions of semidefiniteness, *Cybernetics* **11**, 167–172 (trans. of *Kibernetika* **11** (1975) 145–149).

823. B. M. Stewart [1939], Problem 3918 (k-peg tower of Hanoi), *Amer. Math. Monthly* **46**, 363, solution by J. S. Frame, *ibid.* **48** (1941) 216–217; by the proposer, *ibid.* 217–219.

824. L. Stiller [1988], Massively parallel retrograde analysis, Tech. Rep. BU-CS TR88–014, Comp. Sci. Dept., Boston University.

825. L. Stiller [1989], Parallel analysis of certain endgames, *Internat. Comp. Chess Assoc. J.* **12**, 55–64.

826. L. Stiller [1991], Group graphs and computational symmetry on massively parallel architecture, *J. Supercomputing* **5**, 99–117.

827. L. Stiller [1996], Multilinear algebra and chess endgames, in: *Games of No Chance*, Proc. MSRI Workshop on Combinatorial Games, July, 1994, Berkeley, CA, MSRI Publ. (R. J. Nowakowski, ed.), Vol. 29, Cambridge University Press, Cambridge, pp. 151–192.

828. D. L. Stock [1989], Merlin's magic square revisited, *Amer. Math. Monthly* **96**, 608–610.

829. L. J. Stockmeyer and A. K. Chandra [1979], Provably difficult combinatorial games, *SIAM J. Comput.* **8**, 151–174.

830. J. A. Storer [1983], On the complexity of chess, *J. Comput. System Sci.* **27**, 77–100.

831. W. E. Story [1879], Note on the "15" puzzle, *Amer. J. Math.* **2**, 399–404.

832. P. D. Straffin, Jr. [1985], Three-person winner-take-all games with McCarthy's revenge rule, *College J. Math.* **16**, 386–394.

833. P. D. Straffin [1993], *Game Theory and Strategy*, New Mathematical Library, MAA, Washington, DC.

834. T. Ströhlein and L. Zagler [1977], Analyzing games by Boolean matrix iteration, *Discrete Math.* **19**, 183–193.

835. W. Stromquist and D. Ullman [1993], Sequential compounds of combinatorial games, *Theoret. Comput. Sci.* (Math Games) **119**, 311–321.

836. K. Sugihara and I. Suzuki [1989], Optimal algorithms for a pursuit-evasion problem in grids, *SIAM J. Disc. Math.* **1**, 126–143.

837. K. Sutner [1988], On σ-automata, *Complex Systems* **2**, 1–28.

838. K. Sutner [1989], Linear cellular automata and the Garden-of-Eden, *Math. Intelligencer* **11**, 49–53.

839. K. Sutner [1990], The σ-game and cellular automata, *Amer. Math. Monthly* **97**, 24–34.

840. K. Sutner [1995], On the computational complexity of finite cellular automata, *J. Comput. System Sci.* **50**, 87–97.

841. K. J. Swanepoel [2000], Balancing unit vectors, *J. Combin. Theory Ser. A* **89**, 105–112.

842. J. L. Szwarcfiter and G. Chaty [1994], Enumerating the kernels of a directed graph with no odd circuits, *Inform. Process. Lett.* **51**, 149–153.

843. A. Takahashi, S. Ueno and Y. Kajitani [1995], Mixed searching and proper-path-width, *Theoret. Comput. Sci.* (Math Games) **137**, 253–268.

844. G. Tardos [1988], Polynomial bound for a chip firing game on graphs, *SIAM J. Disc. Math.* **1**, 397–398.

845. M. Tarsi [1983], Optimal search on some game trees, *J. Assoc. Comput. Mach.* **30**, 389–396.

846. R. Telgársky [1987], Topological games: on the 50th anniversary of the Banach-Mazur game, *Rocky Mountain J. Math.* **17**, 227–276.

847. W. F. D. Theron and G. Geldenhuys [1998], Domination by queens on a square beehive, *Discrete Math.* **178**, 213–220.

848. K. Thompson [1986], Retrograde analysis of certain endgames, *Internat. Comp. Chess Assoc. J.* **9**, 131–139.

849. H. Tohyama and A. Adachi [2000], Complexity of path discovery problems, *Theoret. Comput. Sci.* (Math Games) **237**, 381–406.

850. G. P. Tollisen and T. Lengyel [2000], Color switching games, *Ars Combin.* **56**, 223–234.

851. I. Tomescu [1990], Almost all digraphs have a kernel, *Discrete Math.* **84**, 181–192.

852. R. Tošić and S. Šćekić [1983], An analysis of some partizan graph games, *Proc. 4th Yugoslav Seminar on Graph Theory*, Novi Sad, pp. 311–319.

853. A. M. Turing, M. A. Bates, B. V. Bowden and C. Strachey [1953], Digital computers applied to games, in: *Faster Than Thought* (B. V. Bowden, ed.), Pitman, London, pp. 286–310.

854. R. Uehara and S. Iwata [1990], Generalized Hi-Q is NP-complete, *Trans. IEICE* **E73**, 270–273.

855. J. Úlehla [1980], A complete analysis of von Neumann's Hackendot, *Internat. J. Game Theory* **9**, 107–113.

856. D. Ullman [1992], More on the four-numbers game, *Math. Mag.* **65**, 170–174.

857. S. Vajda [1992], *Mathematical Games and How to Play Them*, Ellis Horwood Series in Mathematics and its Applications, Chichester, England.

858. H. J. van den Herik and I. S. Herschberg [1985], The construction of an omniscient endgame database, *Internat. Comp. Chess Assoc. J.* **8**, 66–87.

859. J. van Leeuwen [1976], Having a Grundy-numbering is NP-complete, Report No. 207, Computer Science Dept., Pennsylvania State University, University Park, PA.

860. A. J. van Zanten [1990], The complexity of an optimal algorithm for the generalized tower of Hanoi problem, *Intern. J. Comput. Math.* **36**, 1–8.

861. A. J. van Zanten [1991], An iterative optimal algorithm for the generalized tower of Hanoi problem, *Intern. J. Comput. Math.* **39**, 163–168.

862. I. Vardi [1990], *Computational Recreations in Mathematica*, Addison Wesley.

863. I. P. Varvak [1968], Games on the sum of graphs, *Cybernetics* **4**, 49–51 (trans. of *Kibernetika* **4** (1968) 63–66).

864. J. Veerasamy and I. Page [1994], On the towers of Hanoi problem with multiple spare pegs, *Intern. J. Comput. Math.* **52**, 17–22.

865. H. Venkateswaran and M. Tompa [1989], A new pebble game that characterizes parallel complexity classes, *SIAM J. Comput.* **18**, 533–549.

866. D. Viaud [1987], Une stratégie générale pour jouer au Master-Mind, *RAIRO Recherche opérationelle/Operations Research* **21**, 87–100.

867. J. von Neumann [1928], Zur Theorie der Gesellschaftsspiele, *Math. Ann.* **100**, 295–320.

868. J. von Neumann and O. Morgenstern [1953], *Theory of Games and Economic Behaviour*, Princeton University Press, Princeton, NJ., 3rd edn.

869. J. L. Walsh [1953], The name of the game of Nim, Letter to the Editor, *Math. Gaz.* **37**, 290.

870. T. R. Walsh [1982], The towers of Hanoi revisited: moving the rings by counting the moves, *Inform. Process. Lett.* **15**, 64–67.

871. T. R. Walsh [1983], Iteration strikes back at the cyclic towers of Hanoi, *Inform. Process. Lett.* **16**, 91–93.

872. I. M. Wanless [1997], On the periodicity of graph games, *Australas. J. Combin.* **16**, 113–123.

873. I. M. Wanless [2001], Path achievement games, *Australas. J. Combin.* **23**, 9–18.

874. R. H. Warren [1996], Disks on a chessboard, *Amer. Math. Monthly* **103**, 305–307.

875. A. Washburn [1990], Deterministic graphical games, *J. Math. Anal. Appl.* **153**, 84–96.

876. W. A. Webb [1982], The length of the four-number game, *Fibonacci Quart.* **20**, 33–35.

877. C. P. Welter [1952], The advancing operation in a special abelian group, *Nederl. Akad. Wetensch. Proc.* (Ser. A) **55** = *Indag. Math.***14**, 304-314.

878. C. P. Welter [1954], The theory of a class of games on a sequence of squares, in terms of the advancing operation in a special group, *Nederl. Akad. Wetensch. Proc.* (Ser. A) **57** = *Indag. Math.***16**, 194-200.

879. J. West [1996], Champion-level play of domineering, in: *Games of No Chance,* Proc. MSRI Workshop on Combinatorial Games, July, 1994, Berkeley, CA, MSRI Publ. (R. J. Nowakowski, ed.), Vol. 29, Cambridge University Press, Cambridge, pp. 85–91.

880. J. West [1996], Champion-level play of dots-and-boxes, in: *Games of No Chance,* Proc. MSRI Workshop on Combinatorial Games, July, 1994, Berkeley, CA, MSRI Publ. (R. J. Nowakowski, ed.), Vol. 29, Cambridge University Press, Cambridge, pp. 79–84.

881. J. West [1996], New values for Top Entails, in: *Games of No Chance,* Proc. MSRI Workshop on Combinatorial Games, July, 1994, Berkeley, CA, MSRI Publ. (R. J. Nowakowski, ed.), Vol. 29, Cambridge University Press, Cambridge, pp. 345–350.

882. M. J. Whinihan [1963], Fibonacci Nim, *Fibonacci Quart.* **1**(4), 9–12.

883. R. Wilber [1988], White pebbles help, *J. Comput. System Sci.* **36**, 108–124.

884. R. Wilfong [1991], Motion planning in the presence of movable obstacles. Algorithmic motion planning in robotics, *Ann. Math. Artificial Intelligence* **3**, 131–150.

885. R. M. Wilson [1974], Graph puzzles, homotopy and the alternating group, *J. Combin. Theory* (Ser. B) **16**, 86–96.

886. P. Winkler [2002], Games people don't play, in: *Puzzler's Tribute: a Feast for the Mind,* pp. 301–313, honoring Martin Gardner (D. Wolfe and T. Rodgers, eds.), A K Peters, Natick, MA.

887. D. Wolfe [1993], Snakes in domineering games, *Theoret. Comput. Sci.* (Math Games) **119**, 323–329.

888. D. Wolfe [1996], The gamesman's toolkit, in: *Games of No Chance,* Proc. MSRI Workshop on Combinatorial Games, July, 1994, Berkeley, CA, MSRI Publ. (R. J. Nowakowski, ed.), Vol. 29, Cambridge University Press, Cambridge, pp. 93–98.

889. D. Wolfe [2002], Go endgames are hard, in: *More Games of No Chance,* Proc. MSRI Workshop on Combinatorial Games, July, 2000, Berkeley, CA, MSRI Publ. (R. J. Nowakowski, ed.), Vol. 42, Cambridge University Press, Cambridge, pp. 125–136.

890. D. Wood [1981], The towers of Brahma and Hanoi revisited, *J. Recr. Math.*
14, 17–24.

891. D. Wood [1983], Adjudicating a towers of Hanoi contest, *Intern. J. Comput.
Math.* **14**, 199–207.

892. J.-S. Wu and R.-J. Chen [1992], The towers of Hanoi problem with parallel
moves, *Inform. Process. Lett.* **44**, 241–243.

893. J.-S. Wu and R.-J. Chen [1993], The towers of Hanoi problem with cyclic
parallel moves, *Inform. Process. Lett.* **46**, 1–6.

894. W. A. Wythoff [1907], A modification of the game of Nim, *Nieuw Arch.
Wisk.* **7**, 199–202.

895. A. M. Yaglom and I. M. Yaglom [1967], *Challenging Mathematical Problems
with Elementary Solutions*, Vol. II, Holden-Day, San Francisco, translated
by J. McCawley, Jr., revised and edited by B. Gordon.

896. Y. Yamasaki [1978], Theory of division games, *Publ. Res. Inst. Math. Sciences, Kyoto Univ.* **14**, 337–358.

897. Y. Yamasaki [1980], On misère Nim-type games, *J. Math. Soc. Japan* **32**,
461–475.

898. Y. Yamasaki [1981], The projectivity of Y-games, *Publ. Res. Inst. Math.
Sciences, Kyoto Univ.* **17**, 245–248.

899. Y. Yamasaki [1981], Theory of Connexes I, *Publ. Res. Inst. Math. Sciences,
Kyoto Univ.* **17**, 777–812.

900. Y. Yamasaki [1985], Theory of connexes II, *Publ. Res. Inst. Math. Sciences,
Kyoto Univ.* **21**, 403–410.

901. Y. Yamasaki [1989], *Combinatorial Games: Back and Front*, Springer Verlag, Tokyo (in Japanese).

902. Y. Yamasaki [1989], Shannon switching games without terminals II, *Graphs
Combin.* **5**, 275–282.

903. Y. Yamasaki [1991], A difficulty in particular Shannon-like games, *Discrete
Appl. Math.* **30**, 87–90.

904. Y. Yamasaki [1992], Shannon switching games without terminals III,
Graphs Combin. **8**, 291–297.

905. Y. Yamasaki [1993], Shannon-like games are difficult, *Discrete Math.* **111**,
481–483.

906. Y. Yamasaki [1997], The arithmetic of reversed positional games, *Theoret.
Comput. Sci.* (Math Games) **174**, 247–249, also in *Discrete Math.* **165/166**
(1997) 639–641.

907. L. J. Yedwab [1985] On playing well in a sum of games, M.Sc. Thesis, MIT,
MIT/LCS/TR-348.

908. Y. N. Yeh [1995], A remarkable endofunction involving compositions, *Stud.
Appl. Math* **95**, 419–432.

909. Y. Yesha [1978], Theory of annihilation games, Ph.D. Thesis, Weizmann Institute of Science, Rehovot, Israel.

910. Y. K. Yu and R. B. Banerji [1982], Periodicity of Sprague–Grundy function in graphs with decomposable nodes, *Cybernetics and Systems: An Internat. J.* **13**, 299–310.

911. S. Zachos [1988], Probabilistic quantifiers and games, *J. Comput. System Sci.* **36**, 433–451.

912. D. Zeilberger [2001], Three-rowed Chomp, *Adv. in Appl. Math.* **26**, 168–179.

913. E. Zermelo [1913], Über eine Anwendung der Mengenlehre auf die Theorie des Schachspiels, *Proc. 5th Int. Cong. Math. Cambridge 1912*, Vol. II, Cambridge University Press, pp. 501–504.

914. X. Zhu [1999], The game coloring number of planar graphs, *J. Combin. Theory Ser. B* **75**, 245–258.

915. X. Zhu [2000], The game coloring number of pseudo partial k-trees, *Discrete Math.* **215**, 245–262.

916. M. Zieve [1996], Take-away games, in: *Games of No Chance,* Proc. MSRI Workshop on Combinatorial Games, July, 1994, Berkeley, CA, MSRI Publ. (R. J. Nowakowski, ed.), Vol. 29, Cambridge University Press, Cambridge, pp. 351–361.

917. U. Zwick and M. S. Paterson [1993], The memory game, *Theoret. Comput. Sci.* (Math Games) **110**, 169–196.

918. U. Zwick and M. S. Paterson [1996], The complexity of mean payoff games on graphs, *Theoret. Comput. Sci.* (Math Games) **158**, 343–359.

919. W. S. Zwicker [1987], Playing games with games: the hypergame paradox, *Amer. Math. Monthly* **94**, 507–514.

AVIEZRI S. FRAENKEL
DEPARTMENT OF COMPUTER SCIENCE AND APPLIED MATHEMATICS
THE WEIZMANN INSTITUTE OF SCIENCE
REHOVOT 76100
ISRAEL
  fraenkel@wisdom.weizmann.ac.il
  http://www.wisdom.weizmann.ac.il/~fraenkel